BATCH FERMENTATION

CHEMICAL INDUSTRIES

A Series of Reference Books and Textbooks

Founding Editor

HEINZ HEINEMANN

1. *Fluid Catalytic Cracking with Zeolite Catalysts,* Paul B. Venuto and E. Thomas Habib, Jr.
2. *Ethylene: Keystone to the Petrochemical Industry,* Ludwig Kniel, Olaf Winter, and Karl Stork
3. *The Chemistry and Technology of Petroleum,* James G. Speight
4. *The Desulfurization of Heavy Oils and Residua,* James G. Speight
5. *Catalysis of Organic Reactions,* edited by William R. Moser
6. *Acetylene-Based Chemicals from Coal and Other Natural Resources,* Robert J. Tedeschi
7. *Chemically Resistant Masonry,* Walter Lee Sheppard, Jr.
8. *Compressors and Expanders: Selection and Application for the Process Industry,* Heinz P. Bloch, Joseph A. Cameron, Frank M. Danowski, Jr., Ralph James, Jr., Judson S. Swearingen, and Marilyn E. Weightman
9. *Metering Pumps: Selection and Application,* James P. Poynton
10. *Hydrocarbons from Methanol,* Clarence D. Chang
11. *Form Flotation: Theory and Applications,* Ann N. Clarke and David J. Wilson
12. *The Chemistry and Technology of Coal,* James G. Speight
13. *Pneumatic and Hydraulic Conveying of Solids,* O. A. Williams
14. *Catalyst Manufacture: Laboratory and Commercial Preparations,* Alvin B. Stiles
15. *Characterization of Heterogeneous Catalysts,* edited by Francis Delannay
16. *BASIC Programs for Chemical Engineering Design,* James H. Weber
17. *Catalyst Poisoning,* L. Louis Hegedus and Robert W. McCabe
18. *Catalysis of Organic Reactions,* edited by John R. Kosak
19. *Adsorption Technology: A Step-by-Step Approach to Process Evaluation and Application,* edited by Frank L. Slejko
20. *Deactivation and Poisoning of Catalysts,* edited by Jacques Oudar and Henry Wise
21. *Catalysis and Surface Science: Developments in Chemicals from Methanol, Hydrotreating of Hydrocarbons, Catalyst Preparation, Monomers and Polymers, Photocatalysis and Photovoltaics,* edited by Heinz Heinemann and Gabor A. Somorjai
22. *Catalysis of Organic Reactions,* edited by Robert L. Augustine

23. *Modern Control Techniques for the Processing Industries,* T. H. Tsai, J. W. Lane, and C. S. Lin
24. *Temperature-Programmed Reduction for Solid Materials Characterization,* Alan Jones and Brian McNichol
25. *Catalytic Cracking: Catalysts, Chemistry, and Kinetics,* Bohdan W. Wojciechowski and Avelino Corma
26. *Chemical Reaction and Reactor Engineering,* edited by J. J. Carberry and A. Varma
27. *Filtration: Principles and Practices, Second Edition,* edited by Michael J. Matteson and Clyde Orr
28. *Corrosion Mechanisms,* edited by Florian Mansfeld
29. *Catalysis and Surface Properties of Liquid Metals and Alloys,* Yoshisada Ogino
30. *Catalyst Deactivation,* edited by Eugene E. Petersen and Alexis T. Bell
31. *Hydrogen Effects in Catalysis: Fundamentals and Practical Applications,* edited by Zoltán Paál and P. G. Menon
32. *Flow Management for Engineers and Scientists,* Nicholas P. Cheremisinoff and Paul N. Cheremisinoff
33. *Catalysis of Organic Reactions,* edited by Paul N. Rylander, Harold Greenfield, and Robert L. Augustine
34. *Powder and Bulk Solids Handling Processes: Instrumentation and Control,* Koichi Iinoya, Hiroaki Masuda, and Kinnosuke Watanabe
35. *Reverse Osmosis Technology: Applications for High-Purity-Water Production,* edited by Bipin S. Parekh
36. *Shape Selective Catalysis in Industrial Applications,* N. Y. Chen, William E. Garwood, and Frank G. Dwyer
37. *Alpha Olefins Applications Handbook,* edited by George R. Lappin and Joseph L. Sauer
38. *Process Modeling and Control in Chemical Industries,* edited by Kaddour Najim
39. *Clathrate Hydrates of Natural Gases,* E. Dendy Sloan, Jr.
40. *Catalysis of Organic Reactions,* edited by Dale W. Blackburn
41. *Fuel Science and Technology Handbook,* edited by James G. Speight
42. *Octane-Enhancing Zeolitic FCC Catalysts,* Julius Scherzer
43. *Oxygen in Catalysis,* Adam Bielanski and Jerzy Haber
44. *The Chemistry and Technology of Petroleum: Second Edition, Revised and Expanded,* James G. Speight
45. *Industrial Drying Equipment: Selection and Application,* C. M. van't Land
46. *Novel Production Methods for Ethylene, Light Hydrocarbons, and Aromatics*, edited by Lyle F. Albright, Billy L. Crynes, and Siegfried Nowak
47. *Catalysis of Organic Reactions*, edited by William E. Pascoe
48. *Synthetic Lubricants and High-Performance Functional Fluids*, edited by Ronald L. Shubkin
49. *Acetic Acid and Its Derivatives*, edited by Victor H. Agreda and Joseph R. Zoeller
50. *Properties and Applications of Perovskite-Type Oxides*, edited by L. G. Tejuca and J. L. G. Fierro

51. *Computer-Aided Design of Catalysts*, edited by E. Robert Becker and Carmo J. Pereira
52. *Models for Thermodynamic and Phase Equilibria Calculations*, edited by Stanley I. Sandler
53. *Catalysis of Organic Reactions*, edited by John R. Kosak and Thomas A. Johnson
54. *Composition and Analysis of Heavy Petroleum Fractions*, Klaus H. Altgelt and Mieczyslaw M. Boduszynski
55. *NMR Techniques in Catalysis,* edited by Alexis T. Bell and Alexander Pines
56. *Upgrading Petroleum Residues and Heavy Oils*, Murray R. Gray
57. *Methanol Production and Use*, edited by Wu-Hsun Cheng and Harold H. Kung
58. *Catalytic Hydroprocessing of Petroleum and Distillates*, edited by Michael C. Oballah and Stuart S. Shih
59. *The Chemistry and Technology of Coal: Second Edition, Revised and Expanded,* James G. Speight
60. *Lubricant Base Oil and Wax Processing*, Avilino Sequeira, Jr.
61. *Catalytic Naphtha Reforming: Science and Technology*, edited by George J. Antos, Abdullah M. Aitani, and José M. Parera
62. *Catalysis of Organic Reactions*, edited by Mike G. Scaros and Michael L. Prunier
63. *Catalyst Manufacture,* Alvin B. Stiles and Theodore A. Koch
64. *Handbook of Grignard Reagents,* edited by Gary S. Silverman and Philip E. Rakita
65. *Shape Selective Catalysis in Industrial Applications: Second Edition, Revised and Expanded*, N. Y. Chen, William E. Garwood, and Francis G. Dwyer
66. *Hydrocracking Science and Technology*, Julius Scherzer and A. J. Gruia
67. *Hydrotreating Technology for Pollution Control: Catalysts, Catalysis, and Processes*, edited by Mario L. Occelli and Russell Chianelli
68. *Catalysis of Organic Reactions*, edited by Russell E. Malz, Jr.
69. *Synthesis of Porous Materials: Zeolites, Clays, and Nanostructures,* edited by Mario L. Occelli and Henri Kessler
70. *Methane and Its Derivatives*, Sunggyu Lee
71. *Structured Catalysts and Reactors*, edited by Andrzei Cybulski and Jacob Moulijn
72. *Industrial Gases in Petrochemical Processing*, Harold Gunardson
73. *Clathrate Hydrates of Natural Gases: Second Edition, Revised and Expanded*, E. Dendy Sloan, Jr.
74. *Fluid Cracking Catalysts*, edited by Mario L. Occelli and Paul O'Connor
75. *Catalysis of Organic Reactions*, edited by Frank E. Herkes
76. *The Chemistry and Technology of Petroleum, Third Edition*, *Revised and Expanded*, James G. Speight
77. *Synthetic Lubricants and High-Performance Functional Fluids, Second Edition: Revised and Expanded,* Leslie R. Rudnick and Ronald L. Shubkin

78. *The Desulfurization of Heavy Oils and Residua, Second Edition, Revised and Expanded,* James G. Speight
79. *Reaction Kinetics and Reactor Design: Second Edition, Revised and Expanded,* John B. Butt
80. *Regulatory Chemicals Handbook*, Jennifer M. Spero, Bella Devito, and Louis Theodore
81. *Applied Parameter Estimation for Chemical Engineers*, Peter Englezos and Nicolas Kalogerakis
82. *Catalysis of Organic Reactions,* edited by Michael E. Ford
83. *The Chemical Process Industries Infrastructure: Function and Economics*, James R. Couper, O. Thomas Beasley, and W. Roy Penney
84. *Transport Phenomena Fundamentals*, Joel L. Plawsky
85. *Petroleum Refining Processes*, James G. Speight and Baki Özüm
86. *Health, Safety, and Accident Management in the Chemical Process Industries*, Ann Marie Flynn and Louis Theodore
87. *Plantwide Dynamic Simulators in Chemical Processing and Control*, William L. Luyben
88. *Chemicial Reactor Design*, Peter Harriott
89. *Catalysis of Organic Reactions*, edited by Dennis Morrell
90. *Lubricant Additives: Chemistry and Applications*, edited by Leslie R. Rudnick
91. *Handbook of Fluidization and Fluid-Particle Systems*, edited by Wen-Ching Yang
92. *Conservation Equations and Modeling of Chemical and Biochemical Processes*, Said S. E. H. Elnashaie and Parag Garhyan
93. *Batch Fermentation: Modeling, Monitoring, and Control*, Ali Cinar, Satish J. Parulekar, Cenk Ündey, and Gülnur Birol
94. *Industrial Solvents Handbook: Second Edition*, Nicholas P. Cheremisinoff

ADDITIONAL VOLUMES IN PREPARATION

Chemical Process Engineering: Design and Economics, Harry Silla

Process Engineering Economics, James R. Couper

Petroleum and Gas Field Processing, H. K. Abdel-Aal, Mohamed Aggour, and M. A. Fahim

Thermodynamic Cycles: Computer-Aided Design and Optimization, Chih Wu

Re-Engineering the Chemical Processing Plant: Process Intensification, Andrzej Stankiewicz and Jacob A. Moulijn

BATCH FERMENTATION

Modeling, Monitoring, and Control

Ali Cinar
Satish J. Parulekar
Cenk Ündey

Illinois Institute of Technology
Chicago, Illinois, U.S.A.

Gülnur Birol

Northwestern University
Evanston, Illinois, U.S.A.

MARCEL DEKKER, INC. NEW YORK • BASEL

Library of Congress Cataloging-in-Publication Data
A catalog record for this book is available from the Library of Congress.

ISBN: 0-8247-4034-3

This book is printed on acid-free paper.

Headquarters
Marcel Dekker, Inc.
270 Madison Avenue, New York, NY 10016
tel: 212-696-9000; fax: 212-685-4540

Eastern Hemisphere Distribution
Marcel Dekker AG
Hutgasse 4, Postfach 812, CH-4001 Basel, Switzerland
tel: 41-61-260-6300; fax: 41-61-260-6333

World Wide Web
http://www.dekker.com

The publisher offers discounts on this book when ordered in bulk quantities. For more information, write to Special Sales/Professional Marketing at the headquarters address above.

Current printing (last digit):
10 9 8 7 6 5 4 3 2 1

PRINTED IN THE UNITED STATES OF AMERICA

To Mine and Bedirhan, my heroes and best friends

— AC

To my family, for their encouragement, love, and support

— SJP

To my parents, Gültekin and Gülderen, who gave me life, and mind, and to my sweet Ceylan, my love and inspiration

— CÜ

To İnanç, Uluç and Defne

— GB

Preface

This book deals with batch process modeling, monitoring, fault diagnosis, and control, focusing on batch fermentation processes. Fermentation is one of the main bioprocesses used in pharmaceutical, food, and chemical industries. Most fermentation processes are carried out as batch or fed-batch operations. Batch processes have been around for many millennia, and received increasing attention in the second half of the twentieth century. Although batch processes are simple to set up and operate, modeling, monitoring, and control of these processes is quite challenging. Even in simple fermentation processes, diverse organisms and the large numbers of cells that are produced in various phases of the batch by complex metabolic reactions provide significant challenges to successful process operation. Slight changes in operating conditions during critical phases may have a significant influence on the growth and differentiation of organisms, and impact the quality and yield of the final product. Accurate process models are necessary to monitor and control the progress of the batch, determine transition times to new phases of activity, and diagnose the causes of unacceptable process behavior and product quality. Significant advances have been made in recent years in the development of powerful modeling, monitoring, diagnosis, and control techniques. Various new modeling paradigms have been proposed to develop models of desired accuracy for a specific task. Real-time multivariate process monitoring techniques have been developed to complement quality control based on laboratory analysis of the final product and to permit timely corrective actions to save a batch run destined to produce low quality products during the progress of the run. Control methods that consider desired future trajectories of critical variables, process constraints, and sensor faults have been developed for tighter control of multivariate processes. This book offers a unified presentation of these new methods and illustrates their implementation with a case study of penicillin fermentation.

The book integrates fundamental concepts from biochemical engineering, multivariate statistical theory, model identification, systems theory, and process control, and presents powerful methods for multivariable nonlinear processes with nonstationary and correlated data. Methods are introduced for finding optimal reference trajectories and operating conditions, and for manufacturing the product profitably in spite of variations in the characteristics of raw materials and ambient conditions, malfunctions in equipment, and variations in operator judgment and experience. The book presents both fundamental and data-based empirical modeling methods, several monitoring techniques ranging from simple univariate statistical process control to advanced multivariate process monitoring techniques, many fault diagnosis paradigms and a variety of simple to advanced process control approaches. The integration of techniques in model development, signal processing, data reconciliation, process monitoring, fault detection and diagnosis, quality control, and process control for a comprehensive approach in managing batch process operations by a supervisory knowledge-based system is illustrated. Most of these methods have been presented in various conferences and have been discussed in research journals, but they have not appeared in books for the general technical audience. The focus of the book is on batch fermentation in pharmaceutical processes. However, the methods presented can be used for batch processes in other areas by paying attention to the special characteristics of a specific process.

The book will be a useful resource for engineers and scientists working with fermentation processes, as well as students in biotechnology, modeling, reaction engineering, quality control, and process control courses. One objective of the book is to provide detailed information for understanding, comparing, and implementing new techniques reported in the research literature. Various paradigms are introduced in each subject to provide a balanced view. Some of them are based on the research of the authors, while others have been proposed by other researchers. A well-documented industrial process, penicillin fermentation, is used throughout the book to illustrate the methods, their strengths and limitations. Another objective is to provide a detailed case study to the reader to practice these methods and become comfortable in using them. Data sets, models, and software are provided to encourage the reader to gain hands-on experience. A dynamic simulator for batch penicillin fermentation is available as a web-based application and downloadable material. The fermentation simulator, batch process monitoring software, and software tools for supervision of batch process operations are provided at the website `www.chee.iit.edu/~cinar/batchbook.html`.

Convincing the reader about the strengths and limitations of the techniques discussed in this book would be impossible without reference to proper theory. Theoretical derivations are kept at an appropriate level to enhance the readability of the text, and references are provided for readers seeking more rigorous theoretical treatment. The level of the treatment of methodology in the book requires little background information in various areas such as biotechnology, statistics, system theory, and process control. An outline of the book and various roadmaps to read it are presented in Section 1.4. Introductory books to review the fundamentals are also suggested in Section 1.4, and advanced books are referenced in appropriate chapters in the book. Details of the algorithms are summarized in the text to permit the reader to develop software in his/her favorite environment. Executable software modules are also provided in the aforementioned website for readers who may prefer using our programs.

The book also discusses recent advances that may have an impact on the next generation of modeling, monitoring, and control methods. Metabolic pathway engineering, real-time knowledge-based systems, and nonlinear dynamics are introduced as some of the powerful paradigms that would be of interest.

This book could not have been written without the strong cooperation of the authors and the sacrifices of many family members and friends. The labor and agony of writing a multidisciplinary book tested the strength of several relationships. All four authors are grateful for the encouragement and support they have received from their loved ones. One of the authors, Cenk Undey, has done a magnificent job in coordinating the work of all authors, integrating the manuscript and providing technical support in the use of LaTeX to the others. All four authors are also grateful to Dr. Inanc Birol for contributing an important chapter on System Science Methods for Nonlinear Model Development (Chapter 5). It is certain that the impact of the methods and tools discussed in that chapter will increase in future years in analyzing the dynamics of many nonlinear batch fermentation processes and developing new monitoring and control methods. His insight and knowledge have enhanced the value of the book. It seems that no book can be published free of errors. As time progresses, errors, omissions, and better ways to express the material discussed in the book will be discovered. Each author apologizes for the remaining errors and agrees that they are the fault of the other three.

Batch fermentation operations are abundant in industries that touch many human lives. Pharmaceutical, food, and chemical industries have made significant contributions in improving health and the quality of life. They have also been cited at times for causing challenges to nature and humans. Health, food, comfort, and safety also remind us of disease, limited resources, hunger, and pollution. Advances in technology may play an important role in resolving many conflicts. The authors hope that the methods presented in this book will contribute to the safety and productivity of batch process operations, and ultimately to improving the quality of life and harmony with nature.

Ali Cinar
Satish J. Parulekar
Cenk Ündey
Gülnur Birol

Contents

Nomenclature

a	Alkaloid in Section 2.6.2	
a	Number of equality constraints in Eqs. 7.5 and 7.6	
a_{ij}	Amount of N_i utilized for production of unit amount of P_j in Eq. 2.19	(g/g)
a_{ij}	State transition probabilities of an HMM from state i to state j	
$\underline{a}$	Gas-liquid interfacial area per unit culture volume	(1/m)
$\mathbf{A}$, $\mathbf{B}$	Defined in Eqs. 7.79 and 7.105	
$\mathbf{A} = [a_{ij}]$	State transition matrix of an HMM	
A	Input metabolite in Ch. 2 and 9	
A, R	Number of PCs or LVs in Ch. 3, 4, 6 and 8	
b	Number of inequality constraints in Eqs. 7.5 and 7.6	
$\mathbf{B} = \{b_j(k)\}$	Observation symbol probability distribution	
B	Output metabolite	
$\mathbf{C}$, $\mathbf{E}$	Defined in Eqs. 7.105 and 7.106	
$\mathbf{C} = \{c_i\}$	Initial state distribution (initial state occupancy probability)	
C_{jk}^{D}	Contribution of each element in $x_{\text{new},jk}$ to D-statistic summed over all r components	
$C_i^{J_k}$	FCC of kth reaction affected by enzyme i	
$C_{jk,r}^{t_r}$	Contribution of each element of a new batch run new, jk on the rth score	

C_i	Concentration of specie i in the bulk liquid	(g/L)
C_i^*	Concentration of specie i in the liquid phase at the gas-liquid interface	(g/L)
$C_i^{J_k}$	Flux control coefficient for pathway k with respect to enzyme E_i or reaction i, defined in Eqs. 9.4 and 9.5	
$C_i^{X_j}$	Concentration control coefficient for the intermediate X_j with respect to enzyme E_i, defined in Eq. 9.7	
$C_i^{X_j}$	CCC of intermediate X_j affected by activity of enzyme i	
C_L	Dissolved O_2 concentration in Eqs. 2.47, 2.48, 2.5	(mmole/L)
C_L^*	Dissolved O_2 concentration at maximum saturation	(mmole/L)
C_X	Concentration of biomass (cell mass) in culture	(g/L culture)
C_{iG}	Concentration of specie i in the bulk gas	(g/L)
C_{iG}^*	Concentration of specie i in the gas phase at the gas-liquid interface	(g/L)
C_{jk}^D	Contribution of new observation vector $x_{\text{new},jk}$ to D-statistic	
C_{JK}^Q	Contributions to Q-statistic of J variables over the entire batch run	
C_{jk}^Q	Contributions to Q-statistic of variable j at time k	
$\mathbf{d}$	p or m_d-dimensional vector of disturbance variables	
d	Hyphal diameter	(m)
$d_i(\mathbf{x})$	Linear discriminant score	
$d_i^Q(\mathbf{x})$	Quadratic discrimination score for the ith population	
D	Differential operator in Section 4.4	
D	Dilution rate, defined in Eq. 2.11	(1/h)
D_{iG}	Molecular diffusivity of specie i in the gas phase	(m^2/h)
D_{iL}	Molecular diffusivity of specie i in the liquid phase	(m^2/h)
$\mathbf{e}, \mathbf{E}$	Residuals vector and matrix, respectively, in Ch. 4, 6 and 8	
$\mathbf{e}$	Predicted error vectors, defined in Eqs. 7.152 and 7.157	

$\mathbf{e}$	Output estimation error $\mathbf{y} - \hat{\mathbf{y}}$ in Ch. 4, 6 and 8	
$\mathbf{f}$	Nonlinear function of $\mathbf{d}$, $\mathbf{u}$, $\mathbf{x}$ in Eqs. 2.1 and 7.1	
E_i	Enzyme corresponding to the ith reaction step	
E_j	Enzyme catalyzing reaction j in a metabolic pathway	
F	Volumetric feed rate of nutrient medium	(L/h)
F, f	Bioreactor feed, gas feed or liquid feed as appropriate	
F_s	Volumetric feed rate of nutrient medium in singular control	(L/h)
$f(\phi, \omega)$	Defined in Eqs. 7.88 and 7.89	
f_h	Fraction of hyphal cells that are capable of synthesizing penicillin	
$\mathbf{G}(q)$	Multivariable input-output transfer function matrix	
$\mathbf{G}$	Transfer function matrix defined in Eqs. 7.79 and 7.108	
$\mathbf{G_1}$, $\mathbf{G_2}$	Transfer function matrices for multi-loop feedback control defined in Eq. 7.111	
$\mathbf{G}_C(q)$	Actuator (controlled input) fault TFM	
$\mathbf{G_c}$, $\mathbf{G_m}$	Transfer function matrices associated with feedback controllers and measuring devices, respectively	
$\mathbf{G_d}$	Transfer function matrix defined in Eq. 7.108	
$\mathbf{G_I}$, $\mathbf{K_I}$	Transfer function matrix and gain matrix, respectively, for the decouplers, Eqs. 7.117-7.120	
$\mathbf{G}_M(q)$	Input sensor fault TFM	
$G_1(\mathbf{y}(\mathbf{t_f}))$	Used in the definition of the objective function J in Eq. 7.4	
G_S	Throughput rate of substrate S, involved in the constraint in Eq. 7.7	(g/h)
$G(\mathbf{x}(\mathbf{t_f}))$	Used in the definition of the objective function J in Eq. 7.3	
$\mathbf{g}$	Nonlinear function of $\mathbf{d}$, $\mathbf{u}$, $\mathbf{x}$ in Eq. 2.2	
$g(\mathbf{x}, \mathbf{u})$	Used in the definition of the objective function J in Eq. 7.3	
$g'(\mathbf{y}, \mathbf{u})$	Used in the definition of the objective function J in Eq. 7.4	

$g(i)$	Impulse-response functions in Eq. 7.141	
g, m	Concentrations of glucose and methionine, respectively, in the abiotic phase (Section 2.6.3)	(g/L)
$\mathbf{h}$	Nonlinear function of $\mathbf{x}$ in Eq. 7.2	
$\mathbf{H}$	Hankel matrix in Ch. 4	
$\bar{\mathbf{H}}_{JK}$	Scaled Hankel matrix	
H	Hamiltonian defined in Eq. 7.10	
$h(\mathbf{x}, \mathbf{u})$	Used in the definition of the objective function J in Eq. 7.77	
H_i	Henry's law constant for specie i	
$\mathbf{I}$	Identity matrix	
I	Number of batches in a reference set in Ch. 4, 6 and 8	
$Im(.)$, $Re(.)$	Imaginary and real parts of a complex number or expression	
J	Objective function (performance index) defined in Eq. 7.3	
J_i	Flux of metabolic pathway i	
J_k	Metabolic flux through the kth reaction	
$\mathbf{K}$	Steady-state gain matrix defined in Eq. 7.112	
$\mathbf{K}_\infty$	Kalman filter gain in Ch. 4	
K_{iL}	Overall liquid-based mass transfer coefficient for specie i, defined in Eq. 2.6	(m/h)
k_a, k_s,k_h	Maximum specific growth rates of apical, subapical, and hyphal cells, respectively, in Eq. 2.25	(1/h)
k_{iG}	Gas-side mass transfer coefficient for specie i, defined in Eq. 2.4	(m/h)
k_{iL}	Liquid-side mass transfer coefficient for specie i	(m/h)
k_{u_j}	Kinetic coefficients in Eqs. 2.22-2.24	(1/h)
L, L_0	Nonlinear functions in Eqs. 7.162 and 7.163	
$L[\cdot]$	Simplified log-likelihood function	

M Number of distinct observation symbols per state in HMMs, the alphabet size

m Length of control sequence (controller output) prediction horizon in model predictive control

m_i Maintenance coefficient for specie i in Eq. 2.19 (1/h)

M_S Amount of substrate S supplied in a batch or fed-batch operation, involved in the constraint in Eq. 7.8 (g)

m_{ih}, m_{is}, m_{ia} Intracellular concentrations of methionine in hyphae, swollen hyphal fragments, and arthrospores, respectively (g/g)

$\mathbf{N}_M$ Coefficient matrix of multiplicative modeling faults

$\mathbf{N}_P$ Matrix of time-varying coefficients of multiplicative parametric faults

N Number of states in an HMM

N_i Concentration of nutrient N_i in the abiotic phase (g/L)

N_i Flux of specie i from the gas phase to liquid phase, defined in Eqs. 2.4, 2.5 and 2.6 (g/{m^2.h})

N_i Nutrient i

$\mathbf{O}$ Observation sequence in an HMM ($\mathbf{o}_1 \cdots \mathbf{o}_T$)

P, Q, R Defined in Eqs. 7.79 and 7.132

Q, R, S Positive definite weighting matrices in Eq. 7.163

$\mathbf{p}, \mathbf{P}$ Loading vector and matrix, respectively, in Ch. 4, 6 and 8

P Concentration of target non-biomass product (g/L)

P Target non-biomass product (sections 2.6.4 and 7.2.5)

P_j Concentration of non-biomass product P_j on the basis of the abiotic phase volume (g/L)

P_j Non-biomass product j

$\mathbf{p}$ Parity vector

p Extracellular phosphate concentration in Section 2.6.2 (g/L)

p	Length of output prediction horizon in model predictive control	
p	Target product (cephalosporin C) in Section 2.6.3	
p_i	Intracellular phosphate concentration in Section 2.6.2	(g/g)
pH	Culture pH	
Q	Evolving sequence of states S of an HMM in Section 8.3.3	
Q	Volumetric flow rate of culture in Ch. 2	(L/h)
Q_a	Volumetric flow rate of abiotic phase, defined in Eq. 2.9	(L/h)
Q_b	Volumetric flow rate of biotic phase, defined in Eq. 2.9	(L/h)
Q_{SN}	Quantile of standard Normal distribution	
Q_Y	Quantile of ordered data set	
$[q]$	Intracellular concentration of specie q (Section 2.6.4)	(g/g)
q	Shift operator in Section 4.4	
q_{ij}	Intracellular concentration of specie q in cell type j (Section 2.6.3)	(g/g cell type j)
q_t	Actual state of a discrete-time system at time t	
$\hat{q}_k$	Maximum likelihood estimate	
$\mathbf{R}$	Defined in Eq. 7.113	
$r(\mathbf{K})$	Rank of a matrix $\mathbf{K}$	
r_1, r_2	Amplitudes of periodic variations in u_1 and u_2, respectively (Section 7.3)	
r_d	Specific rate of cell loss due to cell death or cell lysis	(1/h)
r_i	Net rate of generation of specie i in the biotic phase in Ch. 2	(1/h)
r_i	Residual based on the PC model for fault i in Ch. 8	
R_i^{gen}	Rate of generation of specie i due to reactions in the abiotic phase	(g/{L.h})
r_i^{gen}	Net rate of generation of specie i in the biotic phase exclusive of the rate of its loss from the biotic phase due to cell death or cell lysis	(1/h)

r_i^{trans} Biomass-specific rate of transport of specie i from the biotic phase to the abiotic phase (1/h)

r_{qj}^{gen} Specific rate of net generation of specie q in cell type j (g/{g cell type j.h})

r_{qj}^{trans} Specific rate of transport of specie q from the cells of type j to the abiotic phase (g/{g cell type j.h})

$\mathbf{S}$ Riccati transformation matrix in Eqs. 7.136 and 7.137

$\mathbf{S}$ Covariance matrix of scores in Ch. 4, 6 and 8

$\mathbf{S}_B$ Between-class scatter matrix

$\mathbf{S}_f$ Defined in Eq. 7.132

$\mathbf{S}_{PF}(q)$ Plant fault TFM

$\mathbf{S}_{pl}$ Pooled estimate of $\mathbf{\Sigma}$

$\mathbf{S}_{PN}(q)$ Plant noise TFM

$\mathbf{S}_W$ Within-class scatter matrix

$\mathbf{S}_Y$ Total scatter matrix

S Concentration of limiting substrate in the abiotic phase (liquid) (g/L)

S Distinct states of a discrete-time system in Ch. 8

S Limiting substrate

s Laplace transform variable

s_i Score distance based on the PC model for fault i

$\mathbf{t}, \mathbf{T}$ Scores vector and matrix, respectively, in Ch. 4, 6 and 8

t_f Duration of a bioreactor operation (h)

T Sampling period (h)

$\mathbf{u}$ m-dimensional vector of manipulated inputs

$\mathbf{u}, \mathbf{U}$ PLS scores vector and matrix, respectively, in Ch. 4, 6 and 8

u_1, u_2, u_3 Rates of the three metamorphosis reactions in (Eqs. 2.22-2.24) (1/h)

$\mathbf{V}$, $\mathbf{W}$	Scaling matrices in Eq. 7.158	
$\mathbf{V}$, $\mathbf{W}$, $\mathbf{v}_i$, $\mathbf{w}_i$	Defined in Eqs. 7.122, 7.123 and 7.126	
V	Culture volume	(L)
V_a	Volume of abiotic phase, defined in Eq. 2.9	(L)
V_b	Volume of biotic phase, defined in Eq. 2.9	(L)
V_G	Volume of gas phase in the bioreactor (gas phase holdup)	(L)
V_T	Culture volume	(L)
$\overline{V}_{\text{biotic}}$	Specific volume of the cells	(L/g)
$\mathbf{v}_C$	Input actuator noise	
$\mathbf{v}_M$	Input sensor noise	
$\mathbf{v}_P$	Plant noise	
$\mathbf{v}_y$	Output sensor noise	
v_j	Rate of reaction j in a metabolic pathway (Eq. 9.1)	
v, w	Parameters in the definition of J in Eq. 7.95	
$\overline{\mathbf{v}}$	Deviation in $\mathbf{v}(t)$ from its reference, y_r, $\mathbf{v}= \mathbf{d}, \mathbf{u}, \mathbf{x}$ in section 7.4.1	
$\overline{\mathbf{v}}(s)$	Laplace transform of $\overline{\mathbf{v}}(t)$	
$\mathbf{W}$	Projection matrix	
$\mathbf{W}$	Weight matrix in Ch. 4, 6 and 8	
W	Wavelet transform	
$\hat{\mathbf{w}}$	Estimated weight functions in Section 4.4	
$\underline{\mathbf{X}}$	Three-way array of process measurements in Ch. 4, 6 and 8	
$\mathbf{X}$	Unfolded matrix of process measurements in Ch. 4, 6 and 8	
$\mathbf{X}$	Dynamic Matrix defined in Eqs. 7.150b and 7.157	
X	Biomass (cell mass)	
X	Concentration of biomass (cell mass) in culture	(g/L culture)

X_h, X_s, X_a Concentrations of hyphae, swollen hyphal fragments, and arthrospheres, respectively (Section 2.6.3) (g/L culture)

X_h, X_s, X_a Three morphological types of *Cephalosporium acremonium*, hyphae, swollen hyphal fragments, and arthrospheres, respectively (Section 2.6.3)

$\mathbf{x}$ Vector of state variables

$\mathbf{x}$ Vector of process measurements

$\hat{y}$ Predicted output vector

$\mathbf{y}, \mathbf{Y}$ Vector and matrix of quality measurements, respectively, in Ch. 4, 6 and 8

$\mathbf{y}$ l-dimensional vector of output variables

$\mathbf{y_m}$ Vector of measured outputs

$\mathcal{Y}_{k_J}^{+}$, $\mathcal{Y}_{k_K}^{-}$ Stacked vectors of future and past

$Y_{P/X}$ Cell mass phosphate content in Eq. 2.30 (g/g)

Y_{X/N_i} Biomass (cell mass) yield with respect to nutrient N_i in Eq. 2.19 (g/g)

Z Defined in Eq. 7.81

z z-transform variable

Z_a, Z_s, Z_h Mass fractions of apical, subapical, and hyphal cells, respectively, in the total cell population (Eqs. 2.22-2.24)

Z_h, Z_s, Z_a Mass fractions of hyphae, swollen hyphal fragments, and arthrospheres, respectively, in the total cell population (Section 2.6.3)

Greek Letters

α_j Constant characteristic of a particular metabolite P_j in Eq. 2.21

β, $\boldsymbol{\beta}$ Vector and matrix of regression coefficients, respectively

$\beta(i)$ Step-response functions in Eq. 7.142

β_j Constant characteristic of a particular metabolite P_j in Eq. 2.21 (1/h)

$\boldsymbol{\epsilon}$ State estimation error $\mathbf{x} - \hat{\mathbf{x}}$

$\boldsymbol{\Gamma}$ Regressor matrix

$\boldsymbol{\Gamma}$, $\boldsymbol{\Lambda}$ Matrices involved in Eqs. 7.158 and 7.159

$\boldsymbol{\Lambda}$ Relative gain array (RGA)

$\boldsymbol{\lambda}$ Vector of adjoint variables associated with state equations, defined in Eqs. 7.11, 7.81, and 7.136

$\boldsymbol{\phi}'(\mathbf{y}, \mathbf{u})$ Argument vector in the equality constraints in Eq. 7.6

$\boldsymbol{\phi}(\mathbf{x}, \mathbf{u})$ Argument vector in the equality constraints in Eq. 7.5

$\boldsymbol{\phi}_M$ Modeling errors

$\boldsymbol{\phi}_P$ Parametric faults

$\boldsymbol{\Pi}(\omega)$ Hermitian matrix defined in Eq. 7.78

$\boldsymbol{\psi}'(\mathbf{y}, \mathbf{u})$ Argument in the inequality constraints in Eq. 7.6

$\boldsymbol{\psi}(\mathbf{x}, \mathbf{u})$ Argument in the inequality constraints in Eq. 7.5

$\boldsymbol{\Sigma}$ Defined in Eq. 7.127

$\varepsilon(t)$ Error vector, vector of inputs to controllers

$\widehat{\varepsilon}(t)$ Vector of prediction errors

$\boldsymbol{\eta}$, $\boldsymbol{\rho}$ Vectors of adjoint variables associated with integral constraints in Eq. 7.5, defined in Eqs. 7.81

χ Optimum phase difference between u_1 and u_2 in forced periodic operation, defined in Eq. 7.89

$\chi_{ji}(N_i)$ Functions in the expression for ϵ_j in Eq. 2.20 and Table 2.2

$\delta\mathbf{u}_C$ Input actuator faults

$\delta\mathbf{u}_M$ Input sensor faults

$\delta\mathbf{u}_P$ Plant faults

$\delta\mathbf{y}$ Output sensor faults

δ_G Thickness of the gas-side boundary layer (m)

δ_L Thickness of the liquid-side boundary layer (m)

$\epsilon^i_{X_J}$	Elasticity of reaction rate i with respect to concentration of metabolite X_j	
ϵ	Cell-mass specific production rate of P	(1/h)
ϵ	Measurement error	
ϵ_j	Cell-mass specific production rate of P_j	(1/h)
ϵ_{jo}	Characteristic of a particular strain in Eq. 2.20	(1/h)
$\lambda = (\mathbf{A}, \mathbf{B}, \mathbf{C})$	The set of probabilities in an HMM	
$\Lambda_k(r)$	Log-likelihood ratio	
λ_{ij}	(i, j)th element of the relative gain array (RGA), defined in Eq. 7.113	
μ	Specific cell growth rate	(1/h)
μ^{net}	Net specific cell growth rate, defined in Eq. 2.10	(1/h)
μ_a, μ_s, μ_h	Specific growth rates of apical, subapical, and hyphal cells, respectively, defined in Eq. 2.25	(1/h)
μ_o	Characteristic of a particular strain in Eq. 2.17	(1/h)
ν_i	Reaction rate of the ith reaction step	
ω	Forcing frequency in forced periodic operation	(cycles/h)
ω_1, ω_2	Forcing frequencies for u_1 and u_2, respectively, in forced periodic operation	(cycles/h)
Φ	Wavelet scaling function	
$\phi_i(N_i)$	Functions in Eq. 2.17 and Table 2.1	
π_i	Classes of events such as distinct operation modes $i = 1, \cdots, g$	
Ψ	Mother wavelet	
$\psi(X)$	Functions in Eq. 2.17 and Table 2.1	
$\psi_j(P_j)$	Functions in Eq. 2.17 and Table 2.1	
$\psi_{jk}(P_k)$	Functions in the expression for ϵ_j in Eq. 2.20 and Table 2.2	
ρ	Culture density	(g/L)

ρ_b	Density of biomass	(g/L)
ρ_{ij}	(i, j)th element of $\mathbf{\Pi}(\omega)$, section 7.3	
σ	Cell-mass specific uptake rate of limiting substrate S (Section 7.2.5)	(1/h)
σ_i	Cell-mass specific uptake rate of nutrient N_i	(1/h)
σ_j	Standard deviation of summed mean contributions over time instances	
σ_j	jth singular value of a matrix	
σ_k	Standard deviation of summed mean contributions over all process variables	
τ	Cycle period in forced periodic operation	(h)
θ	Model parameters vector	

Subscripts

abiotic	Abiotic phase
biotic	Biotic phase
f	At the end of bioreactor operation ($t = t_f$)
F	Bioreactor feed, gas feed or liquid feed as appropriate
J	Partial derivative with respect to J (Sections 7.2.5 and 7.3.2)
m, max	Maximum values of a variable
min	Minimum value of a variable
r	Reference state/value
sp	Set point
syn, util	Synthesis and utilization, respectively (Section 2.6.3)
0, $\mathbf{0}$	Initial conditions
0, $\mathbf{0}$	Steady-state conditions (Section 7.3)
-	Reciprocal of a scalar or inverse of a matrix

Superscripts

c Complex conjugate

T Transpose of a matrix

$*$ Optimal trajectory/value or desired trajectory/value

Abbreviations

adj **A** Adjoint of a matrix **A**, Eqs. 7.119 and 7.120

AHPCA Adaptive hierarchical principal component analysis

AIC Akaike information criteria

ANN Artificial neural network

AO Additive outlier

AR Auto regressive

ARL Average run length

ARMA Auto regressive moving average

ARMAX Auto regressive moving average with exogenous inputs

ARX Auto regressive model with exogenous inputs

BJ Box-Jenkins

CCC Concentration control coefficient

CPCA Consensus principal components analysis

CUMPRESS Cumulative prediction sum of squares

CUSUM Cumulative sum

CV Canonical variate

CVA Canonical variates analysis

CVSS Canonical variate state space (models)

d.f. Degrees of freedom

diag **A** Diagonal matrix containing the diagonal elements of a matrix **A**

DMC Dynamic-matrix control

DOE Design of experiments

DTW Dynamic time warping

ECM Expected cost of misclassification

EKF Extended Kalman filter

EWMA Exponentially weighted moving average

FCC Flux control coefficient

FDA Fisher's discriminant analysis in Section 8.2.2

FDA Functional data analysis in Section 4.4

FDD Fault detection and diagnosis

FIA Flow injection analysis

FPE Final prediction error

G A specific gene (Section 2.6.4)

GAM Generalized additive model

GC Gas chromatography

GLR Generalized likelihood ratio

GUI Graphical user interface

HMM Hidden Markov model

HPCA Hierarchical principal components analysis

HPLC High pressure liquid chromatography

HPLS Hierarchical partial least squares

ILC Iterative Learning Control

IO Innovational outlier

IV Indicator variable

KBS Knowledge-based System

LCL Lower control limit

LMS Least median squares

LPV Linear parameter varying

LQC Linear quadratic Gaussian control

LTS Least trimmed squares

LTV Linear time varying

LV Latent variable

LWL Lower warning limit

MA Moving average

MAC Model algorithmic control

MARS Multivariate adaptive regression splines

MIMO Multiple-input, multiple-output control/system

MPC Model predictive control

NMPC Nonlinear model predictive control

MBO Model-based optimization

MBPCA Multiblock principal components analysis

MBPLS Multiblock partial least squares

MCA Metabolic control analysis

MFA Metabolic flux analysis

MHBE Moving horizon Bayesian estimator

MIMO Multi-input multi-output

MLE Maximum likelihood estimate

MLR Multiple linear regression

MOBECS Model-Object Based Expert Control System

MPCA Multiway principal component analysis

MPLS Multiway partial least squares

mRNA Messenger ribonucleic acid

MS Mass spectrometer

MSE Least squares mean squared error

MSMPCA Multiscale Multiway principal component analysis

MSPM Multivariate statistical process monitoring

MV Multivariate

NAR Nonlinear auto regressive

NARMAX Nonlinear autoregressive moving average with exogenous inputs

NLTS Nonlinear time series

NO Normal operation

NOC Normal operating conditions

NPETM Nonlinear polynomial models with exponential and trigonometric functions

OD Optical density

OE Output error

OVAT One-variable-at-a-time

PARAFAC Parallel factor analysis

PC Principal component

PCA Principal components analysis

PCD Parameter change detection (method)

PCR Principal components regression

PDA Principal differential analysis

PDF Probability distribution function

PLS Partial least squares (Projection to latent structures)

PRESS Prediction sum of squares

PSSE Penalized sum of squared error

PSSH Pseudo-steady state synthesis

QQ Quantile-Quantile

RGA Relative gain array

RQ Respiratory quotient

RTKBS Real-time knowledge-based systems

RVWLS Recursive variable weighted least squares

RWLS Recursive weighted least-squares

SISO Single-input single-output

SNR Signal-to-noise ratio

SPC Statistical process control

SPE Squared prediction error

SPM Statistical process monitoring

SS Sum of squares

SSE Sum of squares explained

SSR Regression sum of squares

SSY Sum of squares on **Y**-block

STFT Short-time Fourier transform

SV Singular values

SVD Singular value decomposition

TFM Transfer function matrix

UCL Upper control limit

UWL Upper warning limit

VIP Variable influence on projection

BATCH FERMENTATION

1

Introduction

Batch processes have been around for many millennia, probably since the beginning of human civilization. Cooking, bread making, tanning, and wine making are some of the batch processes that humans relied upon for survival and pleasure. The term "batch process" is often used to refer generically to both *batch* and *fed-batch* operations. In the former case, all ingredients used in the operation are fed to the processing vessel at the beginning of the operation and no addition or withdrawal of material takes place during the batch run. In the latter, material can be added during the batch run. For brevity, the term batch is used in this text to refer to both batch and fed-batch operations when there is no need to distinguish between them. The term fed-batch is used to denote addition of material in some portions of an otherwise batch operation.

Batch processes have received increasing attention in the second half of the twentieth century. Specialty chemicals, materials for microelectronics, and pharmaceuticals are usually manufactured using batch processes. One reason for this revival is the advantages of batch operation when there is limited fundamental knowledge and detailed process models are not available. Batch processes are easier to set up and operate with limited knowledge when compared to continuous processes. The performance of the process can be improved by iterative learning from earlier batch runs. A second reason is the increasing pressure to start commercial production of novel materials once patents have been issued to recover research and development costs before competing products affect prices. Another reason is the ability to use the facilities for many products with little or no hardware modification. Many pharmaceutical products are produced in limited quantities and the plant manufactures a specific product for a short period of time before switching to another product. Batch operation is usually more efficient than continuous operation for frequent product changes and small amounts of products.

Although batch processes are simple to set up and operate, modeling, monitoring, and controlling them is quite challenging. Consider a simple operation like cooking spaghetti. The basic steps involved are simple. Heat some water, immerse the spaghetti strings in boiling water, drain the water after the spaghetti is cooked, add oil and sauce, and serve. But the actual process to make good spaghetti is more complex and requires many well-timed decisions. What should be the temperature of the water when the spaghetti strings are added, how long should the spaghetti be cooked in water, how much oil and what other seasoning and ingredients should be added to the spaghetti sauce? This is a process with several *phases* (operations in the same vessel for a specific activity such as cooking or fermentation) and *stages* (operations in different vessels for different activities such as raw material preparation and product separation). The *landmarks* denoting the end of one phase and beginning of the other should be monitored for proper timely actions. For example, spaghetti should not be added to water that is not hot enough, otherwise the strings will stick to each other. A good landmark is boiling of water which can be detected easily as opposed to water temperature reaching 200 ^oF. The latter would work equally well for the cooking operation but will be more difficult to detect, monitor (a thermometer would be needed) and regulate. The duration of keeping the spaghetti in hot water will change because of many factors. These include the relative amounts of water and spaghetti (the initial charge of ingredients), the tenderness of cooked spaghetti (a quality variable that varies with personal taste and weight watching – it is said that absorption of the carbohydrates by the body increases as the spaghetti gets tender), type of spaghetti flour (whole wheat or bleached flour), and the amount of heat provided (one can turn the heat off and keep the strings in hot water longer). Consequently, while developing an optimal reference trajectory for this example process, one may have to take into consideration variations in batch run duration and other factors that influence the degree of cooking. Developing a detailed model of this simple process based on first principles may be even more challenging. A simple empirical model based on data may be accurate enough for most needs. Most industrial batch processes have more process and quality variables, and more stringent operational and financial constraints. Consequently, development of reference trajectories, determination of change point landmark occurrence, quality assessment, and monitoring of process and product safety are much more challenging.

This book focuses on batch process modeling, monitoring, fault diagnosis, and control. The discovery of a new drug such as a new antibiotic or a new manufacturing method that revolutionizes yield and productivity are critical for commercial success. Biology, chemistry, bioinformatics, and biochemical engineering provide the foundations for these advances. But, large-scale commercial production with consistent product quality, stringent process and product safety requirements, and tight production schedules necessitate a different set of skills built upon systems science, statistics, and control theory. The focus then shifts to finding optimal reference trajectories and operating conditions, and manufacturing the product profitably in spite of variations in raw materials and ambient conditions, malfunctions in equipment, and variations in operator judgement and experience. Techniques in model development, signal processing, data reconciliation, process monitoring, fault detection and diagnosis, quality control, and process control need to be integrated and implemented. The book provides a unified source to introduce various techniques in these areas, illustrate many of them, and discuss their advantages and limitations.

The book presents both fundamental and data-based empirical modeling methods, several monitoring techniques ranging from simple univariate statistical process control to advanced multivariate monitoring techniques, many fault diagnosis techniques and a variety of simple to advanced process control approaches. Techniques that address critical issues such as landmark detection, data length adjustment, and advanced paradigms that merge monitoring and diagnosis activities by a supervisory knowledge-based system are discussed. The methods presented can be used in all batch processes by paying attention to the special characteristics of a specific process. The focus of the book is on batch fermentation and pharmaceutical processes. Penicillin fermentation is used as a case study in many chapters throughout the book. Various paradigms are introduced in each subject to provide a balanced view. Some of them are based on the prior research of the authors, others have been proposed by other researchers. Appropriate examples and case studies are presented to illustrate some of the methods discussed. A dynamic simulator for batch penicillin fermentation and batch process monitoring software are provided in the Web. The readers are invited to check the Web site of one of the authors at `www.chee.iit.edu/~cinar/batchbook.html` for the penicillin fermentation simulator and software tools for supervision of batch process operations.

This chapter continues with a discussion of batch process operations in Section 1.1. Section 1.2 provides introductory remarks about the main focus areas of the book: modeling, monitoring, control, and diagnosis. Section 1.3 introduces the penicillin fermentation process that is used in many case studies in various chapters. The last section (1.4) of the chapter provides an outline of the book and provides road maps for readers.

1.1 Characteristics of Batch Processes

Batch operation is characterized by two key elements: (i) the physical configuration that consists of various reactors, tanks, and the network of pipelines available to transfer material between various tanks and production units, and (ii) the sequence of processing tasks. Typically, final products are produced from a number of raw materials and intermediate products through a series of processing tasks. All processing tasks are realized in batch mode of operation, with minimum and maximum batch sizes predetermined by the nature of the processes and the capacity of the reactors [68]. There are number of process specific issues that should be considered in a typical batch processing:

- Because the duration of chemical reactions are fixed, processing times are assumed constant and predetermined irrespective of the particular batch size.
- In many practical applications, the number and size of the individual batches are not known in advance, and hence considered as decision variables. Moreover, merging and splitting of batches are allowed.
- Processing of a single batch is carried out uninterrupted.
- The proportions of input and output materials may be fixed or variable, depending on the particular process.
- Storage conditions depend on availability and capacity of appropriate storage facilities. In extreme cases, reactors themselves can be used as intermediate storage devices.

The inherent advantages of batch processes, such as flexibility to produce multiple related products in the same facility and ability to handle variations in feed stocks, product specifications and market demand pattern, make them well suited for the manufacture of low-volume, high-value products. Ultimate goals of the industry is to reduce time-to-market, lower costs, comply with regulatory requirements, minimize waste and emissions and increase return-on-investment [393].

The main disadvantage of batch processing is the high proportion of unproductive time (down-time) between batches, consisting of times to charge and discharge the reactor, cleaning of vessels and pipes, and restart process.

Batch Bioprocesses

The importance of batch processes in biotech process industries has increased significantly in recent years. Batch processes are extensively used to produce specialty chemicals, biotechnology, pharmaceutical and agricultural products. The production of these high value-added chemicals, as opposed to bulk commodity chemicals, contributes to a significant and growing portion of the revenue and earnings of bioprocess industries. Considering the growing trend in industry towards products with short life cycles and products tailored to specific market needs, rapid process development has become even more significant. With the current pressures of global competition, economic efficiency often dictates whether a manufacturer can compete on a cost basis, an issue of special relevance to the pharmaceutical industry, which is additionally faced with a lengthy government approval process for its production [393]. Environmental concerns are also another key issue faced with batch bioprocesses today.

Batch bioprocesses refer to a partially closed system in which most of the materials required are loaded onto the bioreactor aseptically and are removed at the end of the operation. Contamination of production bioreactors may lead to economic loss and is cause for alarm. Infections by phage are particularly difficult to combat because the virus particles are small enough to escape capture by the filters used to sterilize the air provided to the bioreactors. Phage attacks can be overcome by switching to resistant strains of the microorganisms. In a batch bioprocess, the only material added and removed during the course of operation is air/gas exchange, antifoam and pH controlling agents. For years, batch fermenters were loaded, inoculated, and run to completion with nothing added except air and some agent to control foam. Most modern bioprocesses incorporate adjustments to the medium to control conditions and to supply nutrients and compounds that promote biosynthesis of the desired product. It seems obvious that changes in the batch process should affect formation of the desired product(s) and that these changes can be controlled by additions of certain materials. Also of great interest is interfacing with the bioreactor system with computers to monitor and control it. Since a bioreactor consists of a complicated system of pipes, fittings, wires, and sensors, it is open to malfunctioning. With the aid of on-line monitoring and diagnosis tools, it is now possible to detect many things that can go wrong during the process.

The cultivation broth is assumed to be uniform throughout the reactor at any instant of time in a well-mixed bioreactor. However, these processes exhibit time variant dynamic behavior and are characterized by complex, nonlinear physiological phenomena that are difficult to model.

The stirred tank bioreactor is still the workhorse of bioprocess industries involving microbial cell cultures. Although there are many alternative designs, roughly 85 percent of bioreactors in the world resemble closely to the conventional design. There were already fermentation vats such as those for beer, whiskey, pickles, or sauerkraut, but the conventional design evolved in the 1940's as the pharmaceutical companies scaled up reactors for antibiotics from shake flasks and milk bottles to stirred tanks with features to discourage entry of contaminating organisms. Typical sizes for commercial production bioreactors are 60,000 to 200,000 *liters*, but there are a few that are considerably larger. One famous bioreactor that was known as the Merck hot dog was a cylinder laying on its side with four or five agitators mounted along the top. Its dimensions were 3.6 m diameter by 27 m long. The world's largest industrial bioreactor is still the ICI's air lift system first operated at the Billingham, U.K. plant for producing single-cell protein in 1979. The size of a bioreactor is limited by its ability to remove the heat generated by cellular metabolism. Volume goes up by a dimension cubed while area depends on a dimension squared. This means that the volume of culture fluid overwhelms the heat transfer area when the fermenter is very large. Products based on genetic engineering tend to be produced in small amounts and are suited to much smaller bioreactors. Furthermore, production cultures derived from plant, animal, or insect cells require expensive media which contain many more special nutrients than those present in media employed for synthesis of antibiotics, vitamins, and other products with bulk markets. The microorganisms that make antibiotics, in particular, are relatively easy to cultivate because their products discourage the growth of other microorganisms. Animal cell cultures, in contrast, have no self-protection and cannot compete with hardy, rapidly-growing microorganisms that find the media delectable [133].

1.2 Focus Areas of the Book

Maximizing benefits by optimization of product quality and yield is the ultimate goal of industrial bioprocess operations. In batch processes, this can be achieved by a repeatable process to convert raw materials to final products and tools for supervision of process operation and intervention when needed. Process development involves integration of scientific knowledge with sound design principles. Considering the vast number of permutations in raw materials types and properties, processing methods, and operating conditions, relying exclusively on experimental trial-and-error is not a feasible option for process development. Powerful design of experiments tools are available to reduce the number of experiments, but the cost and length of relying exclusively on experiments are still prohibitive for developing most industrial processes. Use of modeling and simulation tools reduces the cost and the length of the process development period significantly. Process models are also the cornerstone of process monitoring, process control, and fault diagnosis tools that are the building blocks for supervision of process operation and intervention decisions. The book focuses on these four topics: modeling, monitoring, control, and diagnosis. Introductory remarks on these four topics are given in the following subsections.

1.2.1 Batch Process Modeling

Process models can be classified into two groups: *first principles* (fundamental) models and *data-based* (empirical, black box) models. First principles models are based on fundamental theories or laws, such as the conservation of mass, energy and momentum. One of the most important reasons for using fundamental models is the analytical expressions they provide relating key features of the physical system to its dynamic behavior. Data-based models provide relations between measured inputs and outputs that describe how the process responds to changes in various inputs. They can be developed much faster than first principles models, but their accuracy, robustness, and usability are limited. They provide an inexpensive alternative to fundamental models in most monitoring, diagnosis and control tasks.

First Principles Models of Bioprocesses

The central theme of mathematical modeling of bioprocesses is the abstraction of physical phenomena into a suitable simplified mathematical formalism [615]. Even the simplest living cell is a system of such complexity that any mathematical description of it is an extremely modest approximation [32]. For this reason, a fundamental understanding of the phenomena taking place in the cell is needed to develop an acceptable first principles model. In the context of biological systems, this requires the presumption of metabolic intermediates and pathways that are crucial to system behavior and the specific regulatory role they play. This approach may require a number of iterations since the pathways may consist of large number of biochemical reactions [615].

The first step in developing a bioprocess model is to specify model complexity. Model complexity depends primarily on the purpose the model such as description of specific intracellular events or biochemical reactions, effects of environmental variables and effects of bioreactor operating conditions on growth and product formation. Model specifications include the number of biochemical reactions in the model, specification of the stoichiometry for these reactions, and related assumptions and simplifications. In setting up bioprocess models, lumping of biochemical reactions is done while paying attention to the detail level appropriate for the intended use of the model developed. After the model complexity is specified, rates of biochemical reactions are described with appropriate mathematical expressions using linear or nonlinear formulations. The rates are defined as functions of bioprocess variables, namely the concentrations of substrate(s) and metabolic products. These functions are referred to as *kinetic expressions*. Biochemists traditionally use elemental balances as their basic models, formulated as reaction equations. These balances define the biochemical state identifying the components which change considerably during the process, and contain information on the yields of various species with respect to some reference species. Besides stoichiometric relationships, empirical relations discussed in Chapter 2 can also be used as black box expressions.

The second step in modeling is to develop mass, energy, and/or momentum balances based on bioreactor operation mode (batch, fed-batch, or continuous) and combine them with kinetic expressions of the bioprocess. In general, homogeneity is assumed within the bioreactor for the sake of simplicity. Detailed bioreactor models that include spatial non-uniformity are also available. The combination of kinetic and bioreactor models form the complete mathematical description of the bioprocess.

The final step in first principles model development is assigning values to operating and kinetic parameters. The former depend on operating conditions such as volumetric liquid/gas flow rates of inputs and outputs, rotational speed of the impeller, and environmental conditions. The latter are associated with the biological system under consideration. Parameter estimation algorithms are used to assign values to these parameters. The basic steps in developing first principles models of bioprocesses are summarized in Figure 1.1. Detailed discussion of first principles models of bioprocesses and case studies are presented in Chapter 2.

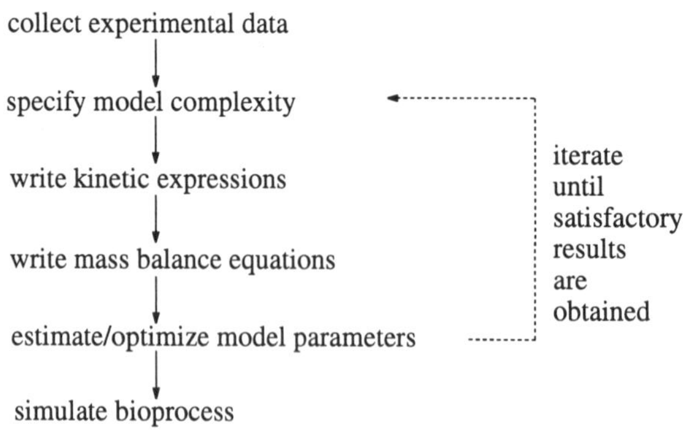

Figure 1.1. Steps in model formulation of bioprocesses.

Data-based Models of Bioprocesses

Process models developed by using fundamental principles *explain and describe* the behavior of the process. This provides the opportunity to assess the importance of various fundamental phenomena such as steps in the metabolic pathway, effects of various modes of mass transfer, or limitations in energy exchange. The user can postulate the existence or lack of some phenomena, modify the model accordingly and compare the predictions of the model with data to determine if the assumptions made could be supported. Often, the process may be too complex or the information may be too limited to develop fundamental models. In addition, the fundamental models developed may be too large to be used in process monitoring, fault diagnosis, and control activities. These activities require fast execution of the models so that regulation of process operation can be made in a timely manner. The alternative model development paradigm is based on developing relations based on process data.

Statistics and system identification literature provide a large number of methods for developing models to represent steady-state and dynamic relations that *describe* process behavior. Powerful model development software is available to build models that relate process inputs to process and quality variables or relate process variables and product properties with ease and speed. There are two important tradeoffs. First, the model describes the process behavior as it is captured in data used for model development. Unlike first principles models, data-based models cannot provide information on behavior that has not been observed (in the sense of capturing in data). Data-based models should not be used for extrapolation, and nonlinear data-based models should be used with caution for interpolation as well. The second tradeoff is the loss of the ability to link physical or biochemical phenomena directly with some part of the model. Hence, the capability to explain the mechanisms in play for the observed behavior is severely limited. One has to rely on sensitivity analysis between inputs and outputs, and expert knowledge to extract information that sheds light on fundamental phenomena that are important in a specific process. However, one should not underestimate the power of good data-based models. In diverse fields such as the stock market, aircraft and ship navigation, oceanography, agriculture, and manufacturing, data-based models have played a significant role. Some data-based modeling methods are built with algorithms that are useful in mining historical databases to elucidate hidden relations in process and product variables. Data-based modeling techniques such as principal components regression and subspace state-space models provide good insight about the largest directions of variation in data and most influential variables. This information can be valuable in enhancing process understanding and developing fundamental models.

Data-based models are frequently used in process monitoring and control, quality control or fault diagnosis activities. Many case studies included in the book illustrate the value and power of data-based models in supervising process operations. Data-based models are discussed in Chapters 4 and 5, their uses in monitoring, control, and fault diagnosis are illustrated in Chapters 6, 7, and 8, respectively.

Applications of Process Models

Various reasons can be listed for making and using mathematical models of bioprocesses, each important in some venue of biotechnology [32]:

- *To organize disparate information into a coherent whole*: Molecular biology and biotechnology have generated and will continue to generate vast amounts of information about components involved in various biological processes and their properties. Models will enable integrated consideration of many interacting components that would shed light on many questions about the genes and their integrated roles.

- *To think logically about components and interactions that are important in a complex system and calculate them*: As more experimental data become available on protein-nucleic acid interactions and theoretical possibilities to predict effects of sequence changes on these parameters improve, genetically structured models can provide a unique resource for predicting the relationship between nucleotide sequence and complex functions of the organism.

- *To discover new strategies in process operation*: Cell metabolism can be controlled by manipulation of environmental parameters such as pH, temperature, dissolved oxygen, aeration rate and other operating conditions. Cell metabolism is reflected in measurable quantities of the culture (measured variables). Models are the crucial link between these two groups of variables, enabling implementation of an algorithm for operating the process effectively based on accessible on-line measurements of the process.

- *To test and modify conventional wisdom*: Many situations encountered in biochemical engineering, and biological science research are extremely complex. It is easy to make erroneous hypotheses or assumptions about a bioprocess. Mathematical modeling and analysis of the resulting model, can aid substantially in avoiding such mistakes or in identifying errors or omissions in earlier thinking and interpretations.

- *To understand the essential, qualitative features*: When analyzing a complex system, it is often sufficient to have certain qualitative results without the need for particular numerical value. Qualitative analysis becomes increasingly important as the system under investigation becomes complex.
- *To build model-based monitoring, control and diagnosis techniques*: Multivariable first principles and data-based models are critical for developing powerful techniques for monitoring and controlling process operations and for diagnosing source causes of faulty operation.

1.2.2 Process Monitoring

Batch processes convert raw materials to products during a finite period of time by following prescribed processing recipes. A high degree of reproducibility is necessary to obtain successful batches. Monitoring and control of batch processes are crucial for detecting deviations from reference trajectories and interfering with undesirable trends to bring the operation to conditions that assure acceptable product quality. The goal of *statistical process monitoring* (SPM) is to detect the existence, magnitude, and time of occurrence of changes that cause a process to deviate from its desired operation.

Traditional statistical process control (SPC) has focused on monitoring quality variables at the end of a batch and if the quality variables are outside the range of their specifications making adjustments (hence control the process) in subsequent batches. An improvement of this approach is to monitor quality variables during the batch run and make adjustments in the same run if they deviate from their expected ranges. Monitoring of quality variables usually involves measurement and reporting delays. Information about quality variations is encoded in process variables, and measurement of process variables is often frequent and highly automated. Hence, monitoring of process variables is useful not only for assessing the status of the process, but also for controlling product quality.

In traditional quality control of multivariable processes, several quality variables are monitored using univariate SPC techniques such as Shewhart charts. This approach considers each variable in isolation. In contrast, multivariate techniques focus on the whole picture and generate an alert when many process variables make small moves from their mean values in a way that indicates a specific trend. They leverage the interaction between variables and monitor changes in the correlation structure of the variables.

The book presents many process monitoring and quality control tools in Chapter 6, starting with simple univariate SPC charts. Several multivariate SPM techniques for end of batch and real-time on-line monitoring are introduced and integrated with quality control. Furthermore, these tools are linked with fault diagnosis to offer an automated process monitoring and diagnosis environment in Chapter 8.

1.2.3 Process Control

Automatic control of batch fermentation processes provides the opportunity to regulate the operation when variations in input conditions such as changes in impurity compositions in feedstock or disturbances during the run such as equipment malfunctions may cause departure from optimal reference trajectories. A simple temperature control loop or stirrer speed controller can save a 80,000 liter batch from getting ruined.

Control of batch fermentation processes can be defined as a sequence of problems. The first problem is the determination of optimal trajectories to be followed during a batch run. Given a good model, this can be cast as an open-loop optimization problem. Another approach for determining these trajectories is to extract them from historical data bases of good batches by using statistical techniques such as principal components analysis. The second problem is the low level closed-loop control of critical process variables. This may be achieved by using several single-input single-output (SISO) control loops to regulate each controlled variable by manipulating an influential manipulated variable paired with it. The third problem is higher level control that can be addressed by selecting a multi-loop or a multivariable control approach. The former necessitates the coordination of the operation of SISO loops, the latter focuses on the development of a single controller that regulates all controlled variables by all manipulated inputs. While such a controller can be built without using any low level SISO loops, practice in other areas has favored the use of SISO loops for redundancy and reliability. In that case, the multivariable controller supplies the set-points to SISO loops. The multivariable control system can be based on linear quadratic optimal control theory or model predictive control (MPC). The optimal control theory has many success stories in various fields ranging from aerospace to manufacturing and power generation. In recent years MPC has become appealing because it can handle process constraints, disturbances, and modeling errors very effectively. MPC involves the solution of a real-time constrained optimization problem at each sampling time. While this is a limiting factor, the increase of computation speed and reduction of computation cost over the years works in favor of MPC. Techniques for addressing these three problems are discussed in Chapter 7.

1.2.4 Fault Diagnosis

When process monitoring indicates abnormal process operation, *diagnosis* activities are initiated to determine the source causes of this abnormal behavior. Experienced plant personnel have good insight in integrating various pieces of information provided by process measurements to determine the cause(s) of a fault. Various fault diagnosis paradigms can automate this effort and provide timely information to plant personnel about the most likely causes for abnormal operation. Reduction of down time, fixing the actual problem (as opposed to a secondary fault), and scheduling regular maintenance as opposed to doing emergency repairs contribute significantly over time to the profitability of the process.

The book introduces many alternative fault diagnosis paradigms and illustrates some in more detail through case studies. It also proposes the integration of monitoring and diagnosis activities by linking SPM tools with a real-time knowledge-based system that acts as a supervisor of process operations and fault diagnosis agent. Fault diagnosis techniques and knowledge-based systems are covered in Chapter 8. A forward-looking proposal for further integration of monitoring and diagnosis with control system performance assessment, supervisory control of batch fermentation process operation, and plantwide decision making systems is presented in Chapter 9.

1.3 Penicillin Fermentation

In September 1928, Alexander Fleming, a professor of bacteriology at St. Mary's Medical School in London, observed that mould had developed accidentally on a *Staphylococcus aureus* culture plate that was left on the laboratory bench and that the mould had created a bacteria-free circle around itself. He was inspired to further experiment and he found that a mould culture prevented growth of *Staphylococcus*, even when diluted 800 times. He named the active substance *penicillin* [154]. In December 1945, he and his colleagues (Florey and Chain) received the Nobel Price in medicine for the discovery of penicillin and its curative effect in various infectious diseases [424]. This accidental discovery saved thousands of lives in later years and had a major impact on pharmaceutical production of various antibiotics.

Industrial Scale Penicillin Production

There are basically two major kinds of antibiotics, namely, *narrow-spectrum antibiotics* and *broad-spectrum antibiotics.* Narrow-spectrum antibiotics control a narrow range of microorganisms, such as Gram-positive

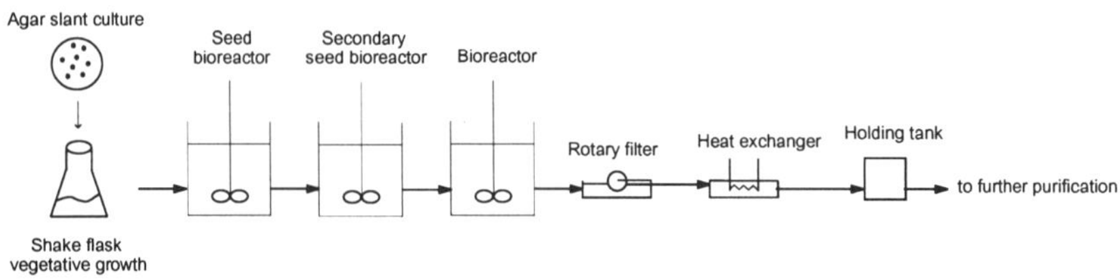

Figure 1.2. Penicillin cultivation process.

or Gram-negative bacteria but not both. Penicillin is an example of narrow spectrum antibiotic. Broad-spectrum antibiotics are active against a wide range of microorganisms such as both Gram-negative and Gram-positive bacteria. Tetracycline is an example of a broad-spectrum antibiotic.

The penicillin family includes penicillin G, penicillin V, penicillin O, and many synthetic and semisynthetic derivatives such as ampicillin, amoxicillin, nafcillin and ticarcilin. β-lactams are the largest group of antibiotics covering approximately 65% of the World market. More than 60% of these antibiotics are penicillin derivatives (either penicillin V or penicillin G). The total worldwide production of bulk penicillin is about 25,000 tons of which about 70% is sold either directly, *i.e.* penicillin V for oral administration and penicillin G for use in animal feed mixtures or a sterile salt, or it is converted to amoxicillin and ampicillin via 6-APA [424].

Although penicillin is produced by many *Penicillium* and *Aspergillus* strains, industrial penicillin is completely produced by *Penicillium chrysogenum*. Highly developed mutants are also used in industry. The medium for penicillin production typically contains an organic nitrogen source (e.g. corn steep liquor), fermentable carbohydrate (e.g. sucrose, lactose, fructose, or glucose), calcium carbonate as a buffer and other inorganic salts as necessary. Oxygen should also be added not to exceed 40% of saturation which corresponds to a volumetric oxygen uptake rate of 0.4-0.8 $mmol/l/min$. Although it is strain-specific, pH and temperature of cultivation broth are typically between 5-7 and 23-28$°C$, respectively. The culture volume is 40,000-200,000 l and is vigorously agitated using turbine agitators. Under these circumstances, a maximum theoretical yield of penicillin on glucose is estimated to be 0.12 $g\ penicillin/g\ glucose$ [26]. A typical industrial scale process for penicillin production is shown in Figure 1.2 starting with culture growth. *Penicillium chrysogenum* strains are used to inoculate 100 ml of the medium in a 500 ml flask at 25$°C$. After incubating for four days, the contents of the flask are transferred to a new 2 l medium which is further incubated for two more days. Depending on the size of the final

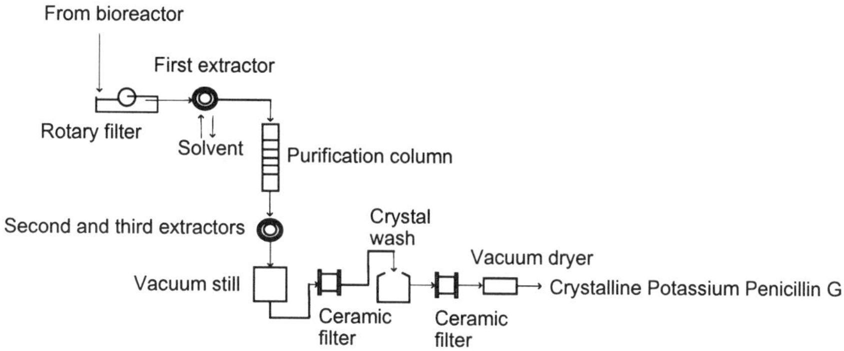

Figure 1.3. Downstream processes in industrial scale penicillin production.

culture volume, sequential growth is followed in this manner. Inoculum size is typically around 10% of the total culture volume. Since formation of secondary metabolites (in this case, penicillin) is usually not associated with cell growth, it is a common practice to grow the cells in a batch culture followed by a fed-batch operation to promote synthesis of the antibiotic. When inoculum at the desired concentration is obtained, an industrial size bioreactor (40,000-200,000 l) is inoculated. The bioreactor is operated for five to six days in fed-batch mode. After the cultivation stage, a series of product recovery techniques are applied depending on the required purity of the final product. Flow diagram for penicillin recovery process is given in Figure 1.3.

A typical time course of penicillin cultivation is represented in Figure 1.4. First, the cells are grown batchwise until they enter early stationary phase of batch growth which is also associated with the depletion of substrate. Then, the process is switched to fed-batch operation that is accompanied by penicillin production. At this stage, process is said to be in the *production phase.* Experimental data are displayed in [26]. Detailed discussion of various physiological phases is presented in Section 2.7.

1.4 Outline of the Book

The book consists of nine chapters. Chapters 2-5 focus on modeling. Chapter 6 presents a variety of process monitoring techniques. Chapter 7 presents control techniques for batch process operation and Chapter 8 discusses various fault diagnosis paradigms. Chapter 9 outlines recent developments that will impact fermentation process modeling, monitoring,

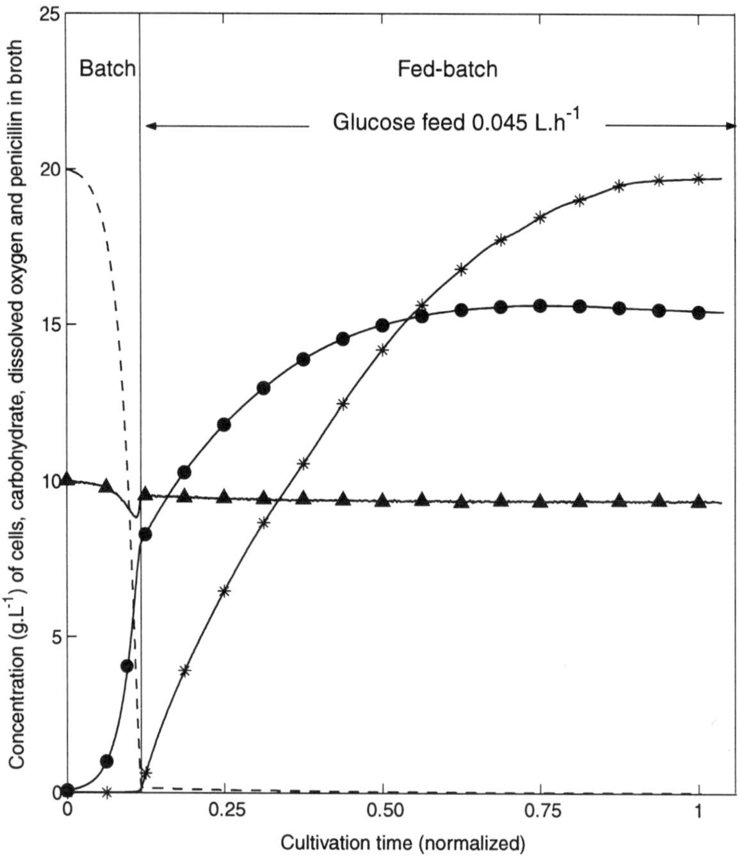

Figure 1.4. Time course of changes in carbohydrate (glucose, – –), dissolved oxygen (% saturation/10) (▲), penicillin (g.L^{-1} × 10) (∗) and biomass (•) concentrations in a penicillin fermentation simulation [61].

and control, and speculates about the future.

Chapter 2 focuses on the development of process models based on first principles. Considering the uncertainty in some reaction and metabolic pathways, and in various parameters, both unstructured and structured kinetic models are discussed. Case studies for penicillin fermentation are presented for both types of models along with simulation results. Chapter 3 presents various concepts and techniques that deal with experimental data collection and pretreatment. Sensors and computer-based data acquisition is discussed first. Then, statistical design of experiments techniques are introduced for preliminary screening experiments. Factorial and fractional factorial designs are summarized and statistical analysis tools are presented for interpretation of results. Data pretreatment issue is divided into outlier detection and data reconciliation, and signal noise reduction. Wavelets are introduced in this section for use in noise reduction. Finally, techniques for theoretical confirmation of data such as stoichiometric balances and thermodynamics of cellular growth are presented to provide a reality check of experimental data. Chapter 4 tackles the modeling problem by focusing on data-based models. First, theoretical foundations in multivariate statistics, such as principal components analysis (PCA), multivariable regression techniques, and functional data analysis, are summarized. Then, various statistical techniques for batch process modeling (multiway PCA, multivariate covariates regression, and three-way techniques) are presented. Extensions to nonlinear model development are discussed and artificial neural networks, nonlinear input-output modeling, and nonlinear partial least squares (PLS) modeling are introduced as alternative techniques for developing nonlinear models. Chapter 5 focuses on nonlinear model development from systems science and chaos point of view. It illustrates how the concept of correlation can be extended to the nonlinear framework and used for model development and reduction.

Chapter 6 deals with batch process monitoring problem. It starts with discussion of statistical process monitoring (SPM) tools for univariate problems (Section 6.1). Shewhart, cumulative sum (CUSUM), and exponentially weighted moving average (EWMA) charts are presented. Then, multivariate tools (PCA and PLS) for SPM of continuous processes are discussed in Section 6.2. The phase change point (landmark) detection problem and data length adjustment are discussed in Section 6.3, introducing indicator variable, dynamic time warping (DTW) and curve registration techniques. In Section 6.4, SPM of multivariable batch processes is discussed and multiway PCA, multiway PLS, multiscale SPM with wavelets techniques are introduced. Finally in Section 6.5, on-line SPM of batch processes is addressed using multiway PCA and hierarchical PCA techniques, and Kalman filters for final product quality estimation.

Chapter 7 presents various control problems in batch process operations. The first problem is the determination of the optimal reference trajectories that should be followed during the batch run. This is an optimal open-loop control problem. A related problem is the determination of the benefits, if any, of forced periodic operation of the fermentation system and the variables and operating conditions that will maximize productivity and selectivity. The other control problems focus on closed-loop control using multi-loop, linear quadratic Gaussian, and model predictive control techniques.

Chapter 8 discusses various fault diagnosis techniques. One approach is based on determining first the variables that contribute to the increase in the statistic that indicates an out-of-control signal and then using process knowledge to reason about the source causes that will affect those variables to identify the likely causes of faulty operation. The contribution plots method is presented in the first part of the chapter. Automating the integration of the variables indicated by contribution plots and process knowledge with a knowledge-based system (KBS) is discussed in the last section of the chapter. Section 8.2 of the chapter is devoted to multivariate statistical classification techniques such as discriminant analysis and Fisher's discriminant function, and their integration with PCA. Section 8.3 focuses on a variety of model-based techniques from systems science for fault diagnosis. Generalized likelihood ratio, parity relations, observers, Kalman filter banks, and hidden Markov models are presented. Section 8.4 is devoted to model-free fault diagnosis techniques such as limit checking, hardware redundancy and KBSs. The last section outlines real-time supervisory KBSs that integrate SPM, contribution plots and KBS rules to provide powerful fault diagnosis systems.

Chapter 9 introduces some related developments in modeling, dynamic optimization, and integration of various tasks in batch process operations management. Metabolic engineering, metabolic flux analysis and metabolic control analysis concepts are introduced and their potential contributions to modeling is discussed. Dynamic optimization and its potential in industrial applications is discussed and compared with classical and advanced automatic control approaches. The integration of various tasks in process operation using a supervisory knowledge-based system is outlined for on-line process supervision.

Background Information and Road Maps to Use the Book

The book is written for professionals and students interested batch fermentation process operations. It requires little background information in various areas such as biotechnology, statistics, system theory, and process control. Introductory materials in biotechnology can be found in various process engineering books [35, 426, 546]. Applied statistics books for engineers and scientists [78, 167, 400, 626] provide the basic theory and techniques. A reference for multivariate statistics [262] would be useful for Chapters 6 and 8. Several good textbooks are available for basic concepts in process control [366, 438, 541]. Advanced books in all these areas are referenced in appropriate chapters in the book.

Ideally, the chapters in this book should be read in the sequence they appear. However, allowing for potential diversity in background in fundamentals of various readers, we suggest here alternate roadmaps. Chapters 1 and 9, being the first and last chapters in this book, have been intentionally kept descriptive and can be followed by readers with diverse fundamental background and technical expertise with relative ease. The remaining chapters could be followed in the order in which they appear or in an alternate sequence. Readers with much more familiarity with process modeling and control than with statistical methods may start with Chapters 2, 5 and 7. This sequence would then be followed by Chapters 3 and 4, which deal with data collection and pretreatment and data-based modeling, This would then set the stage for statistical techniques for process monitoring and fault diagnosis, which are the subjects of Chapters 6 and 8. In this alternate sequence, the readers may need to refer to appropriate sections in Chapters 3 and 4 while going over certain sections of Chapters 5 and 7. Finally, readers very conversant with statistics and statistical methods but not familiar with engineering processes may start with Chapters 3 and 4 and may transit to Chapters 6 and 8, followed by Chapters 2, 5 and 7. Many chapters rely on sufficient knowledge of linear algebra and systems theory. The readers will be well served by referring to appropriate help material, as needed, on these, which can be found in many undergraduate and graduate texts in engineering and mathematics. Access to the help material will permit the readers to focus on differences in methodologies/techniques discussed in individual sections of the appropriate chapters. At the start of each chapter, we have provided a brief layout of the chapter. Depending on the level of familiarity with different sections in a chapter, the readers may make their own menu for going over the chapter, reading perhaps sections that they are more familiar with first, followed by reading the sections with which they are less familiar or unfamiliar. The book is intended to be a valuable resource guide. For further in-depth review of particular topics, the readers should access suggested references.

2

Kinetics and Process Models

2.1 Introduction and Background

Growth of living (viable) cells requires intimate contact of a small quantity of living cells with a liquid solution (medium) containing appropriate levels of nutrients at a suitable pH and temperature. Depending on the morphology of cells under consideration, one needs to consider two different manifestations of cell growth. For unicellular organisms which divide as they grow, an increase in biomass (mass of viable cells) is accompanied by an increase in the number of cells present in the culture (cell-medium suspension). The situation is very different in the case of growth of molds, which are popular organisms for industrial production of a variety of antibiotics. In the case of molds, the length and number of mycelia increase as the growth proceeds. The growing mold therefore increases in size and density (concentration) but not necessarily in numbers. (There isn't a one-to-one relation between the number of distinct multicellular units and amount of biomass.)

The extent of complexity of the kinetic description to be considered depends on the complexity of the physical situation under consideration and the intended application of the kinetics (fundamental understanding of cellular processes, design and simulation of bioprocesses, optimization and control of bioprocesses). However simple or however complex the kinetic description be, it must incorporate certain key cellular processes, such as cell replication (cell growth), consumption of essential nutrients, synthesis of end products (followed by intracellular accumulation or excretion of these), and cell death/lysis.

Biological reactors employed for production of commercially significant metabolites using living cells involve two or more phases (a single gas phase, at least one liquid phase, and at least one solid phase). The cells are usually in contact with a liquid phase. Whether the cells are suspended in the liquid phase (suspension culture) or attached to a suitable solid support (immobilized) and in contact with the liquid phase, the interactions between the two phases [biotic phase (cell population) and abiotic phase (liquid)] must be considered and fully accounted for. Both phases are multicomponent systems. The abiotic phase usually contains all of the nutrients essential for cell growth and various end products of cellular metabolism that are excreted. Some of the end products may undergo further reactions in this phase. A classic example is the hydrolysis of antibiotics such as penicillin in the liquid medium. Transport of nutrients from abiotic phase to biotic phase is essential for utilization of these for cell growth and maintenance and for formation of a host of metabolic intermediates and end products. Some of the end products are retained within the cells (intracellular metabolites), while others are excreted by the cells (transport from biotic phase to abiotic phase). The large number of chemical reactions occurring within a cell result in accumulation or depletion of energy. Exchange of energy between abiotic and biotic phases must be accounted for to determine the culture temperature. The temperature of the abiotic phase usually determines the temperature of the biotic phase. Some of the cellular reactions impact the acid-base equilibria in the biotic phase and in turn the pH of the abiotic phase, which in turn influences cellular activities and transport processes across the abiotic - biotic two-phase interface. In addition to transport of essential nutrients and end products of cellular metabolism between the two phases, one must also consider transport of ionic species (such as protons and cations). As a result of cellular reactions, the properties of the abiotic phase, such as viscosity, may change during the course of cell cultivation.

An individual cell is a complex multicomponent system in which a large number of independent enzyme-catalyzed chemical reactions occur simultaneously, subject to a variety of constraints. In a growing cell population, there is cell-to-cell variation as concerns cell age and cell function (cell activity). Thus, at a given time and in a sufficiently small region of physical space in a culture, some cells may be newly born, others may be of intermediate age and dividing, while still others may be much older and subject to death or lysis. In the case of molds, in an individual multicellular unit, there may be significant variation as concerns cell age. There is also differentiation among different cells as concerns replication, utilization of essential nutrients and formation of the target end product (for example, antibiotics such as cephalosporin and penicillin). Some of the cells thus may be actively dividing but incapable of or less efficient in synthesizing the target metabolite, while some others may be fully capable of synthesis of the target end product.

2.2 Mathematical Representation of Bioreactor Operation

A popular form of operation of bioreactors employing living cells involves the use of a well-mixed reactor, the mixing being accomplished by mechanical agitation and/or fluid motion. The uniformity of composition and temperature in the reactor allows its representation as a lumped parameter system. Although well-mixed reactors are almost a norm in cell cultivation, tubular reactors are used in bioprocesses involving immobilized cells and immobilized enzymes, these bioreactors being distributed parameter systems. In view of this, the focus in this chapter is on lumped parameter systems. The dynamics of bioreactors that can be viewed as lumped parameter systems can be described succinctly as

$$\frac{d\mathbf{x}}{dt} = \mathbf{f}(\mathbf{x}, \mathbf{u}, \mathbf{d}), \qquad \mathbf{x}(t_0) = \mathbf{x}_0, \tag{2.1}$$

with $\mathbf{x}$ denoting the set of variables which represent the status of the cell culture in the bioreactor, the so-called state variables, and $\mathbf{u}$ and $\mathbf{d}$ representing the set of external variables which indirectly influence the status of the cell culture, the so-called input variables. The input variables are further classified into inputs that are manipulated ($\mathbf{u}$) and inputs that are not manipulated ($\mathbf{d}$), the so-called disturbance variables. Let n, m and p denote the number of state variables, manipulated inputs and disturbance variables. The right hand side in Eq. 2.1 contains information on how the temporal variations in state variables are influenced by the state and input variables, these influences in general being nonlinear ($\mathbf{f}$ - a nonlinear function of $\mathbf{x}$, $\mathbf{u}$ and $\mathbf{d}$). In the subsequent sections, illustrations will be provided on how descriptions of different operating modes for bioreactors and bioprocesses with varying levels of complexity as concerns description of kinetics of cellular and extracellular processes can be concisely represented in the form of Eq. 2.1. As these illustrations are discussed, it will be evident that not all state variables can be measured or estimated. There can be a variety of reasons for not monitoring variations in a state variable, including lack of availability of an assay (procedure for analysis)/ measuring device/sensor, monitoring difficulties due to rapid fluctuations in the state variable, and costs associated with frequent measurement of the same. The specifics of an assay (analytical procedure) or measuring device may impose limitations on frequency of measurement of certain state variables. In such situations, the measurements of the state variable at discrete times may have to be supplanted by estimations of the same (based on these measurements) at times when no measurements were made. Some of the state variables, which cannot be measured as frequently as desired, can be estimated from measurement of certain other parameters, which in turn can be measured as frequently as desired. An example of such a state variable is the biomass concentration in the culture (usually expressed as dry mass of cells per unit culture volume). Direct estimation of this variable requires time-intensive separation (of the abiotic and biotic phases via centrifugation or filtration) and gravimetric procedures. A reliable procedure for monitoring biomass concentration involves determination of turbidity of cell culture via measurement of optical density (OD) of the culture and estimating the biomass concentration using a predetermined correlation between the biomass concentration and OD. The advantage of this estimation is rendered by one's ability to measure OD as frequently as possible. It must therefore be realized that only some of the state variables may be monitored or estimated. The set of variables which can be measured will be referred to as bioreactor outputs, $\mathbf{y}$, with the number of outputs being l. The relations among the state variables, the input variables and the output (measured) variables can then be succinctly stated as

$$\mathbf{y} = \mathbf{g}(\mathbf{x}, \mathbf{u}, \mathbf{d}). \tag{2.2}$$

2.3 Bioreactor Operation Modes

The three popular modes of operation of mechanically agitated reactors for cell cultivation are batch, fed-batch and continuous operations. The mechanically agitated reactors are equipped with capabilities for on-line and off-line sensing of a variety of culture characteristics, such as pH, temperature, concentrations of dissolved gases (such as CO_2 and O_2), biomass concentration, cell morphology, concentrations of various components of the nutrient medium, total protein content of cells, activities of certain proteins, and concentrations of certain metabolites, and tighter control of some of these using appropriate controllers. The classification of the reactor operation is based on culture [suspension of cells in a liquid medium or a composite of liquid medium and cells attached to a suitable solid support (immobilized cells)]. Besides the liquid and solid phases, these reactors also have a gas phase, to provide oxygen to liquid culture in an aerobic bioprocess, to provide a blanket of an inert such as nitrogen in an anaerobic bioprocess, and to remove carbon dioxide (from the culture) generated as a product of cellular metabolism. Irrespective of the mode of operation with respect to culture, the bioreactors are always operated in continuous mode with respect to gas phase (gas phase continuously entering and leaving the reactor).

2.3.1 Batch Operation

A batch culture operation is characterized by no addition to and withdrawal from the culture of biomass, fresh nutrient medium and culture broth (with the exception of gas phase). The batch operation is initiated by addition of a small amount (with respect to sterile nutrient medium) of a cell culture (the so-called "inoculum") to a sterile nutrient medium. The inoculum is derived from serial batch cultures, the so-called starter cultures or subcultures. Some of the culture conditions (such as pH and dissolved oxygen level) during each subculture are usually left uncontrolled. A typical batch culture operation is strictly not a batch operation since it may involve addition of an acid/base for pH control and antifoam to suppress foaming in the culture and withdrawal of small portions of culture for assessing the status of the culture. Any net volume changes due to these additions and withdrawals are usually minimized by using concentrated acid/base and antifoam solutions and by keeping the number and volume of samples withdrawn within limits. As concerns cell mass accumulation resulting from uptake and utilization of nutrients, the batch culture is characterized by a lag phase (during which period cells in the inoculum adjust to the shock in their environment and accelerate synthesis of enzymes needed to utilize nutrients in the liquid medium), which is followed by an active growth phase (both the cell number and cell mass usually increase exponentially with time in this phase). Cell growth continues in the next phase, albeit at a slower rate since substantial consumption of nutrients has already occurred and such growth is usually referred to as non-exponential growth. Production of some of the metabolites (including a variety of antibiotics and enzymes of commercial importance), whose synthesis is not necessarily directly proportional to cell growth, is accelerated (and in some cases initiated) in this second growth phase. These metabolites are usually referred to as secondary metabolites. The growth phase is followed by a stationary phase upon near complete exhaustion of one or more nutrients essential for cell growth. Synthesis of secondary metabolites is usually promoted in the stationary phase. The stationary phase is followed by the death phase, which is characterized by a significant decline in the cell number density (or the viable cell concentration). A batch operation is usually terminated near the end of the growth phase or during the stationary phase. In industrial bioprocesses, serial batch culture operations are very common. In a typical operation, a portion of the culture from the previous batch is used as inoculum for the next batch. Since there is in a sense recycle of culture from batch to batch, these operations are referred to as repeated batch operations with recycle. If there is no transfer of culture from a batch to the next, the serial operations are referred to as repeated batch operations without recycle and are indistinguishable from a single (once-through) batch culture or parallel batch cultures (conducted simultaneously).

2.3.2 Fed-Batch Operation

A fed-batch culture operation is characterized by predetermined or controlled addition of nutrient medium in an otherwise batch operation (no withdrawal of culture). This operation allows for temporal variation in the supply of nutrients, thereby allowing tighter control of various cellular processes such as cell growth, nutrient uptake and production of target metabolites. As mentioned earlier, synthesis of secondary metabolites, including a variety of antibiotics and enzymes, is promoted under culture conditions where cell growth is discouraged. Controlled addition of nutrients in a fed-batch operation allows for control of cell growth and thereby promotes production of secondary metabolites. The total mass of culture increases during a fed-batch operation and so does the culture volume unless nutrients in highly concentrated form (such as solid powders) are fed to the culture. The feed rate can be varied in a predetermined fashion or by using feedback control. The addition of nutrient feed is terminated upon reaching the maximum permissible volume. A fed-batch operation may be followed by a terminal batch operation, with culture volume being equal to maximum permissible volume, to utilize the nutrients remaining in the culture at the end of fed-batch operation. A fed-batch operation is usually preceded by a batch operation. A typical run involving fed-batch operation therefore very often consists of the fed-batch operation sandwiched between two batch operations. This entire sequence (batch→fed-batch→batch) may be repeated many times leading to serial (or repeated) fed-batch operation. As in the case of repeated batch operation with recycle, transfer of culture from one sequence to the next to inoculate the next sequence is common in these serial operations.

2.3.3 Continuous Operation

In a continuous culture operation, nutrients essential for growth are continuously fed and a portion of the culture is continuously withdrawn. The culture volume is controlled using a level controller. A continuous culture is usually preceded by a batch or fed-batch culture. If the mass flow rates of the bioreactor feed and bioreactor effluent are identical and time-invariant, a time-invariant (steady state) operation can be realized after sufficient time since the start of continuous culture operation. The status of the culture can be determined easily by analysis of the bioreactor effluent, thereby causing no interference with bioreactor operation, which is certainly not the case with batch and fed-batch operations. As in a fed-batch culture, the feed rate to a continuous bioreactor can be varied in a temporal sense in a predetermined fashion or using feedback control. Since the culture conditions (in a global sense) can be kept time-invariant, continuous cultures are easier to monitor and control. When culture conditions which promote biomass growth are substantially different from those which promote production of a target metabolite (on a per cell basis), a simple continuous culture operation described here may not yield the best productivity of the target metabolite. Two-staged continuous culture operations where cell growth is promoted in the first stage and synthesis of a target metabolite is promoted in the second stage have been shown to yield much higher productivity when compared to the highest productivity attainable in single-stage continuous culture. Such two-staged operations may be attained spatially in two continuous cultures in series and temporally in a single continuous culture by switching from a growth promoting medium to a production medium and vice versa[451]. These are some of the advantages and flexibility that a continuous culture offers over batch and fed-batch cultures. Unlike the operation of continuous processes employed for production of chemicals, long-term operation of continuous cultures is subject to many operating difficulties, including risks of contamination and loss in productivity due to cell washout in case of unanticipated disturbances and substantial changes in characteristics of the biotic phase.

2.4 Conservation Equations for a Single Bioreactor

Irrespective of the type of operation, a description of the behavior of a suspension culture requires applying the principle of conservation for each of the three distinct phases (gas, liquid and solid phases) and constituents of each phase. When the nutrients are in the liquid phase, as is the case with submerged cultures and which are the primary focus here, the solid phase is comprised essentially entirely of cell mass. In cases involving nutrients (such as cellulose) which are insoluble in the liquid medium, the solid phase is comprised of cell mass and solid nutrients. Where a particular specie is present in more than one phase, the conservation equations for that specie in each of these phases must account for interphase transport of that specie. As an illustration, the conservation equations for a single well-mixed bioreactor are presented here, with the feed to the bioreactor (in the case of fed-batch and continuous cultures) being considered to be sterile. Each of the three phases are considered to be well-mixed. The bioreactor operation is considered to be isothermal. Appendages/modifications to these conservation equations due to operational modifications such as recycle of cell mass, downstream separations, and two-stage continuous cultures will be addressed briefly at the end of this chapter.

2.4.1 Conservation Equations for the Gas Phase

For the three bioreactor operation modes (batch, fed-batch and continuous), the continuously flowing gas phase is ubiquitous. The conservation equation for a specie i in the gas phase (e.g., $i = O_2$, CO_2) can then be expressed as

$$d(C_{iG}V_G)/dt = Q_{GF}C_{iGF} - Q_GC_{iG} - N_i\underline{a}V, \quad V_G = V_T - V \tag{2.3}$$

with C_{iG} denoting concentration of specie i in the gas phase, Q_G the volumetric gas phase flow rate, V_G the gas phase holdup in the bioreactor (volume of gas phase in bubbles and head space), the subscript F the gas feed, N_i the flux of specie i from the gas phase to liquid phase, $\underline{a}$ the gas-liquid interfacial area per unit culture volume, V the culture volume, and V_T the volume of empty reactor. Equation 2.3 provides a *volume-averaged* description of the gas phase. In a bioreactor, the gas phase is introduced at the bottom of the bioreactor using spargers. As is evident from Equation 2.3, the rate of transport of a specie i depends on the gas-liquid interfacial area, which is higher, the smaller the size of gas bubbles (say bubble diameter). Although the bubble size near the sparger is more or less uniform, there are variations in characteristics of gas bubbles (such as bubble shape and size, gas-liquid interfacial area and concentrations of various gaseous species) during their ascent through the culture. Further, these characteristics of the gas phase in the head space above the culture may be substantially different from those of the gas phase in ascending gas bubbles. A detailed accounting of (bubble-to-bubble and gas bubbles to headspace) heterogeneity in the gas phase, although possible, can certainly divert one's attention from description of culture behavior. For this reason, Equation 2.3 is commonly used for representation of events in the gas phase. It must be realized that while the volume of the head space in batch and continuous cultures is essentially time-invariant [and therefore so is V_G as per Eq. 2.3] since the culture volume is essentially unchanged, an increase in culture volume (V) during a fed-batch operation implies a reduction in the head space and V_G [Eq. 2.3].

The transport of a specie i from the gas phase to the liquid phase and vice versa occurs through boundary layers on each side of the gas-liquid interface in the two phases. The dominant mechanism for transport of specie i in each boundary layer in a direction orthogonal to the gas-liquid interface is molecular diffusion. Assuming that specie i does not participate in any chemical reactions in the gas-side boundary layer, the flux N_i in Eq. 2.3 can be expressed as

$$N_i = k_{iG}(C_{iG} - C^*_{iG}), \quad k_{iG} = \frac{D_{iG}}{\delta_G}, \tag{2.4}$$

with C^*_{iG} denoting the concentration of specie i in the gas phase at the gas-liquid interface, k_{iG} the gas-side mass transfer coefficient for specie i, D_{iG} the molecular diffusivity of i in the gas phase, and δ_G the thickness of the gas-side boundary layer. On the other side of the gas-liquid interface, one must consider in the liquid-side boundary layer the transport of specie i by molecular diffusion. Such transport occurs in parallel with consumption or generation, as appropriate, of specie i as a result of cellular metabolism by cells present in the liquid-side boundary layer and is therefore influenced by the latter. Precise description of events occurring in the liquid-side boundary layer then requires solution of conservation equations which account for diffusion of specie i and its participation in one or more reactions within cells leading to its consumption or generation. Similar conservation equations must also be considered for all species that are non-volatile and participate in cellular reactions. These conservation equations are typically nonlinear second-order (spatially) ordinary (partial) differential equations and simultaneous solution of these can be a computationally challenging task. For this reason, it is assumed that cellular reactions occur to negligible extents in the liquid-side boundary layer. Since the gas-liquid interface has infinitesimal capacity to retain specie i, flux of specie i must be continuous at the gas-liquid interface and the gas and liquid phases must be at equilibrium with respect to species i at the gas-liquid interface. When the two phases are dilute with respect to i, the equilibrium is described by Henry's law. The following relations then apply at the gas-liquid interface.

$$N_i = k_{iG}(C_{iG} - C^*_{iG}) = k_{iL}(C^*_i - C_i), \quad C^*_{iG} = H_i C^*_i, \tag{2.5}$$

In Eq. 2.5, C_i and C^*_i denote the concentrations of specie i in the bulk liquid and the liquid phase at the gas-liquid interface, H_i the Henry's law constant for specie i, k_{iL} the liquid-side mass transfer coefficient for specie i ($k_{iL} = D_{iL}/\delta_L$), D_{iL} the molecular diffusivity of i in the liquid phase, and δ_L the thickness of the liquid-side boundary layer. The thicknesses of the boundary layers in the gas and liquid phases are dependent, among other things, on flow patterns in the two phases. Since the interfacial concentrations are not easily tractable, in view of Eq. 2.5, the flux of specie i across the gas-liquid interface can be expressed in terms of easily tractable variables, C_{iG} and C_i, as

$$N_i = K_{iL}(C_{iG} - H_i C_i), \quad K_{iL} = \frac{k_{iG} k_{iL}}{k_{iL} + H_i k_{iG}}. \tag{2.6}$$

The volumetric flow rate of gas feed is usually many-fold greater than that of liquid feed. A pseudo-steady state hypothesis is therefore invoked as concerns Eq. 2.3, with the term on the left side (accumulation term) being much less than any of the three terms on the right side (as concerns absolute magnitudes). Eq. 2.3 then reduces to an algebraic relation. The conservation equations for the culture and its constituents are discussed next.

2.4.2 Conservation Equations for Cell Culture

The conservation equation for the culture can be stated as

$$\frac{d(\rho V)}{dt} = \rho_F Q_F - \rho Q, \tag{2.7}$$

with ρ and Q denoting the density and volumetric effluent rate, respectively, of the culture and ρ_F and Q_F the density and volumetric flow rate, respectively, of the sterile feed (usually liquid). Q is trivial in batch and fed-batch operations, while Q_F is trivial in a batch operation. The mass balance in Eq. 2.7 is simplified via a customary assumption that ρ_F and ρ are not significantly different. This assumption is reasonable since the densities of nutrient medium and biomass are not substantially different. The simplified form of Eq. 2.7 is

$$\frac{dV}{dt} = Q_F - Q. \tag{2.8}$$

The culture is comprised of the biotic phase (cell mass) and the abiotic (extracellular) phase. Let the concentration of biomass (X, cell mass) be denoted as C_X ($C_X = X$, the notation popular in the biochemical engineering literature is X) and the density of biomass as ρ_b. It is then not difficult to deduce that the volume fractions of the biotic and abiotic phases in the culture are C_X/ρ_b and (1 - C_X/ρ_b), respectively. The volumes of biotic and abiotic phases (V_b and V_a, respectively) and the volumetric flow rates of these phases in the bioreactor effluent (in the case of continuous culture, Q_b and Q_a, respectively) can then be expressed as

$$V_b = \frac{C_X V}{\rho_b}, \quad V_a = (1 - \frac{C_X}{\rho_b})V, \quad Q_b = \frac{C_X Q}{\rho_b} \quad Q_a = (1 - \frac{C_X}{\rho_b})Q. \tag{2.9}$$

The volume fraction of the biotic phase is commonly considered to be negligible. While this consideration is valid in cultures that are dilute in biomass

($C_X \ll \rho_b$), in cultures that are concentrated in biomass, neglecting the volume fraction of the biotic phase may lead to certain pitfalls.

Before proceeding further, some comments are in order regarding bases for different species. For a particular specie, the basis for choice is based on how the specie is monitored. Thus, the biomass concentration is expressed on the basis of unit culture volume, concentrations of species present in the abiotic phase are expressed on the basis of unit abiotic phase volume, and concentrations of intracellular species are usually expressed on the basis of unit biomass amount. For rate processes occurring entirely in the abiotic phase, the basis is the volume of the abiotic phase (V_a), while for rate processes occurring in the biotic phase (metabolic reactions) and at the interface between the abiotic and biotic phases (such as species transport), the basis is the amount of the biotic phase. On a larger scale, a single cell is viewed also as a catalyst (hence the name biocatalyst), or in a stricter sense, an autocatalyst, since resource utilization and generation of end products of cellular metabolism are promoted by the cell. The rates of proliferation/replication of a living species and other processes associated with it (utilization of resources and synthesis of end products) are as a result proportional to the amount of the living species.

Approaches to representation of the biotic phase according to the number of components (species) used for such representation and whether or not the biotic phase is viewed as a heterogeneous collection of discrete cells have been succinctly classified by Fredrickson and Tsuchiya [166]. Representations which view each cell as a multicomponent mixture are referred to as structured representations, while those which view the biotic phase as a single component (like any specie in the abiotic phase) are termed unstructured representations. An unsegregated representation is based on use of average cellular properties and does not account for cell-to-cell heterogeneity. A segregated representation, on the other hand, involves description of behavior of discrete, heterogeneous cells suspended in the culture and thereby accounts for cell-to-cell variations. The segregated-structured representation is most suitable for a bioreactor. In order to have tractable representations of biotic phase, it is often assumed that the cell-to-cell variations do not substantially influence the kinetic processes in the biotic phase. The segregated representation can then be reduced to an unsegregated representation based on average cell properties. The discussion in this chapter is limited to the latter perspective. With this in mind, the conservation equation for the biotic phase can be stated as

$$\frac{d(C_X V)}{dt} = \mu^{\text{net}} C_X V - Q C_X, \quad \mu^{\text{net}} = \mu - r_d, \tag{2.10}$$

with μ denoting the specific cell growth rate, r_d the specific rate of cell loss due to cell death or cell lysis, and μ^{net} the net specific growth rate, respectively, the basis for each being unit biomass amount. It must be kept in mind that C_X (later referred to also as X) represents the concentration of viable cell mass. The mass fraction of dead cells in the total cell population is considered to be negligible. In view of Eqs. 2.8 and 2.10, the temporal variation in C_X can be described as

$$\frac{dC_X}{dt} = \mu^{\text{net}} C_X - DC_X, \quad D = \frac{Q_F}{V}, \tag{2.11}$$

with D denoting the dilution rate for the culture. The mass balance above applies for all three reactor operations under consideration, with D being trivial for a batch operation.

The conservation equation for a specie i in the abiotic phase in its general form can be expressed as

$$\frac{d(C_i V_a)}{dt} = Q_F C_{iF} + N_i \underline{a} V + R_i^{\text{gen}} V_a + r_i^{\text{trans}} C_X V - Q_a C_i, \tag{2.12}$$

with C_{iF} denoting the concentration of specie i in the nutrient feed, R_i^{gen} the rate of generation of specie i due to any reactions in the abiotic phase, r_i^{trans} the biomass-specific rate of transport of specie i from the biotic phase to the abiotic phase, and Q_a being trivial in batch and fed-batch operations. When the specie is supplied in the feed gas (as is the case with oxygen), C_{iF} is trivial and N_i is non-trivial. For species which are not transported into the biotic phase (for example, macromolecules like starch), r_i^{trans} is trivial. Although bulk of the chemical transformations occur in the biotic phase, some species may undergo reactions in the abiotic phase and for these species R_i^{gen} is non-trivial. Two examples of this situation are acidic or enzymatic hydrolysis of starch to generate readily metabolizable carbohydrates and degradation of antibiotics and enzymes in abiotic phase. This sets the stage for accounting for the intracellular components (components of the biotic phase).

The conservation equation for a specie i in biotic phase can be succinctly stated as

$$\frac{d(c_i C_X V)}{dt} = (r_i - r_i^{\text{trans}}) C_X V - Q c_i C_X \tag{2.13}$$

with c_i denoting intracellular concentration of specie i in the biotic phase (mass of i per unit biomass, i.e., mass fraction of specie i in the biotic phase) and r_i the net rate of generation of specie i in the biotic phase. In view of the biomass balance [Eq. 2.10], Eq. 2.13 can be restated as

$$\frac{dc_i}{dt} = r_i - r_i^{\text{trans}} - \mu^{\text{net}} c_i, \quad r_i = r_i^{\text{gen}} - r_d c_i \tag{2.14a}$$

It should be observed that Eq. 2.14a is valid irrespective of the mode of operation of the bioreactor and the last term on the right side represents the effect of dilution due to net cell growth. For species which are retained inside the cells (e.g., large macromolecules), r_i^{trans} is trivial. The net rate of generation of specie i in the biotic phase, r_i, includes the rate of loss of specie i from the biotic phase due to cell death or cell lysis. In view of Eq. 2.10, Eq. 2.14a can be stated alternately as

$$\frac{dc_i}{dt} = r_i^{\text{gen}} - r_i^{\text{trans}} - \mu c_i, \tag{2.14b}$$

with r_i^{gen} being the net rate of generation of specie i in the biotic phase exclusive of the rate of its loss from the biotic phase due to cell death or cell lysis. In the case of cell lysis, specie i will be introduced into the abiotic phase at the rate equal to $r_d C_X V$ and this must be accounted for in R_i^{gen} in Eq. 2.12.

Conservation equations for intracellular species (and therefore temporal variations in quantities of these) are not accounted for in an unstructured representation of kinetics of cellular processes (the so-called unstructured kinetic models). For species that are present in both abiotic and biotic phases, examples of which include readily metabolizable nutrients and metabolites that are excreted by the cells, no differentiation is (can be) made between r_i^{gen} (specie synthesis in the biotic phase) and r_i^{trans} (specie transport across the outer cell membrane into the abiotic phase). The conservation equation for a specie i in the abiotic phase is then based on Eq. 2.12 and is expressed as

$$\frac{d(C_i V_a)}{dt} = Q_F C_{iF} + N_i \underline{a} V + R_i^{gen} V_a + \mathrm{r}_\mathrm{i}^{\text{gen}} C_X V - Q_a C_i. \tag{2.15}$$

For a target product metabolite (specie i) that is retained in the cells, an unstructured kinetic model must still include the conservation equation for the intracellular product, viz., Eq. 2.13, with r_i^{trans} of course being trivial.

2.5 Unstructured Kinetic Models

An unstructured kinetic representation provides a simplistic global (with respect to cell mass) view of the net result of metabolic rate processes occurring in the living cells. In brief, the unstructured representation can be described as

$$Nutrients\ (\sum N_i) + \ Cell\ mass\ \rightarrow \tag{2.16}$$
$$more\ Cell\ mass\ +\ Non\text{-}biomass\ Products(\sum P_j)$$

This representation then involves conservation equations for cell mass, key nutrients, and metabolites of interest (target products), the rates of generation/consumption of the individual species being expressed in general in terms of concentrations of nutrients in abiotic phase, N_i ($N_i = C_i$ for nutrient N_i as per the notation commonly used in literature in biochemical engineering and biotechnology), cell mass concentration, X ($X = C_X$), concentrations of metabolites of interest (P_j, $P_j = C_j$ for product P_j for an extracellular metabolite and $P_j = c_jC_X$ for an intracellular metabolite as per the notation commonly used in literature in biochemical engineering and biotechnology), and other parameters such as culture pH and temperature (T). For biomass (cell mass), the specific cell growth rate is therefore expressed as $\mu = \mu(N_1, N_2, \ldots, P_1, P_2, \ldots, X, pH, T)$. Consumption of a nutrient N_i implies that the rate of its generation in the biotic phase, r_i^{gen}, is negative [Eq. 2.15], with the consumption rate usually being expressed as $(- r_i^{\text{gen}}) = \sigma_i(N_1, N_2, \ldots, P_1, P_2, \ldots, X, \mu, pH, T)$ and being referred to as the cell mass-specific uptake rate of nutrient N_i. Similarly, for a target metabolite P_j, the rate of its generation in the biotic phase, r_j^{gen} (whether or not the metabolite is excreted) [Eqs. 2.14a and 2.15], is expressed as $r_j^{\text{gen}} = \varepsilon_j(N_1, N_2, ..., P_1, P_2, \ldots, X, \mu, pH, T)$, with ε_j being referred to as the cell mass-specific production rate of metabolite P_j.

2.5.1 Rate Expressions for Cell Growth

More commonly, the dependence of specific cell growth rate on concentrations of various nutrients, cell mass and target metabolites is expressed in uncoupled, multiplicative form as per the relation

$$\mu = \mu_o\phi_1(N_1)\phi_2(N_2)...\phi_i(N_i)...\varphi_1(P_1)\varphi_2(P_2)...\varphi_j(P_j)...\varphi(X). \tag{2.17}$$

with μ_o being constant characteristic of a particular strain. The popular forms of $\phi_i(N_i)$ and $\varphi_j(P_j)$ are based on the following common experimental observations. Cell growth is usually promoted by increased presence of some nutrients N_i (i.e., with increasing concentrations of these) at least up to some threshold levels and may be discouraged at high concentrations of some of these (the so-called "substrate inhibition"). Such nutrients are called (growth) rate-limiting nutrients. Nutrients which do not influence cell growth are not rate-limiting [$\phi_i(N_i) = 1$ for these nutrients]. Cell growth may be unaffected by the presence of a target metabolite [$\varphi_j(P_j) = 1$ for a metabolite P_j] or may be discouraged as the product P_j accumulates in the culture [i.e., $\varphi_j(P_j)$ decreases with increasing P_j]. The former is often the case when the amount of P_j in the culture is significantly less than that of cell mass (X), an example of this situation being production of many antibiotics and enzymes. The latter is often the case when the amount of P_j in the culture is comparable or greater than that of cell mass (X), an example of this situation being production of alcohols (such as ethanol and butanol) and organic acids (such as acetic, citric, formic, lactic and succinic acids) by a variety of microbial species. One of the determinants of cell proliferation is accessibility of nutrients in the abiotic phase to cells. This accessibility is reduced with increasing biomass concentration in the culture and as a result, cell growth may be discouraged as the biotic fraction of the culture is increased [i.e., $\varphi(X)$ may decrease with increasing X, $\varphi(X) \leq 1$]. The popular forms of $\phi_i(N_i)$, $\varphi_j(P_j)$ and $\varphi(X)$ are provided in Table 2.1.

In the classical chemical literature, rate expressions for homogeneous (fluid-based) reactions are of the power-law type, i.e., the rate of a reaction is proportional to some power of concentration of a reactant or a product for that reaction, the power (exponent) being referred to as the order of the reaction with respect to that species. For the large number of expressions available in the literature for rate of cell growth (see [35, 426, 545] for several examples of these), the orders of reactions with respect to nutrients are less than unity and positive those with respect to end-products are non-positive (not surprising since synthesis of building blocks for cellular material and synthesis of end-products are competing processes as concerns utilization of nutrient and energy resources within the living species) (Table 2.1).

The activity of each cell is a net result of thousands of molecular level chemical reactions occurring inside the cell that are promoted by a large number of enzymes (biological catalysts). These reactions are therefore surface-based reactions. Following the classical literature in chemistry on catalytic reactions, the rate expressions for individual reactions are usually of the Langmuir-Hinshelwood type, Michaelis-Menten expression being one example [126]. Depending on the positive (activation, induction) and negative (inhibition, repression) effects of various chemicals on the activity of an enzyme and the rate of the reaction it catalyzes, expressions with varying degrees of complexity have been proposed in the literature, all of these bearing a strong resemblance to the Langmuir-Hinshelwood type rate expressions used for chemical catalytic reactions. Due to enzyme-based

Table 2.1. Dependence of the specific cell growth rate μ on concentrations of nutrients, products and cell mass [35, 447, 451]

Function	Form	Reference
$\phi_j(N_j)$	$\frac{N_j}{K_j+N_j+N_j^2/K_{Ij}}, K_{Ij} > 0$	[19, 190, 600]
	$\frac{N_j}{K_j+N_j}$ if $K_{Ij} \to \infty$	[8, 9, 35]
	$1 - e^{-N_j/K_j}$	[35]
	$(1 + K_j N_j^{-\lambda_j})^{-1}$	[35]
	$N_j e^{-\gamma_j N_j}$	[7, 453]
$\varphi_j(P_j)$	$\frac{K_j}{(K_j+P_j)}$	[8, 9]
	$e^{-\alpha_j P_j}$	[8, 9]
	$(1 - \frac{P_j}{P_{jm}})^{\alpha}$	[47, 190, 239, 326, 338, 395] [544, 600]
$\varphi(X)$	$(1 - \frac{X}{X_m})$	[23, 24]

nature of cellular metabolism, rate expressions for cell replication, nutrient uptake and synthesis of metabolites of interest are usually analogous to those for individual enzyme-catalyzed reactions (Langmuir-Hinshelwood type). Some of the entries in Tables 2.1 and 2.2 are reflective of this.

The uncoupled, multiplicative form in Eq. 2.17, although by far the most popular, is not the only way of relating the specific cell growth rate (μ) to concentrations of various nutrients, cell mass and target metabolites. Alternate expressions for μ do not uncouple the effects of concentrations of individual species influencing cell growth. One such expression, used previously for description of growth of *P. chrysogenum*, the strain used for synthesis of penicillin, is the Contois kinetics, viz.,

$$\mu = \frac{\mu_o S}{K_s X + S}, \tag{2.18}$$

with μ_o and K_s being kinetic coefficients independent of S and X, and S the concentration of the rate-limiting nutrient.

2.5.2 Rate Expressions for Nutrient Uptake

The cell mass-specific uptake rate of nutrient N_i, σ_i, is usually related to activities accounting for consumption of that nutrient, namely cell growth (an aggregate of all reactions occurring in an individual cell), cell maintenance, and production (in significant amounts, amounts comparable to amount of cell mass) of certain metabolites of interest and is therefore expressed as

$$\sigma_i = \frac{\mu}{Y_{X/N_i}} + m_i + \Sigma a_{ij}\varepsilon_j \tag{2.19}$$

with Y_{X/N_i} being the biomass yield with respect to nutrient N_i, m_i (referred to as the maintenance coefficient) accounting for consumption of N_i for maintenance activities of cells, and a_{ij} being the amount of N_i utilized for production of unit amount of metabolite P_j. (The reciprocal of a_{ij} is commonly referred to as the yield of P_j with respect to N_i.) Since production of metabolites P_j is a part of cellular metabolism, the biomass yield in Eq. 2.19 is the apparent biomass yield (and not the true biomass yield) when consumption of N_i for production of P_j is accounted for directly [as in the last term in Eq. 2.19] and also indirectly via uptake of N_i for biomass production. When the last term in Eq. 2.19 is trivial, the cell mass yield (Y_{X/N_i}) in this relation represents the true cell mass yield. A direct accounting of utilization of a nutrient N_i for production of metabolite P_j, as represented by the last term in Eq. 2.19, is justified only when the amount of P_j is substantial, so that utilization of N_i for synthesis of P_j is comparable to that for cell growth. The significance of utilization of N_i for cell maintenance relative to utilization of the same nutrient for cell growth increases as the specific cell growth rate is reduced.

2.5.3 Rate Expressions for Metabolite Production

The dependence of specific formation rate of metabolite P_j (ε_j) on concentrations of various nutrients, cell mass and target metabolites is usually expressed in two different ways. The first approach involves expressing the dependence in an uncoupled, multiplicative form as per the relation

$$\varepsilon_j = \varepsilon_{jo}\chi_{j1}(N_1)\chi_{j2}(N_2)...\chi_{ji}(N_i)...\psi_{j1}(P_1)\psi_{j2}(P_2)...\psi_{jk}(P_k)...\psi(X) \tag{2.20}$$

with ε_{jo} being constant characteristic of the metabolite P_j. The popular forms of $\chi_{ji}(N_i)$ and $\psi_{jk}(P_k)$ are based on the experimental observations for a particular strain and the metabolite P_j. Synthesis of a metabolite P_j may be

Table 2.2. Dependence of the specific formation rate P_j, ϵ_j on concentrations of nutrients, products and cell mass [35, 447, 451]

Function	Form	Reference
$\chi_{ji}(N_j)$	$\frac{N_i}{N_i+K'_{ji}}$	[8, 9, 35, 453, 575]
	$\frac{N_i}{K'_{ji}+N_i+N_i^2/K'_{Iji}}$	[190, 600]
$\psi_{jk}(P_k)$	$\frac{K'_{jk}}{(K'_{jk}+P_k)}$	[8, 9]
	$e^{-\alpha_{jk}P_{jk}}$	[8, 9]
	$(1-\frac{P_k}{P'_{km}})^\beta$	[47, 190, 239, 326, 395, 544, 600]
$\psi(X)$	$(1-\frac{X}{X'_m})$	[325, 470]

- promoted by increased presence of a nutrient N_i (χ_{ji} increases with increasing concentration of N_i) at least up to some threshold levels and may be discouraged at high concentrations of N_i (the so-called "substrate inhibition"),
- discouraged by increased presence of the nutrient N_i (χ_{ji} decreases with increasing concentration of N_i, $\chi_{ji} \leq 1$), or
- unaffected by variations in concentration of N_i ($\chi_{ji} = 1$ for all values of N_i) (Table 2.2).

Production of a metabolite P_j is usually (i) unaffected by the presence of a target metabolite P_k [$\psi_{jk}(P_k) = 1$] or may be discouraged as the product P_k accumulates in the culture [i.e., $\psi_{jk}(P_k)$ decreases with increasing P_k]. Accessibility of nutrients in the abiotic phase to cell mass is reduced with increasing biomass concentration in the culture (X) at high levels of the same and as a result, cellular metabolism (including synthesis of metabolite P_j) may be affected negatively as the biotic fraction of the culture is increased [i.e., $\psi(X)$ may decrease with increasing X, $\psi(X) \leq 1$]. The popular forms of $\chi_{ji}(N_i)$, $\psi_{jk}(P_k)$ and $\psi(X)$ are provided in Table 2.2.

The second approach in relating the specific formation rate of metabolite P_j (ε_j) to concentrations of various nutrients, cell mass and target metabolites involves expressing ε_j as a function of the specific cell growth rate, viz., $\varepsilon_j = \varepsilon_j(\mu)$. A popular relation that follows this approach is the Leudeking-Piret rate expression used to represent kinetics of synthesis of metabolite P_j, viz.,

$$\varepsilon_j = \alpha_j \mu + \beta_j, \tag{2.21}$$

with α_j and β_j being constants characteristic of the particular metabolite P_j.

2.5.4 Miscellaneous Remarks

It was mentioned at the beginning of section 2.5 that the cell mass-specific rates of cell growth, uptake of various nutrients, and synthesis of various metabolites are influenced by two additional culture parameters, viz., pH and temperature. Many of the parameters in the rate expressions discussed in Eqs. 2.17 - 2.21 and Tables 2.1 and 2.2 are functions of pH and temperature. These two culture parameters are tightly controlled in bioreactor operations in research laboratories and industrial bioprocesses. This is the reason why we do not dwell much on the effects of pH and temperature on culture dynamics. Before concluding this section, it is pertinent to relate the discussion in this section to the compact representation of bioreactor dynamics provided in Eqs. 2.1 and 2.2. When an unstructured representation is used for cellular metabolism, the state variables ($\mathbf{x}$) would include concentrations of gaseous species such as O_2 and CO_2, concentrations of nutrients (including dissolved O_2) and extracellular products of interest (including dissolved CO_2), cell mass concentration, culture volume, and concentrations of intracellular metabolites of interest. If pH and temperature are not controlled, the state variables would also include these two additional variables. The inputs to the bioreactor ($\mathbf{u}$ and $\mathbf{d}$) typically would include volumetric flow rates and composition of gas feed and liquid feed (for fed-batch and continuous operations). The outputs from the bioprocess ($\mathbf{y}$) typically would include all bioreactor variables that are monitored (including culture parameters that are measured on-line or off-line, composition of the effluent gas, and volumetric flow rates of effluent gas and culture withdrawal).

2.6 Structured Kinetic Models

Structured kinetic representations are warranted in situations involving significant changes in composition of the biotic phase and the kinetics of cellular rate processes is significantly sensitive to changes in cell composition. Since it is not practical to account for variations in every component of the biotic phase, the structured kinetic model for a particular bioprocess must focus on carefully selected key components and rate processes of major interest for that bioprocess. Depending on a particular application of interest, a variety of structured kinetic representations have been reported in the literature. These can be classified as (1) morphologically structured models, (2) chemically structured models, (3) genetically structured models, and (4) metabolically structured models, and some binary and higher combinations of (1)-(4). Since antibiotic (such as penicillin) production is the example industrial bioprocess considered throughout this book, the illustrations provided below will pertain more often to antibiotic production processes. In these illustrations, which are imported from the literature, the notation used in the source will be followed as much as possible so that interested readers will have little difficulty in accessing additional details in the source references.

2.6.1 Morphologically Structured Models

The kinetics of nutrient utilization and product (for example, an antibiotic) formation by filamentous microorganisms and molds is usually quite complex. A characteristic of growth of these living species is cellular differentiation, with different cell types differing from one another in terms of functions (cell growth, uptake of different nutrients, and synthesis of a target metabolite such as an antibiotic) they perform. The illustration provided here pertains to penicillin production. Morphologically structured models for penicillin production by *P. chrysogenum* have been proposed earlier by Megee et al. [382], Nielsen [423], and Paul and Thomas [459]. The models due to Megee et al. [382] and Paul and Thomas [459], although very comprehensive, involve large number of parameters, which make their identification and validation difficult. In comparison, the model proposed by Nielsen [423, 424, 425] is simplified and flexible and is used here as an illustration. This is not to say that we prefer one structured model over the others. Three cell types are considered in the Nielsen [423, 424, 425] model, apical cells, subapical cells, and hyphal cells (denoted as a, s and h, respectively). Uptake of nutrients and formation of biomass occurs only in apical and subapical compartments of an hyphal element (a multicellular unit). In an hyphal element, the apical compartment is located between a tip and the first septum. The cells in the interior (with respect to apical compartment) have an intracellular composition similar to that of apical cells and form the subapical compartment. Three metamorphosis reactions are considered in the Nielsen [423, 424, 425] model: branching, tip extension and differentiation. During *tip extension*, some apical cells become subapical cells. *Branching* refers to formation of new apical compartments from the cells in the subapical compartment. The subapical cells further away from the tip become more and more vacuolated as their age increases. As a result, cells further away from the tip contain large vacuoles. These cells, which form the hyphal compartment, play an important role in transport of protoplasm toward the tip section. Formation of vacuolated hyphal cells from the subapical cells is referred to as *differentiation*. The transition from active subapical cells to completely vacuolated hyphal cells takes place gradually. The hyphal cells located in the vicinity of the subapical compartment are therefore assumed to retain the metabolic activity and ability to grow as do the subapical cells. This has been accounted for in the formulation of the model by considering that a fraction f_h of the hyphal cells is metabolically active ([423, 424, 425]).

The kinetic expressions for branching, tip extension and differentiation are considered to be first order in cell type being transformed, which leads to the following rate expressions for the metamorphosis reactions under consideration.

Branching (1):

$$Z_s \rightarrow Z_a, \quad u_1 = k_{u_1} Z_s \tag{2.22}$$

Extension (2):

$$Z_a \rightarrow Z_s, \quad u_2 = k_{u_2} Z_a \tag{2.23}$$

Differentiation (3):

$$Z_s \rightarrow Z_h, \quad u_3 = k_{u_3} \phi_3(S) Z_s, \quad \phi_3(S) = \frac{1}{1 + K_{u_3} S}. \tag{2.24}$$

In Eqs. 2.22 - 2.24, Z_a, Z_s, and Z_h represent the mass fractions of apical, subapical, and hyphal cells, respectively, in the total cell population, u_j's (j = 1, 2, 3) the rates of the three metamorphosis reactions and k_{u_j}'s (j = 1, 2, 3) the kinetic coefficients for these. *Differentiation* is assumed to be inhibited by the carbon source (usually glucose, S = glucose concentration in the abiotic phase). The form of $\phi_3(S)$ in Eq. 2.24 is a special case of the form of $\phi_i(N_i)$ ($N_i = S$) in Table 2.1. The specific growth rates of each cell type have been represented by the Monod kinetics, viz.,

$$\mu_j = k_j \phi(S), \quad \phi(S) = \frac{S}{K_s + S}, \quad j = a, s, h. \tag{2.25}$$

In Eq. 2.25, k_j's (j = a, s, h) represent the maximum values of specific growth rate of each cell type. The mass balances for the three cell types then can be described by Eqs. 2.14a and 2.14b ($c_i = Z_i$, $i = a, s, h$, $r_i^{\text{trans}} = 0$) with r_i^{gen} ($i = a, s, h$) being

$$r_a^{\text{gen}} = u_1 - u_2 + \mu_a Z_a, \quad r_s^{\text{gen}} = u_2 - u_1 - u_3 + \mu_s Z_s, \qquad (2.26)$$
$$r_h^{\text{gen}} = u_3 + f_h \mu_h Z_h.$$

Via addition of Eq. 2.14b for the three cells types and in view of the identity $Z_a + Z_s + Z_h = 1$, the expression for the specific cell growth rate μ can be deduced to be

$$\mu = \mu_a Z_a + \mu_s Z_s + f_h \mu_h Z_h. \qquad (2.27)$$

Cell death and cell lysis have not been considered in the model by Nielsen ([423, 424, 425]). One must note that the specific cell growth rate in this structured kinetic representation is dependent not only on the extracellular glucose concentration (S), but also on the fractions of the three cell types (intracellular variables as concerns the total cell population). The following expressions have been employed for the other two key processes, viz., glucose (S) uptake and penicillin (P) synthesis (specific rates denoted as σ_S and ε_P, respectively).

$$\sigma_s = \alpha_1 \mu + m_s + \alpha_2 \varepsilon_P, \quad \varepsilon_P = k_2 (Z_s + f_h Z_h) \chi(S), \qquad (2.28)$$
$$\chi(S) = \frac{S}{K_2 + S + S^2/K_I}.$$

In Eq. 2.28, the parameters α_1, α_2, k_2, K_2, and K_I are independent of S, X and Z_i's ($i = a,s,h$). The rate expression for glucose uptake in Eq. 2.28 is similar to Eq. 2.19, with α_1 being the reciprocal of the cell mass yield with respect to glucose ($Y_{X/S}$). Before leaving this illustration, the functional segregation of the three cell types must be commented upon. While all three cell types are considered to be capable of growth [Eq. 2.25], only the subapical cells and a fraction (f_h) of the hyphal cells are capable of synthesizing penicillin [Eq. 2.28], and the three cell types participate in different metamorphosis reactions [Eqs. 2.22-2.24]. The conservation equations for glucose (S) and the extracellular product (penicillin, P) are provided by Eq. 2.15 with $N_i = 0$ and $R_i^{\text{gen}} = 0$ ($i = S$, P, products of hydrolysis of penicillin pooled together with penicillin), $r_S^{\text{gen}} = -\sigma_S$, $r_P^{\text{gen}} = \varepsilon_P$, and $C_J = J$, $J = P, S, X$. Conservation equations for oxygen in the gas phase and the abiotic phase have not been considered in the Nielsen model [423, 424, 425], since it is assumed that the culture is not oxygen limited (with dissolved oxygen level assumed to be in excess of 45% saturation). The supply of sufficient amounts of oxygen may indeed become a critical problem at high biomass concentrations. The state variables [Eq. 2.1] in this model therefore are $\mathbf{x} = [X\ S\ P\ Z_s\ Z_h]^T$ for batch and continuous cultures and $\mathbf{x} = [X\ S\ P\ Z_s\ Z_h\ V]^T$ for a fed-batch culture. The identity $Z_a + Z_s + Z_h = 1$ implies that only two of the three fractions of the cell population are independent state variables.

2.6.2 Chemically Structured Models

The effects of key chemicals on the key rate processes are accounted for in the chemically structured models. All viable cells in the cell population are considered to be functionally similar, with conservation equations in the abiotic and biotic phases being considered for those species that are present in both phases. For such species, generation in the abiotic and biotic phases and transport across the interface between the abiotic and biotic phases must be fully accounted for. Synthesis of several antibiotics and other secondary metabolites by a host of microorganisms is inhibited by high concentrations of phosphate. Since cell growth is promoted by phosphate and the production of the secondary metabolite depends on both the cell mass concentration and production of the secondary metabolite per unit cell mass (specific production of the target metabolite), an optimum phosphate level which leads to maximum production of the secondary metabolite exists, as has been shown in previous experimental studies (see [461]). The illustration provided here pertains to a structured model for alkaloid production by *Claviceps purpurea* [461]. Let p and p_{int} denote the concentrations of extracellular and intracellular phosphate (KH_2PO_4), respectively, the dimensions for the respective variables being g phosphate/L abiotic phase and g phosphate/g biomass.

Phosphate is considered to be the rate-limiting nutrient as concerns cell growth. Following expressions have been employed for specific cell growth rate (μ) and specific cell lysis rate (r_d) [Eq. 2.10].

$$\mu = k_1[1 - exp(-K_1 p_{\text{int}})], \quad r_d = k_2 X. \tag{2.29}$$

In the relations above, k_1, k_2, and K_1 are the kinetic parameters. The dependence of μ on p_{int} is expressed by the Tessier equation. Cell lysis releases phosphate into the abiotic phase in quantity proportional to the cell mass phosphate content ($Y_{P/X}$) and the intracellular phosphate concentration (p_i). This release must be accounted for in the mass balance for extracellular phosphate, which is described by Eq. 2.12 with $C_p = p$ and $N_p = 0$, and R_p^{gen} and r_p^{trans} being

$$R_p^{\text{gen}} = (Y_{P/X} + p_{\text{int}})k_2 X^2 V/V_a, \quad r_p^{\text{trans}} = -\frac{k_3 p}{K_2 + p}. \tag{2.30}$$

In Eq. 2.30, k_3 and K_2 are kinetic parameters. The conservation equation for intracellular phosphate is provided by Eq. 2.14b with $c_i = c_p = p_{\text{int}}$ and r_p^{gen} = - $Y_{P/X}\mu$. The specific phosphate utilization rate for generation of biomass is therefore considered to be proportional to specific cell growth rate. It should be noted that the expression for r_p^{trans} in Eq. 2.30 is similar to the Monod expression. The transport of many species from the abiotic phase to the biotic phase and vice versa is facilitated by transport proteins (such as permeases). The rate expressions for transport of species across the outer cell membrane therefore bear close resemblance to the rate expressions for enzyme-catalyzed reactions (such as the Michaelis-Menten expression). Finally, the mass balance for the alkaloid (target metabolite) is provided by Eq. 2.15 with $C_i = C_a = a$, $N_a = 0$, $R_a^{\text{gen}} = 0$, and r_a^{gen} being provided by

$$r_a^{\text{gen}} = \varepsilon_a = k_4 K_3/(K_3 + p_i^2), \tag{2.31}$$

with K_3 and k_4 being the kinetic parameters. It is evident from Eqs. 2.29 and 2.31 that while increasing intracellular phosphate content is conducive for cell growth, it inhibits alkaloid synthesis due to repression of phosphatase activity. In this chemically structured model, the rates of all key kinetic activities, viz., cell growth, phosphate consumption, (- r_p^{gen}), and alkaloid production are expressed in terms of the conditions prevailing in the biotic phase, viz., p_i in the present case. The state variables [Eq. 2.1] in this model therefore are $\mathbf{x} = [X \;\; p \;\; p_i \;\; a]^T$ for batch and continuous cultures and $\mathbf{x} = [X \;\; p \;\; p_i \;\; a \;\; V]^T$ for a fed-batch culture.

2.6.3 Chemically and Morphologically Structured Models

This illustration pertains to production of the antibiotic cephalosporin C (CPC) by the mold *Cephalosporin acremonium*. As in the case of penicillin production (section 2.6.1), experimental observations have revealed that the cell population is comprised of three different morphological types, viz., hyphae (h), swollen hyphal fragments (s), and arthrospores (a). These three cell types undergo two metamorphosis reactions shown below.

$$(i) \;\; X_h \rightarrow X_s, \quad (ii) \;\; X_s \rightarrow X_a. \tag{2.32}$$

Transformation of hyphae into swollen hyphal fragments involves assimilation of glucose (a carbon source, denoted as g) and methionine (a nitrogen source, denoted as m). The uptake of these two nutrients is confined mainly to hyphae and swollen hyphal fragments. Of the three cell types, only the swollen hyphal fragments are primarily capable of synthesizing CPC. Experimental studies have indicated that the rate of CPC synthesis is directly related to the activity of enzymes responsible for this. These enzymes are induced by intracellular methionine and are repressed by glucose. Only hyphae are capable of replication (growth). The rate of reaction (i) is expressed as a function of concentrations of hyphae, glucose and methionine, while the rate of reaction (ii) is expressed as a function of concentrations of glucose and swollen hyphal fragments. Let Z_h, Z_s and Z_a denote the mass fractions of hyphae, swollen hyphal fragments, and arthrospores, respectively, in the total cell population. Then the conservation equations for the three cell types can be expressed as in Eq. 2.14a with $c_i = Z_i$, $r_i^{\text{trans}} = 0$, $i = h$, s, a. The net rates of generation of the three cell types, r_i $(i = h, s, a)$, are expressed as ([35, 374])

$$r_h = (\mu' - \beta - k_D)Z_h, \; r_s = \beta Z_h - (\gamma + k_D)Z_s, \; r_a = \gamma Z_s - k_D Z_a \quad (2.33)$$

with k_D being the kinetic coefficient for cell death or cell lysis and the specific rates μ', β and γ being expressed as

$$\mu' = \mu_m \phi_1(g), \; \beta = (k_{11} + k_{12}m/(K_m + m))\phi_2(g), \; \gamma = k_{21} + k_{22}\phi_1(g),$$
$$\phi_1(g) = g/(K_g + g), \; \phi_2(g) = g/(K_G + g), \quad (2.34)$$

with g and m denoting concentrations of glucose and methionine, respectively, in the abiotic phase, and $\mu_m, k_{11}, k_{12}, k_{21}, k_{22}, K_g, K_G$ and K_m being the kinetic parameters. In view of the identities $Z_h + Z_s + Z_a = 1$ and $X_h/Z_h = X_s/Z_s = X_a/Z_a = X$, it can be deduced from Eqs. 2.14a and 2.33 that

$$\mu^{\text{net}} = \mu - k_D, \quad \mu = \mu' Z_h. \quad (2.35)$$

The structured model by Matsumura et al. [374] incorporates conservation equations for glucose and methionine in the abiotic phase. A separate mass balance for intracellular glucose is not considered. Hence, one needs to employ Eq. 2.15 for glucose $(i = g)$ and Eq. 2.12 for methionine $(i = m)$ with $N_i = 0$ and $R_i^{\text{gen}} = 0$ $(i = g, m)$. Glucose uptake occurs predominantly in hyphae and swollen hyphal fragments and as a result

$$r_g^{\text{gen}} = -(\mu_m Z_h / Y_{H/G} + v_m Z_s)\phi_1(g), \quad (2.36)$$

in Eq. 2.15 with $Y_{H/G}$ (yield of hyphae based on glucose consumption) and v_m being kinetic parameters. For methionine in the abiotic phase,

$$r_m^{\text{trans}} = -(U_h Z_h + U_s Z_s), \quad U_h = U_{mh}\phi_3(m),$$
$$U_s = U_{ms}\phi_3(m), \quad \phi_3(m) = m/(K_m + m), \quad (2.37)$$

with U_{mh}, U_{ms}, and K_m being the associated kinetic parameters.

Owing to the considerable significance of intracellular methionine in regulating expression of CPC-synthesizing enzymes, conservation equations for methionine in the three cell types are also part of the structured model. Let m_{ih}, m_{is} and m_{ie} denote concentrations of methionine inside the hyphae, swollen hyphal fragments, and arthrospores, respectively. The conservation equation for the cell type j ($j = h, s, a$), which is analogous to that for total cell mass, Eq. 2.11, has the form

$$dX_j/dt = (\mu_j - k_D)X_j - DX_j, \quad j = h, s, a, \tag{2.38}$$

with the effective specific growth rates μ_h, μ_s, and μ_a being expressed based on Eq. 2.33 as

$$\mu_h = (\mu' - \beta), \quad \mu_s = \beta Z_h/Z_s - \gamma, \quad \mu_a = \gamma Z_s/Z_a. \tag{2.39}$$

The conservation equations for an intracellular species in one or more of the cell types can be expressed in a form similar to those in Eqs. 2.13 and 2.14b. For a specie q, the temporal variation in its intracellular concentration in cell type j, q_{ij}, can therefore be expressed as

$$dq_{ij}/dt = r_{sj}^{\text{gen}} - r_{sj}^{\text{trans}} - \mu_j q_{ij}, \tag{2.40}$$

with r_{sj}^{gen} and r_{sj}^{trans} representing the specific rate of net generation of specie q in cell type j and the specific rate of transport of specie q from cells of type j into the abiotic phase [both in units mass (moles) of q /{time. mass of cells of type j}]. For methionine ($s = m$), these rates have the following forms for the three cell types under consideration [35, 374].

$$\begin{aligned}
r_h^{\text{gen}} &= V_{h,syn}^{\max} - k_{3h}m_{ih} - V_{h,util}^{\max}\phi_1(g)m_{ih} - \beta m_{ih},\\
r_h^{\text{trans}} &= -U_h, \quad r_s^{\text{trans}} = -U_s,\\
r_s^{\text{gen}} &= V_{s,syn}^{\max} - k_{3s}m_{is} - V_{s,util}^{\max}\phi_1(g)m_{is} + \beta m_{ih}Z_h/Z_s - \gamma m_{is},\\
r_a^{\text{gen}} &= -k_{3a}m_{ia} + \gamma m_{is}Z_s/Z_a, \; r_a^{\text{trans}} = 0.
\end{aligned} \tag{2.41}$$

The first terms on the right sides of expressions for r_h^{gen} and r_s^{gen} account for biosynthesis of methionine in hyphae and swollen hyphal fragments, respectively. The terms in the expressions above containing β and γ represent (dis)appearance of methionine in a particular cell type population associated with interconversion between two cell types. The presence of glucose in the abiotic medium is considered to increase the rate of methionine utilization for protein synthesis. The estimation of the kinetic parameters in Eq. 2.41 has been based on comparison of the experimentally measured and model predicted values of the average intracellular methionine concentration, $(m_i)_{\text{avg}}$, the value predicted by the model being $(m_i)_{\text{avg}} = m_{ih}Z_h + m_{is}Z_s + m_{ia}Z_a$.

The kinetic parameters in Eq. 2.41 have been considered to be constant. The rate of synthesis of CPC is considered to depend on activity of enzymes responsible for synthesis of CPC. The conservation equation for this enzyme pool (denoted as e) in swollen hyphal fragments is provided by Eq. 2.40 with $q_{ih} = e_{ih} = e$, with

$$r_{eh}^{\text{trans}} = 0, \quad r_{eh}^{\text{gen}} = (1/X_s)(V_{mE} m_{is} X_s / \{m_{is} + K_E\})_{t-t_I} Q - \gamma e,$$
$$Q = \{1 + (\kappa/\alpha^n) g\} / \{1 + (\kappa/\alpha^n)(1+\eta) g^n\} \tag{2.42}$$

and $V_{mE}, K_E, \kappa, \alpha, n,$ and η being the kinetic parameters. The effect of catabolite repression by glucose is included in Q $(n > 1)$. The subscript $(t - t_I)$ denotes evaluation at time $t' = t - t_I$, with t_I representing the time lag between induction and gene expression. Finally, the mass balance for the target product (p), cephalosporin C ($C_p = p$), is expressed as in Eq. 2.15 with

$$r_p^{\text{gen}} = e Z_s, \quad R_p^{\text{gen}} = -k_{PD} p. \tag{2.43}$$

The second expression in Eq. 2.43 accounts for degradation of cephalosporin C in the abiotic phase. The magnitudes of various kinetic parameters for the structured model are reported in [35] and [374]. The state variables (Eq. 2.1) in this model therefore are $\mathbf{x} = [X\ g\ m\ Z_h\ Z_s\ m_{ih}\ m_{is}\ m_{ia}\ e\ p]^{\text{T}}$ for batch and continuous cultures and $\mathbf{x} = [X\ g\ m\ Z_h\ Z_s\ m_{ih}\ m_{is}\ m_{ia}\ e\ p\ V]^{\text{T}}$ for a fed-batch culture. The identity $Z_h + Z_s + Z_a = 1$ implies that only two of the three fractions of the cell population are independent state variables.

2.6.4 Genetically Structured Models

In the case of protein synthesis, the knowledge of pertinent mechanisms at the molecular level allows formulation of genetically structured models. Protein synthesis assumes particular importance in manufacture of enzymes of industrial importance, hormones, and other commercial polypeptides. These models are robust, that is, these can be used for reliable prediction at conditions different from those used for estimation of model parameters and model evaluation and as such are very useful for optimization with respect to environmental and genetic parameters. A simple illustration is considered here that focuses on transcription and translation processes involved in synthesis of a target protein. One must consider conservation equations for the messenger RNA (mRNA) obtained by transcription of a particular gene G and the product of translation of the message carried by the mRNA, viz., the target protein (P). Let [G], [mRNA], and [P] denote the molar intracellular concentrations (moles per unit cell mass) of G, mRNA, and P, respectively. The conservation equations for mRNA and P then are provided by Eq. 2.14b [329, 330] with $c_i = [i]$, $i =$ mRNA, P and

$$r^{\text{gen}}_{\text{mRNA}} = k_p \eta [G] - k_d [\text{mRNA}], \quad r^{\text{gen}}_P = k_q \xi [\text{mRNA}] - k_e [P],$$
$$r^{\text{trans}}_{\text{mRNA}} = 0. \tag{2.44}$$

In Eqs. (2.44), k_p and k_q are the kinetic coefficients for transcription of the gene and translation of mRNA, η the efficiency of promoter utilization, ξ the efficiency of utilization of the mRNA at the ribosomes, and k_d and k_e the kinetic coefficients for deactivation of the mRNA and the active protein, respectively. For intracellular proteins, r^{trans}_P is trivial, while for proteins partially excreted from living cells, r^{trans}_P is non-trivial and positive. In balanced growth, pseudo-steady state hypothesis (PSSH) is often invoked for the specific mRNA and the target protein, i.e., the rate of intracellular accumulation of each species (left side of Eq. 2.14b) is considered to be insignificant compared to rates of other processes (the non-trivial terms on the right side of Eq. 2.14b). Application of PSSH for an intracellular protein results in the following algebraic relations.

$$[\text{mRNA}] = k_p \eta [G]/(k_d + \mu), \quad [P] = k_q \xi [\text{mRNA}]/(k_e + \mu). \tag{2.45}$$

The cell mass-specific rate of synthesis of the target protein, r^{gen}_P, therefore can be deduced to be

$$r^{\text{gen}}_P = k_p k_q \eta \xi [G] \mu / \{(k_d + \mu)(k_e + \mu)\}. \tag{2.46}$$

From this the cell mass-specific production rate of the target protein (ε_P, total rate of protein production in the culture $= \varepsilon_P X V$) can be obtained as follows. If the cells are subject to death, then $\varepsilon_P = r^{\text{gen}}_P - r_d$, while if the cells are subject to lysis, then assuming total release of protein from the cells undergoing lysis into the abiotic phase, $R^{\text{gen}}_P V_a = r_d [P] X V$ and in that case $\varepsilon_P = r^{\text{gen}}_P$. If the target protein is partially excreted, then one must consider mass balances for it in both biotic and abiotic phases with r^{trans}_P providing the linkage between the two balances.

The rate of expression of an operator-regulated gene depends on the efficiency of transcription of that gene (η), which in turn is determined by interactions of modulating species at operator sites and RNA polymerase binding. This efficiency is thus proportional to the probability that the operator site O is not bound to repressor protein R. The genetically structured model involves a large number of model parameters representing various molecular interactions. A specific genetic change would affect only certain interactions and therefore specific model parameters. Further details on this model [329, 330] are spared here and interested readers are referred to the source references ([33, 329, 330]). For *Escherichia coli*, the kinetic parameters for the transcription and translation processes, viz., k_p and k_q, have been correlated to the specific cell growth rate (μ), with both parameters increasing with increasing μ [329, 330]. Such kinetic models will allow mapping of nucleotide sequence into cell population productivity and therefore afford the user capability for systematic optimization of cloned DNA inserts and in the long run the genetic makeup of the organism.

2.7 Case Studies

2.7.1 An Unstructured Model for Penicillin Production

In this case study, the mechanistic model of Bajpai and Reuss [36] was used as starting point for model development. The original model has been extended by including additional input variables such as agitation power, and aeration rate. Functional relationships among the process variables are summarized in Table 2.3 and all inputs and outputs are listed in Figure 2.1. A variety of mathematical representations have been suggested for describing certain biological behaviors by researchers referenced earlier in the text and others. We used the representations by Bajpai and Reuss [36] but readers are cautioned that several other representations may also be used to describe the penicillin fermentation process as we discussed earlier.

Unstructured Models

Mass balance equations can be summarized as follows.

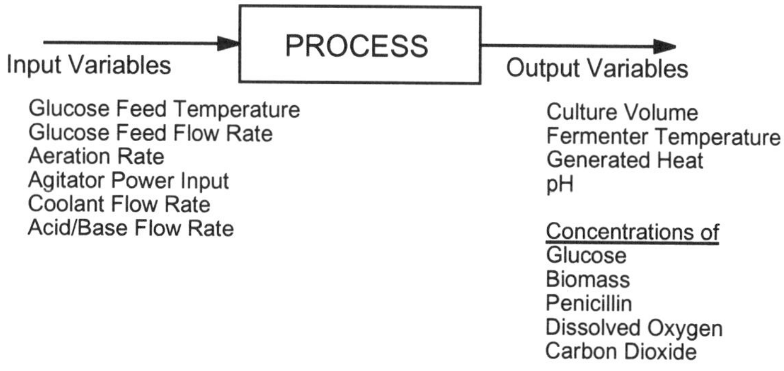

Figure 2.1. Input/output structure of the process.

Table 2.3. Functional relationship among the process variables

Model Structure
X = f (X, S, C_L, H, T)
S = f (X, S, C_L, H, T)
C_L = f (X, S, C_L, H, T)
P = f (X, S, C_L, H, T, P)
CO_2 = f (X, H, T)
H = f (X, H, T)

Biomass: The dependence of specific growth rate on carbon and oxygen substrates was assumed to follow Contois kinetics [36] to consider the biomass inhibition. The biomass growth has been described by Eq. 2.11 with $C_x = X$ and $\mu^{\text{net}} = \mu$ and the specific growth rate μ being

$$\mu = \mu_x \frac{S}{(K_x X + S)} \frac{C_L}{(K_{ox} X + C_L)} \tag{2.47}$$

in the original model [36]. The variables and parameters used are defined in Table 2.3 and 2.4.

In order to include the effects of environmental variables such as pH and temperature, biomass formation can be related to these variables by introducing their effects in the specific growth rate expression [61] to give:

$$\mu = \left[\frac{\mu_x}{1 + \frac{K_1}{[H^+]} + \frac{[H^+]}{K_2}} \right] \frac{S}{(K_x X + S)} \frac{C_L}{(K_{ox} X + C_L)} \times \left[k_g e^{-\frac{E_g}{(RT)}} - k_d e^{-\frac{E_d}{(RT)}} \right]. \tag{2.48}$$

This would in turn affect the utilization of substrate and the production of penicillin. Direct effects of pH and temperature on penicillin production

Table 2.4. Initial conditions, kinetic and controller parameters for normal operation (adapted from [61])

Time: t (h)	
Initial Conditions	**Value**
Biomass concentration: X (g/L)	0.1
Carbon dioxide concentration: CO_2 (mmole/L)	0.5
Culture volume: V (L)	100
Dissolved oxygen concentration: C_L (= C_L^* at saturation) (g/L)	1.16
Heat generation: Q_{rxn} (cal)	0
Hydrogen ion concentration: $[H^+]$ (mole/L)	$10^{-5.5}$
Penicillin concentration: P (g/L)	0
Substrate concentration: S (g/L)	15
Temperature: T (K)	297
Activation energy for growth: E_g (cal/mole)	5100
Arrhenius constant for growth: k_g	1×60^3
Activation energy for cell death: E_d (cal/mole)	52000
Arrhenius constant for cell death: k_d	10^{33}
Constant: K_1 (mole /L)	10^{-10}
Constant: K_2 (mole /L)	7×10^{-5}
Constant relating CO_2 to growth: α_1 (mmole CO_2/ g biomass)	0.143
Constant relating CO_2 to maintenance energy:	
Constant relating CO_2 to penicillin production:	
Constant: p	9
Constant: b	0.60
Constants in K_{la}: α, β	72, 0.5
Constant in F_{loss}: λ (h^{-1})	2.5×10^{-4}
Constant in heat generation: r_{q2} (cal/g biomass.h)	1.6783×10^{-1}
Cooling water flow rate: F_c (L/h)	
Contois saturation constant: K_x (g/L)	0.15
Density × heat capacity of medium: $\rho\ C_p$ (1/L°C)	1/1580
Density × heat capacity of cooling liquid: $\rho_c C_{pc}$ (1/L°C)	5/2000
Feed substrate concentration: s_f (g/L)	600
Feed flow rate of substrate: F (L/h)	
Feed temperature of substrate: T_f (K)	298
Heat transfer coefficient of cooling/heating liquid: a (cal/h°C)	1050
Inhibition constant: K_p (g/L)	0.0002
Inhibition constant for product formation: K_I (g/L)	0.10
Maintenance coefficient on substrate: m_x (h^{-1})	0.014
Maintenance coefficient on oxygen: m_o (h^{-1})	0.467
Maximum specific growth rate: μ_x (h^{-1})	0.092
Oxygen limitation constant: K_{ox}, K_{op} (no limitation)	0
Oxygen limitation constant: K_{ox}, K_{op} (with limitation)	2×10^{-2}, 5×10^{-4}
Penicillin hydrolysis rate constant: K (h^{-1})	0.04
pH : (Base)K_c,τ_I:(h),τ_D:(h)	8×40^{-4}, 4.2, 0.2655
(Acid)K_c,τ_I:(h),τ_D:(h)	1×10^{-4}, 8.8, 0.125
Specific rate of penicillin production: μ_p (h^{-1})	0.005
Temperature: (Cooling)K_c,τ_I:(h),τ_D:(h)	70, 0.5, 1.6
(Heating)K_c,τ_I:(h),τ_D:(h)	5, 0.8, 0.05
Yield constant: $Y_{x/s}$ (g biomass/g glucose)	0.45
Yield constant: $Y_{x/o}$ (g biomass/g oxygen)	0.04
Yield constant: $Y_{p/s}$ (g penicillin/g glucose)	0.90
Yield constant: $Y_{p/o}$ (g penicillin/g oxygen)	0.20
Yield of heat generation: r_{q1} (cal/g biomass)	60
α_2 (mmole CO_2/ g biomass h)	4×10^{-7}
α_3 (mmole CO_2/ L h)	10^{-4}

are not considered due to the complex nature of the phenomenon, and unavailability of the experimental data.

A typical inhibition term that includes hydrogen ion concentration $[H^+]$ is introduced into the specific growth rate expression. It has been found that the $[H^+]$-dependent term in the rectangular parentheses in Eq. 2.48. The values of K_1 and K_2 are chosen to be in the range of their typical values in the literature [426, 545].

The influence of temperature on the specific growth rate of a microorganism shows an increasing tendency with an increase in temperature up to a certain value which is microorganism specific and a rapid decrease is observed beyond this value. This decrease might be treated as a death rate [545]. These effects are reflected in the temperature-dependent term in Eq. 2.48 with k_g and E_g being the pre-exponential constant and activation energy for cell growth, and k_d and E_d being the pre-exponential constant and activation energy for cell death, respectively. Typical values for these parameters were taken from the literature [545]. An adjustment has been made so that an increase in temperature enhanced the biomass formation up to 35°C.

Penicillin:

The production of penicillin is described by non-growth associated product formation kinetics. The hydrolysis of penicillin is also included in the rate expression [36] for completeness.

$$\frac{dP}{dt} = \epsilon_p X - KP - \frac{P}{V}\frac{dV}{dt} \tag{2.49}$$

where, ϵ_p is the specific penicillin production rate defined as:

$$\epsilon_p = \epsilon_o \frac{S}{(K_p + S + S^2/K_I)} \frac{C_L^p}{(K_{op}X + C_L^p)} \tag{2.50}$$

Substrate inhibition kinetics for penicillin production was originally proposed by Bajpai and Reuss [36] to successfully represent the observed experimental behavior. They commented that the proposed mechanism should not be considered to throw any light upon the nature of phenomena involved. Others point out that industrial strains of penicillin production are tolerant to high levels of glucose and question the use of substrate inhibition terms in Eq. 2.50. Large quantities of substrate results in only little improvement in penicillin production.

Substrates:

The utilization of each of the two substrate (glucose and oxygen) largely for biomass growth, and penicillin formation, and cell maintenance [36]. The mass balances for glucose and oxygen for the variable volume fed-batch operation therefore are

Glucose:

$$\frac{dS}{dt} = -\frac{\mu}{Y_{x/s}}X - \frac{\epsilon_p}{Y_{p/s}}X - m_x X + \frac{Fs_f}{V} - \frac{S}{V}\frac{dV}{dt} \tag{2.51}$$

Dissolved Oxygen:

$$\frac{dC_L}{dt} = -\frac{\mu}{Y_{x/o}}X - \frac{\epsilon_p}{Y_{p/o}}X - m_o X + K_{la}(C_L^* - C_L) - \frac{C_L}{V}\frac{dV}{dt} \tag{2.52}$$

with yield coefficients $Y_{x/s}$, $Y_{p/s}$, $Y_{x/o}$, and $Y_{p/o}$ and maintenance coefficients m_x and m_o being constants characteristic of a particular penicillin producing strain. Whereas Bajpai and Reuss [36] have considered the overall mass transfer coefficient K_{la} to be constant, we have assumed K_{la} to be a function of agitation power input P_w and flow rate of oxygen f_g (as $f_g = Q_G = Q_{GF}$ suggested by [35].

$$K_{la} = \alpha\sqrt{f_g}\left(\frac{P_w}{V}\right)^{\beta}. \tag{2.53}$$

The values of α and β are assigned so that the dependence of penicillin concentration on K_{la} showed behavior very similar to the predictions of [36].

Volume Change:

The change in the bioreactor volume during fed-batch process operation is provided by a modified form of Eq. 2.8, which is

$$\frac{dV}{dt} = F + F_{a/b} - F_{\text{loss}} \tag{2.54}$$

The effect of acid/base addition on the bioreactor volume is accounted for by $F_{a/b}$ (volumetric feed rate of acid/base addition) The term F_{loss} accounts for evaporative loss during fermentation. The loss in volume due to evaporation is in fact more significant than the acid/base addition term in industrial cultivations. Normally the air entering the bioreactor is fairly dry and after bubbling through the broth it is at about 90 - 100 % relative humidity. Typically, 10 to 20 % of the total broth can be lost due to evaporation during one week of fermentation, the actual amount depending on the temperature of the fermentation. F_{loss} is a function of temperature and culture volume V of the cultivation broth. An accurate relationship can be developed by carrying out a set of experiments at different temperatures and measuring the humidity of the inlet and exit gas and the volume of the culture broth at different times during each experiment.

Culture Temperature:

Neglecting all other sources of heat generation except that caused by microbial reactions, the volumetric heat production rate is given as:

$$\frac{dQ_{rxn}}{dt} = r_{q_1}\frac{dXV}{dt} + r_{q_2}XV \tag{2.55}$$

where r_{q_1} is assumed to be constant and might be treated as a yield coefficient [426]. During the product synthesis phase, when the rate of biomass formation is rather low, there is still significant heat generation associated with metabolic maintenance activities. Therefore, we have included the second term on the right hand side of Eq. 2.55 to account for the heat production during maintenance. Because the heat generation and CO_2 evolution show similar profiles, their production rate due to growth (dX/dt) and biomass (X) should have the same ratio as a first approximation. Based on this observation, r_{q_2} is calculated and tabulated in Table 2.4. The energy balance is written based on a coiled type heat exchanger which is suitable for a laboratory scale fermentor [424]:

$$\frac{dT}{dt} = \frac{F}{s_f}(T_f - T) + \frac{1}{V\rho c_p}\left[Q_{rxn} - \frac{aF_c^{b+1}}{F_c + \dfrac{aF_c^b}{2\rho_c c_{pc}}}\right] \tag{2.56}$$

Carbon Dioxide:

The introduction of variables which are easy to measure yet important in terms of their information content has been very helpful in predicting other important process variables. One such variable is CO_2 from which biomass may be predicted with high accuracy. In this work, CO_2 evolution is assumed to be due to growth, penicillin biosynthesis and maintenance requirements as suggested by [398]. The CO_2 evolution is:

$$\frac{dCO_2}{dt} = \alpha_1\frac{dX}{dt} + \alpha_2 X + \alpha_3 \tag{2.57}$$

Here, the values of α_1, α_2 and α_3 are chosen to give CO_2 profiles similar to the predictions of [398].

The extended model developed consists of differential equations 2.11, 2.49, and 2.51-2.57 that are solved simultaneously.

Simulation Results of the Unstructured Model

Simulations have been carried out to check the performance of the simulator. In all runs, a batch operation has been followed by a fed-batch operation upon near complete depletion of the carbon source (glucose). This has been done by assigning a threshold value to glucose concentration which was chosen to be 0.3 g/L. The system switches to the fed-batch mode of operation when the level of glucose concentration reaches this threshold value. The predictions of the model under different conditions are compared with experimental data of Pirt and Righelato [471] and the simulation results of Bajpai and Reuss [36]. Note that most of the parameters are functions of the strain, nature of the substrate and the environmental conditions like pH and temperature. The additional terms that were introduced increased the stiffness of the ordinary differential equations. For that reason, some of the parameter values are readjusted. These readjusted parameters are listed in Table 2.4.

Figures 2.2, 2.3, 2.4, 2.5, and 2.6 show the simulation results under normal operating conditions with the pH and temperature being controlled at 5.0 and 25°C, respectively. The model successfully predicted the concentration profiles of biomass (Figure 2.2), glucose (Figure 2.3), penicillin (Figure 2.4), dissolved oxygen (Figure 2.5), and carbon dioxide (Figure 2.6). Typical experimental data are also shown in Figure 1.4 for comparison. In the batch operation, glucose and oxygen are mainly used for biomass growth. In Figures 2.2, 2.3, 2.4, 2.5, and 2.6, phase I represents the lag phase where no biomass production is observed. Phase II represents the exponential growth phase where the specific growth rate is maximum and so is the substrate utilization rate. Phase III is the late exponential or early stationary phase where the operation is switched to fed-batch mode and penicillin production starts. At this stage, glucose and oxygen are used for both biomass growth and penicillin production. Phase IV is the stationary phase where biomass production is essentially negligible and penicillin production is high. When the concentration of penicillin reaches its high value and levels off, it is common practice to stop the operation. All phases are simulated successfully via the unstructured model.

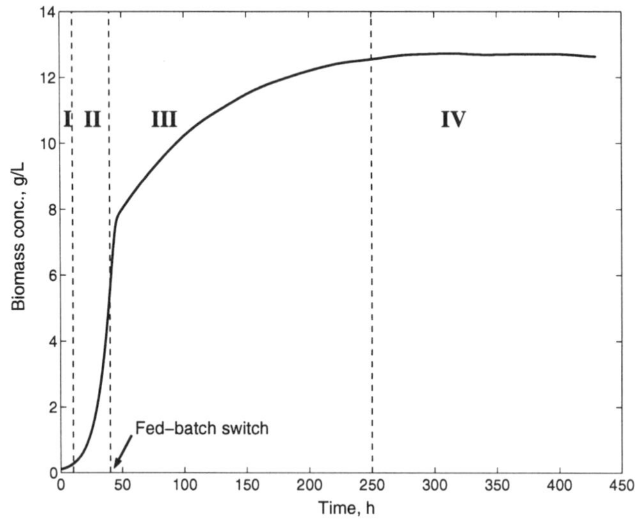

Figure 2.2. Time course of biomass concentration based on unstructured model.

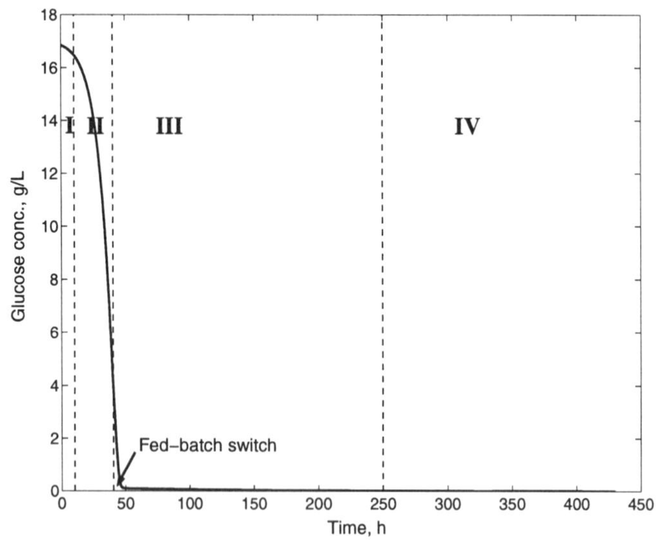

Figure 2.3. Time course of glucose concentration based on unstructured model.

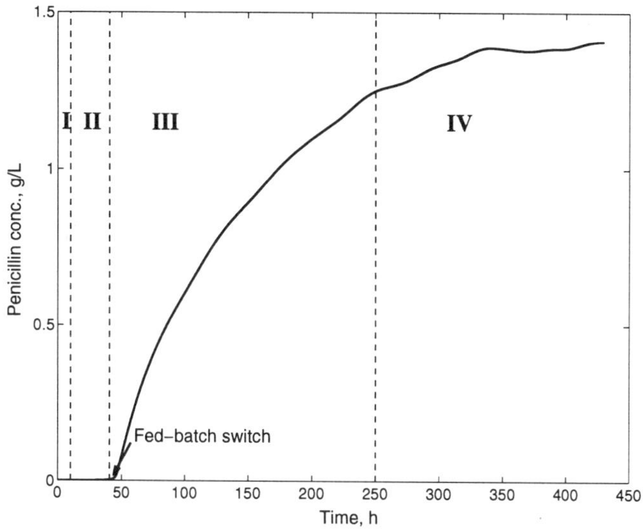

Figure 2.4. Time course of penicillin concentration based on unstructured model.

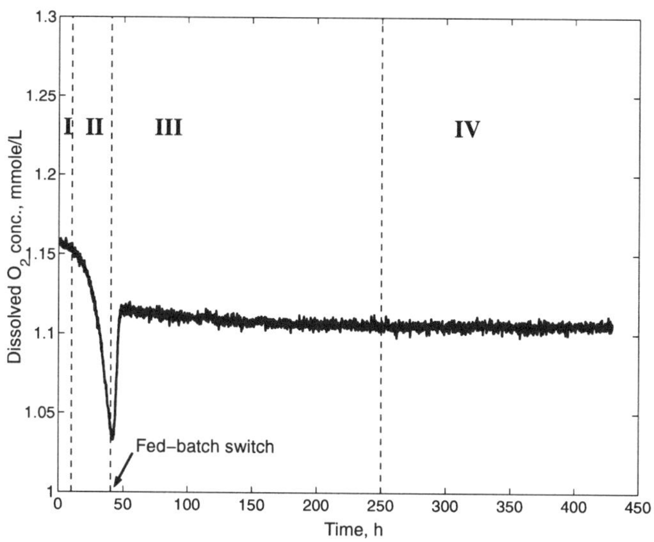

Figure 2.5. Time course of dissolved oxygen concentration based on unstructured model.

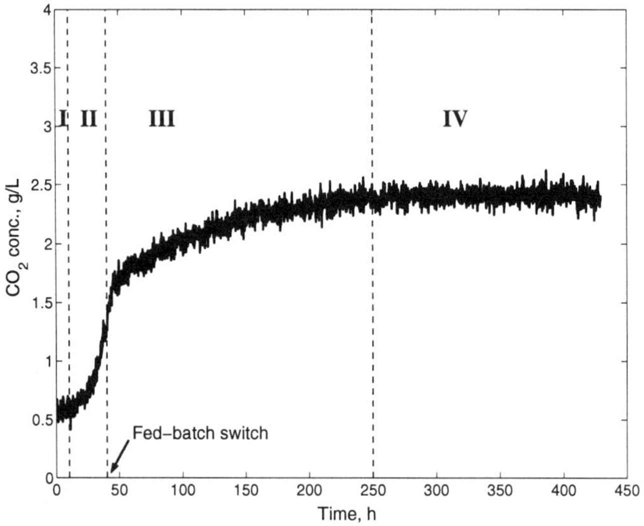

Figure 2.6. Time course of carbon dioxide concentration based on unstructured model.

2.7.2 A Structured Model for Penicillin Production

The model proposed in this case is a derivative of the morphologically structured model by Nielsen [424] and accounts for effects of dissolved oxygen on cell growth and penicillin production and variations in volume fractions of abiotic and biotic phases due to biomass formation [63]. Penicillin production is considered to occur in the subapical hyphal cell compartment and to be affected by glucose and oxygen.

Morphology and Metamorphosis

The morphological structure of the model is described in detail elsewhere [423, 424]. Each hyphal element is divided into three cell compartments/regions: apical (Z_a), subapical (Z_s) and hyphal (Z_h). Branching, tip extension and differentiation are the three metamorphosis processes considered [423]. Let Z_a, Z_s and Z_h denote the mass fractions of apical, subapical and hyphal portions, respectively in the cell population. Both branching and tip extension are considered to be first order processes, viz.

Branching:

$$Z_s \rightarrow Z_a \qquad u_1 = k_1 Z_s \tag{2.58}$$

Extension:

$$Z_a \rightarrow Z_s \qquad\qquad u_2 = k_2 Z_a \tag{2.59}$$

Differentiation:

$$Z_s \rightarrow Z_h \qquad\qquad u_3 = \frac{k_3 Z_s}{S K_3 + 1}. \tag{2.60}$$

Mass Balance Equations

Growth of apical and subapical cells is described by saturation type kinetics including effects of both glucose and oxygen in multiplicative form. The motivation for this is an earlier modeling work on penicillin production by Bajpai and Reuss [36] where growth has been described by Contois kinetics. Here, in order to reduce the model complexity, Monod kinetics has been used for describing the growth as suggested by Nielsen [423].

$$\mu_a = k_a \left(\frac{S}{K_{sa} + S}\right)\left(\frac{C_L}{K_{oa} + C_L}\right) \tag{2.61}$$

$$\mu_s = k_s \left(\frac{S}{K_{ss} + S}\right)\left(\frac{C_L}{K_{os} + C_L}\right) \tag{2.62}$$

Zangirolami et al. [685] suggest that hyphal cells may still retain the same metabolic activity and growth ability exhibited in the subapical compartment to some extent and considers a growing fraction (f_h) of hyphal cells in their model. On the other hand, Nielsen [423] suggests that hyphal cells have a metabolism completely different from the actively growing apical and subapical cells, and hence, they are believed not to contribute to the overall growth process and assumes μ_h to be zero. For simplicity, the growth rate of hyphal cells (μ_h) is also considered to be trivial based on Nielsen's work [423]. The overall specific growth rate (μ), which is an average of the growth rates of individual compartments, is then obtained as

$$\mu = \mu_a Z_a + \mu_s Z_s. \tag{2.63}$$

In view of the above, the conservation equations for the three compartments (components) of the cell population can be expressed as ($Z_a + Z_s + Z_h = 1$)

$$\frac{dZ_a}{dt} = u_1 - u_2 + (\mu_a - \mu) Z_a \qquad \text{apical cells} \tag{2.64}$$

$$\frac{dZ_s}{dt} = -u_1 + u_2 - u_3 + (\mu_s - \mu) Z_s \qquad \text{subapical cells} \tag{2.65}$$

$$\frac{dZ_h}{dn} = u_3 - \mu Z_h \qquad \text{hyphal cells.} \tag{2.66}$$

The terms μZ_i ($i = a,\ s,\ h$) in Eqs. 2.64, 2.65, and 2.66 account for dilution associated with biomass formation. The random fragmentation at different positions in individual hyphal elements leads to distribution in characteristics of the population such as the mass and numbers of total tips and actively growing tips [423]. Estimation of the average properties of the hyphal elements has been addressed theoretically by Nielsen [423] and experimentally using image analysis by Yang et. al. [674, 675]. In this case, we have made use of this population model based on the average properties of the hyphal elements [423]. In summary,

Hyphal element balance:

$$\frac{de}{dt} = (\varphi\mu - D)e \tag{2.67}$$

where D is the dilution rate.
Hyphal mass balance:

$$\frac{d\bar{m}}{dt} = (\mu - \varphi\bar{m})\bar{m} \tag{2.68}$$

Actively growing tips balance:

$$\frac{d\bar{n}}{dt} = (\phi - \varphi\bar{n})\bar{n} \tag{2.69}$$

where ϕ ($= a_2 u_1, a_2$ is constant) is the specific branching frequency $(1/\{g.h\})$ and is a function of Z_s. φ is the specific rate of fragmentation $(1/\{g.h\})$ and is assumed to be constant. The number of actively growing tips is less than the total number of tips ($\bar{n}_{\text{total}}$) due to formation of non-growing tips by fragmentation. Two inactive tips are formed as a result of fragmentation resulting in formation of an additional hyphal element. The average number of inactive tips on each element is exactly 2. The total number of tips then is:

$$\bar{n}_{\text{total}} = \bar{n} + 2. \tag{2.70}$$

The mass of the average hyphal growth unit ($\bar{m}_{hgu}$) and total hyphal growth unit length (l_{hgu}) are then obtained as [85]

$$\bar{m}_{hgu} = \frac{\bar{m}}{\bar{n}_{\text{total}}}, \qquad l_{hgu} = \frac{4\bar{m}_{hgu}}{\pi\rho(1-w)d^2}. \tag{2.71}$$

Mass balances are as follows:

For glucose:

$$\frac{dS}{dt} = \frac{F}{V}(S_f - S) - \sigma_s \bar{m} e \frac{V}{V_{\text{abiotic}}} - \frac{S}{V_{\text{abiotic}}}\frac{dV_{\text{abiotic}}}{dt} \tag{2.72}$$

$$\text{where} \qquad \sigma_s = \alpha_a \mu_a + \alpha_s \mu_s + m_s + \frac{1}{Y_{p/s}} \varepsilon_p Z_s. \tag{2.73}$$

For oxygen:

$$\frac{dC_L}{dt} = K_{la}(C_L^* - C_L) - \frac{F}{V} C_L - \sigma_o \bar{m} e \frac{V}{V_{\text{abiotic}}} - \frac{C_L}{V_{\text{abiotic}}} \frac{dV_{\text{abiotic}}}{dt} \tag{2.74}$$

$$\text{where} \quad \sigma_o = \frac{1}{Y_{x/o}} \mu + \frac{1}{Y_{p/o}} \varepsilon_p Z_s + m_o. \tag{2.75}$$

For penicillin:

$$\frac{dP}{dt} = \varepsilon_p Z_s \bar{m} e \frac{V}{V_{\text{abiotic}}} - KP - \frac{F}{V} P - \frac{P}{V_{\text{abiotic}}} \frac{dV_{\text{abiotic}}}{dt} \tag{2.76}$$

$$\text{where} \quad \varepsilon_p = \varepsilon_m \frac{S}{(K_p + S)(1 + \frac{S}{K_i})} \frac{C_L^p}{(K_{op} \bar{m} e + C_L^p)}. \tag{2.77}$$

The last term in Eqs. 2.72, 2.74 and 2.76 is due to volume correction that is applied to glucose, penicillin and oxygen concentrations since these are based on liquid volume (V_{abiotic}). Biomass concentration, X ($= \bar{m}e$, e = number of elements per culture volume, and $\bar{m}$ = average mass per element) is on the other hand based on culture volume (V).

In Eq. 2.73, m_s and ε_p are the maintenance on glucose and the specific rate of product formation, respectively and α_a and α_s are the stoichiometric biomass yield coefficients for apical and subapical cell compartments, respectively. The last term on the right hand side of σ_s (Eq. 2.73) reflects the fact that the target antibiotic is synthesized only by subapical cells. The dissolved oxygen balance (Eq. 2.74) can similarly be expressed after accounting for oxygen consumption due to cell growth, cell maintenance and product formation and m_o is the maintenance on oxygen in Eq. 2.75. The mass balance for penicillin in Eq. 2.76 accounts for hydrolysis/degradation for the antibiotic with K being the degradation/hydrolysis coefficient. The form of ε_p in Eq. 2.77 is chosen so as to reflect the inhibitory effects observed at high biomass and glucose concentrations.

These balances [Eqs. 2.72, 2.74 and 2.76] reduce to standard balances without volume correction when $X << 1/\bar{V}_{\text{biotic}}$, since $V_{\text{abiotic}} = V$ in that case.

Simulation Results of the Structured Model

For simplicity, at all times, all the hyphal elements were assumed to have the same composition of the three compartments with the same number of actively growing tips and mass. The model parameters are presented in Table 2.5. Parameters related to growth and substrate consumption were taken from Nielsen [423]. Again for simplicity, the growth kinetics of apical and subapical compartments are assumed to be the same resulting in the same stoichiometric yield coefficients for the two compartments and the same maximum specific growth rates ($k_a = k_s$).

In all simulations, a batch operation is considered to be followed by a fed-batch operation. The transition from batch culture to fed-batch culture occurs when the level of glucose concentration reaches a threshold value (10 g/L); such threshold values are commonly used in industrial scale penicillin production. The predictions of the model presented here under different operating conditions were compared with various experimental data. Note that most of the parameters are specific to the strain employed, substrate used and culture parameters such as pH, and temperature. Hence, this work focuses on capturing the general dynamic behavior of penicillin production rather than concentrating on strain or medium specific conditions. A set of simulation results are illustrated through Figures 2.7 and 2.13. Similar to the unstructured model, it is obvious from the simulated results that there are four distinct phases based on growth and are shown in Figures 2.7 through 2.13.

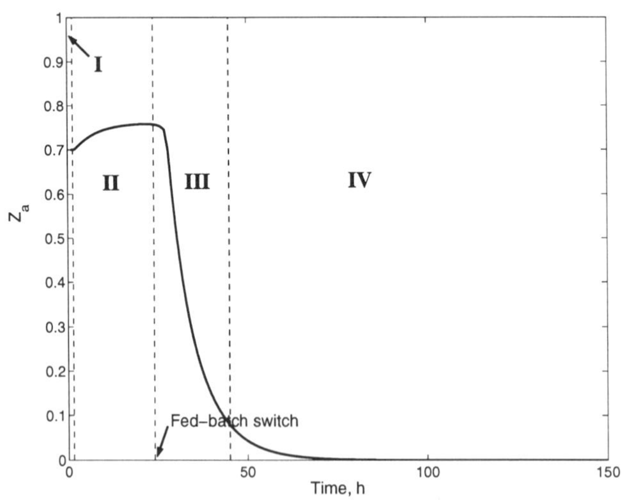

Figure 2.7. Time course of the apical fraction of the cells based on the structured model.

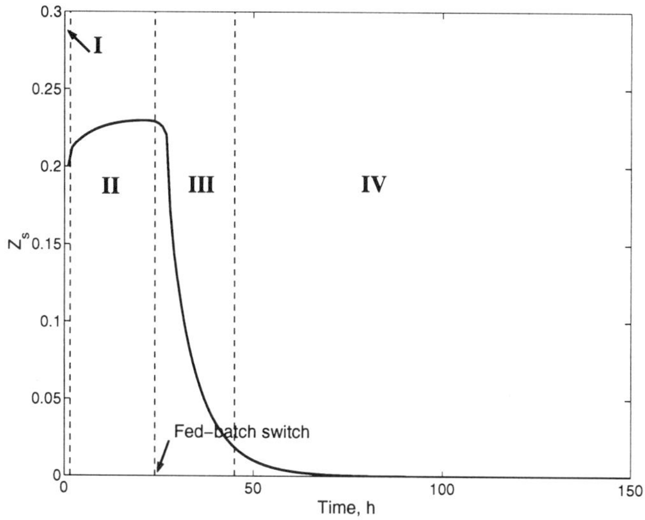

Figure 2.8. Time course of the subapical fraction of the cells based on the structured model.

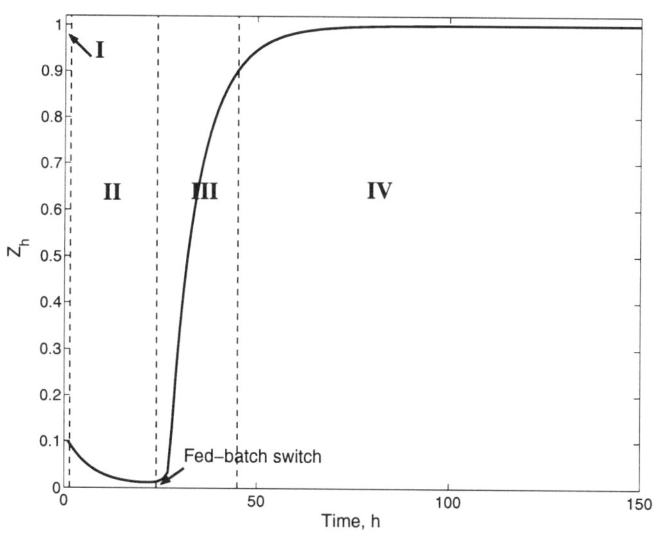

Figure 2.9. Time course of the hyphal fraction of the cells based on the structured model.

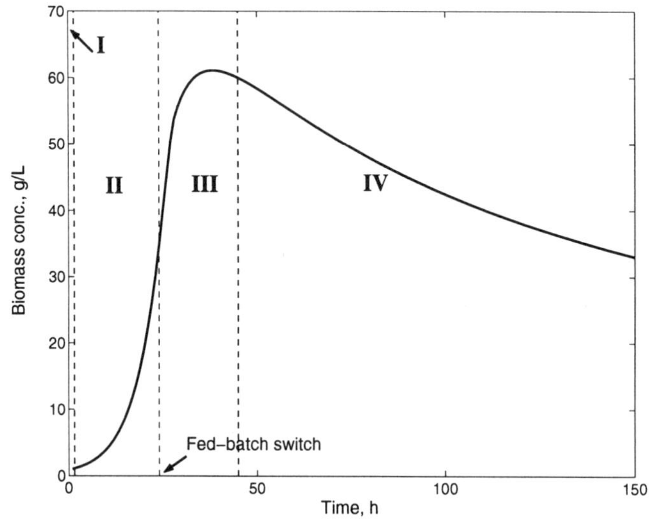

Figure 2.10. Time course of biomass concentration based on the structured model.

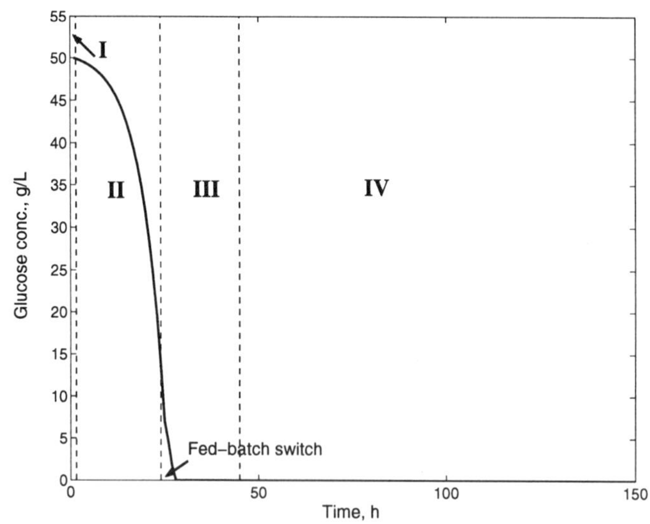

Figure 2.11. Time course of glucose concentration based on the structured model.

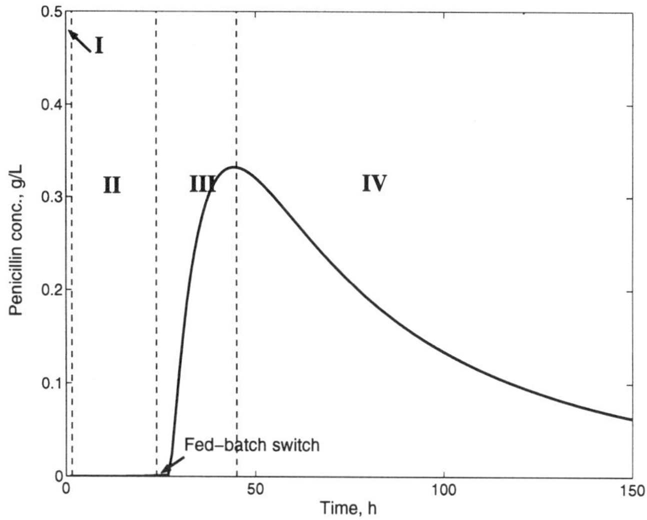

Figure 2.12. Time course of penicillin concentration based on the structured model.

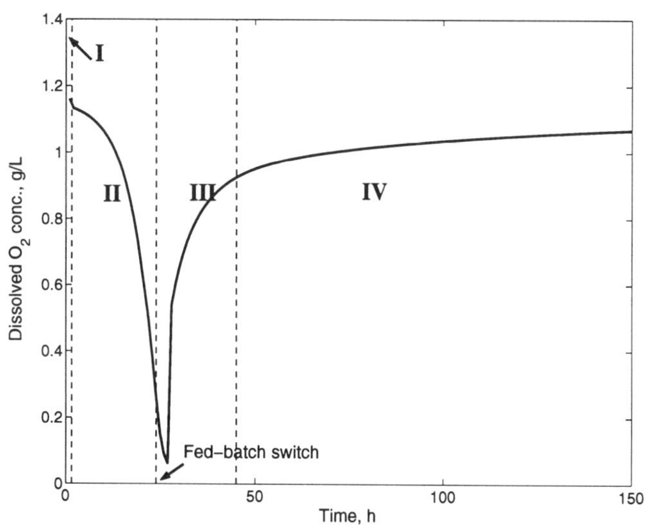

Figure 2.13. Time course of dissolved oxygen concentration based on the structured model.

Table 2.5. Parameter set for structured model

Parameter	Value	Reference
$Y_{x/o}$	0.04 g/g	[36]
$Y_{p/s}$	0.90 g/g	[36]
$Y_{p/o}$	0.20 g/g	[36]
k_1	2.30 1/h	[423]
k_2	0.70 1/h	[423]
k_3	0.85 1/h	[423]
K_3	4 g/L	[423]
k_a	0.16 1/h	[423]
k_s	0.16 1/h	[423]
K	0.004 1/h	modified from [36]
K_{sa}	0.03 g/L	[423]
K_{oa}	$0.02\times C_L^*$ g/L (under O_2 limitation, otherwise 0)	modified from [36]
K_{ss}	0.03 g/L	[423]
K_{os}	$0.02\times C_L^*$ g/L (under O_2 limitation, otherwise 0)	modified from [36]
a_2	5.2×10^8 tips/g	[423]
φ	6×10^7 $1/\{gh\}$	[423]
p	3	modified from [36]
ρ	1×10^6 g/m^3	assigned
w	0.67	[423]
α_a, α_s	0.45	[423]
m_s	0.015 1/h	[36]
m_o	0.467 1/h	[36]
ε_m	0.005 1/h	[36]
K_p	0.0002 g/L	[36]
K_i	0.1 g/L	[36]
K_{op}	5×10^{-4} (under O_2 limitation, otherwise 0)	modified from [36]
ρ_b	294 g/L	[264]
K_{la}	200 1/h	[36]

3

Experimental Data Collection and Pretreatment

Data collection during the progress of a batch is necessary for monitoring and controlling process operation for optimal cultivation and production. An essential element of effective monitoring and control is high quality data obtained during the progress of the batch via appropriate instrumentation/sensors. The information collected may also be used for modeling the process or improving the process design and production policies.

This chapter focuses on a number of important topics about experimental data collection. Section 3.1 outlines desirable properties of sensors and discusses various on-line and off-line sensors used in fermentation processes. Section 3.2 presents data acquisition systems and describes computer-based data collection and control. Section 3.3 presents introductory concepts in statistical design of experiments. Outliers in data and signal noise may have strong influence on the parameter values and structure of models developed, decisions about process status in monitoring, and regulatory action selected in process control. Consequently, outlier detection and data pretreatment such as reconciliation and denoising of data are critical for developing accurate models. Section 3.4 introduces various techniques on outlier detection and data reconciliation. Process data may contain various levels of signal noise. Section 3.5 introduces wavelets and discusses various noise reduction techniques. Section 3.6 outlines methods used in theoretical confirmation of data, in particular stoichiometric balances and thermodynamics of cell growth.

3.1 Sensors

Sensors may be categorized as on-line and off-line sensors. On-line sensors are preferred since they provide process information quickly without any disruption in the process and any sampling and cultivating delays, and fewer human errors, and allow for arbitrary frequencies of measurement. Off-line analysis techniques are used because of the difficulty and expense of developing sterilizable probes or constructing a sampling system for some process variables and product properties.

Sensors must have several characteristics that must meet the specifications for use in a particular application [366, 439, 475]:

Accuracy is the degree of conformity to standard when the device is operated under specified conditions. This is typically described in terms of maximum percentage of deviation expected based on a full-scale reading on the device specification sheet.

Precision (Repeatability) is the exactness with which a measuring instrument repeats indications when it measures the same property under the same conditions. Sensors display a drift in time which can be corrected by periodic calibration.

Range is the difference between the minimum and the maximum values of the sensor output in the intended operating limits. It is essential that accuracy and precision improve as the range is reduced, which implies that a small range would be preferred. However, the range must be large enough to span the expected variation of the process variable under typical operational conditions, including disturbances and set point changes.

Durability refers to the endurance of a sensor under the exposure to different operational conditions (pH, temperature, acidity). Since most of the industrial scale cultivations require extensive periods of operation time for completion (2-20 days), the sensor response should be stable for extended periods.

Reliability is the degree of how well a sensor maintains both precision and accuracy over its expected lifetime. Reliability is a function of the failure rate, failure type, ease of maintenance, and robustness of the sensor.

Response Time is the time it takes for the sensor output to reach its final value. It indicates how quickly the sensor will respond to changes in the environment. This parameter indicates the speed of the sensor and must be compared with the speed of the process.

Sensor technology is a rapidly growing field that has significant potential to improve the operation, reliability, serviceability, and utility of many engineering systems. Advances in chemistry, biochemistry, materials science and engineering have accelerated the development of new and more capable sensors. However, as the complexity and the capabilities of the sensors increase, there is a significant amount of associated cost. Cost may be a critical consideration in the selection of a sensor and the way a measurement should be made (on-line or off-line). There may always be a trade-off between a high-cost on-line frequent measurement or a relatively low-cost off-line infrequent measurement.

On-line Sensors

On-line sensors are crucial for monitoring and controlling a process for its safe and optimal performance. These instruments can be classified as

1. sensors that do not come in contact with the cultivation broth, (e.g., in a thermocouple)
2. *in-situ* sensors that are immersed directly into the cultivation broth and hence are in contact with it (e.g., pH meter, dissolved oxygen probe and level sensor).
3. other sensors, such as tachometer and rotameter.

When the sensors/probes directly come in contact with the cultivation broth, one potential problem is to maintain aseptic conditions. Under these conditions, probe should be sterilizable and should be placed in a way so as to avoid any possible leakage from/ to the bioreactor through the connections. The seal is usually accomplished by elastomer "O" rings that also provide an easy insertion of the probe.

The location of the sensor in the fermenter is very important since the contents of the bioreactor are usually heterogeneous. As a result, the measurements of variables that are critical for control action will be dependent on the location of the sensor. Conventionally, sensors are placed in the midsection of the vessel, though placement somewhere else may also be considered depending on the design of the bioreactor. A sensor should be placed in a region with sufficient turbulence to maintain the surface of the sensor clean and avoid build-up of material on it. Besides corrupting the sensor output, such build-up may lead to fouling of the sensor.

In the absence of *in-situ* sensors, on-line analysis of medium components is preferred. The main idea is to sample the medium automatically by collecting it in a loop that has a relatively small volume compared to the cultivation broth and to analyze it. Automatic sampling can be performed in two ways: (1) direct withdrawal of sample by using a syringe or a catheter into a loop, (2) use of a membrane module that separates the sample from the tip of the sensor. After collecting the sample, microbial activity has to be stopped to achieve precise measurement by either adding a deactivating agent or by cooling the sample [424]. Direct sampling allows for the measurement of the biomass and intracellular and extracellular components. The membrane modules used can be categorized either as membranes placed in a recycle loop connected to the bioreactor or *in-situ* membrane modules. The membrane may be replaced during the operation without any interruption of the process if it is placed in a recycle loop.

Flow Injection Analysis (FIA) has proven to be a valuable tool for on-line measurement of medium components due to its high speed (frequent analysis ability), good precision, and reliability. For the same purpose, other analytical systems are also used such as Mass Spectrometer (MS), High Pressure Liquid Chromatography (HPLC), Gas Chromatography (GC), with somewhat less efficiency. Among the *in-situ* sensors, Pt-resistance thermometers are commonly used for temperature measurement. Temperature control is typically implemented by manipulating flow rate of coolant circulating in coils if the temperature exceeds the control limits and by steam injection if the temperature goes below the minimum acceptable limit. For pH measurement, glass electrodes are used and the pH is regulated by the addition of acid or alkali. Dissolved Oxygen (DO_2) is measured by Pt, Ag/AgCl, Ag and Pb sensors. They could be either *polarographic* which are expensive but reliable or *galvanic* types. DO_2 is kept within the desired control limits by changing the agitator speed, inlet air flow rate and gas composition. The level of foam is determined by using conductance or capacitance sensors that trigger the addition of aliquots of antifoaming agent when there is excessive amount of foam formation.

Agitation speed and its power requirement are measured by a tachometer and watt-meter, respectively. Variable speed drives perform the control action. Air flow rate is measured by rotameters and mass flow meters and regulated by flow control valves. The pressure built inside the bioreactor is measured by spring and oil-filled diaphragms and regulated by pressure control valves. Feed flow rate is measured by electro-magnetic flow meters, vortex devices and electronic balances, it is controlled by upstream flow control valve and peristaltic pumps. On-line gas analysis (O_2 and CO_2) is performed by gas analyzers (paramagnetic and infrared, respectively) and by mass spectrometer.

Off-line Sensors

Off-line analysis becomes a viable option especially when there is a need to measure a large number of medium components in order to improve the understanding of the process. Disadvantages of off-line analysis include infrequent and time-delayed process information. This may be caused by the speed of the instrument or preliminary treatment of the sample. In some cases, the amount of sample necessary may be sufficiently high to cause a volume change in the bioreactor if sampling is done frequently. In a typical cultivation, substrates, precursors, intermediate metabolites and products are measured in addition to other biomass related material (*biotic phase*) and other components of the cell-free medium (*abiotic phase*).

Dry weight and optical density measurements are used to determine the mass of the cells. For homogeneous cell populations, usually optical density is correlated with the weight of the sample. Microscopy and plate counts are used to measure the number of cell colonies present in an aliquot of cell suspension and on an agar-plate, respectively. Coulter counter provides a means for counting the number of particles passing through an orifice hence giving size distribution. But, this instrument is very expensive and has a limited usage due to the problems associated with small cells and inability to measure fungal organisms. Flow cytometry is used to determine the protein, DNA and RNA contents of the biotic phase. Although this is a very powerful technique, it can only be applied to unicellular cultures. There are also chemical methods, such as enzymatic assays and colorimetric analysis for the measurement of these compounds, but some of them may be quite labor intensive. Image analysis systems are very useful especially in performing detailed morphological studies.

For the measurement of medium components, HPLC, being less selective, offers a wide range of advantages over FIA. GC is another widely used instrument with limited capacity. For certain components such as glucose, lactate and ethanol, analyzers specifically designed for these components are also available (e.g., YSI Glucose Analyzer). Physical properties of the cultivation medium such as viscosity, density and turbidity are also measured off-line in most of the cultivations.

The interested readers may find detailed information about sensors in many references [366, 396, 424, 439, 475].

3.2 Computer-Based Data Acquisition

Data collection and process control in most modern fermentation systems are computer-based. The computer is interfaced to process sensors by analog to digital converters and to final control elements such as control valves with digital to analog converters (Figure 3.1). This interface provides a link between hardware signals and software variables. Some analyzers may have their own microprocessors to refine and interpret data that they col-

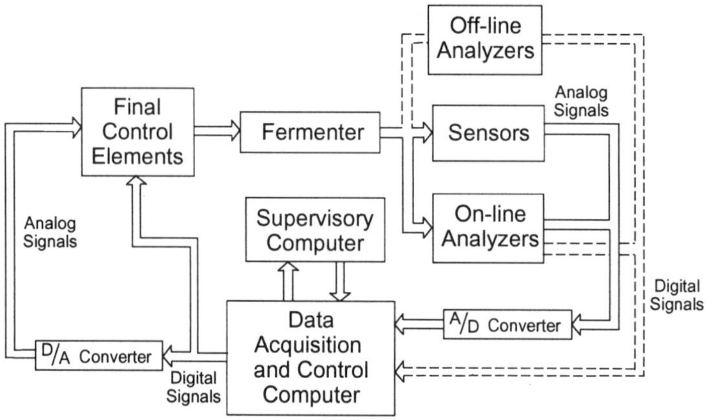

Figure 3.1. A typical data acquisition system.

lect. A supervisory level computer may also be included to perform higher level monitoring, diagnosis, optimization and advanced control tasks (Figure 3.1). In the data-acquisition system, the input to the system is a physical variable such as temperature, pressure or flow rate. Such a physical variable is first converted to an electrical signal (a voltage or current signal) by a suitable transducer. Then it is transmitted to A/D converter. The digitized signal is sent to the computer. The computer records the data, monitors process status and generates control commands and messages to plant personnel. The control commands are converted to analog signals by the D/A converter and sent to the final control elements.

The Analog-to-Digital (A/D) converter, also called as *encoder*, is a device that converts an analog signal into a digital signal, usually a numerically coded signal. A/D converter is needed as an interface between an analog component such as a sensor or transducer and a digital component such as a computer.

The Digital-to-Analog (D/A) converter, also called as *decoder*, is a device that converts a digital signal into an analog signal. This converter is needed as an interface between a digital component and an analog component (a physical device) to operate the physical device, such as a control valve or a pump.

The conversion of an analog signal into a digital signal (binary number) is an approximation since an analog signal can take on an infinite number of values, whereas the numbers that can be formed by a finite set of digits are limited. This approximation process is called *quantization*. In other words, "quantizing" means transforming a continuous or analog signal into a set of discrete states. Since the number of bits in the digital code is finite, A/D conversion results in a finite resolution, rounding off an analog number to the nearest digital level and producing a *quantization error*.

The functionality and ease of use of commercially available data collection and processing packages have significantly improved over the years. Most commercial data acquisition software in the market are capable of

- capturing and recording process data over time
- data reconciliation and outlier detection
- custom tailoring data treatment according to the user's needs
- transferring the data to other software
- sending out commands or data to control instruments and final control elements
- alarm generation and handling
- inputting time series data from any device into any application program
- creating charts and graphs that automatically uptake real-time data from serial devices
- performing real time analysis of data
- storing and compressing the data

Most software work with popular operating systems such as Windows 2000 and Unix. User-friendly graphical user interface (GUI) of software provides a convenient environment for the user. Simple, menu driven, step by step set-up is possible in most commercial software due to the interactive nature of the GUI. Hierarchical password protection personalizes the user. In most applications, controllers can be designed and set points can be changed as a function of any parameter using simple pictorial function blocks avoiding any programming.

3.3 Statistical Design of Experiments

Experiments are frequently performed to assess the effects of inputs, operating conditions, and changes in the process on the outputs. For example, the effects of variations in fermentation temperature, air flow rate, or strain type used on the attributes of a product would provide valuable information for optimizing productivity. Experiments are costly since they consume time and raw materials. Properly planned experiments minimize unnecessary duplications, generate more information with fewer experiments, and reduce the effects of measurement noise on data used for analysis. Statistically designed experiments provide unambiguous results at a minimum cost and provide information about interactions among variables [213]. An experiment is a means for drawing inferences about the real world and care must be exercised to define the scope of the experiment broad enough to include all conditions that the experimenter wishes to consider.

Most technical personnel focus on generating information from data, an activity that is called *statistical analysis*. However, equal attention should be given to generate informative data. The process of planning the experiments to generate the maximum amount of information with the minimum number of experiments is called statistical *design of experiments* (DOE). The objectives of DOE can be:

- To compare a set of treatments to determine whether their effects differ on some response (output) of interest
- To establish cause and effect relationships between *outputs* (*responses, dependent variables*) and *inputs* (*factors, independent variables*)
- To identify the most important inputs
- To identify interactions between inputs
- To identify improved settings for inputs to optimize the outputs
- To estimate empirical relationships between inputs and outputs.

The amount of data needed to cover this wide range of objectives that include comparison, screening, regression and optimization varies from one process to another. Consequently, the methods selected to design the experiments depend on the objective. Exploratory experiments can be designed as an iterative process where additional experiments are designed based on insight gained from analyzing the results of prior experiments. The literature on design of experiments is quite rich. Some of the popular books include [78, 370, 401]. More sophisticated design and analysis techniques include response surface analysis [77, 407], multivariate design of process experiments [277], and various types of advanced designs used for example in the pharmaceutical industries for drug discovery where very large numbers of configurations must be screened rapidly. The discussion in this section will be limited to screening experiments where the most influential inputs and interactions are determined. Two-level factorial designs are of great practical importance for comparison and screening studies. They are discussed in this section to underline the wealth of information that can be extracted from a process by a proper design and to contrast with the favorite approach of most technical people, the one-variable-at-a-time (OVAT) experimentation.

The OVAT approach involves variation in the level of an input (with levels of all other inputs being fixed) to find the input level that yields an optimal response. This procedure is then repeated for each of the remaining inputs. The OVAT procedure can be carried out for several iterations. The inputs that were varied in previous sets of experiments are kept at levels that gave optimal responses. The OVAT approach necessitates more experiments than the factorial design based experimental plans. Experiments must be duplicated and the results must be averaged to reduce the effects of measurements errors. This increases further the number of experiments conducted based on the OVAT approach. As illustrated later, the averaging process is an integral part of the analysis of data collected by factorial design based experimental plans. Furthermore, the OVAT approach does not provide information on the impact of the interaction of inputs on the response. Consequently, the OVAT approach must be avoided as much as possible.

Design of experiments to collect data for building empirical dynamic models of processes is another challenging problem. Here the focus is on designing input sequences that have specific characteristics so that the process is excited properly to generate data rich in information. This problem has been studied in the systems science, system identification and process control communities. The interested reader is referred to [346, 558].

3.3.1 Factorial Design

In any process, there may be a large number of input variables (factors) that may be assumed *a priori* to affect the process. Screening experiments are conducted to determine the inputs and interactions of inputs that influence the process significantly. In general the relationship between the inputs and outputs can be represented as

$$y = f(x_1, x_2, \cdots, x_p) + \epsilon \; . \tag{3.1}$$

where $x_i, i = 1 : p$ are the factors, (ϵ) is the random and systematic error and y is the response variable. Approximating this equation by using Taylor series expansion:

$$\begin{aligned} y \quad = \quad & b_0 + b_1x_1 + b_2x_2 + \cdots + b_px_p + b_{12}x_1x_2 + \cdots + b_{ij}x_ix_j + \cdots \\ & + b_{jp}x_jx_p + \cdots + b_{11}x_1^2 + \cdots + b_{ii}x_i^2 + \cdots + b_{pp}x_p^2 \\ & + \text{Higher Order Terms} + \epsilon \end{aligned} \tag{3.2}$$

a polynomial response surface model is obtained where b_i denotes the parameters of the model. The first task is to determine the factors (x_i) and the interactions (x_ix_j, $x_ix_jx_k$ and higher order interactions) that influence y. Then, the coefficients like b_i, b_{ij}, b_{ijk} of the influential inputs and interactions are computed. These parameters of the response surface models can be determined by least squares fitting of the model to experimental data. Several decisions have to be made before designing the experiments. A detailed summary of the decision making process is discussed in [88].

This section presents the two-level factorial design approach to select the conditions for conducting screening experiments that determine the significant factors and interactions. To perform a general factorial design, the investigator selects a fixed number of levels (two in most screening experiments) for each factor and then runs experiments with all possible combinations of levels and variables. If there are p factors, 2^p experiments must be conducted to cover all combinations. The number of experiments to be conducted grows rapidly with increasing number of factors. While 8 experiments are needed for 3 factors, 64 experiments are necessary for 6 factors. The factors may be continuous variables such as substrate feed rate (R) or bioreactor temperature (T) or discrete variables such as the strain (S) of the *inoculum.* The low and high levels of continuous variables may be coded using the $-$ and $+$ signs or 0 and 1, respectively. Qualitative (discrete) variables limited to two choices are coded using the same nomenclature. The levels of inputs to be used in each experiment are listed in a design matrix (Table 3.1).

Two-level factorial designs are appealing for a number of reasons. They require a few experiments to indicate major trends in process operation and determine promising directions for further experiments. They form the basis for two-level *fractional* factorial designs. They can be readily augmented to form composite designs, hence they are building blocks to construct efficient data collection strategies that match the complexity of the problem studied. The results of the experiments can be interpreted using simple algebra and computations. The interpretation of experimental results by discovering the significant factors and interaction effects is illustrated below by an example.

Example. A set of screening experiments are conducted in a laboratory scale fermenter and separation system to determine the effects of substrate

Table 3.1. Three alternative notations for 2^3 full factorial designs

Run	R	T	S	R	T	S	
1	−	−	−	0	0	0	1
2	+	−	−	1	0	0	r
3	−	+	−	0	1	0	t
4	+	+	−	1	1	0	rt
5	−	−	+	0	0	1	s
6	+	−	+	1	0	1	rs
7	−	+	+	0	1	1	ts
8	+	+	+	1	1	1	rts

feed rate (R), bioreactor temperature (T) and two different strains (S) of the *inoculum* on the total amount of product (yield Y) in a fed-batch run. The high (+) and low (−) settings for feed rate R (L/h) and temperature T ^oC are 0.08, 0.02 and 35, 17, respectively. Two strains A (−) and B (+) are used. It is assumed that approximately 5% higher production is reached when strain A is used (first four runs in Table 3.2). The fictitious experiments and penicillin production information are listed in a tabular form (Table 3.2).

Table 3.2. Data from a 2^3 full factorial design for investigating the effects of substrate feed rate (R L/h), bioreactor temperature (T ^oC) and *inoculum* strains (S) on the total amount of product (Y *grams*) in a fed-batch run.

Run	R	T	S	Y
1	−	−	−	69.24
2	+	−	−	214.82
3	−	+	−	59.45
4	+	+	−	133.49
5	−	−	+	65.78
6	+	−	+	201.93
7	−	+	+	57.07
8	+	+	+	126.82

The first group of information to extract is the effect of each variable on yield. For example, with everything else besides the experimental error remaining the same, what is the effect of temperature on the product (penicillin) yield? Consider for example runs 1 and 3 in Table 3.2: The variation in the yield is due to a variation in T and experimental error. In fact, there are four pairs of runs in Tables 3.1 and 3.2 where R and S have identical values in each pair and T is at two different levels. The variations in yield with variation in temperature for the four pairs and the corresponding R and S settings are listed in Table 3.3.

Table 3.3. The effect of temperature on penicillin yield

Individual Measure of the effect of changing T from 17^oC to 35^oC		Level of other factors	
		R (L/h)	S (Type)
$y_3 - y_1$	$=59.45 - 69.24 = -9.79$	0.02	A
$y_4 - y_2$	$=133.49 - 214.82 = -81.33$	0.08	A
$y_7 - y_5$	$=57.07 - 64.41 = -4.69$	0.02	B
$y_8 - y_6$	$=133.71 - 213.61 = -79.9$	0.08	B

The *main effect* of temperature is the average of these four differences (-43.93). It is denoted by T (not to be confused by T that is the symbol for the input) and it indicates the *average effect* of temperature *over all conditions of the other factors.* The main effects of the other factors can be computed similarly. A more efficient computation can be made by noting that the main effect is the difference between two averages:

$$\text{main effect of factor } i = \overline{y}_{i+} - \overline{y}_{i-} \tag{3.3}$$

where $\overline{y}_{i+}$ and $\overline{y}_{i-}$ are the average responses for the + and − levels of variable i, respectively. Hence, for T:

$$T = \frac{y_3 + y_4 + y_7 + y_8}{4} - \frac{y_1 + y_2 + y_5 + y_6}{4} \tag{3.4}$$

Similar equations can be developed for other main effects. The main effects of all three factors are $T = -43.73$, $R = 106.38$, and $S = -6.35$. □

All eight observations are used to compute the information on each of the main effects, providing a fourfold replicate of the differences. To secure the same precision in the OVAT approach for estimating the main effect of temperature, eight experiments have to be conducted, four at each level of temperature, while the other two inputs are *fixed* at one of their respective levels. A total of 24 experiments (a threefold increase) is needed to obtain the estimates of the three main effects. In general, a p-fold (p =number of factors) increase in the number of experiments are needed for OVAT over the full factorial approach. Even if all changes are made with respect to a common experimental condition in OVAT design, $(p+1)/2$ times more experiments are needed than the full factorial designs [78].

The implicit assumption in the OVAT design is that the main effect observed for one factor will remain the same at different settings of the other factors. In other words, the variables act on the response additively. If this assumption is correct, the results based on the OVAT design will provide complete information about the effects of various factors on the response even though the OVAT design would necessitate more experiments to match the precision of factorial design. If the assumption is not appropriate, data based on factorial design (unlike the OVAT design) can detect and estimate interactions between factors that lead to nonadditivity [78].

Interaction Effects. The effect of a factor may be much greater at one level of another factor than its other level. If the factors do not behave additively, they *interact.* A geometric representation of contrasts corresponding to main effects and interactions is given in Figure 3.2. The interaction between two factors is called *two-factor interaction.* Most of the interactions between a larger number of factors are usually smaller. The experimental data collected provide the opportunity to compute and assess the significance of these interactions. A measure of two-factor interaction with R_1 and R_2 as factors is provided by the difference between the average effect of one factor at one level of the second factor and its average effect at the other level of the second factor. Two factor interactions are denoted as $R_1 \times R_2$ or $R_1 R_2$ (when the $\times$ can be dropped without causing ambiguity). Thus, the temperature and inoculum strain interaction is denoted by $T \times S$ or TS.

Consider the interaction between the first and third factors in Table 3.1. The average effect of the first factor for one level of the third factor (R_3=+) is $(y_6 - y_5)/2 + (y_8 - y_7)/2$, and for the other level of the third factor ($R_3 = -$) is $(y_2 - y_1)/2 + (y_4 - y_3)/2$. The first and third factor interaction thus is

$$R_1 \times R_3 = \frac{1}{2}\left[\frac{(y_6 - y_5 + y_8 - y_7)}{2} - \frac{(y_2 - y_1 + y_4 - y_3)}{2}\right] \tag{3.5}$$

This equation can be rearranged to give:

$$\begin{aligned} R_1 \times R_3 &= \frac{1}{4}(y_1 - y_2 + y_3 - y_4 - y_5 + y_6 - y_7 + y_8) \\ &= \frac{y_1 + y_3 + y_6 + y_8}{4} - \frac{y_2 + y_4 + y_5 + y_7}{4}. \end{aligned} \tag{3.6}$$

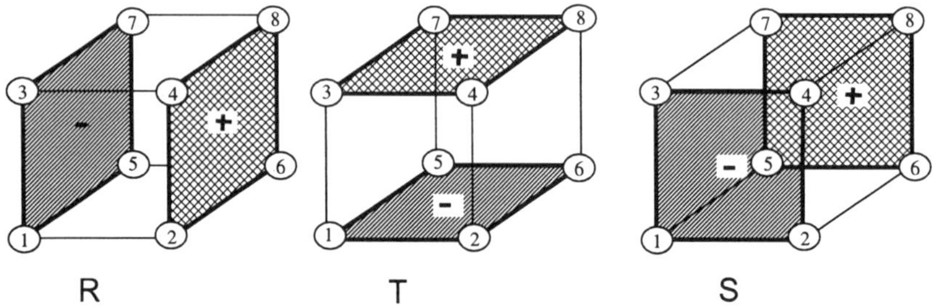

(a) Main effects

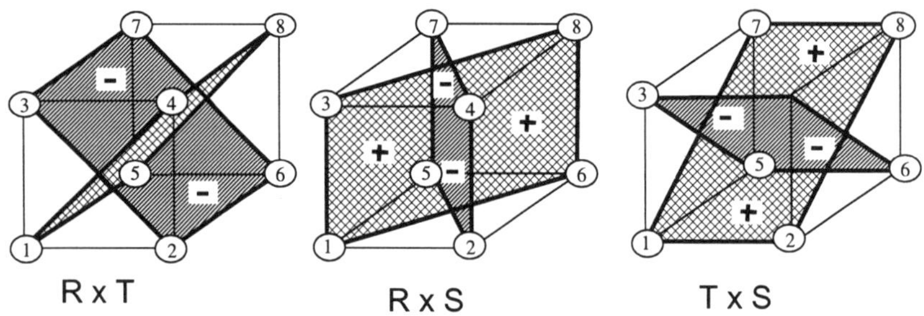

(b) Two-factor interactions

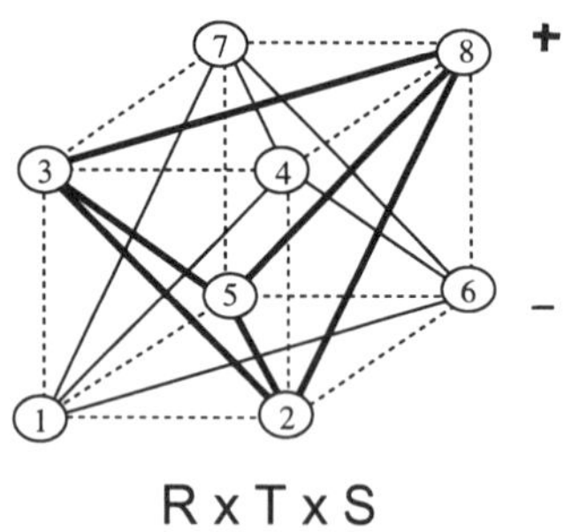

(c) Three-factor interactions

Figure 3.2. Geometric representation of contrasts corresponding to main effects and interactions [78].

Similarly,

$$\begin{aligned} R_1 \times R_2 &= \frac{y_1 + y_4 + y_5 + y_8}{4} - \frac{y_2 + y_3 + y_6 + y_7}{4} \\ R_2 \times R_3 &= \frac{y_1 + y_2 + y_7 + y_8}{4} - \frac{y_3 + y_4 + y_5 + y_6}{4}. \end{aligned} \tag{3.7}$$

The interactions of higher number of factors are denoted using the same convention. For example, the three-factor interaction between concentration, temperature, and strain is denoted by $R \times T \times S$. Three-factor interactions are computed using similar equations. The interaction between the three factors (factor levels as listed in Table 3.1) and illustrated in Figure 3.2 is computed by using two factor interactions. The interaction between R_1 and R_2 for one level of $R_3(-)$ is $[(y_4 - y_3) - (y_2 - y_1)]/2$ and for the other level of $R_3(+)$ $[(y_8 - y_7) - (y_6 - y_5)]/2$. Half of their difference (for $R_3[+] - R_3[-]$) is defined as the three factor interaction:

$$\begin{aligned} R_1 \times R_2 \times R_3 &= \frac{1}{4}\big(- y_1 + y_2 + y_3 - y_4 + y_5 - y_6 - y_7 + y_8 \big) \\ &= \frac{y_2 + y_3 + y_5 + y_8}{4} - \frac{y_1 + y_4 + y_6 + y_7}{4}. \end{aligned} \tag{3.8}$$

Example. Computation of the two-factor and three-factor interactions for the penicillin fermentation data.

The two-factor interactions are computed using Eqs. 3.6 and 3.7. The three-factor interaction is computed using Eq. 3.8:

$$\begin{aligned} R \times T &= -36.69 \\ R \times S &= 1.63 \\ T \times S &= 0.89 \\ R \times T \times S &= -0.92 \end{aligned} \tag{3.9}$$

□

The levels of factors such as those displayed in Table 3.2 can be used to generate a table of contrast coefficients that facilitates the computation of the effects (Table 3.4). The signs of the main effects are generated using the signs indicating the factor levels. The signs of the interactions are generated by multiplying the signs of the corresponding experiment levels (main effect signs). For example, the main effect T is calculated by using the signs of the third column:

$$\begin{aligned} T &= \frac{-69.24 - 214.82 + 59.45 + 133.49 - 64.41 - 213.61 + 59.72 + 133.71}{4} \\ &= \frac{-175.71}{4} = -43.9275 \end{aligned}$$

Table 3.4. Signs for calculating the effects from a 2^3 full factorial design. Last column (product) for use in the fermentation example

Run	mean	R	T	S	RT	RS	TS	RTS	yield
1	+	−	−	−	+	+	+	−	69.24
2	+	+	−	−	−	−	+	+	214.82
3	+	−	+	−	−	+	−	+	59.45
4	+	+	+	−	+	−	−	−	133.49
5	+	−	−	+	+	−	−	+	64.41
6	+	+	−	+	−	+	−	−	213.61
7	+	−	+	+	−	−	+	−	59.72
8	+	+	+	+	+	+	+	+	133.71
	8	4	4	4	4	4	4	4	divisor

Similarly, Eqs. 3.6-3.8 can be readily obtained from the information in columns 5 through 8 of Table 3.4.

Randomization and Blocking. *Randomization* of experiments is desired to reduce the inferential validity of data collected in spite of unspecified disturbances. For example, the run numbers in Table 3.2 are written on pieces of paper and a drawing is made to pick the sequence of experiments to be conducted. *Blocking* is used to eliminate unwanted variability. The variability may be introduced by changes in raw materials. For example, the substrate may be prepared in batches that may be enough for 4 batch runs, necessitating the use of two batches of raw materials to run the eight experiments of the 2^3 factorial design of the fermentation example. The experimental design can be arranged in two blocks of four runs to minimize the effect of variations due to two different batches of substrate. If runs 1, 4, 6, 7 use one substrate batch, and runs 2, 3, 5, 8 use the second substrate batch, two data points from each substrate batch are used in the computation of the main effects. This eliminates the additive effect associated with substrate batches from each main effect. There is a tradeoff in this experimental design. The RTS interaction and the experiments using a specific substrate batch are *confounded* (Table 3.4). All experiments with one of the substrate batches correspond to experiments where RTS is $-$, and all experiments with the other substrate batch correspond to RTS being $+$. Therefore, one cannot estimate the effect of the three-factor interaction separately from the effect of substrate batch change. Fortunately, the three-factor interaction is usually less important than the main effect and the two-factor interactions which are measured more precisely by this design.

The concept of confounding is discussed further in the section on fractional factorial design. More detailed treatment of blocking and confounding is available [78].

3.3.2 Fractional Factorial Design

The number of experiments required in a full 2^k factorial design increases geometrically with k. For a process with 7 factors, for example $2^7 = 128$ experiments are needed to estimate the 7 main effects, and 120 interactions. There are 21 two-factor, 35 three-factor, 35 four-factor, 21 five-factor, 7 six-factor and 1 seven-factor interactions [78]. Fortunately, not all of these interactions are important. Furthermore, the main effects tend to be larger in absolute magnitude than the two-factor interactions, which in turn are greater in absolute magnitude than the three-factor interactions, and so on. This then permits neglecting higher order terms in the Taylor series expansion in Eq. 3.2. Consequently, the information collected by a full factorial design will have redundancies, and fewer experiments than the required number (2^p) may be enough to extract all the relevant information. A popular experimental design approach that plans only part of the full factorial design is the *fractional* factorial design. The fractional factorial designs are named according to the fraction of the full design used, half-fraction indicating that only half of the experiments are conducted.

Consider a process where the effects of five factors are investigated. A full factorial design would necessitate $2^5 = 32$ runs as listed in Table 3.5. One possible half-fraction design (16 runs) includes all runs indicated by an asterisk in the half-fraction column of Table 3.5. The half-fraction design for an experiment with five factors is designated as 2^{5-1} to underline that the design has five variables, each at two levels, and only $2^4 = 16$ runs are used

$$\frac{1}{2}2^5 = 2^{-1}2^5 = 2^{5-1} \tag{3.10}$$

The selection of the specific 16 runs is important. One way to make the selection is to start with a full 2^4 design for the first four variables 1, 2, 3, and 4. Then, derive the column of signs for the 1234 interaction and use it to define the levels of variable 5. Thus, 5=1234 as displayed in Table 3.5. Because only 16 runs are carried out, 16 quantities can be estimated: the mean, 5 main factors and 10 two-factor interactions. But there are 10 three-factor interactions, 5 four-factor interactions and 1 five-factor interaction as well. Consider the three-factor interaction 123 written for the 16 runs in Table 3.5. They are identical to the two-factor interaction 45, hence 123 = 45. The remaining 16 runs not included in the half-fraction

Table 3.5. Full and half-fraction 2^5 factorial design

Run	1	2	3	4	5	1234	half-fraction	123	45
1	−	−	−	−	−	+		−	+
2	+	−	−	−	−	−	*	+	+
3	−	+	−	−	−	−	*	+	+
4	+	+	−	−	−	+		−	+
5	−	−	+	−	−	−	*	+	+
6	+	−	+	−	−	+		−	+
7	−	+	+	−	−	+		−	+
8	+	+	+	−	−	−	*	+	+
9	−	−	−	+	−	−	*	−	−
10	+	−	−	+	−	+		+	−
11	−	+	−	+	−	+		+	−
12	+	+	−	+	−	−	*	−	−
13	−	−	+	+	−	+		+	−
14	+	−	+	+	−	−	*	−	−
15	−	+	+	+	−	−	*	−	−
16	+	+	+	+	−	+		+	−
17	−	−	−	−	+	+	*	−	−
18	+	−	−	−	+	−		+	−
19	−	+	−	−	+	−		+	−
20	+	+	−	−	+	+	*	−	−
21	−	−	+	−	+	−		+	−
22	+	−	+	−	+	+	*	−	−
23	−	+	+	−	+	+	*	−	−
24	+	+	+	−	+	−		+	−
25	−	−	−	+	+	−		−	+
26	+	−	−	+	+	+	*	+	+
27	−	+	−	+	+	+	*	+	+
28	+	+	−	+	+	−		−	+
29	−	−	+	+	+	+	*	+	+
30	+	−	+	+	+	−		−	+
31	−	+	+	+	+	−		−	+
32	+	+	+	+	+	+	*	+	+

Table 3.6. The confounding patterns for the 2^{5-1} design with the defining relation I=12345

1	=	2345
2	=	1345
3	=	1245
4	=	1235
5	=	1234
12	=	345
13	=	245
14	=	235
15	=	234
23	=	145
24	=	135
25	=	134
34	=	125
35	=	124
45	=	123

design selected satisfy the relationship $123 = -45$. Consequently, the 123 and 45 interactions are *confounded.* The individual interactions 123 and 45 are called *aliases* of each other. A relationship such as $5 = 1234$ used to construct the 2^{5-1} design is called the *generator* of the design. Recall that the numbers 1 to 5 used above or the uppercase letters used in Section 3.3.1 denote a factor and a column of $-$ and $+$ signs indicate its level. The multiplication of the elements of a column by another column having identical elements is represented as $1 \times 1 = 1^2 = \mathrm{I}$. Similarly $2 \times 2 = \mathrm{I}$ and $\mathrm{T} \times \mathrm{T} = \mathrm{I}$. Furthermore, $2 \times \mathrm{I} = 2$. Hence,

$$5 \times 5 = 5^2 = 1234 \times 5 = 12345 \quad \text{or} \quad \mathrm{I} = 12345 \tag{3.11}$$

The relation I=12345 is called the *defining relation* of the design and is the key for determining all confoundings. For example, multiplying both sides of the defining relation with 1 yields 1= 2345, indicating that the main effect 1 is confounded with the four-factor interaction 2345. All confounding patterns for the 2^{5-1} design with the defining relation I=12345 are given in Table 3.6.

The complementary half-fraction design for 2^{5-1} is made up by all the entries in Table 3.5 without the asterisk in the "half-fraction" column. Its defining relation is I=−12345, the "−" sign indicating that the − level of 1234 interaction is used. Higher fractions such as 1/4 or 1/8 may also be of interest because of limited resources to conduct experiments. Then, additional defining relations must be used to design the experiment plan. The selection of the defining contrasts and confounded effects becomes more challenging as the number of factors and level of the fractions increase, necessitating systematic procedures such as the algorithm proposed by Franklin [164].

Design Resolution. Fractional factorial designs are classified according to their resolutions as well. A design of resolution R has no p-factor effect confounded with any other effect containing less than $R-p$ factors. Usually the resolution of the design is indicated with Roman numerals as subscripts such as 2_{V}^{5-1} for the design with the defining relation I=12345. Referring to Table 3.6, main effects are confounded only with 4-factor interactions ($R-4=1$, hence $R=$V) and 3-factor interactions are confounded only with 2-factor interactions ($R-3=2$, hence $R=$V again). In general, the resolution of a two-level fractional design is equal to the length of the shortest word in the defining relation (I=12345 has $R=$V) [78]. A design of resolution $R=$III does not confound main effects with one another, but confounds main effects with two-factor interactions. A design of resolution $R=$IV does not confound main factors and two-factor interactions, but confounds two-factor interactions with other two-factor interactions. Consequently, given the number of experiments that will be performed, the design with the highest resolution is sought. The selection of the defining relation plays a critical role in the resolution. In general, to construct a 2^{p-1} fractional factorial design of highest possible resolution, one writes a full factorial design for the first $p-1$ variables and associates the pth variable with the interaction $123\cdots(p-1)$ [78].

Exploratory experimentation is an iterative process where results from a small number of experiments are used to obtain some insight about the process and use that information to plan additional experiments for learning more about the process. It is better to conduct sequential experimentation in exploratory studies using fractional factorial designs and use these fractions as the building blocks to design more complete sets of experiments as needed.

3.3.3 Analysis of Data from Screening Experiments

Once the numerical values of main and interaction effects are computed, one must decide which effects are significant. Comparison of the estimates of effects and standard errors indicates the dominant effects. Consequently, the standard errors must be computed. If replicate runs are made at each set of experimental conditions, the variation between their outcomes may be used to estimate the standard deviation of a single observation and consequently the standard deviation of the effects [78]. For a specific combination of experimental conditions, n_i replicate runs made at the ith set of experimental conditions yield an estimate s_i^2 of the variance σ^2 having $\nu_i = n_i - 1$ degrees of freedom. In general, the pooled estimate of the run variance for g sets of experimental conditions is

$$s^2 = \frac{\nu_1 s_1^2 + \nu_2 s_2^2 + \cdots + \nu_g s_g^2}{\nu_1 + \nu_2 + \cdots + \nu_g} \tag{3.12}$$

with $\nu = \nu_1 + \nu_2 + \cdots + \nu_g$ degrees of freedom.

A direct estimate of the variance σ^2 is not available if there are no replicate runs. An alternate way to estimate σ^2 may be based on the assumption that the interactions of large number of factors would be negligible and the numerical values computed would measure differences caused by experimental error.

Example. Inspection of the interactions computed for the example problem (Eq. 3.9) shows that RT=-36.69, RS=1.63, TS=0.89, and RTS=-0.92. Since there are only three factors and RT is more than an order of (absolute) magnitude greater than the other interactions, one may either use RTS or pool RS, TS, and RTS to compute an estimate of σ^2. The exclusion of RT may seem arbitrary, but its large magnitude would justify its exclusion. The first approach will yield $s = \sqrt{(-0.92)^2} = 0.92$. The second will give $s = \sqrt{[1.63^2 + 0.89^2 + (-0.92)^2]/3} = 1.19$. □

Once a standard error is estimated, it can be compared with the magnitude of the effects of various factors and interactions to assess their significance.

Example. Determine the dominant effects of the example problem. Recall the main effect values $T = -43.93$, $R = 110.71$ and $S = -1.39$. Either estimate of s indicates that T and R, and their interaction (RT) are more influential than all other factors and interactions. An increase in temperature reduces the product yield while an increase in substrate feed rate increases it. The two strains have no significant influence on the yield. The interaction of temperature and feed rate is such that a joint increase in both variables causes a reduction in the yield. □

Quantile-Quantile plots. A more systematic approach for the assessment of effects is based on comparison of the magnitudes of actual effects to what might be expected from a Normal distribution. This may be done

Table 3.7. Data and computed values for Q-Q plot of the main effects and interactions for the experimental data in Tables 3.1 and 3.2 and standard Normal distribution

i	Ordered $y_{(i)}$	effect	$Q_Y(f_i)$	f_i	$Q_{SN}(f_i)$
1	-43.93	T	-43.93	0.0862	-1.3646
2	-36.69	RT	-36.69	0.2241	-0.7561
3	-1.39	S	-1.39	0.3621	-0.3515
4	-0.92	RTS	-0.92	0.5000	0
5	0.89	TS	0.89	0.6379	0.3515
6	1.63	RS	1.63	0.7759	0.7561
7	110.71	S	110.71	0.9138	1.3646

using normal probability paper [78] or quantile-quantile (Q-Q) plots. Since the Q-Q plots can easily be generated by many software packages, the procedure to use them in assessing the importance of effects will be outlined. Assume that the data (effects of all factors and interactions in this case) are represented as a set of values y_i, $i = 1, 2, \ldots, n$.

1. Order the data according to magnitude: $y_{(1)} \leq y_{(2)} \leq \cdots \leq y_{(n)}$.

2. For each ordered data, set $Q_Y(f_i) = y_{(i)}$, $i = 1, 2, \ldots, n$ where Q denotes a quantile.

3. Calculate the quantiles for a standard Normal distribution (Q_{SN}) using the empirical relation

$$Q_{SN}(f_i) = 4.91\left[f^{0.14} - (1-f)^{0.14}\right] \quad \text{where} \quad f_i = \frac{i - 3/8}{n + 1/4} \qquad (3.13)$$

4. Plot $Q_Y(f_i)$ versus $Q_{SN}(f_i)$. Approximate linearity indicates that data are consistent with standard Normal distribution. Significant deviation of specific effects from linearity indicates that they are important effects.

Example. Develop the Q-Q plot for the main effects and interactions computed for the penicillin fermentation data in Tables 3.1 and 3.2.

The second and fourth columns of Table 3.7 provide the quantiles of the main effects and interactions. The last column of the table displays the corresponding quantiles of the standard Normal distribution computed using Eq. 3.13. The two sets of quantiles are plotted in Figure 3.3. The

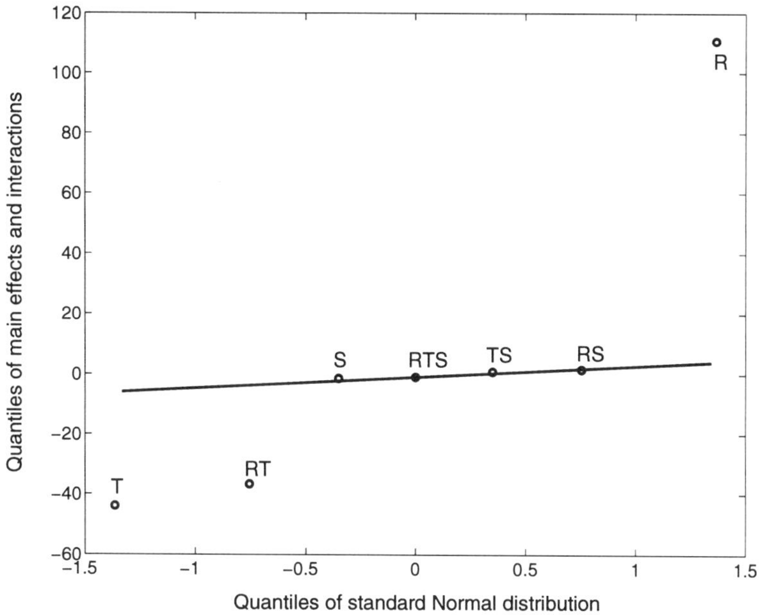

Figure 3.3. Quantile-quantile plot of main effects and interactions against standard Normal distribution.

same main effects and interactions noted earlier (R, T, RT) deviate substantially from the Normal distribution (i.e., from the line of unit slope passing through the origin in Figure 3.3).

3.4 Data Pretreatment: Outliers and Data Reconciliation

Data pretreatment is necessary to assure that data used in modeling, monitoring and control activities provide an accurate representation of what is happening in a process. Data corruption may be caused by failures in sensors or transmission lines, process equipment malfunctions, erroneous recording of measurement and analysis results, or external disturbances. These faults would cause data to have spikes, jumps, or excessive oscillations. For example, sensor faults cause bias change, drift or increase in signal noise and result in abnormal patterns in data. The general strategy is to detect data that are not likely based on other process information (*outlier detection*) and to substitute these data with estimated values that are in agreement with other process information (*data reconciliation*). The implementation of this simple strategy is not straightforward. A significant change in a variable reading may be caused by a process equipment fault or a sensor fault. If the change in signal magnitude is due to an equipment fault, this change reflects what is really happening in the process and the signal value should not be modified. Corrective action should be taken to eliminate the effect of this disturbance. However, if the magnitude of the signal has changed because of a sensor fault, the process is most likely behaving the way it should, but the information about the process (measured value) is wrong. In this case, the signal value must be modified. Otherwise, any action taken based on the erroneous reading would cause an unwarranted process upset. The challenge is to decide when the significant change is caused by something that is happening in the process and when it is caused by erroneous reporting of measurements. This necessitates a comprehensive effort that includes signal noise reduction, fault detection, fault diagnosis, and data reconciliation. However, many of these activities rely on the accuracy of measurement information as well.

This section focuses on detection of outliers and gross errors, and data reconciliation. Data reconciliation involves both the elimination of gross errors and the resolution of the contradictions between the measurements and their constraints such as predictions from the model equations of the process. Detection and reduction of random signal noise are discussed in Section 3.5. Techniques for fault diagnosis and sensor fault detection are presented in Chapter 8. The implementation of these techniques must be well coordinated because of the interactions among signal conditioning, fault detection and diagnosis, process monitoring and control activities. The use of an integrated supervisory knowledge-based system for on-line process supervision is discussed in Chapter 8.

Most outlier detection and data reconciliation techniques are developed for continuous processes where the desired values of the important process variables are constant for extended periods of time. Consequently, these techniques are often based on the existence of stationary signals (constant mean value over time) and many of them have focused on assessment and reconciliation of steady state data. The time-dependent nonstationary data that batch processes generate may necessitate modification of the techniques developed for continuous processes prior to their application for batch processes.

3.4.1 Data Reconciliation

The objective of data reconciliation is to convert contaminated process data into consistent information by resolving contradictions between measurements and constraints imposed on them by process knowledge. The simplest reconciliation case is steady-state linear data reconciliation which can be set as a quadratic programming problem. Given the process measurements vector $\tilde{\mathbf{x}}$, the vector of unmeasured variables $\mathbf{v}$ and the data adjustments $\mathbf{a}$ (their final values will be the solution of the optimization problem), the data reconciliation problem is formulated as

$$\begin{aligned} &\min_{\mathbf{a}} \quad F(\mathbf{a}) = \mathbf{a}^T \boldsymbol{\Sigma}_1^{-1} \mathbf{a} \\ &\textit{such that} \quad \mathbf{B}_1(\tilde{\mathbf{x}} + \mathbf{a}) + \mathbf{P}\mathbf{v} = \mathbf{0} \end{aligned} \tag{3.14}$$

where $\mathbf{B}_1$ and $\mathbf{P}$ are the matrices of coefficients corresponding to $\tilde{\mathbf{x}}$ and $\mathbf{v}$ in Eq. 3.14 and $\boldsymbol{\Sigma}_1$ is the covariance matrix of $\tilde{\mathbf{x}}$. For example if a leak detection problem is being formulated, $\tilde{\mathbf{x}}$ will consist of component flow rates and the constraint equations will be the material balances. Matrix projections can be used to remove the unmeasured variables [113] such that the constraints in Eq. 3.14 are transformed to a reduced set of process constraints that retain only the measured variables. The covariance matrix of the reduced constraints is

$$\mathbf{H_e} = cov(\mathbf{e}) = \mathbf{B}_1^R \boldsymbol{\Sigma}_1^{-1} (\mathbf{B}_1^R)^T \tag{3.15}$$

where $\mathbf{B}_1^R$ is the "material balance" coefficient matrix of the reduced constraints with a residual vector (reduced balance residuals)

$$\mathbf{e} = \mathbf{B}_1^R \tilde{\mathbf{x}} \ . \tag{3.16}$$

The optimal value of the objective function F is

$$F = \mathbf{e}^T \mathbf{H_e}^{-1} \mathbf{e} \ \sim \ \chi_m^2 \tag{3.17}$$

which follows a chi-squared (χ^2) distribution with m degrees of freedom where m is the rank of $\mathbf{H}_e$ [113]. Additional restrictions such as flow rates being positive or zero may be introduced so that $(\tilde{\mathbf{x}} + \mathbf{a})$ is not negative. This framework can be combined with principal components analysis for gross error detection and reconciliation [259, 591].

Other data reconciliation and gross error detection paradigms have been proposed for linear processes operating at steady state. A serial strategy for detecting and identifying multiple gross errors eliminates sequentially measurements susceptible to gross errors, recomputes a test statistic, and compares it against a critical value [258, 519]. The use of generalized likelihood ratio (Section 8.3) method for identifying abrupt changes [651] has been proposed to discriminate between gross measurement errors and process faults (for example between malfunctions of flow rate sensors and leaks) [409]. The approach has been extended to dynamic linear processes [410]. Techniques based on Kalman filters [563], maximum likelihood functions [516, 517], successively linearized horizons [491], orthogonal collocation [341], neural networks [269], and discretization of differential-algebraic equation systems [12] have been developed. Recent books [29, 518] provide details of many techniques for data reconciliation and gross error detection.

The first critical step in data reconciliation is the detection, identification and elimination of gross errors. Some strategies and methods to carry out this task are presented in Section 3.4.2.

3.4.2 Outlier Detection

Outliers or gross errors corrupt process data. Spikes in data that are candidates for outliers can be detected easily by visual inspection. Statistical tools or heuristics can then be used to assess validity of the spikes as outliers, and based on this assessment, data analysis, modeling, and/or monitoring activities may be undertaken. Detection of outliers is critical for having reliable data to make decisions about the operation of a process and to develop empirical (data based) models. Consequently, the literature on outliers is dispersed in statistics, process engineering and systems science as robust estimation, regression, system identification, and data analysis. Since many references from the process engineering literature have been provided in Section 3.4.1, outlier detection methods developed by statisticians are outlined first in this section. This is followed by a discussion of outlier detection in multivariable systems by principal components analysis (PCA).

Various books [42, 225, 599, 523] and survey papers [20, 41, 201, 476, 512] in the statistics literature provide a good account of many techniques used in outlier detection. Outlier detection in time series has received significant attention [96, 281, 538]. Fox [159] distinguished two types of outliers: Type I, the *additive outlier* (AO), consisting of an error that affects only a single observation, and Type II, the *innovational outlier* (IO), consisting of an error that affects a particular observation and all subsequent observations in the series. Abraham and Chuang [3] considered regression analysis for detection of outliers in time series. Lefrancois [332] developed a tool for identifying over-influential observations in time series, and presented a method for obtaining various measures of influence for the autocorrelation function, as well as thresholds for declaring an observation over-influential.

A popular outlier detection technique in time series is the *leave-one-out* diagnostic idea for linear regression where one deletes a single observation at a time, and for each deletion computes a Gaussian maximum likelihood estimate (MLE) (Section 8.3) for the missing datum [80, 222, 283]. Because more than one outlier may exist in data, some outliers may be masked by other dominating outliers in their vicinity. A patch of outlying successive measurements is common in time series data, and masking of outliers by other outliers is a problem that must be addressed. One approach for determining patches of outliers is the generalization of the leave-one-out technique to the leave-k-out diagnostics. However, at times the presence of a gross outlier will have sufficient influence such that deletion of aberrant values elsewhere in the data has little effect on the estimate. More subtle types of masking occur when moderate outliers exist close to one another [379]. These types of masking can often be effectively uncovered by an iterative deletion process that removes suspected outlier(s) from the data and recomputes the diagnostics.

Several modeling methods have been proposed to develop empirical models when outliers may exist in data [91, 595]. The strategy used in some of these methods first detects and deletes the outlier(s), then identifies the time series models. A more effective approach is to accommodate the possibility of outliers by suitable modifications of the model and/or method of analysis. For example, mixture models can be used to accommodate certain types of outliers [10]. Another alternative is the use of *robust estimators* that yield models (regression equations) that represent the data accurately in spite of outliers in data [69, 219, 524]. One robust estimator, the L_1 estimator, involves the use of the least *absolute values* regression estimator rather than the traditional least *sum of squares* of the residuals (the least squares approach). The magnitudes of the residuals (the differences between the measured values and the values estimated by the model equation) have a strong influence on the model coefficients. Usually an outlier yields a large residual. Because the least squares approach takes the square of the residuals (hence it is called the L_2 regression indicating that the residual is squared), the outliers distort the model coefficients more than L_1 regression that uses the absolute values of the residuals [523]. An improved group of robust estimators includes the M estimator [216, 391, 245] that substitutes a function of the residual for the square of the residual and the *Generalized M* estimator [15, 217] that includes a weight function based on the regressor variables as well. An innovative approach, the *least trimmed squares* (LTS) estimator uses the first h ordered squared residuals in the sum of squares ($h < n$, where n is the number of data points), thereby excluding the $n-h$ largest squared residuals from the sum and consequently allowing the fit to stay away from the influence of potential outliers [523]. A different robust estimator, the *least median squares* (LMS) estimator, is based on the *medians* of the residuals and tolerates better outliers in both dependent and independent (regressor) variables [523].

Subspace modeling techniques such as principal components analysis

(PCA) provide another framework for outlier detection [224, 591, 592] and data reconciliation. PCA is discussed in detail in Section 4.1. One advantage of PCA based methods is their ability to make use of the correlations among process variables, while most univariate techniques are of limited use because they are ignoring variable correlations. A method that integrates PCA and sequential analysis [592] to detect outliers in linear processes operated at steady state is outlined in the following paragraphs. Then, PCA based outlier detection and data reconciliation approach for batch processes is discussed.

PCA can be used to build the model of the process when it is operating properly and the data collected do not have any outliers. In practice, the data sets from good process runs are collected, inspected and cleaned first. Then the PCA model is constructed to provide the reference information. When a new batch is completed, its data are transformed using the same PCs and its scores (see Section 4.1) are compared to those of the reference model. Significant increases in the scores indicate potential outliers. Since the increases in scores may be caused by abnormalities in process operation, the outlier detection activities should be integrated with fault detection activities. The PCA framework can also be used for data reconciliation as illustrated in the example given in this section.

Consider a set of linear combinations of the reduced balance residuals $\mathbf{e}$ defined in Eq. 3.16:

$$\mathbf{y_e} = \mathbf{W_e}^T\mathbf{e} = \mathbf{\Lambda_e}^{-1/2}\mathbf{U_e}^T\mathbf{e} \tag{3.18}$$

where $\mathbf{\Lambda_e}$ is a diagonal matrix whose elements are the magnitude ordered eigenvalues of $\mathbf{H_e}$ (Eq. 3.15). Matrix $\mathbf{U_e}$ contains the orthonormalized eigenvectors of $\mathbf{H_e}$ (detailed discussion of PCA computations are presented in Section 4.1). The elements of vector $\mathbf{y_e}$ are called PC scores and correspond to individual principal components (PC). The random variable $\mathbf{e}$ has a statistical distribution with the mean $\mathbf{0}$ and covariance matrix $\mathbf{H_e}$ ($\mathbf{e} \sim (\mathbf{0}, \mathbf{H_e})$). Consequently, $\mathbf{y_e} \sim (\mathbf{0}, \mathbf{I})$ where $\mathbf{I}$ denotes the identity matrix (a diagonal matrix with 1s in the main diagonal), and the correlated variables $\mathbf{e}$ are transformed into an uncorrelated set ($\mathbf{y_e}$) with unit variances. Often the measured variables are Normally distributed about their mean values. Furthermore, the central limit theorem would be applicable to the PCs. Consequently, $\mathbf{y_e}$ is assumed to follow Normal distribution ($\mathbf{y_e} \sim N(\mathbf{0}, \mathbf{I})$) and the test statistic for each PC is

$$y_{\mathbf{e},i} = (\mathbf{W}_\mathbf{e}^T\mathbf{e})_i \sim N(0,1), \qquad i = 1, \cdots, m \tag{3.19}$$

which can be tested against tabulated threshold values. When an outlier is detected by noting that one or more $y_{\mathbf{e},i}$ are greater than their threshold values, Tong and Crowe [592] proposed the use of contribution plots

(Section 8.1) to identify the cause of the outlier detected. They have also advocated the use of sequential analysis approach [625] to make statistical inferences for testing with fewer observations whether the mean values of the PCs are zero.

Outlier detection in batch processes can be done by extending the PCA based approach by using the multiway PCA (MPCA) framework discussed in Section 4.5.1. The MPCA model or reference models based on other paradigms such as functional data analysis (Section 4.4) representing the reference trajectories can also be used for data reconciliation by substituting "reasonable" estimated values for outliers or missing observations. The example that follows illustrates how the MPCA models can be used for outlier detection and data reconciliation.

Example Consider a data set collected from a fed-batch penicillin fermentation process. Assume that there are a few outliers in some of the variables such as glucose feed rate and dissolved oxygen concentration due to sensor probe failures. This scenario is realized by adding small and large outliers to the values of these variables as shown in Figure 3.4. Locations of the outliers for the two variables are shown in Table 3.8.

In this example, a multiway PCA (MPCA) model with four principal components is developed out of a reference set (60 batches, 14 variables, 2000 samples) for this purpose. A number of multivariate charts are then constructed to unveil the variables that might contain outlying data points and the locations of the outliers in those variables. The first group of charts one might inspect is the SPE, T^2 charts and the charts showing variable contributions to these statistics. Contribution plots are discussed in detail in Section 8.1. Both SPE and T^2 charts signal the outliers and their locations correctly (Figure 3.5), but they do not give any information about which variable or variables have outliers. At this point of the analysis, contribution (to SPE and T^2 values) plots are inspected to find out the variables responsible for inflating SPE and T^2. Since outliers will be projected farther from the plane defined by MPCA model, their SPE values are expected to be very high. Consistently, SPE contribution plot indicates two variables

Table 3.8. Locations of the outliers

Variable	Locations of the outliers (sample no.)
Glucose feed rate (no.3)	500, 750, 800, 1000, 1500, 1505, 1510
Dissolved O_2 conc. (no.6)	450, 700, 900, 1400, 1405, 1410

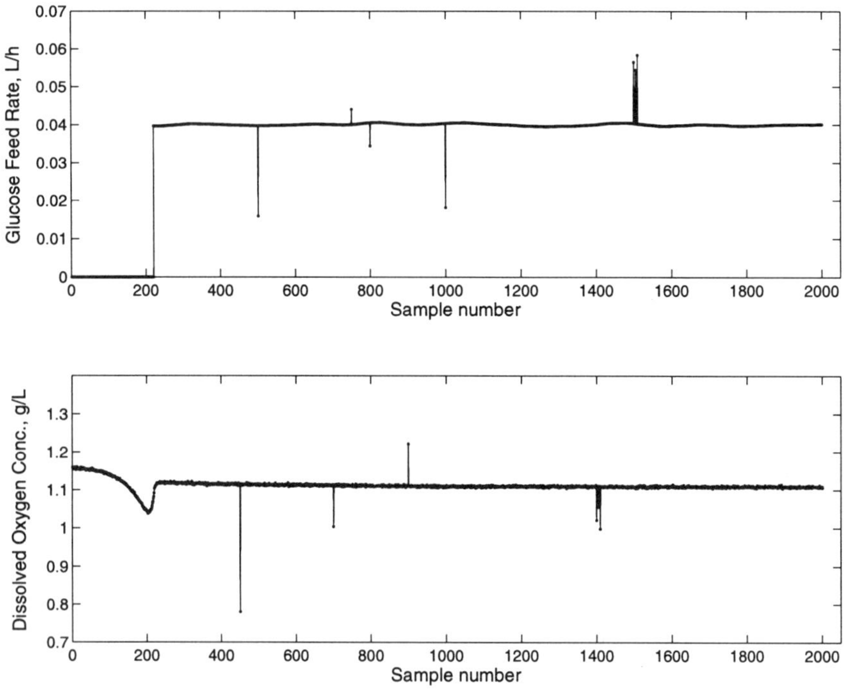

Figure 3.4. Raw data profiles containing outliers.

(variables 3 and 6, glucose feed rate and dissolved oxygen concentration, respectively). T^2 contributions represent similar information. Variable 3 is not that obvious from that chart but variable 6 can be clearly distinguished. Now that the variables with outliers and the overall locations of outliers are identified, the locations of outliers in each variable need to be found. This task can be accomplished either by directly inspecting individual variable trajectories (Figure 3.4) or by using multivariate temporal contribution plots (to both SPE and T^2) for the identified variables (Figure 3.6). Some of the critical change points on these curves (Figure 3.6) indicate the multivariate nature of the process, and the important events that take place. For instance, the sudden drop in variable 3 at sample 450 is due to corresponding outlier in variable 6 and the dip in variable 6 around sample 300 indicates the switch from batch to fed-batch operation. In this example, all of the outliers are clearly detected by using PCA. To prevent multivariate charts from signaling that the process is out-of-control, these outliers are marked for removal. However, for further analysis and modeling purposes, they should be replaced with estimates. The ability of the PCA technique

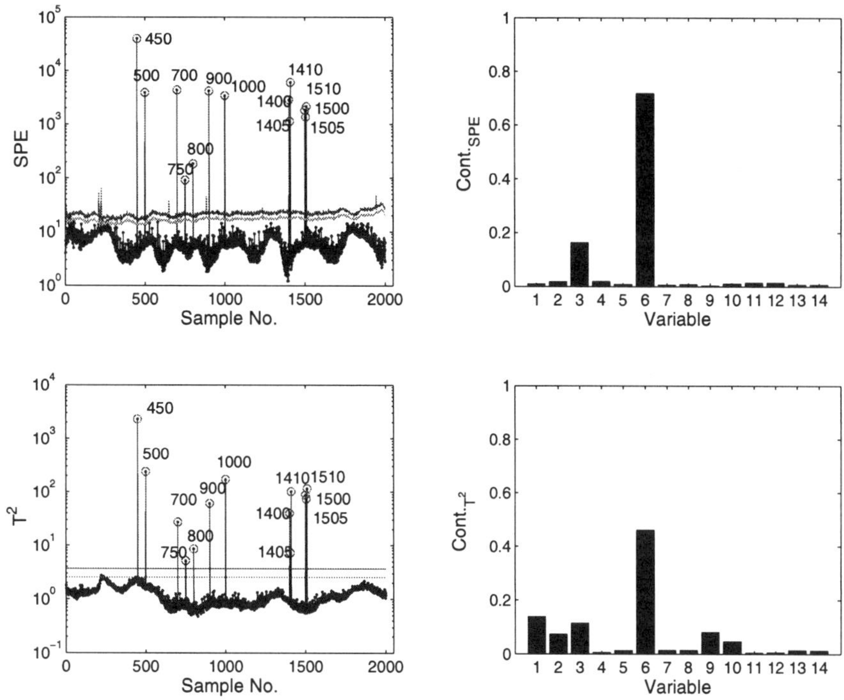

Figure 3.5. Multivariate charts for detecting and diagnosing outliers. Variable 3 is glucose feed rate and variable 6 is dissolved oxygen concentration.

to handle missing data is used for estimation of these by restricting the estimates with observed values up to time interval k and the correlation structure of the reference set variables as defined by the loading matrix $\mathbf{P}$ of the MPCA model. The procedure followed in this example is based on locating the first outlier, replacing it with a PCA-based estimate and repeating this procedure with the next outlier. Projection of the already known observations made on J variables $[x_k(kJ \times 1)]$ into the reduced space is performed by calculating the $\mathbf{t}_{R,k}$ scores (Eq. 3.20). This scores vector $\mathbf{t}_{R,k}$ is then used to predict the next observation set $[x_{k+1}((k+1)J \times 1)]$ with the outlier becoming the missing value (old value replaced by a 0) as shown in Eq. 3.21:

$$\mathbf{t}_{R,k} = (\mathbf{P}_k^T \mathbf{P}_k)^{-1} \mathbf{P}_k^T \mathbf{x}_k \tag{3.20}$$

$$\mathbf{x}_{(k+1)J\times 1} = \mathbf{P}_{k+1} \mathbf{t}_{R,k} \tag{3.21}$$

Note that $\mathbf{P}_k$ is a $(kJ \times R)$ matrix having as columns the elements of p-

loading vectors ($\mathbf{p}_r$) from all R principal components up to time k just before the outlier is observed. The matrix ($\mathbf{P}_k^T \mathbf{P}_k$) is well conditioned because all $\mathbf{p}_r$ vectors are orthogonal to each other [435].

All these techniques use data sets that have already been collected. This is acceptable for model development and analysis of completed batches, but it is not satisfactory for real time process monitoring and control activities. Outlier detection during the progress of the batch is more challenging and a compromise must be made between speed and sophistication of the outlier detection technique used. As a first step, simple upper and lower limits for each measured variable may be used to identify major outliers. More sophisticated tools based on multivariate process monitoring techniques can also be used, noting the need to discriminate between real process faults (that may persist for a number of sampling times) and outliers that usually have shorter durations. Signal processing tools such as change detection techniques based on maximum likelihood ratios can be considered for critical variables [45]. Outlier detection and data reconciliation during the

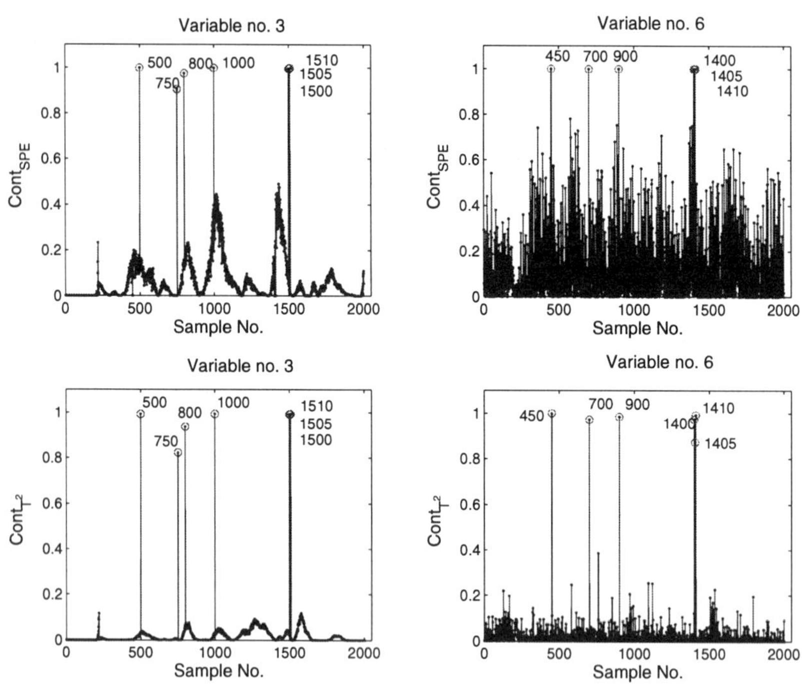

Figure 3.6. Temporal contribution plots for locating outliers in identified variables.

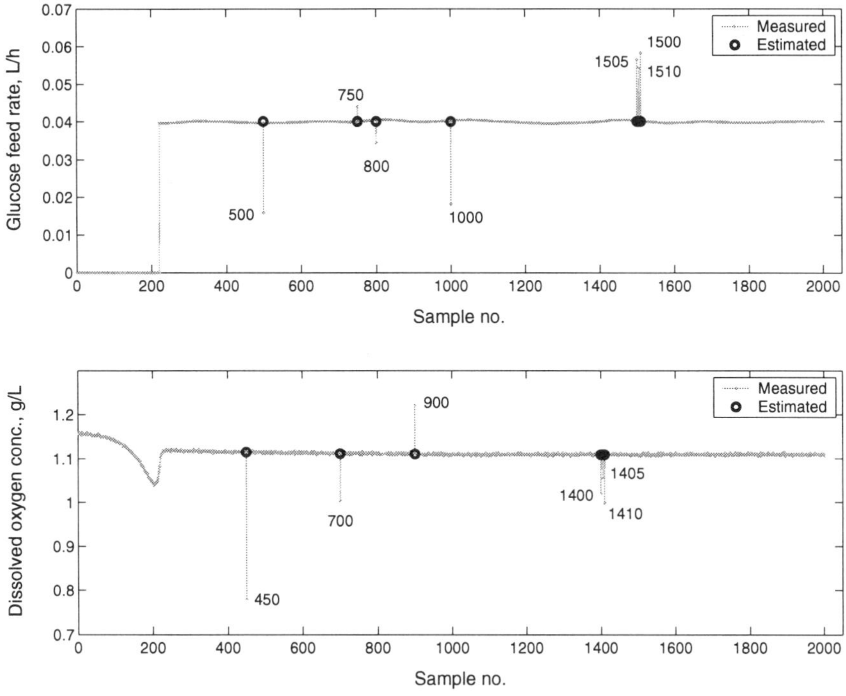

Figure 3.7. Outlier removal and principal components estimates of the removed samples.

progress of the batch in real time is an active research and development area [607, 647].

A different paradigm is the use of *smart sensors* that include in the sensor system functionality to detect outliers, reduce noise and conduct self diagnosis of operation status. This will shift the burden of information reliability assessment from the data collection and process operations computer to the individual sensors. In coming years, cost reductions and improvements in reliability would increase the feasibility of this option.

3.5 Data Pretreatment: Signal Noise Reduction

Sensors and transmission lines can pick up noise from the environment and report compromised readings. Various signal processing and filtering paradigms have been proposed to reduce signal noise. Some simple remedies can be developed by boosting the signal strength or selecting a variable that is less susceptible to noise. An example of the former is noise reduction in thermocouple signals that are at millivolt level. If the signal is amplified to volt level before transmitting, the signal to noise ratio during transmission is significantly improved. Converting a voltage signal to current that is less noise prone is an example of the latter remedy.

If several measurements are made for the same variable at a specific sampling time, measurement error is reduced by averaging these. This is the approach used in traditional statistical quality control in discrete manufacturing processes. It may also be used for quality variables at the end of a batch if several measurements can be made easily and at low cost. The problem is more challenging if a single measurement is made for each variable at each sampling instant. The paradigms used in this case include averaging over time, decomposing the signal into low and high frequency components, and use of multivariate data analysis techniques such as PCA to separate signal information from noise. A general framework for data averaging for signal filtering is presented and the use of PCA for noise reduction is illustrated in Section 3.5.1. The PCA method is discussed in detail in Section 4.1. Signal decomposition can be implemented by various techniques described in the signal processing literature [45, 207, 271]. Most of these techniques are available in commercial software such as Matlab Signal Processing Toolbox [373]. A particular signal decomposition approach that has captured attention in recent years is the wavelet decomposition, which can implement time-frequency decomposition simultaneously. Wavelets and their use in noise reduction are discussed in Section 3.5.2. Multivariate data analysis techniques use a different premise: If a coordinate transformation can be made to explain the major variation in data, what has not been extracted from the measurements would be mostly random noise. If the signals are reconstructed by using only the information retained in the new coordinate system, then the noise will be filtered out.

3.5.1 Signal Noise Reduction Using Statistical Techniques

Simple noise filtering tools can be developed by using time series model representation (Section 4.3.1) where the filtered signal $y(n)$ at sampling time n is related to the signal $x(n)$:

$$\begin{aligned} y(n) &= b_1 x(n) + b_2 x(n-1) + \cdots + b_{n_b+1} x(n-n_b) \\ &\quad -a_2 y(n-1) \cdots - a_{n_a+1} y(n-n_a) \end{aligned} \tag{3.22}$$

where n is the current sampling time and n_a and n_b are the lengths of the past sampling time windows for y and x signals, respectively. This is the standard time-domain representation of a digital filter. Assuming zero initial conditions and starting with $y(1)$, the progression of this representation is:

$$\begin{aligned} y(1) &= b_1x(1) \\ y(2) &= b_1x(2) + b_2x(1) - a_2y(1) \\ y(3) &= b_1x(3) + b_2x(2) + b_3x(1) - a_2y(2) - a_3y(1) \; . \end{aligned} \tag{3.23}$$

For example, if $n_b = 2$ and $n_a = 2$, then $y(6)$ is computed by writing Eq. 3.22 for this set of index values:

$$y(6) = b_1x(6) + b_2x(5) + b_3x(4) - a_2y(5) - a_3y(4) \; . \tag{3.24}$$

To compute the estimated value $y(6)$ for sensor reading $x(6)$, the weighted sum of the three most recent sensor readings and two most recent estimated (filtered) values are used. Two limiting cases may be considered to generate the estimates by using a time series model: the use of sensor readings only (all $a_i = 0$) and the use of previous estimates only (all $b_i = 0$). The former is called *moving average* (MA) since all readings included in a sliding time window (the window width is determined by the value of $n_b + 1$) are used to estimate the filtered signal. One option is to assign all values equal weights (all b_i are equal). Another option is to assign them different weights, perhaps to give a higher emphasis to more recent readings. If only past estimated values are used, then the current value is regressed over the previous estimates, yielding an *auto regressive* (AR) model. If an MA model is used, the last few readings are averaged to eliminate random noise. This is reasonable because the noise is "random" and consequently sometimes it will be positive and at other times it will be negative with a mean value that is zero in theory. By averaging a few readings, it is hoped that the noise components in each measurement cancel out in the averaging process. A pure AR model is not appealing because the estimation is based only on past estimated values and the actual measurements are ignored. The filter is called autoregressive moving average (ARMA) when both AR and MA terms are included. The reduction of noise by using ARMA and MA filters is illustrated in the following example.

Example Consider a low frequency, high amplitude sinusoidal signal containing high frequency, low amplitude noise (Figure 3.8(a)). Moving average (MA) and autoregressive moving average (ARMA) filtering are applied to denoise this signal. Filter design and fine tuning the parameters are important issues that are influential as concerns the result (Figure 3.8). After filtering (especially with good performance filters), increase in the signal-to-noise ratio is obvious. □

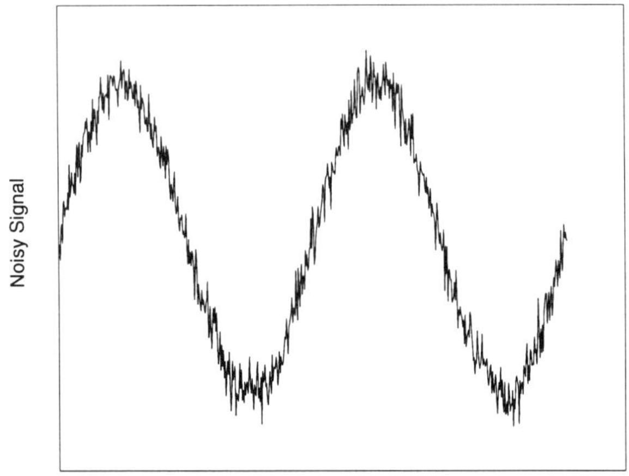

(a) Noisy raw signal.

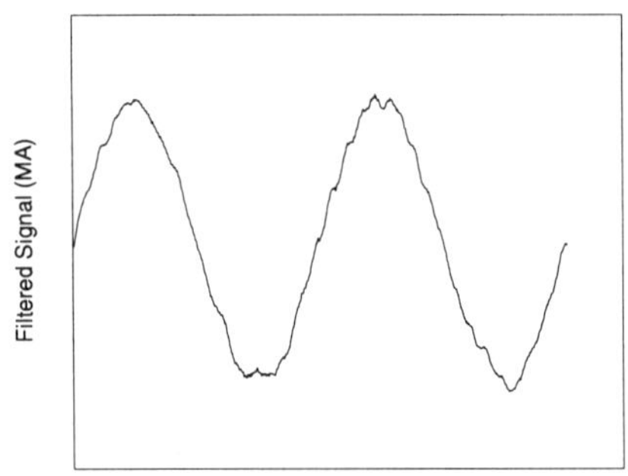

(b) Filtered signal with a poor MA filter.

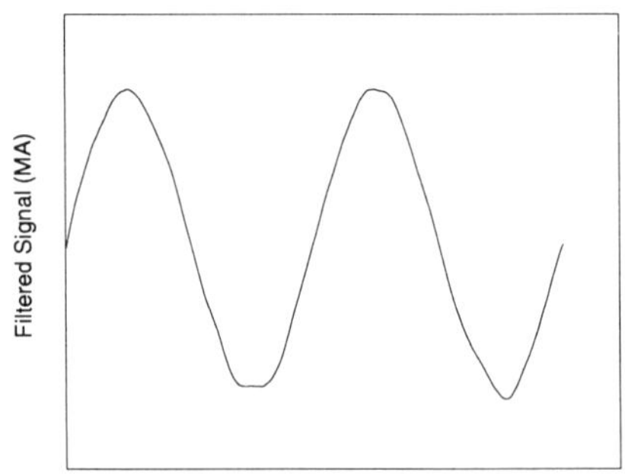

(c) Filtered signal with a good MA filter.

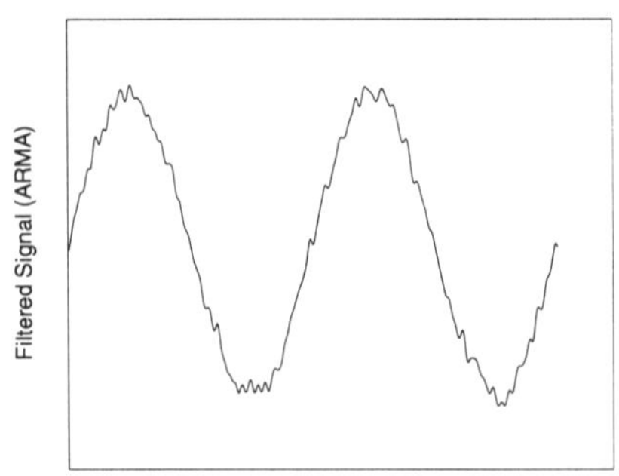

(d) Filtered signal with a poor ARMA filter.

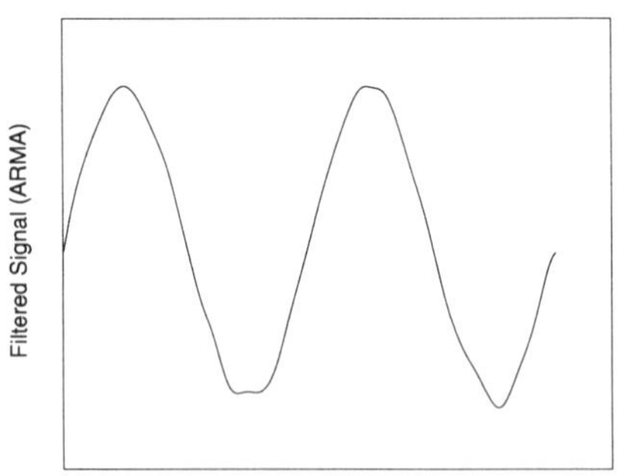

(e) Filtered signal with a good ARMA filter.

Figure 3.8. Noise reduction using MA and ARMA filters.

Noise Reduction by PCA. Multivariate statistical analysis tools process the entire data set and make use of the relations between all measured variables. Consequently, the noise reduction philosophy is different than the filtering techniques discussed earlier where each measured variable is treated separately. The multivariable statistical technique is utilized to separate process information from the noise by decomposing relevant process information from random variations. This is followed by reconstruction of the signals using only the process information. PCA determines the new coordinates and extracts the essential information from a data set. Principal Components (PC) are a new set of coordinates that are orthogonal to each other. The first PC indicates the direction of largest variation in data, the second PC indicates the largest variation not explained by the first PC in a direction orthogonal to the first PC (Fig. 4.1). Consequently, the first few PCs describe mostly the actual variations in data while the portion of the variation not explained by these PCs contains most of the random measurement noise. By decomposing the data to their PCs and reconstructing the measurement information by using only the PCs that contain process information, measurement noise can be filtered. The methods for computing PCs and determining the number of PCs that contain significant process information are described in Section 4.1. The random noise filtering by PCA is illustrated in the following example.

Example An MPCA model with 4 principal components is developed using a reference data set containing 60 batches, 14 variables and 2000 samples of each variable over the batches. Scores of the first two principal components are plotted in Figure 3.9, for the reference data set representing normal operating conditions. As a result of PCA modeling, noise level is decreased as shown in Figure 3.10. This is expected since a PC model extracts the most relevant correlation information among the variables, unmodeled part of the data being mostly noise. □

3.5.2 Wavelets and Signal Noise Reduction

Wavelets were developed as an alternative to Short Time Fourier Transform (STFT) for characterizing non-stationary signals. Wavelets provide an opportunity to localize events in both time and frequency by using windows of different lengths while in STFT the window length is fixed. A wavelet transform can be represented as

$$W(a,b) = \frac{1}{\sqrt{|a|}} \int x(t)\Psi\left(\frac{t-b}{a}\right)dt \tag{3.25}$$

where Ψ represents the *mother wavelet*, $x(t)$ is the original signal, a and b are scale and translation parameters, respectively, and the factor $1/\sqrt{|a|}$

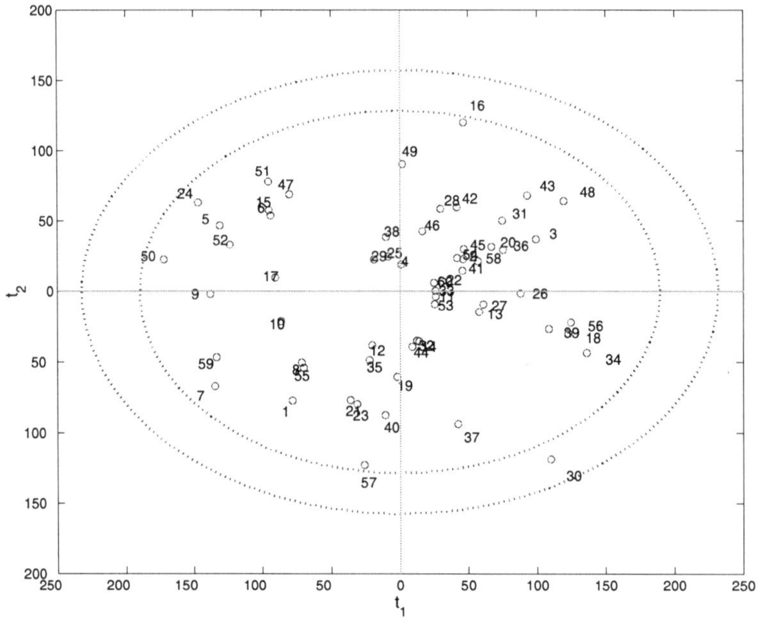

Figure 3.9. Biplot of first two score vectors ($\mathbf{t}_1$ and $\mathbf{t}_2$, respectively) of the MPCA model representing normal operation with 95 and 99 % control limits.

is used to ensure that the energy of the scaled and translated signals are the same as the mother wavelet. Scale parameter specifies the location in frequency domain and translation parameter determines the location in time domain. This equation can be interpreted as the inner product of $x(t)$ with the scaled and translated versions of the basis function Ψ [116]:

$$W(a,b) = \int x(t)\Psi_{(a,b)}(t)dt \tag{3.26}$$

$$\Psi_{(a,b)}(t) = \frac{1}{\sqrt{|a|}}\Psi\left(\frac{t-b}{a}\right). \tag{3.27}$$

Scaled and translated versions of the basis functions are obtained from the mother wavelet (Eq. 3.27). The discrete wavelet transform is used to reduce the computational burden without losing significant information. To obtain the discretized wavelet transform, scale and translation parameters are discretized as $a = 2^j$ and $b = 2^j \times k$. Then, there exists Ψ with good time-frequency localization properties such that the discretized wavelets

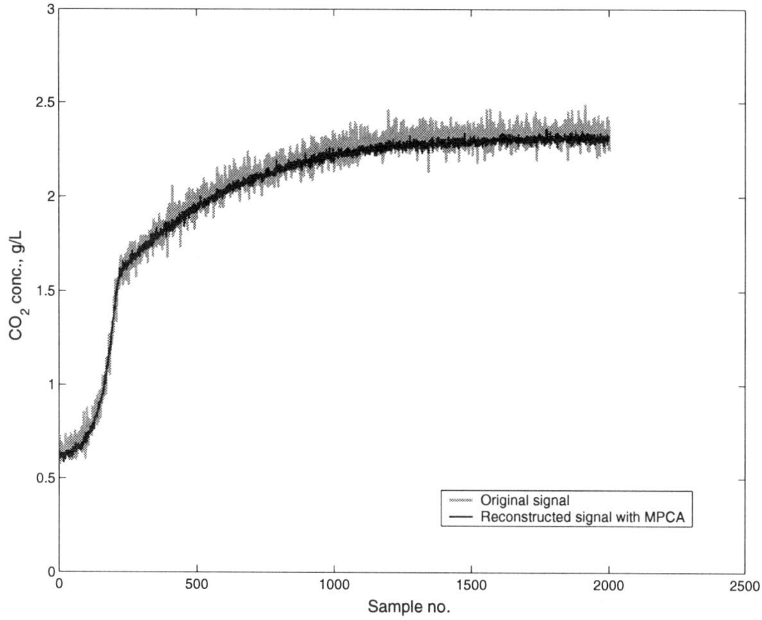

Figure 3.10. CO_2 concentration profile before and after MPCA based denoising.

constitute an orthonormal basis. For this reason, although there are other choices for discretization, dyadic discretization is used frequently [116]. The discretized wavelet function becomes

$$\Psi_{(j,k)}(t) = 2^{-j/2}\Psi(2^{-j}t - k) \tag{3.28}$$

j and k are the scale and translation parameters, respectively.

According to Mallat's multiresolution theory [362], any square integrable signal can be represented by successively projecting it on scaling and wavelet functions. The scaling function is shown as

$$\Phi_{(j,k)}(t) = 2^{-j/2}\Phi(2^{-j}t - k) \ . \tag{3.29}$$

The scaling coefficients $a_{j,k}$ (Eq. 3.30) which are the low frequency content of the signal are obtained by the inner product of the signal with the scaling function Φ. The wavelet coefficients $d_{j,k}$ (Eq.3.31) which are the high frequency content of the signal are obtained by the inner product of the original signal with the wavelet function Ψ.

$$a_{j,k} = < x, \Phi_{j,k} > = \int x(t)\Phi_{(j,k)}(t)dt \tag{3.30}$$

$$d_{j,k} = < x, \Psi_{j,k} > = \int x(t)\Psi_{(j,k)}(t)dt \tag{3.31}$$

where $< \cdot, \cdot >$ indicates the inner product operation. Mallat [362] developed a fast pyramid algorithm for wavelet decomposition based on successive filtering and dyadic downsampling. Figure 3.11 represents this process for one scale. The input signal X is filtered by a low pass filter $L(n)$ and a high pass filter $H(n)$ in parallel obtaining the projection of the original signal onto wavelet function and scaling function. Dyadic downsampling is applied to the filtered signal by taking every other coefficient of the filtered output. The same procedure is repeated for the next scale to the downsampled output of $L(n)$ shown as A1, since the low pass output includes most of the original signal content. By applying this algorithm successively, scaling coefficients a_j and wavelet coefficients d_j at different scales j can be found as

$$a_j = La_{j-1}, \quad d_j = Ha_{j-1} \ . \tag{3.32}$$

Increasing the scale yields scaling coefficients that become increasingly smoother versions of the original signal. The original signal can be computed recursively by adding the wavelet coefficients at each scale and the scaling coefficients at the last scale.

Haar wavelet [116] is the simplest wavelet function that can be used as a basis function to decompose the data into its scaling and wavelet coefficients. It is defined as

$$\Psi(t) = \begin{cases} 1 & , \quad 0 \le t \le 1/2 \\ -1 & , \quad 1/2 < t \le 1 \\ 0 & , \quad otherwise \end{cases} \tag{3.33}$$

and its graphical representation is shown in Figure 3.12. The scaling and wavelet coefficients for the Haar wavelet are [1,1] and [1,-1], respectively. Haar wavelet transform gives better results if the process data contain jump discontinuities. Most of batch process data by nature contain such discontinuities which make Haar wavelet a suitable basis function for decomposing

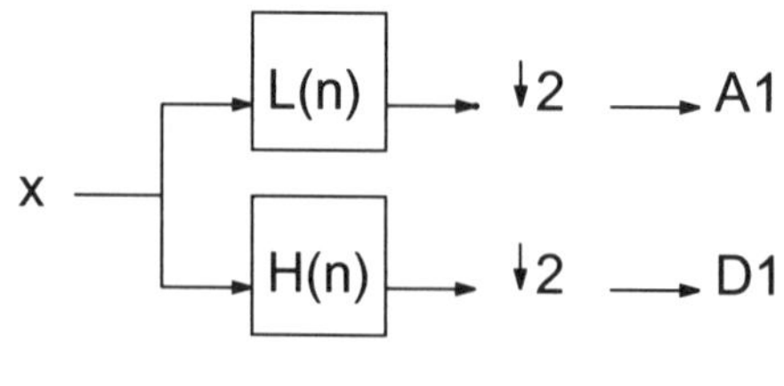

Figure 3.11. Discrete wavelet decomposition.

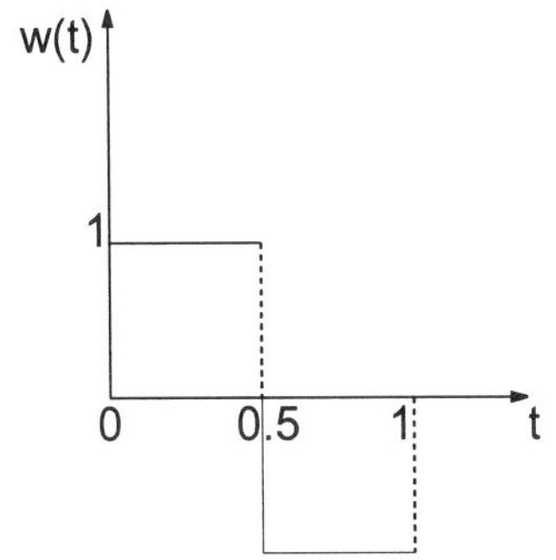

Figure 3.12. Haar wavelet.

batch process data. A noisy process signal (CO_2 evolution rate) was decomposed in four scales using Haar wavelet in Figure 3.13. The low frequency component (dominating nonlinear dynamics) of the original signal (uppermost figure) is found in the scaling coefficients at the last scale whereas the high frequency components that are mostly comprised of noise appear at wavelet coefficients at different scales.

Wavelets are widely used to remove the noise from signals by extracting the low frequency content and removing the high frequency content above a threshold value. The denoised signal is obtained by reconstructing the signal by applying inverse wavelet transform to the scaling and thresholded wavelet coefficients. Thresholding, a crucial step of wavelet denoising, can be applied either as soft or hard thresholding. Hard thresholding (Eq. 3.34) removes the wavelet coefficients smaller than the threshold and replaces them with zero:

$$\delta_h(x) = \begin{cases} x & , \quad |x| > \lambda \\ 0 & , \quad otherwise \end{cases} \tag{3.34}$$

where $\delta_h(x)$ denotes the threshold value of x. Soft thresholding shrinks the wavelet coefficients which are greater than the threshold value towards zero by subtracting the threshold value from the wavelet coefficients as well:

$$\delta_s(x) = \begin{cases} x - \lambda & , \quad x > \lambda \\ 0 & , \quad |x| \leq \lambda \\ x + \lambda & , \quad x < -\lambda \, . \end{cases} \tag{3.35}$$

Different methods for selecting the threshold value have been suggested in the literature by Donoho and co-workers [132]. These methods are grouped

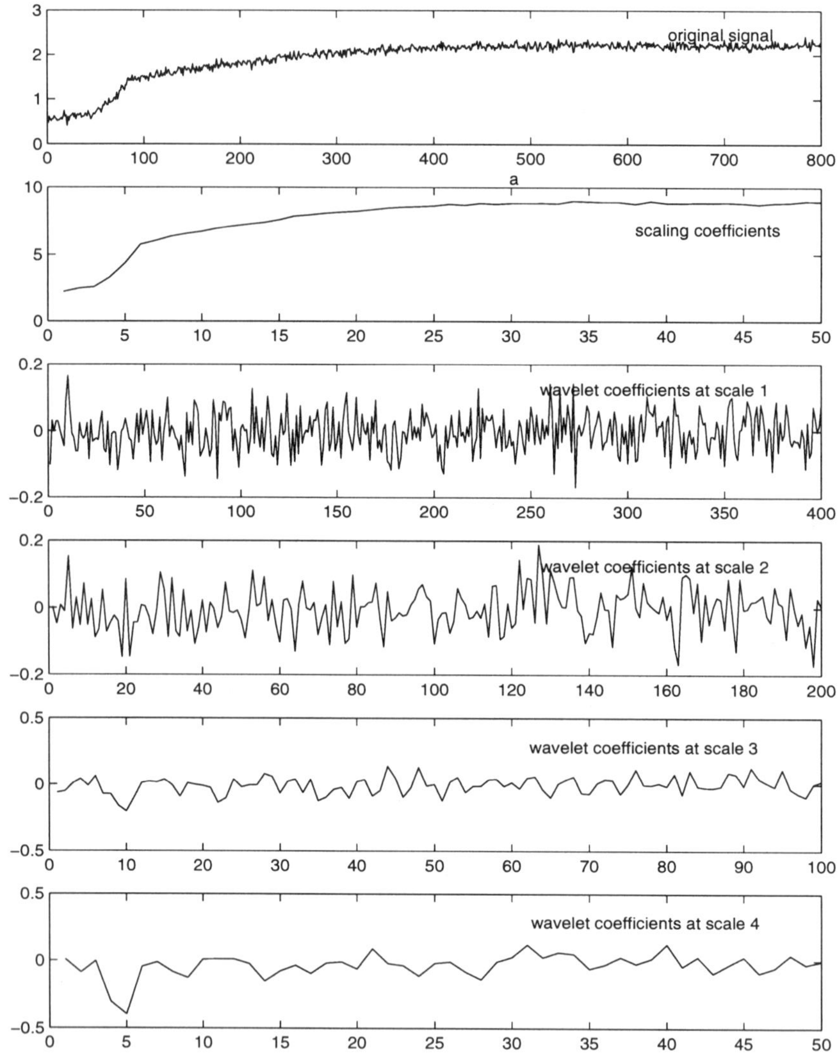

Figure 3.13. Wavelet decomposition of a process signal (CO_2 evolution rate).

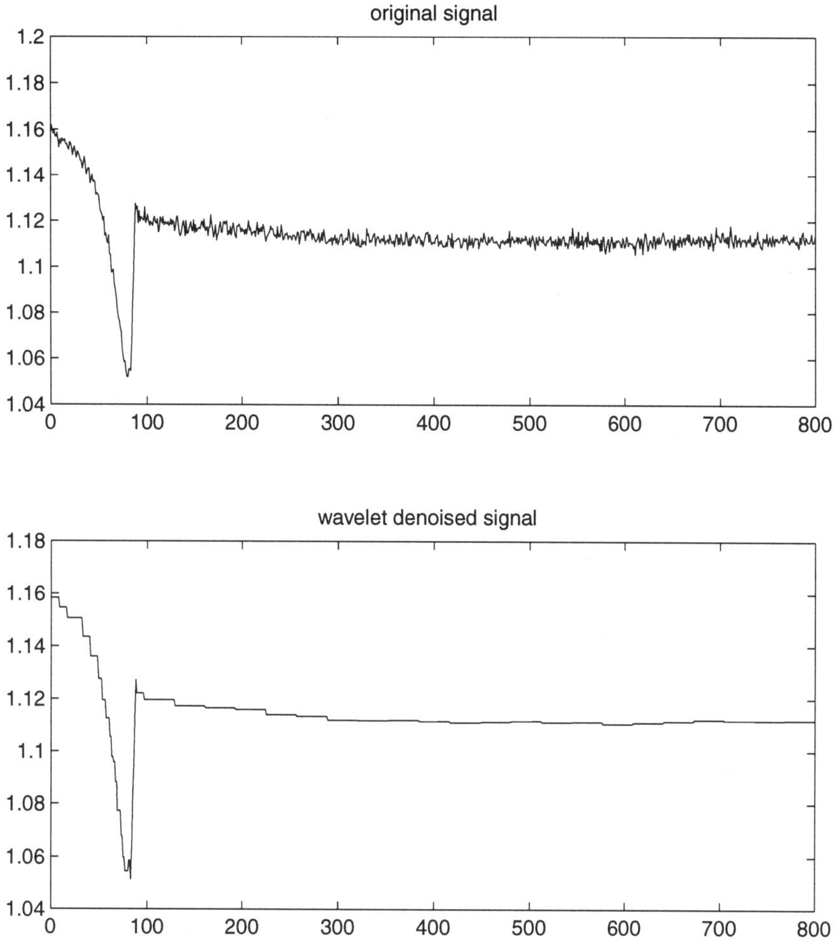

Figure 3.14. Wavelet denoising of a process signal.

into two categories as global thresholding and level-dependent thresholding. A single threshold value is applied for all scales in global thresholding whereas for level-dependent thresholding, a different threshold value is selected for each scale. Level-dependent thresholding is suitable especially for data with non-stationary noise. Figure 3.14 illustrates the wavelet denoising of a process variable using level-dependent hard thresholding. Haar wavelet was used for de-noising the data in three scales.

3.6 Theoretical Confirmation/Stoichiometry and Energetics of Growth

Data collected from a process should also be checked for consistency by using fundamental process knowledge such as stoichiometry and the energetics of growth. Cell growth involves consumption of nutrients for the synthesis of additional biomass. The nutrients that supply energy and raw materials for the biosynthesis should be compatible with the enzymatic machinery of the cell. Knowledge of the reaction stoichiometry provides a convenient way of obtaining various yield coefficients and consequently provides information for formulating a growth medium that will supply all the required nutrients in balanced amounts. This will in turn be very useful for (i) determining the other quantities by expressing one in terms of the others, (ii) monitoring the bioprocess, (iii) eliminating the measurements of compounds that are difficult to measure while keeping track of the easy to measure ones.

For many microorganisms, the energy and carbon requirements for growth and product formation can be met by the same organic compound. This considerably simplifies the analysis of cellular kinetics.

3.6.1 Stoichiometric Balances

To examine cell growth, it is important to know what the cells are made of, that is their chemical composition. Although there are many different biological species, it turns out that a very large fraction of their mass is made of a few elements, namely carbon (C), oxygen (O), nitrogen (N) and hydrogen (H). Minor elements in the cell include phosphorus, sulfur, calcium, potassium and sodium. Typically, 70% of cell mass is water and the remainder is dry matter. Therefore it is conventional to express cell composition on a dry basis. Nearly half of the dry matter in cells is carbon and the elements carbon, oxygen, nitrogen and hydrogen make up about 92% of the total dry mass. In different microbes, the carbon content varies from 46 to 50%, hydrogen from 6 to 7%, nitrogen from 8 to 14% and oxygen from 29 to 35%. These are small variations and they appear to depend on substrate and growth conditions. For many engineering calculations, it is reasonable to consider the cell as a chemical species having the formula of $CH_{1.8}O_{0.5}N_{0.2}$. This engineering approximation is a good starting point for many quantitative analyses while a more carefully formulated empirical formula based on gravimetric techniques may be necessary for complete material flow analysis. The cell "molecular weight" for the generic molecular formula stated above is then $12+1.8 + 0.5(16) +0.2\ (14) = 24.6$. More generally, the elemental composition of the cell can be represented as $CH_aO_bN_c$. Elemental composition of selected microorganisms can be found in [26].

When the cells grow in a medium (a source of all elements needed by the cells) in the presence of oxygen, they oxidize or respire some of the carbon to produce energy for biosynthesis and maintenance of cellular metabolic machinery. Furthermore, cells may produce extracellular products that accumulate in the broth. The overall growth process may therefore be represented simplistically as:

$$Cell + Source\ (Carbon,\ \ Nitrogen,\ \ etc.) + O_2 \rightarrow$$
$$More\ \ Cells + Extracellular\ \ Products + CO_2 + H_2O \qquad (3.36)$$

Carbon dioxide and water on the product side of the reaction (overall growth process) result from oxidation of carbon source (such as glucose) in the medium. Assuming that glucose and ammonia are the sole C and N sources, and the cell composition is represented as $CH_{1.8}O_{0.5}N_{0.2}$, the overall cell growth may be described by

$$C_6H_{12}O_6 + aNH_3 + bO_2 \rightarrow$$
$$\alpha CH_{1.8}O_{0.5}N_{0.2} + \beta CH_xO_yN_z + \gamma CO_2 + \delta H_2O\ . \qquad (3.37)$$

Here, $CH_xO_yN_z$ is the elemental composition of extracellular product and a, b, x, y, z ,α, β, γ and δ are the parameters to be determined. In order to calculate these parameters, some additional information, such as yield coefficient ($Y_{x/s}$), respiratory quotient (RQ) and degree of reductance (γ_D), is needed.

Elemental balances, when applied to Eq. 3.37, lead to the following algebraic relations

$$6 = \alpha + \beta + \gamma \qquad (3.38)$$
$$12 + 3a = 1.8\alpha + x\beta + 2\delta \qquad (3.39)$$
$$6 + 2b = 0.5\alpha + y\beta + 2\gamma + \delta \qquad (3.40)$$
$$a = 0.2\alpha + z\beta \qquad (3.41)$$

Eqs. 3.38-3.41 reduce the degrees of freedom (parameters to be determined, such as a, b, x, y, z, α, β, γ, and δ by four. If the elemental composition of the extracellular product is available a priori, then additional information on this variables, such as cell mass yield ($Y_{x/s}$ and respiratory quotient (RQ), is needed. If the elemental composition of the extracellular product is not available, then information on five variables must be available from experiments for complete specification of Eq. 3.37.

Cell Mass Yield can be defined as the amount of cell mass produced per unit amount of substrate consumed,

$$Y_{x/s} = \frac{amount\ \ of\ \ cell\ \ mass\ \ produced}{amount\ \ of\ \ substrate\ \ consumed} = \frac{\triangle X}{\triangle S} \qquad (3.42)$$

The subscript x/s denotes that, cell yield (X) is based on substrate (S). This notation is especially important when there is more than one substrate which significantly influences cell mass yield. This definition of yield can be extended to non-biomass products (P) with the basis being substrate consumed or biomass produced:

$$Y_{p/s} = \frac{amount\ of\ product\ produced}{amount\ of\ substrate\ consumed} = \frac{\triangle P}{\triangle S} \tag{3.43}$$

or

$$Y_{p/x} = \frac{amount\ of\ product\ produced}{amount\ of\ cell\ mass\ produced} = \frac{\triangle P}{\triangle X}. \tag{3.44}$$

The cell mass yield based on oxygen ($Y_{x/o}$) and yield of ATP (Adenosine triphosphate, $Y_{ATP/x}$) can be obtained in analogous manner.

Respiratory Quotient, RQ, is defined as the rate of carbon dioxide formation divided by the rate of oxygen consumption in aerobic growth.

$$RQ = \frac{rate\ of\ CO_2\ formation}{rate\ of\ O_2\ consumption} \tag{3.45}$$

This ratio can be calculated from on-line measurements of feed and exit CO_2 and O_2 using CO_2 and O_2 analyzers. If the nature of the major extracellular product(s) is known (i.e., x, y, z of $CH_xO_yN_z$), then it is possible to calculate the parameters α, β, γ and δ in Eq. 3.37 from experimental measurement of RQ and one other measurement. If no significant amount of extracellular product is formed, as in some cell growth processes, then it is evident from Eqs. 3.38-3.41 ($\beta = 0$) only one measurement such as RQ is needed to calculate stoichiometric coefficients.

Degree of Reductance of an organic compound is defined as the number of electrons available for transfer to oxygen upon combustion of the compound to CO_2, N_2 and H_2O. It is also defined as the number of equivalents of available electrons per g atom of the compound. The number of equivalents for carbon, hydrogen, oxygen and nitrogen are 4, 1, -2 and -3 respectively. In view of this, for different cell compositions, degree of reductance can be calculated. Examples of the degree of reductance values for a wide range of compounds can be found in [514].

3.6.2 Thermodynamics of Cellular Growth

A complex network of metabolic reactions in microbial growth is involved. These reactions are either *catabolic* or *anabolic*. The former type releases energy while the latter consumes energy. However, some energy is always lost as heat. For this reason, in large-scale processes, it is necessary to remove this heat so that the culture is maintained at its optimum temperature. When there is negligible amount of extracellular product formation under aerobic conditions, the growth reaction (Eq. 3.36) may be rewritten as,

$$\begin{aligned} &Cell + Source\ (Carbon,\ Nitrogen,\ etc.) + O_2 \rightarrow \\ &More\ Cells + CO_2 + H_2O \end{aligned} \tag{3.46}$$

and assuming cell composition is $CH_{1.8}O_{0.5}N_{0.2}$, Eq. 3.37 becomes

$$\begin{aligned} &C_6H_{12}O_6 + aNH_3 + bO_2 \rightarrow \\ &\quad \alpha CH_{1.8}O_{0.5}N_{0.2} + \beta CO_2 + \gamma H_2O \end{aligned} \tag{3.47}$$

The total heat evolved ($\triangle Q$) during growth can be calculated from an enthalpy balance

$$\Delta Q = (-\Delta H_s)(-\Delta S) + (-\Delta H_N)(-\Delta N) - (-\Delta H_x)(-\Delta X) \tag{3.48}$$

where, ΔH_s, ΔH_N, ΔH_x are the heats of combustion of carbon substrate, nitrogen substrate and cells in kcal/g respectively. ΔS, ΔN, ΔX are the amounts of the corresponding materials consumed or produced. The value of the heat of combustion of cells can be estimated using a modification of the Dulong equation [67] using cell composition data that is experimentally determined,

$$(-\Delta H_x) = 8.076C + 34.462(H - \frac{O}{8}) \tag{3.49}$$

while heats of combustion of carbon substrate, and nitrogen substrate can be found in standard chemistry books. The heats of combustion for a variety of organisms are in a narrow range around 22 kJ/g. Note that there are several empirical correlations for ΔQ as a function of the amount of oxygen consumed during aerobic growth as well [67].

It is also common to define a yield term, Y_{kcal}, based on the heat evolution by cell growth as

$$Y_{kcal} = \frac{\Delta X}{\Delta Q} \ . \tag{3.50}$$

Typical values of Y_{kcal} range between 0.096 and 0.126 $g/kcal$ for many microorganisms.

When significant amount of product is present, based on the stoichiometric description of cell growth (Eq. 3.36), total heat evolved (Eq. 3.48) should be modified to

$$\Delta Q = (-\Delta H_s)(-\Delta S) + (-\Delta H_N)(-\Delta N) - \tag{3.51}$$

$$(-\Delta H_x)(-\Delta X) - \sum(-\Delta H_{P_i})(-\Delta P_i) \tag{3.52}$$

where $(-\Delta H_{P_i})$ and $(-\Delta P_i)$ represent the heats of combustion of products P_i and the amount of those products respectively.

Example

In this example, stoichiometric balances and calculation of yield coefficients will be illustrated for growth of *Penicillium chrysogenum* and penicillin production. For growth, a simple stoichiometric model can be used that is based on the theoretical analysis of biosynthesis and polymerization by Nielsen [424] and is given by:

$$CH_{1.81}O_{0.58}N_{0.20}S_{0.004}P_{0.010} + 0.139CO_2 + 0.458NADH \rightarrow$$
$$1.139CH_2O + 0.20NH_3 + 0.004HSO_4^- + 0.010H_2PO_4^- + Y_{x,ATP}ATP$$
$$+ 0.243NADPH. \tag{3.53}$$

The stoichiometry (Eq. 3.53) is based on a cell with the composition given in Table 3.9. C-source is glucose, N-source is ammonia, S-source is sulfate and P-source is phosphate. The stoichiometry is given on a C-mole basis and the elemental composition of the biomass is calculated from the content of the various building blocks [424]. The required ATP and NADPH for biomass synthesis are supplied by the catabolic pathways, and excess NADH formed in the biosynthetic reactions is, together with NADH formed in the catabolic pathways, reoxidized to oxygen via the electron transport chain. Based on the above stoichiometry, the yield coefficient of biomass on glucose can be easily calculated as $Y_{x/s} = 1.139$ *C-mole glucose/C-mole biomass.* In a similar manner, the yield coefficient of biomass on other nutrient sources such as $Y_{x/\text{ammonia}}$, and $Y_{x/\text{phosphate}}$. can also be calculated. The yield coefficient is usually given as *g per g dry weight* in the literature. To convert the literature data to a *C-mole* basis a molecular weight of 24 *g/C-mole* and an ash content of 5% can be assumed.

For the calculation of the theoretical yield of penicillin, a simple stoichiometric balance proposed by Cooney and Acevedo [112] can be used:

$$1.5C_6H_{12}O_6 + 2NH_3 + H_2SO_4 + 2NADH_2 + PAA + 5ATP \rightarrow$$
$$penicillin\ G + 2FADH_2 + CO_2 + 7H_2O. \tag{3.54}$$

α-AAA (α-Aminoadipic acid) is the starting compound in the pathway for penicillin biosynthesis and acts as a carrier. If it is recycled, its net synthesis is of little concern in the overall material and energy requirements for formation of large quantities of penicillin. But if it is used once and then degraded, the net demand for energy will contribute to the energy demands of penicillin and cell synthesis. If the synthesis of α-AAA cannot be neglected and if α-AAA is used once and discarded, then the overall stoichiometry is obtained as:

$$3C_6H_{12}O_6 + 3NH_3 + H_2SO_4 + PAA + 3ATP \rightarrow penicillin\ G$$
$$+ 2FADH_2 + 4CO_2 + 7H_2O + 4NADH_2 + \alpha - AAA. \quad (3.55)$$

Both penicillin G and α-AAA would accumulate. From the above stoichiometry (Eqs. 3.54 and 3.55), the theoretical yield of penicillin on either glucose, or ammonia or sulfate can be calculated based on the definition of yield coefficient for the two cases (in which α-AAA is either recycled or discarded) [112]. Theoretical yield coefficients are presented in Table 3.10, [112] where the stoichiometry of Eq. 3.54 is used in case 1 and the stoichiometry of Eq. 3.55 is used in case 2.

Table 3.9. Composition, ATP and NADPH requirements of a *Penicillium chrysogenum*. Adapted from [424].

Macromolecule	Content[a]	ATP[b] Defined[c]	ATP[b] Complex[d]	NADPH[b] Defined	NADPH[b] Complex
Protein	0.45	19.918	17.505	8.295	0
RNA	0.08	3.315	3.315	-0.266	-0.266
DNA	0.01	0.389	0.389	0.016	0.016
Lipid	0.05				
Phospholipids	0.035	1.652	1.652	0.655	0.655
Sterolesters	0.010	0.805	0.805	0.029	0.029
Tryacylglycerol	0.005	0.295	0.295	0.119	0.119
Carbohydrates	0.25				
Cell Wall	0.22	2.901	2.901	-0.356	-0.356
Glycogen	0.03	0.370	0.370	0	0
Soluble Pool[e]	0.08				
Amino Acids	0.04				
Nucleotides	0.02				
Metabolites etc.	0.02				
Ash	0.08				
Transport[f]					
Ammonia		7.101	1.736		
Amino Acids		0	4.341		
Sulfate		0.137	0		
Phosphate[g]		2.116	2.116		
Total	1	38.999	35.425	8.492	197

[a] : The macromolecular composition is given in *g* per *g DW* for balanced growth at a specific growth rate of about 0.1 h^{-1}.
[b] : The calculation of ATP and NADPH requirements (in *mmole* per *DW*) are shown in [424].
[c] : Data for growth on a defined medium of glucose and inorganic salts.
[d] : Data for growth on a complex medium containing glucose, inorganic salts and amino acids.
[e] : The metabolic costs for biosynthesis of building blocks present in the soluble pool are included in the cost for the macromolecules.
[f] : In estimation of the ATP requirement for transport it is assumed that glucose is transported by facilitated diffusion.
[g] : For the phosphate requirements it is assumed that 3 *moles* of phosphate are needed for synthesis of each nucleotide.

Table 3.10. Theoretical conversion yield of penicillin G. Adapted from [112].

Product	Case 1 α-AAA recycled	Case 2 α-AAA discarded
$\frac{\text{g penicillin G}^{a}}{\text{g glucose}}$	1.10	0.66
$\frac{\text{g 6-APA}^{b}}{\text{g glucose}}$	0.67	0.40
$\frac{10^6 \text{ units penicillin G}^{c}}{\text{g glucose}}$	1.80	1.10
$\frac{\text{g penicillin G}}{\text{g } NH_3}$	10.5	7.0
$\frac{\text{g penicillin G}}{\text{g } H_2SO_4}$	3.60	3.60

[a] : The macromolecular weight of sodium salt of benzylpenicillin is 356.4. The molecular formula is $C_{16}H_{17}N_2O_4SNa$.
[b] : The molecular weight of 6-aminopenicillanic acid (6-APA) is 216.28.
[c] : One international unit of penicillin is equal to 0.6 μg of benzyl sodium penicillin.

4

Methods for Linear Data-Based Model Development

Process models may be developed by using either first principles such as material and energy balances, or process input and output information. The advantages of *first principle models* include the ability to incorporate the scientist's view of the process into the model, describe the internal dynamics of the process, and *explain* the behavior of the process. Their disadvantages are the high cost of model development, the bias that they may have because of the model developer's decisions, and the limitations on including the details due to lack of information about specific model parameters. Often, some physical, chemical or transport parameters are computed using empirical relations, or they are derived from experimental data. In either case, there is some uncertainty about the actual value of the parameter. As details are added to the model, it may become too complex and too large to run model computations on the computer within an acceptable amount of time. However, this constraint has a moving upper limit, since new developments in computer hardware and software technologies permit faster execution. Fundamental models developed may be too large for faster execution to be used in process monitoring and control activities. These activities require fast execution of the models so that regulation of process operation can be made in a timely manner. The alternative model development paradigm is based on developing relations based on process data.

Input-output models are much less expensive to develop. However, they only *describe* the relationships between the process inputs and outputs, and their utility is limited to features that are included in the data set collected for model development. They can be used for interpolation but they should not be used for extrapolation. There are numerous well established techniques for *linear* input-output model development. *Nonlinear* input-output model development techniques have been proposed during the last four decades, but they have not been widely accepted. There are more than twenty different paradigms, and depending on the type of nonlinearities in the data, some paradigms work better than others for describing a specific process. The design of experiments to collect data and the amount of data available have an impact on the accuracy and predictive capability of the model developed. Data collection experiments should be designed such that all key features of the process are excited in the frequency ranges of interest. Since, the model may have terms that are composed of combinations of inputs and/or outputs, exciting and capturing the interactions among variables is crucial. Hence, the use of routine operational data for model development, without any consideration of exciting the key features of the model, may yield good fits to the data, but provide models that have poor predictive ability. The amount of data needed for model development increases with the order of first principle models, linear input-output models, and nonlinear input-output models.

Biochemical processes have become increasingly instrumented in recent years. More variables are being measured and data are being recorded more frequently [304, 655]. This creates a data overload, and most of the useful information gets hidden in large data sets. There is a large amount of correlated or redundant information in these process measurements. This information must be compressed in a manner that retains the essential information about the process, extracts process knowledge from measurement information, and presents it in a form that is easy to display and interpret. A number of methods from multivariate statistics, systems theory and artificial intelligence for data based model development are presented in this chapter.

Model development may have various goals. These goals warrant consideration of the following cases. One case is the interpretation and modeling of one block of data such as measurements of process variables. Principal components analysis (PCA) may be useful for this to retain essential process information while reducing the size of the data set. A second case is the development of a relationship between two groups of data such as process variables and product variables, the regression problem. PCA regression or partial least squares (PLS) regression techniques would be good candidates for addressing this problem. Discrimination and classification are activities related to process monitoring that lead to fault diagnosis. PCA and PLS based techniques as well as artificial neural networks (ANN) and knowledge-based systems may be considered for such problems. Since all these techniques are based on process data, the reliability of data is critical for obtaining dependable results from the implementation of these techniques.

Data-based models may be linear or nonlinear and *describe* only the process behavior captured by the data collected. Methods for development of linear models are easier to implement and more popular. Since most monitoring and control techniques are based on the linear framework, use of linear models is a natural choice. However, nonlinear empirical models that are more accurate over a wider range of operating conditions are desirable for processes with strong nonlinearities. ANNs provide one framework for nonlinear model development. Extensions of PCA and PLS to develop nonlinear models have also been proposed. Several nonlinear time series modeling techniques have been reported. Nonlinear system science methods provide a different framework for nonlinear model development and model reduction. This chapter will focus on linear data-based modeling techniques. References will be provided for their extensions to the nonlinear framework. ANNs will also be discussed in the context of model development. Chapter 5 will introduce nonlinear modeling techniques based on systems science methods.

Section 4.1 introduces PCA. Various multivariate regression techniques are outlined in Section 4.2. Input-output modeling of dynamic processes with time series and state-space modeling techniques, state estimation with Kalman filters and batch process modeling with local model systems are introduced in Section 4.3. Functional data analysis that treats data as representation of continuous functions is discussed in Section 4.4. Statistical methods for modeling batch processes such as multivariate PCA and multivariate PLS, multivariate covariates regression and three-way techniques like PARAFAC and Tucker are introduced in Section 4.5. ANNs and their use in dynamic model development are presented in Section 4.6. Finally, Section 4.7 introduces extensions of linear techniques to nonlinear model development, nonlinear time series modeling methods, and nonlinear PLS techniques.

4.1 Principal Components Analysis

Principal Components Analysis (PCA) is a multivariable statistical technique that can extract the essential information from a data set reported as a single block of data such as process measurements. It was originally developed by Pearson [462] and became a standard multivariate statistical technique [18, 254, 262, 263]. PCA techniques are used to develop models describing the expected variation under normal operation (NO). A reference data set is chosen to define the NO for a particular process based on the data collected from various periods of plant operation when the performance is good. The PCA model development is based on this data set. This model can be used to detect outliers in data, data reconciliation, and deviations from NO that indicate excessive variation from normal target or unusual patterns of variation. Operation under various known upsets can also be modelled if sufficient historical data are available to develop automated diagnosis of source causes of abnormal process behavior [488].

Principal Components (PC) are a new set of coordinate axes that are orthogonal to each other. The first PC indicates the direction of largest variation in data, the second PC indicates the largest variation unexplained by the first PC in a direction orthogonal to the first PC (Fig. 4.1). The number of PCs is usually less than the number of measured variables.

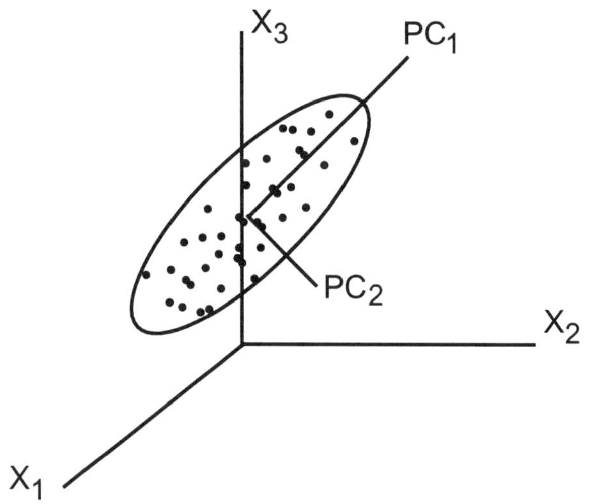

Figure 4.1. PCs of three–dimensional data set projected on a single plane [488].

PCA involves the orthogonal decomposition of the set of process measurements along the directions that explain the maximum variation in the data. For a continuous process, the elements of the data matrix ($\mathbf{X}$) are x_{ij} where $i = 1, \cdots, n$ indicates the number of samples and $j = 1, \cdots, m$ indicates the number of variables. The directions extracted by the orthogonal decomposition of $\mathbf{X}$ are the eigenvectors $\mathbf{p}_i$ of $\mathbf{X}^T\mathbf{X}$ or the PC loadings

$$\mathbf{X} = \mathbf{t}_1\mathbf{p}_1^T + \mathbf{t}_2\mathbf{p}_2^T + \cdots + \mathbf{t}_A\mathbf{p}_A^T + \mathbf{E} \tag{4.1}$$

where $\mathbf{X}$ is an $n \times m$ data matrix with n observations of m variables, $\mathbf{E}$

is $n \times m$ matrix of residuals, and the superscript T denotes the transpose of a matrix. Ideally the dimension A is chosen such that there is no significant process information left in $\mathbf{E}$, and $\mathbf{E}$ represents random error. The eigenvalues of the covariance matrix of $\mathbf{X}$ define the corresponding amount of variance explained by each eigenvector. The projection of the measurements (observations) onto the eigenvectors define new points in the measurement space. These points constitute the score matrix, $\mathbf{T}$ whose columns are $\mathbf{t_i}$ given in Eq. 4.1. The relationship between $\mathbf{T}, \mathbf{P}$, and $\mathbf{X}$ can also be expressed as

$$\mathbf{T} = \mathbf{XP} \,, \qquad \mathbf{X} = \mathbf{TP}^T + \mathbf{E} \tag{4.2}$$

where $\mathbf{P}$ is an $m \times A$ matrix whose jth column is the jth eigenvector of $\mathbf{X}^T\mathbf{X}$, and $\mathbf{T}$ is an $n \times A$ score matrix.

The PCs can be computed by spectral decomposition [262], computation of eigenvalues and eigenvectors, or singular value decomposition. The covariance matrix $\mathbf{S}$ $(\mathbf{S}{=}\mathbf{X}^T\mathbf{X}/(m-1))$ of data matrix $\mathbf{X}$ can be decomposed by *spectral decomposition* as

$$\mathbf{S} = \mathbf{PLP}^T \tag{4.3}$$

where $\mathbf{P}$ is a unitary matrix[1] whose columns are the normalized eigenvectors of $\mathbf{S}$ and $\mathbf{L}$ is a diagonal matrix that contains the ordered eigenvalues l_i of $\mathbf{S}$. The scores $\mathbf{T}$ are computed by using the relation $\mathbf{T} = \mathbf{XP}$.

Singular value decomposition is

$$\mathbf{X} = \mathbf{U\Lambda V}^T \tag{4.4}$$

where the columns of $\mathbf{U}$ are the normalized eigenvectors of $\mathbf{XX}^T$, the columns of $\mathbf{V}$ are the normalized eigenvectors of $\mathbf{X}^T\mathbf{X}$, and $\mathbf{\Lambda}$ is a 'diagonal' matrix having as its elements the positive square roots of the magnitude ordered eigenvalues of $\mathbf{X}^T\mathbf{X}$. For an $n \times m$ matrix $\mathbf{X}$, $\mathbf{U}$ is $n \times n$, $\mathbf{V}$ is $m \times m$ and $\mathbf{\Lambda}$ is $n \times m$. Let the rank of $\mathbf{X}$ be denoted as p, $p \leq \min(m, n)$. The first p rows of $\mathbf{\Lambda}$ make a $p \times p$ diagonal matrix, the remaining $n - p$ rows are filled with zeros. Term by term comparison of the last two equations yields

$$\mathbf{P} = \mathbf{V} \qquad \text{and} \qquad \mathbf{T} = \mathbf{U\Lambda} \,. \tag{4.5}$$

For a data set that is described well by two PCs, the data can be displayed in a plane. The data are scattered as an ellipse whose axes are in

[1] A unitary matrix $\mathbf{A}$ is a complex matrix in which the inverse is equal to the conjugate of the transpose: $\mathbf{A}^{-1} = \mathbf{A}^*$. Orthogonal matrices are unitary. If $\mathbf{A}$ is a *real* unitary matrix then $\mathbf{A}^{-1} = \mathbf{A}^T$.

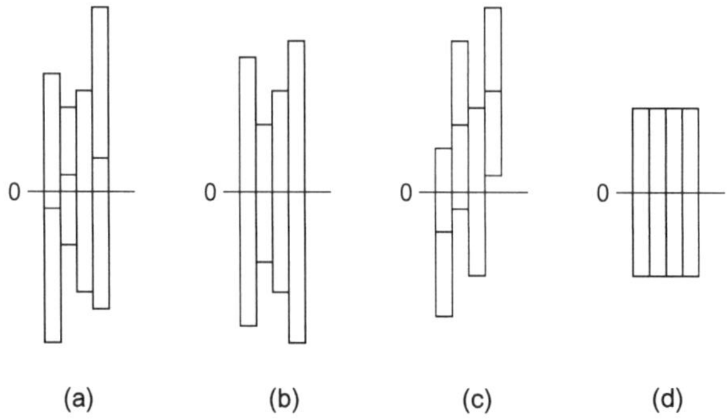

Figure 4.2. Data preprocessing: Scaling of the variables. (a) Raw data, (b) After mean-centering only, (c) After variance-scaling only, (d) After autoscaling (mean-centering and variance-scaling) [145, 181].

the direction of PC loadings in Figure 4.1. For higher number of variables data will be scattered as an ellipsoid.

PCA is sensitive to scaling and outliers. The process data matrix should be mean-centered and scaled properly before the analysis. Scaling is usually performed by dividing all the values for a certain variable by the standard deviation for that variable so that the variance in each variable is unity (Figure 4.2(d)) corresponding to assumption that all variables are equally important. If *a priori* knowledge about the relative importance about the variables is available, important variables can be given a slightly higher scaling weight than that corresponding to unit variance scaling [82, 206].

The selection of appropriate number of PCs or the maximum significant dimension A is critical for developing a parsimonious PCA model [253, 262, 528]. A quick method for computing an approximate value for A is to add PCs to the model until the percent of the variation explained by adding additional PCs becomes small. Inspect the ratio $\sum_{i=1}^{A} l_i / \sum_{i=1}^{p} l_i$ where $\mathbf{L}$ is the diagonal matrix of ordered eigenvalues of $\mathbf{S}$, the covariance matrix. The sum of the variances of the original variables is equal to the *trace* ($tr(\mathbf{S})$), the sum of the diagonal elements of $\mathbf{S}$:

$$S_1^2 + S_2^2 + \cdots + S_p^2 = tr(\mathbf{S}) \ . \tag{4.6}$$

where $tr(\mathbf{S}) = tr(\mathbf{L})$. A more precise method that requires large computational time is cross-validation [309, 659]. Cross-validation is implemented by excluding part of the data, performing PCA on the remaining data, and computing the prediction error sum of squares (PRESS) using the data retained (excluded from model development). The process is repeated until every observation is left out once. The order A is selected as that minimizes the overall PRESS. Two additional criteria for choosing the optimal number of PCs have also been proposed by Wold [659] and Krzanowski [309], related to cross-validation. Wold [659] proposed checking the following ratio

$$R = \frac{\mathrm{PRESS}_A}{\mathrm{RSS}_{A-1}} \tag{4.7}$$

where RSS_A is the residual sum of squares after Ath principal component based on the PCA model. When R exceeds unity upon addition of another PC, it suggests that the Ath component did not improve the prediction power of the model and it is better to use $A-1$ components. Krzanowski [309] suggested the ratio

$$W = \frac{\left(\mathrm{PRESS}_{A-1} - \mathrm{PRESS}_A\right)/D_m}{\mathrm{PRESS}_A/D_A} \tag{4.8}$$

$$D_m = I + JK - 2A, \qquad D_A = JK(I-1) - \sum_{i=1}^{A} I + JK - 2i$$

where D_m and D_A denote the degrees of freedom required to fit the Ath component and the degrees of freedom after fitting the Ath component, respectively. If W exceeds unity, then this criterion suggests that the Ath component could be included in the model [435].

4.2 Multivariable Regression Techniques

Several regression techniques can be used to relate two groups of variables such as process measurements $\mathbf{X}$ and quality variables $\mathbf{Y}$. The availability of a model provides the opportunity to predict process or product variables and compare the measured and predicted values. The residuals between the predicted and measured values of the variables can be used to develop various SPM techniques and tools for identification of variables that have contributed to the out-of-control signal.

Multivariable linear regression is the most popular technique for model development. The model equation is

$$\mathbf{Y} = \mathbf{X}\boldsymbol{\beta} + \mathbf{E} \quad \text{where} \quad \boldsymbol{\beta} = (\mathbf{X}^T\mathbf{X})^{-1}\mathbf{X}^T\mathbf{Y} \tag{4.10}$$

where $\mathbf{E}$ is the residual which is equal to 0 for the estimate $\tilde{\mathbf{Y}} = \mathbf{X}\boldsymbol{\beta}$. A critical issue in using this approach for modeling multivariable processes is the colinearity among process variables. Colinearity causes numerical difficulties in computing the inverse $(\mathbf{X}^T\mathbf{X})^{-1}$. Hence, the computation of the regression coefficients $\boldsymbol{\beta}$ by the least-squares approach may not be possible. Even if $\boldsymbol{\beta}$ is computed, the standard errors of the estimates of the $\boldsymbol{\beta}$ coefficients associated with the colinear regressors become very large. This causes uncertainty and sensitivity in these $\boldsymbol{\beta}$ estimates.

Colinearity can be detected by standardizing all predictor variables (mean centered, unit variance) and computing correlations and coefficients of determination.

$$z_{ij} = \frac{x_{ij} - \bar{x}_j}{d_j} \quad d_j^2 = \sum_{i=1}^{m} (x_{ij} - \bar{x}_j)^2 , \quad i = 1, \cdots , m , j = 1, \cdots , p . \tag{4.11}$$

There is significant colinearity among some predictor variables if:

- The correlation between any two predictors exceeds 0.95 (only colinearity between *two* predictors can be assessed).
- The coefficient of determination R_j^2 of each predictor variable j regressed on all the other predictor variables exceeds 0.90, or the variance inflation factor $VIF_j = (1 - R_j^2)^{-1}$ is less than 10 (variable j is colinear with one or more of the other predictors). VIF_j is the (j, j) *th* diagonal element of the matrix $\mathbf{Z}^T\mathbf{Z}^{-1}$ where $\mathbf{Z} = [z_{ij}]$. R_j^2 can be computed from the relationship between R_j^2 and VIF_j.
- Some of the *eigenvalues* of the correlation matrix $\mathbf{Z}^T\mathbf{Z}$ are less than 0.05. Large elements of the corresponding *eigenvectors* identify the predictor variables involved in the colinearity.

Remedies in regression with colinear data include

- Stepwise regression
- Ridge regression
- Principal components regression
- Partial least squares (PLS) regression

These techniques will be introduced in the sections that follow.

4.2.1 Stepwise Regression

Predictor variables are added to or deleted from the prediction (regression) equation one at a time. Stepwise variable selection procedures are useful when a large number of candidate predictors is available. It is expected that only one of the strongly colinear variables will be included in the model. Major disadvantages of stepwise regression are the limitations in identifying alternative candidate subsets of predictors, and the inability to guarantee the optimality of the final model. The procedure is:

- Fit p single variable regression models, calculate the *overall* model F-statistic for each model. Select the model with the largest F-statistic. If the model is significant, retain the predictor variable and set $r = 1$.
- Fit $p-r$ reduced models, each having the r predictor variables selected in the previous stages of variable selection and one of the remaining candidate predictors. Select the model with the largest overall F-statistic. Check the significance of the model by using the *partial* F-statistic.
- If the partial F-statistic is not significant, terminate the procedure. Otherwise, increment r by 1 and return to step 2.

Computation of F-statistics:
Regression sum of squares: $SSR = \sum(\hat{y}_i - \bar{y})^2$, with p degrees of freedom (d.f.), Error sum of squares: $SSE = \sum(y_i - \bar{y})^2$, with d.f.$= m - p - 1$. Denote a model of order r by M_2 and a model of order $r+1$ by M_1, and their error sum of squares by SSE_2 and SSE_1, respectively. Then

$$\text{Overall F-statistic}: \quad F_{p,n-p-1} = \frac{SSR/p}{SSE/n-p-1} \tag{4.12}$$

$$\text{Partial F-statistic}: \quad F_{1,m-r-2} = \frac{MSR(M_1|M_2)}{MSE_1} \tag{4.13}$$

where

$$MSR(M_1|M_2) = \frac{SSE_2 - SSE_1}{r+1-r} \qquad MSE_1 = \frac{SSE_1}{m-r-2} \; . \tag{4.14}$$

4.2.2 Ridge Regression

The computation of regression coefficients $\boldsymbol{\beta}$ in Eq. 4.10 is modified by introducing a ridge parameter k such that

$$\boldsymbol{\beta} = \left[\mathbf{Z}^T\mathbf{Z} + k\mathbf{I}\right]^{-1}\mathbf{Z}^T\mathbf{Y} \; . \tag{4.15}$$

Standardized ridge estimates β_j $j = 1, \cdots, p$ are calculated for a range of values of k and are plotted versus k. This plot is called a *ridge trace.* The $\boldsymbol{\beta}$ estimates usually change dramatically when k is initially incremented by a small amount from 0. Some $\boldsymbol{\beta}$ coefficients may even change sign. As k is increased, the trace stabilizes. A k value that stabilizes all $\boldsymbol{\beta}$ coefficients is selected and the final values of $\boldsymbol{\beta}$ are estimated.

A good estimate of the k value is obtained as

$$k = \frac{p\ MSE}{\sum_{j=1}^{p}(\beta_j^*)^2} \tag{4.16}$$

where β_j^*s are the least-squares estimates for the standardized predictor variables, and MSE is the least squares mean squared error, $SSE/(m-p-1)$.

Ridge regression estimators are biased. The tradeoff for stabilization and variance reduction in regression coefficient estimators is the bias in the estimators and the increase in the squared error.

4.2.3 Principal Components Regression

Principal components regression (PCR) is one of the techniques to deal with ill-conditioned data matrices by regressing the system properties (e.g. quality measurements) on the principal components scores of the measured variables (e.g. flow rates, temperature). The implementation starts by representing the data matrix $\mathbf{X}$ with its scores matrix $\mathbf{T}$ using the transformation $\mathbf{T} = \mathbf{XP}$. The number of principal components to retain in the model must be determined as in the PCA such that it optimizes the predictive power of the PCR model. This is generally done by using cross validation. Then, the regression equation becomes

$$\mathbf{Y} = \mathbf{TB} + \mathbf{E} \tag{4.17}$$

where the optimum matrix of regression coefficients $\mathbf{B}$ is obtained as

$$\hat{\mathbf{B}} = (\mathbf{T}^T\mathbf{T})^{-1}\mathbf{T}^T\mathbf{Y}\ . \tag{4.18}$$

Substitution of Eq. 4.18 into Eq. 4.17 leads to trivial $\mathbf{E}$'s. The inversion of $\mathbf{T}^T\mathbf{T}$ should not cause any problems due to the mutual orthogonality of the scores. Score vectors corresponding to small eigenvalues can be left out in order to avoid colinearity problems. Since principal components regression is a two-step method, there is a risk that useful predictive information would be discarded with a principal component that is excluded. Hence caution must be exercised while leaving out vectors corresponding to small eigenvalues.

4.2.4 Partial Least Squares

Partial Least Squares (PLS), also known as Projection to Latent Structures, develops a biased regression model between **X** and **Y**. It selects latent variables so that variation in **X** which is most predictive of the product quality data **Y** is extracted. PLS works on the sample covariance matrix $(\mathbf{X}^T\mathbf{Y})(\mathbf{Y}^T\mathbf{X})$ [180, 181, 243, 349, 368, 661, 667]. Measurements on k process variables taken at n different times are arranged into a $(n \times m)$ process data matrix **X**. The p quality variables are given by the corresponding $(n \times p)$ matrix **Y**. Data (both **X** and **Y** blocks) are usually preprocessed prior to PLS analysis. PLS modeling works better when the data are fairly symmetrically distributed and have fairly constant "error variance" [145]. Data are usually centered and scaled to unit variance because in PLS any given variable will have the influence on the model parameters that increases with the variance of the variable. Centering and scaling issues were discussed earlier in Section 4.1. The PLS model can be built by using the non-linear iterative partial least-squares algorithm (NIPALS). The PLS model consists of outer relations (**X** and **Y** blocks individually) and an inner relation (linking both blocks). The outer relations for the **X** and **Y** blocks are respectively

$$\mathbf{X} = \mathbf{T}\mathbf{P}^T + \mathbf{E} = \sum_{a=1}^{A} \mathbf{t}_a \mathbf{p}_a^T + \mathbf{E} \tag{4.19}$$

$$\mathbf{Y} = \mathbf{U}\mathbf{Q}^T + \mathbf{F} = \sum_{a=1}^{A} \mathbf{u}_a \mathbf{q}_a^T + \mathbf{F} \tag{4.20}$$

where **E** and **F** represent the residuals matrices. Linear combinations of **x** vectors are calculated from the latent variable $\mathbf{t}_a = \mathbf{w}_a^T \mathbf{x}$ and those for the **y** vectors from $\mathbf{u}_a = \mathbf{q}_a^T \mathbf{y}$ so that they maximize the covariance between **X** and **Y** explained at each dimension. $\mathbf{w}_a$ and $\mathbf{q}_a$ are loading vectors. The number of latent variables can be determined by cross-validation [659].

For the first latent variable, PLS decomposition is started by selecting one column of $\mathbf{Y}, \mathbf{y}_j$, as the starting estimate for $\mathbf{u}_1$. (Usually, the column of **Y** with greatest variance is chosen.) Starting in the **X** data block (for the first latent variable):

$$\mathbf{w}_1^T = \frac{\mathbf{u}_1^T \mathbf{X}}{\|\mathbf{u}_1^T \mathbf{u}_1\|} \quad , \quad \mathbf{t}_1 = \frac{\mathbf{X}\mathbf{w}_1}{\|\mathbf{w}_1^T \mathbf{w}_1\|} \ . \tag{4.21}$$

In the **Y** data:

$$\mathbf{q}_1^T = \frac{\mathbf{t}_1^T \mathbf{Y}}{\|\mathbf{t}_1^T \mathbf{t}_1\|} \quad , \quad \mathbf{u}_1 = \frac{\mathbf{Y}\mathbf{q}_1}{\|\mathbf{q}_1^T \mathbf{q}_1\|} \ . \tag{4.22}$$

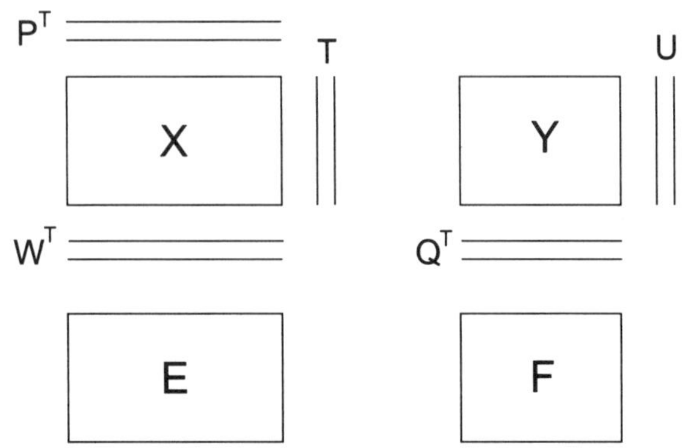

Figure 4.3. The matrix relationships in PLS [145]. **T** and **U** show PLS scores matrices on **X** and **Y** blocks, respectively, **P**, **X** loadings, **W** and **Q** represent weight matrices for each block, **E** and **F** are residual matrices formed by the variation in the data that were left out of modeling.

At this point, the convergence is checked by comparing $\mathbf{t}_1$ in Eq. 4.21 with the $\mathbf{t}_1$ from the preceding iteration. If they are equal within rounding error, one proceeds to Eq. 4.23 to calculate **X** data block loadings $\mathbf{p}_1$ and weights $\mathbf{w}_1$ are rescaled using the converged $\mathbf{u}_1$. Otherwise, $\mathbf{u}_1$ from Eq. 4.22 is used.

$$\mathbf{p}_1^T = \frac{\mathbf{t}_1^T \mathbf{X}}{\|\mathbf{t}_1^T \mathbf{t}_1\|} . \tag{4.23}$$

The regression coefficient b for the inner relation is computed as

$$b_1 = \frac{\mathbf{t}_1^T \mathbf{u}_1}{\|\mathbf{t}_1^T \mathbf{t}_1\|} . \tag{4.24}$$

Once the scores and loadings have been calculated for the first latent variable, **X**- and **Y**-block residuals are computed as

$$\mathbf{E}_1 = \mathbf{X} - \mathbf{t}_1 \mathbf{p}_1^T \tag{4.25}$$

$$\mathbf{F}_1 = \mathbf{Y} - b_1 \mathbf{t}_1 \mathbf{q}_1^T . \tag{4.26}$$

The entire procedure is now repeated for the next latent variable starting with Eq. 4.21. **X** and **Y** are replaced with the residuals $\mathbf{E}_1$ and $\mathbf{F}_1$, respectively, and all subscripts are incremented by 1. Hence, the variability explained by the earlier latent variables is filtered out from **X** and **Y** by replacing them in the next iteration with their residuals that contain unexplained variation.

Several enhancements have been made to the PLS algorithm [118, 198, 343, 363, 664, 660, 668] and software is available for developing PLS models [472, 548].

4.3 Input-Output Modeling of Dynamic Processes

Methods for developing models to describe steady state relationships of processes are presented in Sections 4.1 and 4.2. The description of batch fermentation processes and the general form of their model equations in Chapter 2 (for example Eq. 2.1 or Eq. 2.3) indicate that dynamic input-output models are more appropriate for representing the behavior of these processes. Two types of dynamic models are introduced in this section: time series models (Section 4.3.1) and state space models (Section 4.3.2). State estimators are also presented in conjunction with state space models. The linear model structures are discussed in this section. They can handle mild nonlinearities. They can also result from linearization around an operating point. Their extensions to nonlinear models are discussed in Section 4.7. Use of these modeling paradigms to develop more complex models of batch and semi-batch processes is reported in Section 4.3.4.

Inputs, outputs, disturbances and state variables will be denoted as $\mathbf{u}$, $\mathbf{y}$, $\mathbf{d}$ and $\mathbf{x}$, respectively. The models can be in continuous time (differential equations) or discrete time (difference equations). For multivariable processes where $u_1(t)$, $u_2(t)$, $\cdots$, $u_m(t)$ are the m inputs, the input vector $\mathbf{u}(t)$ at time t is written as a column vector. Similarly, the p outputs, and the n state variables are defined by column vectors:

$$\mathbf{y}(t) = \begin{pmatrix} y_1(t) \\ \vdots \\ y_p(t) \end{pmatrix}, \quad \mathbf{u}(t) = \begin{pmatrix} u_1(t) \\ \vdots \\ u_m(t) \end{pmatrix}, \quad \mathbf{x}(t) = \begin{pmatrix} x_1(t) \\ \vdots \\ x_n(t) \end{pmatrix} \tag{4.27}$$

Disturbances $\mathbf{d}(t)$, residuals $\mathbf{e}(t) = \mathbf{y}(t) - \hat{\mathbf{y}}(t)$, and random noise attributed to inputs, outputs and state variables are also represented by column vectors with appropriate dimensions in a similar manner.

4.3.1 Time Series Models

Time series models have been popular in many fields ranging from modeling stock prices to climate. They could be cast as a regression problem where the regressor variables are the previous values of the same variable and past values of inputs and disturbances. They are also called black box models because they describe the relationship of the present value of the output to external variables but do not provide any knowledge about the physical description of the processes they represent.

A general linear discrete time model for a single variable $y(t)$ can be written as

$$y(t) = \eta(t) + w(t) \tag{4.28}$$

where $w(t)$ is a disturbance term such as measurement noise and $\eta(t)$ is the noise-free output

$$\eta(t) = G(q, \theta)u(t) \tag{4.29}$$

with the rational function $G(q, \theta)$ and input $u(t)$. The function $G(q, \theta)$ relates the inputs to noise-free outputs whose values are not known because the measurements of the outputs are corrupted by disturbances such as measurement noise. The parameters of $G(q, \theta)$ (such as b_i in Eq. 4.30) are represented by the vector θ, and q is called the shift operator (Eq. 4.31). Assume that relevant information for the current value of output $y(t)$ is provided by past values of $y(t)$ for n_y previous time instances and past values of $u(t)$ for n_u previous instances. The relationship between these variables is

$$\begin{aligned} &\eta(t) + f_1\eta(t-1) + \cdots + f_{n_y}\eta(t-n_y) \\ &\quad = b_1u(t) + b_2u(t-1) + \cdots + b_{n_u}u(t-(n_u-1)) \end{aligned} \tag{4.30}$$

where f_i, $i = 1, 2, \ldots, n_y$ and b_i, $i = 1, 2, \ldots, n_u$ are parameters to be determined from data. Defining the shift operator q as

$$y(t-1) = q^{-1}y(t) \tag{4.31}$$

Eq. (4.30) can be written using two polynomials in q

$$\begin{aligned} &\eta(t)\left(1 + f_1q^{-1} + \cdots + f_{n_y}q^{-n_y}\right) \\ &\quad = u(t)\left(b_1 + b_2q^{-1} + \cdots + b_{n_u}q^{-n_u}\right) . \end{aligned} \tag{4.32}$$

This equation can be written in a compact form by defining the polynomials

$$\begin{aligned} F(q) &= (1 + f_1q^{-1} + \cdots + f_{n_y}q^{-n_y}) \\ B(q) &= (b_1 + b_2q^{-1} + \cdots + b_{n_u}q^{-(n_u-1)}) \end{aligned} \tag{4.33}$$

where

$$\eta(t) = G(q, \theta)\, u(t) \quad \text{with} \quad G(q, \theta) = \frac{B(q)}{F(q)}. \tag{4.34}$$

Often the inputs may have a delayed effect on the output. If there is a delay of n_k sampling times, Eq. (4.30) is modified as

$$\begin{aligned}\eta(t) + f_1\eta(t-1) + \cdots + f_{n_y}\eta(t-n_y) \\ = b_1u(t-n_k) + b_2u(t-(n_k+1)) + \cdots + b_{n_u}u(t-(n_u+n_k-1)) \; .\end{aligned} \tag{4.35}$$

The disturbance term can be expressed in the same way

$$w(t) = H(q,\theta)e(t) \tag{4.36}$$

where $e(t)$ is white noise and

$$H(q,\theta) = \frac{C(q)}{D(d)} = \frac{1 + c_1q^{-1} + \cdots + c_{n_c}q^{-n_c}}{1 + d_1q^{-1} + \cdots + d_{n_d}q^{-n_d}}. \tag{4.37}$$

The model (Eq. 4.28) can be written as

$$y(t) = G(q,\theta)u(t) + H(q,\theta)e(t) \tag{4.38}$$

where the parameter vector θ contains the coefficients b_i, c_i, d_i and f_i of the transfer functions $G(q,\theta)$ and $H(q,\theta)$. The model structure is described by five parameters n_y, n_u, n_k, n_c, and n_d. Since the model is based on polynomials, its structure is finalized when the parameter values are selected. These parameters and the coefficients are determined by fitting candidate models to data and minimizing some criteria based on reduction of prediction error and parsimony of the model.

The model represented by Eq. (4.38) is known as the **Box-Jenkins (BJ) model**, named after the statisticians who have proposed it [79]. It has several special cases:

- **Output error (OE) model.** When the properties of disturbances are not modeled and the noise model $H(q)$ is chosen to be identity ($n_c = 0$ and $n_d = 0$), the noise source $w(t)$ is equal to $e(t)$, the difference (error) between the actual output and the noise-free output.

- **AutoRegressive Moving Average model with eXogenous inputs (ARMAX).** If the same denominator is used for G and H

  $$A(q) = F(q) = D(q) = 1 + a_1q^{-1} + \cdots + a_{n_a}q^{-n_a} \; . \tag{4.39}$$

  Hence Eq. (4.38) becomes

  $$A(q)y(t) = B(q)u(t) + C(q)e(t) \tag{4.40}$$

  where $A(q)y(t)$ is the autoregressive (regressing on previous values of the same variable $y(t)$) term, $C(q)e(t)$ is the moving average of white noise $e(t)$, and $B(q)u(t)$ represents the contribution of external inputs. Use of a common denominator is reasonable if the dominating disturbances enter the process together with the inputs.

- **AutoRegressive model with eXogenous inputs (ARX).** A special case of ARMAX is obtained by letting $C(q) = 1$ ($n_c = 0$).

These models are used for prediction of the output given the values of inputs and outputs in previous sampling times. Since white noise cannot be predicted, its current value $e(t)$ is excluded from prediction equations. Predicted values are denoted by a ^ over the variable symbol, for example $\hat{y}(t)$. To emphasize that predictions are based on a specific parameter set θ, the nomenclature is further extended to $\hat{y}(t \mid \theta)$.

The computation of parameters θ is usually cast as a minimization problem: select the values for θ that minimize the prediction errors $\varepsilon(t, \theta) = y(t) - \hat{y}(t \mid \theta)$ for given sets of data over a time period. For N data points

$$\hat{\theta}_N = \arg \min_{\theta} \frac{1}{N} \sum_{t=1}^{N} \varepsilon^2(t, \theta) \tag{4.41}$$

where "arg min" denotes the minimizing argument. This criteria has to be extended to prevent overfit of data. A larger model with many parameters may fit data used for model development very well, but it may give large prediction errors when new data are used. Several criteria have been proposed to balance model fit and model complexity. Two of them are given here to illustrate how they balance accuracy and parsimony:

- **Akaike's Information Criterion (AIC)**

$$\min_{d,\theta} \left(1 + \frac{2d}{N}\right) \sum_{t=1}^{N} \varepsilon^2(t, \theta) \tag{4.42}$$

where d is the number of parameters estimated (dimension of θ).

- **Final Prediction Error (FPE)**

$$\min_{d,\theta} \left(\frac{1 + d/N}{1 - d/N} \frac{1}{N}\right) \sum_{t=1}^{N} \varepsilon^2(t, \theta) \, . \tag{4.43}$$

Model development (also called system identification) involves several critical activities including design of experiments and collection of data, data pretreatment, model fitting, model validation and acceptability of the model for its use. A vast literature has been developed over the last 50 years in various aspects of model identification [212, 354, 346, 500, 558]. A schematic diagram in Figure 4.4 [347] where the ovals represent human activities and decision making steps and the rectangles represent computer-based computations and decisions illustrates the links between critical activities.

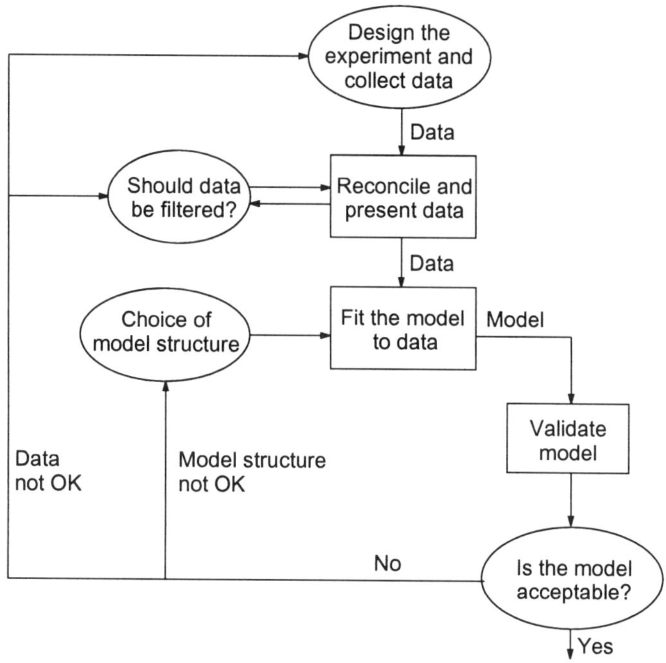

Figure 4.4. Model identification [347].

4.3.2 State-Space Models

State variables are the minimum set of variables that are necessary to describe completely the state of a system. In quantitative terms, given the values of state variables $x(t)$ at time t_0 and the values of inputs $u(t)$ for $t > t_0$, the values of outputs $y(t)$ can be computed for $t > t_0$. Various types of state-space models are introduced in this section. Recall the models derived from first principles in Chapter 2. The process variables used in these models can be subdivided into measured and unmeasured variables, and all process variables can be included in the set of state variables while the measured variables can form the set of output variables. This way, the model can be used to compute all process variables based on measured values of output variables and the state-space model.

Classical state-space models are discussed first. They provide a versatile modeling framework that can be linear or nonlinear, continuous or discrete time, to describe a wide variety of processes. State variables can be defined based on physical variables, mathematical solution convenience or ordered importance of describing the process. Subspace models are discussed in the second part of this section. They order state variables according to the magnitude of their contributions in explaining the variation in data. State-space models also provide the structure for developing state estimators where one can estimate corrected values of state variables, given process input and output variables and estimated values of process outputs. State estimators are discussed in the last part of this section.

Classical State-Space Models

State space models relate the variation in state variables over time to their values in the immediate past and to inputs with differential or difference equations. Algebraic equations are then used to relate output variables to state variables and inputs at the same time instant. Consider a system of first-order differential equations (Eq. 4.44) describing the change in state variables and a system of output equations (Eq. 4.45) relating the outputs to state variables

$$\frac{d\mathbf{x}}{dt} = \dot{\mathbf{x}}(t) = \mathbf{f}(\mathbf{x}(t), \mathbf{u}(t)) \tag{4.44}$$

$$\mathbf{y}(t) = \mathbf{h}(\mathbf{x}(t), \mathbf{u}(t)) \ . \tag{4.45}$$

For a specific time t_0, $\dot{\mathbf{x}}(\mathbf{t_0})$ can be computed using Eq. 4.44 if $x(t)$ and $u(t)$ are known at time t_0. For an infinitesimally small interval δt, one can compute $\mathbf{x}(t_0 + \delta t)$ using

$$\mathbf{x}(t_0 + \delta t) = \mathbf{x}(t_0) + \delta t \cdot \mathbf{f}(\mathbf{x}(t_0), \mathbf{u}(t_0)). \tag{4.46}$$

Then, the output $\mathbf{y}(t_0 + \delta t)$ can be computed using $\mathbf{x}(t_0 + \delta t)$ and Eq. (4.45). Equation (4.46) is the Euler's method for the solution of Eq. (4.44) if δt is a small number. This computation sequence can be repeated to compute values of $\mathbf{x}(t)$ and $\mathbf{y}(t)$ for $t > t_0$ if the corresponding values of $\mathbf{u}(t)$ are given for future values of time such as $t_0 + 2\delta t, \cdots, t_0 + k\delta t$. The model composed of Eqs. (4.44)-(4.45) is called the *state-space model*, the vector $\mathbf{x}(t)$, the *state vector*, and its components $x_i(t)$ the *state variables*. The dimension of $\mathbf{x}(t)$, n (Eq. (4.27)) is the model order.

State-space models can also be developed for discrete time systems. Let the current time be denoted as t_k and the next time instant where input values become available as t_{k+1}. The equivalents of Eqs. (4.44)-(4.45) in discrete time are

$$\mathbf{x}(t_{k+1}) = \mathbf{f}(\mathbf{x}(t_k), \mathbf{u}(t_k)) \quad k = 0, 1, 2, \cdots \tag{4.47}$$

$$\mathbf{y}(t_k) = \mathbf{h}(\mathbf{x}(t_k), \mathbf{u}(t_k)) \ . \tag{4.48}$$

For the current time $t_0 = t_k$, the state at time $t_{k+1} = t_0 + \delta t$ is now computed by using the difference equations (4.47)-(4.48). Usually, the time interval between the two discrete times $\delta t = t_{k+1} - t_k$ is a constant value equal to the sampling time.

Linear State-Space Models

The functional relations $\mathbf{f}(\mathbf{x}, \mathbf{u})$ and $\mathbf{h}(\mathbf{x}, \mathbf{u})$ in Eqs. (4.44)-(4.45) or Eqs. (4.47)-(4.48) were not restricted so far. They could be nonlinear. For the sake of easier mathematical solutions, if justifiable, they can be restricted to be linear. The linear continuous models are represented as

$$\begin{aligned} \dot{\mathbf{x}}(t) &= \mathbf{A}\mathbf{x}(t) + \mathbf{B}\mathbf{u}(t) \\ \mathbf{y}(t) &= \mathbf{C}\mathbf{x}(t) + \mathbf{D}\mathbf{u}(t) \ . \end{aligned} \tag{4.49}$$

The linear discrete time model is

$$\begin{aligned} \mathbf{x}(t_{k+1}) &= \mathbf{F}\mathbf{x}(t_k) + \mathbf{G}\mathbf{u}(t_k) \quad k = 0, 1, 2, \cdots \\ \mathbf{y}(t_k) &= \mathbf{C}\mathbf{x}(t_k) + \mathbf{D}\mathbf{u}(t_k) \ . \end{aligned} \tag{4.50}$$

The notation in Eq. (4.50) can be simplified by using $\mathbf{x}_k$ or $\mathbf{x}(k)$ to denote $\mathbf{x}(t_k)$:

$$\begin{aligned} \mathbf{x}_{k+1} &= \mathbf{F}\mathbf{x}_k + \mathbf{G}\mathbf{u}_k \quad k = 0, 1, 2, \cdots \\ \mathbf{y}_k &= \mathbf{C}\mathbf{x}_k + \mathbf{D}\mathbf{u}_k \ . \end{aligned} \tag{4.51}$$

Matrices $\mathbf{A}$ and $\mathbf{B}$ are related to matrices $\mathbf{F}$ and $\mathbf{G}$ as

$$\mathbf{F} = e^{\mathbf{A}T} \qquad \mathbf{G} = \int_0^T e^{\mathbf{A}\tau}\mathbf{B}d\tau \tag{4.52}$$

where the sampling interval $T = t_{k+1} - t_k$ is assumed to be equal for all values of k. The dimensions of these matrices are

$$\begin{array}{ccccccccccc} \mathbf{A} & : & n & \times & n & \qquad & \mathbf{B} & : & n & \times & m \\ \mathbf{C} & : & p & \times & n & \qquad & \mathbf{D} & : & p & \times & m \end{array}$$

These models are called linear *time-invariant* models. Mild nonlinearities in the process can often be described better by making the matrices in model equations (4.49) or (4.50) time dependent. This is indicated by symbols such as $\mathbf{A}(t)$ or $\mathbf{F}_k$.

Disturbances

Disturbances are inputs to a process. Some disturbances can be measured, others arise and their presence is only recognized because of their influence on process and/or output variables. The state-space model needs to be augmented to incorporate the effects of disturbances on state variables and outputs. Following Eq. (4.28), the state-space equation can be written as

$$\begin{aligned} \dot{\mathbf{x}}(t) &= \mathbf{f}(\mathbf{x}(t), \mathbf{u}(t), \mathbf{w}(t)) \\ \mathbf{y}(t) &= \mathbf{h}(\mathbf{x}(t), \mathbf{u}(t), \mathbf{w}(t)) \end{aligned} \tag{4.53}$$

where $\mathbf{w}(t)$ denotes disturbances. It is necessary to describe $\mathbf{w}(t)$ in order to compute how the state variables and outputs behave. If the disturbances are known and measured, their description can be appended to the model. For example, the linear state-space model can be written as

$$\begin{aligned} \dot{\mathbf{x}}(t) &= \mathbf{A}\mathbf{x}(t) + \mathbf{B}\mathbf{u}(t) + \mathbf{E}_1\mathbf{w}_1(t) \\ \mathbf{y}(t) &= \mathbf{C}\mathbf{x}(t) + \mathbf{D}\mathbf{u}(t) + \mathbf{E}_2\mathbf{w}_2(t) \ . \end{aligned} \tag{4.54}$$

where $\mathbf{w}_1(t)$ and $\mathbf{w}_2(t)$ are disturbances affecting the state variables and outputs, respectively, and $\mathbf{E}_1$ and $\mathbf{E}_2$ are the corresponding coefficient matrices. This model structure can also be used to incorporate modeling uncertainties (represented by $\mathbf{w}_1(t)$) and measurement noise (represented by $\mathbf{w}_2(t)$).

Another alternative is to develop a model for *unknown* disturbances to describe $\mathbf{w}(t)$ as the output from a dynamic system with a known input $\mathbf{u}_w(t)$ that has a simple functional form.

$$\begin{aligned} \dot{\mathbf{x}}_w(t) &= \mathbf{f}_\mathbf{w}(\mathbf{x}_w(t), \mathbf{u}_w(t)) \\ \mathbf{w}(t) &= \mathbf{h}_w(\mathbf{x}_w(t), \mathbf{u}_w(t)) \end{aligned} \tag{4.55}$$

where the subscript w indicates state variables, inputs and functions of the disturbance(s). Typical choices for input forms may be an impulse, white noise, or infrequent random step changes. Use of fixed impulse and step changes lead to deterministic models, while white noise or random impulse and step changes yield stochastic models [347]. The disturbance model is appended to the state and output model to build an augmented dynamic model with *known* inputs.

Linearization of Nonlinear Systems

Sometimes a nonlinear process can be modeled by linearizing it around a known operating point. If the nonlinear terms are expanded using the linear terms of Taylor series and equations are written in terms of deviations of process variables (the so-called *deviation variables*) from the operating point, a linear model is obtained. The model can then be expressed in state-space form [438, 541].

Consider the general state-space equation Eqs. (4.44-4.45) and assume that there is a stable stationary solution (a steady state) at $\mathbf{x} = \mathbf{x}_{ss}, \mathbf{u} =$

$\mathbf{u}_{ss}$:

$$\mathbf{f}(\mathbf{x}_{ss}, \mathbf{u}_{ss}) = 0 \; . \tag{4.56}$$

If $\mathbf{f}(\mathbf{x}, \mathbf{u})$ has continuous partial derivatives in the neighborhood of the stationary solution $\mathbf{x} = \mathbf{x}_{ss}, \mathbf{u} = \mathbf{u}_{ss}$, then for $\ell = 1, \cdots, n$:

$$\begin{aligned} f_\ell(x, u) &= f_\ell(\mathbf{x}_{ss}, \mathbf{u}_{ss}) + \frac{\partial f_\ell}{\partial x_1}(\mathbf{x}_{ss}, \mathbf{u}_{ss})(x_1 - x_{ss,1}) + \cdots \\ &+ \frac{\partial f_\ell}{\partial x_n}(\mathbf{x}_{ss}, \mathbf{u}_{ss})(x_n - x_{ss,n}) + \frac{\partial f_\ell}{\partial u_1}(\mathbf{x}_{ss}, \mathbf{u}_{ss})(u_1 - u_{ss,1}) \\ &+ \cdots + \frac{\partial f_\ell}{\partial u_m}(\mathbf{x}_{ss}, \mathbf{u}_{ss})(u_m - u_{ss,m}) + r_k(\mathbf{x} - \mathbf{x}_{ss}, \mathbf{u} - \mathbf{u}_{ss}) \end{aligned} \tag{4.57}$$

where all partial derivatives are evaluated at $(\mathbf{x}_{ss}, \mathbf{u}_{ss})$ and $\mathbf{r}_k$ denotes the higher order terms that yield nonlinear expressions, which are assumed to be negligible. Consider the Jacobian matrices $\mathbf{A}$ and $\mathbf{B}$ that have the partial derivatives in Eq. (4.57) as their elements:

$$A = \begin{pmatrix} \frac{\partial f_1}{\partial x_1} & \cdots & \frac{\partial f_1}{\partial x_n} \\ \vdots & \ddots & \vdots \\ \frac{\partial f_n}{\partial x_1} & \cdots & \frac{\partial f_n}{\partial x_n} \end{pmatrix}, \quad B = \begin{pmatrix} \frac{\partial f_1}{\partial u_1} & \cdots & \frac{\partial f_1}{\partial u_m} \\ \vdots & \ddots & \vdots \\ \frac{\partial f_n}{\partial u_1} & \cdots & \frac{\partial f_n}{\partial u_m} \end{pmatrix} \tag{4.58}$$

with the partial derivatives being evaluated at $(\mathbf{x}_{ss}, \mathbf{u}_{ss})$. In view of Eq. 4.56, Eq. 4.57 can now be written in a compact form as

$$\mathbf{f}(\mathbf{x}, \mathbf{u}) = \mathbf{A}(\mathbf{x} - \mathbf{x}_{ss}) + \mathbf{B}(\mathbf{u} - \mathbf{u}_{ss}) + \mathbf{r}(\mathbf{x} - \mathbf{x}_{ss}, \mathbf{u} - \mathbf{u}_{ss}) \; . \tag{4.59}$$

Neglecting the higher order terms $\mathbf{r}_k(\mathbf{x} - \mathbf{x}_{ss}, \mathbf{u} - \mathbf{u}_{ss})$ and defining the deviation variables

$$\bar{\mathbf{x}} = \mathbf{x} - \mathbf{x}_{ss}, \qquad \bar{\mathbf{u}} = \mathbf{u} - \mathbf{u}_{ss} \tag{4.60}$$

Eq. (4.44) can be written as

$$\dot{\bar{\mathbf{x}}} = \mathbf{A}\bar{\mathbf{x}} + \mathbf{B}\bar{\mathbf{u}}, \tag{4.61}$$

The output equation is developed in a similar manner:

$$\bar{\mathbf{y}} = \mathbf{C}\bar{\mathbf{x}} + \mathbf{D}\bar{\mathbf{u}} \tag{4.62}$$

where the elements of $\mathbf{C}$ and $\mathbf{D}$ are the partial derivatives $\partial h_i / \partial x_j$ with $i = 1, \cdots, p$ and $j = 1, \cdots, n$ and $\partial h_i / \partial u_j$ with $i = 1, \cdots, p$ and $j = 1, \cdots, m$, respectively. Hence, the linearized equations are of the same form as the original state-space equations in Eq. 4.49. Linearization of discrete time nonlinear models follows the same procedure and yields linear difference equations similar to Eq. (4.50).

Subspace State-Space Models

Subspace state-space models are developed by using techniques that determine the largest directions of variation in the data and build models that describe the data. PCA and PLS are subspace methods used for steady state data. They could be used to develop models for dynamic relations by augmenting the appropriate data matrices with lagged values of the variables. In recent years, dynamic model development techniques that rely on subspace concepts have been proposed [315, 316, 613, 621]. Subspace methods are introduced in this section to develop state-space models for process monitoring and closed-loop control.

Consider a simple state-space model without external inputs u_k

$$\begin{aligned} \mathbf{x}_{k+1} &= \mathbf{F}\mathbf{x}_k + \mathbf{H}\epsilon_k \\ \mathbf{y}_k &= \mathbf{C}\mathbf{x}_k + \epsilon_k \end{aligned} \tag{4.63}$$

where $\mathbf{x}_k$ is the state variable vector of dimension n at time k and $\mathbf{y}_k$ is the observation vector with p output measurements. The stochastic input ϵ_k is the serially uncorrelated innovation vector having the same dimension as $\mathbf{y}_k$ and covariance $\mathbf{E}[\epsilon_k \epsilon_{k+l}^T] = \mathbf{\Delta}$ if $l = 0$, and $\mathbf{0}$ otherwise. This representation would be useful for process monitoring activities where "appropriate" state variables (usually the first few state variables) are used to determine if the process is operating as expected. The statistics used in statistical process monitoring (SPM) charts assume no correlation over time between measurements. If state-space models are developed such that the state variables and residuals are uncorrelated at zero lag, the statistics can be safely applied to these calculated variables instead of measured process outputs. Several techniques, balanced realization [21], PLS realization [416], and the canonical variate realization [315, 413] can be used for developing these models. Negiz and Cinar [413] have proposed the use of state variables developed with *canonical variate analysis* based realization to implement such SPM techniques to multivariable continuous processes.

Subspace algorithms generate the process model by successive approximation of the memory or the state variables of the process by determining successively functions of the past that have the most information for predicting the future [316]. Canonical variates analysis (CVA), a subspace algorithm, is used to develop state space models [315] where the first state variable contains the largest amount of information about the process dynamics, the second state variable is orthogonal to the first (does not repeat the information explained in the previous state variable) and describes the largest amount of the remaining process variation. The first few significant state variables can often be used to describe the greatest variation in the process. The system order n is determined by inspecting the dominant singular values (SV) of a covariance matrix (the ratio of the specific SV to the sum of all the SVs [21] generated by singular value decomposition (SVD) or an information theoretic approach such as the Akaike Information Criterion (AIC) [315].

The Hankel matrix (Eq. 4.65) is used to develop subspace models. It expresses the covariance between future and past stacked vectors of output measurements. If the stacked vectors of future ($\mathcal{Y}_{k_J}^{+}$) and past ($\mathcal{Y}_{k_K}^{-}$) data are given as

$$\mathcal{Y}_{k_J}^{+} = \begin{bmatrix} \mathbf{y}_k \\ \mathbf{y}_{k+1} \\ \vdots \\ \mathbf{y}_{k+J-1} \end{bmatrix} \quad \text{and} \quad \mathcal{Y}_{k_K}^{-} = \begin{bmatrix} \mathbf{y}_{k-1} \\ \mathbf{y}_{k-2} \\ \vdots \\ \mathbf{y}_{k-K} \end{bmatrix} \tag{4.64}$$

the Hankel matrix (note that $\mathbf{H}_{KJ}$ is different than the $\mathbf{H}$ matrix in Eq. (4.63)) is

$$\mathbf{H}_{KJ} = E\left[\mathcal{Y}_{k_J}^{+} \mathcal{Y}_{k-1_K}^{-^T}\right] = \begin{bmatrix} \mathbf{\Lambda}_1 & \mathbf{\Lambda}_2 & \cdots & \mathbf{\Lambda}_K \\ \mathbf{\Lambda}_2 & \mathbf{\Lambda}_3 & \cdots & \mathbf{\Lambda}_{K+1} \\ \vdots & \vdots & \vdots & \vdots \\ \mathbf{\Lambda}_J & \mathbf{\Lambda}_{J+1} & \cdots & \mathbf{\Lambda}_{J+K-1} \end{bmatrix} \tag{4.65}$$

where $\mathbf{\Lambda}_\ell$ is the autocovariance of $\mathbf{y}_k$'s which are ℓ time period apart and $E[\cdot]$ denotes the expected value of a stochastic variable. K and J are past and future window lengths. The non-zero singular values of the Hankel matrix determine the order of the system, i.e., the dimension of the state variables vector. The non-zero and dominant singular values of $\mathbf{H}_{JK}$ are chosen by inspection of singular values or metrics such as AIC.

CV (canonical variate) realization requires that covariances of future and past stacked observations be conditioned against any singularities by taking their square roots. The Hankel matrix is scaled by using $\mathbf{R}_K^-$ and $\mathbf{R}_J^+$ defined in Eq. (4.67). The scaled Hankel matrix ($\overline{\mathbf{H}}_{JK}$) and its singular value decomposition is given as

$$\bar{\mathbf{H}}_{JK} = \left[\mathbf{R}_J^+\right]^{-1/2} \mathbf{H}_{JK} \left[\mathbf{R}_K^-\right]^{-1/2} = \mathbf{U}\mathbf{\Sigma}\mathbf{V}^T \tag{4.66}$$

where

$$\begin{aligned} (\mathbf{R}_J^+) &= E\left(\mathcal{Y}_{k_J}^{+} \mathcal{Y}_{k_J}^{+^T}\right) \\ (\mathbf{R}_K^-) &= E\left(\mathcal{Y}_{k-1_K}^{-} \mathcal{Y}_{k-1_K}^{-^T}\right). \end{aligned} \tag{4.67}$$

$\mathbf{U}_{pJ\times n}$ contains the n left eigenvectors of $\overline{\mathbf{H}}_{JK}$, $\mathbf{\Sigma}_{n\times n}$ contains the singular values (SV), and $\mathbf{V}_{Kp\times n}$ contains the n right eigenvectors of the decomposition. The subscripts associated with $\mathbf{U}$, $\mathbf{\Sigma}$ and $\mathbf{V}$ denote the dimensions of these matrices. The SVD matrices in Eq. 4.66 include only the SVs and eigenvectors corresponding to the n state variables retained in the model. The full SV matrix $\mathbf{\Sigma}$ has dimension $Jp \times Kp$ and it contains the SVs in a descending order. If the process noise is small, all SVs smaller than the nth SV are effectively zero and the corresponding state variables are excluded from the model.

The state variables are given as

$$\mathbf{x}_k = \mathbf{\Sigma}^{1/2}\mathbf{V}^T \left(\mathbf{R}_K^-\right)^{-1/2} \mathcal{Y}_{k-1_K}^- \ . \tag{4.68}$$

Once $\mathbf{x}_k$ (or $\mathbf{x}(t)$) is known, $\mathbf{F}$, $\mathbf{G}$ (or $\mathbf{A}$, $\mathbf{B}$), $\mathbf{C}$, and $\mathbf{\Delta}$ can be constructed [413]. The covariance matrix of the state vector based on CV decomposition $E[\mathbf{x}_k\mathbf{x}_k^T] = \mathbf{\Sigma}$ reveals that $\mathbf{x}_k$ are independent at zero-lag.

The second subspace state-space model includes external inputs:

$$\begin{aligned} \mathbf{x}_{k+1} &= \mathbf{F}\mathbf{x}_k + \mathbf{G}\mathbf{u}_k + \mathbf{H}_1\mathbf{w}_k \\ \mathbf{y}_k &= \mathbf{C}\mathbf{x}_k + \mathbf{D}\mathbf{u}_k + \mathbf{H}_2\mathbf{v}_k \end{aligned} \tag{4.69}$$

where $\mathbf{F}, \mathbf{G}, \mathbf{C}, \mathbf{D}, \mathbf{H}_1$ and $\mathbf{H}_2$ are system matrices, and $\mathbf{w}$ and $\mathbf{v}$ are Normally distributed, zero-mean noise vectors. It can be developed using CV or other methods such as N4SID [613].

The subspace state-space modeling framework has been used to develop batch-to-batch process monitoring and control techniques that utilize information from previous batches along with measurements from the ongoing batch (Section 6.6).

4.3.3 State Estimators

A state estimator is a computational algorithm that deduces the state of a system by utilizing knowledge about system and measurement dynamics, initial conditions of the system, and assumed statistics of measurement and system noises [182]. State estimators can be classified according to the set of state variables estimated. *Full-order state estimators* estimate all n state variables of the process. *Minimum-order state estimators* estimate only the unmeasurable state variables. *Reduced-order state estimators* estimate some of the measured state variables in addition to all unmeasurable state variables.

The estimator is designed to minimize the estimation error in a well-defined statistical sense by using all measurement information and prior knowledge about the process. The accuracy of the estimates is affected by errors in process models used. Three estimation problems can be listed: filtering, smoothing, and prediction (Fig. 4.5). In *filtering*, the time at which the estimate is desired coincides with the latest measurement time.

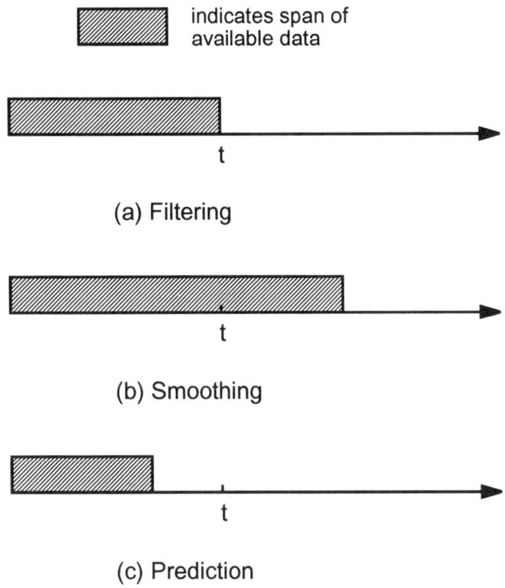

Figure 4.5. Estimation problems.

In *smoothing*, the time of the estimate falls within the span of measurement data available. The state of the process at some prior time is estimated based on all measurements collected up to the current time. In *prediction*, the time of the estimate occurs after the last available measurement. The state of the process in some future time is estimated. The discussion in this section focuses on filtering (Fig. 4.6), and in particular on Kalman filtering technique.

An estimate $\hat{x}$ of a state variable x is computed using the measured outputs y. An *unbiased* estimate $\hat{x}$ has the same expected value as that of

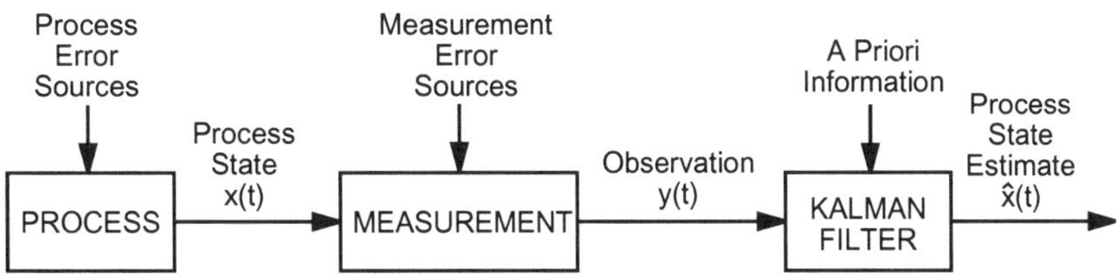

Figure 4.6. Kalman filtering technique.

the variable being estimated (x). A *minimum variance* (unbiased) estimate has its error variance that is less than or equal to the variance of any other unbiased estimate. A *consistent* estimate $\hat{x}$ converges to the true value of x as the number of measurements increase. Kalman filters are unbiased, minimum variance, consistent estimators. They are also *optimal* estimators in the sense that they minimize the mean square estimation error.

Discrete Kalman Filter

Consider a discrete time system with a state equation

$$\mathbf{x}_k = \mathbf{F}_{k-1}\mathbf{x}_{k-1} + \mathbf{w}_{k-1} \tag{4.70}$$

where $\mathbf{x}_k$ is an abbreviation for $\mathbf{x}(t_k)$, and the subscript of $\mathbf{F}_{k-1}$ indicates that it is time dependent ($\mathbf{F}(t_{k-1})$). Note that the time index is shifted back by 1 with respect to the discrete time state-space model description in Eq. (4.50) to emphasize the filtering problem. $\mathbf{w}_k$ is a zero mean, white (Gaussian) sequence with covariance $\mathbf{Q}_k$, and the system is not subjected to external inputs (unforced system) ($\mathbf{G}(t_k) = 0$). The measured output equation is

$$\mathbf{y}_k = \mathbf{C}_k\mathbf{x}_k + \mathbf{v}_k \tag{4.71}$$

where $\mathbf{v}_k$ is a vector of random noise with zero mean and covariance $\mathbf{R}_k$ corrupting the output measurements $\mathbf{y}_k$. Given the *prior estimate* of $\mathbf{x}_k$ denoted by $\hat{\mathbf{x}}_k^-$, a recursive estimator is sought to compute an *updated estimate* $\hat{\mathbf{x}}_k^+$ based on measurements $\mathbf{y}_k$. The recursive estimator uses only the most recent values of measurements and prior estimates, avoiding the need for a growing storage of past values. The updated estimate is a weighted sum of $\hat{\mathbf{x}}_k^-$ and $\mathbf{y}_k$:

$$\hat{\mathbf{x}}_k^+ = \mathbf{K}'_k\hat{\mathbf{x}}_k^- + \mathbf{K}_k\mathbf{y}_k \tag{4.72}$$

where $\mathbf{K}'_k$ and $\mathbf{K}_k$ are unspecified (yet) time-varying weighting matrices. Expressing the estimates as the sum of unknown real values and estimation errors denoted by $\tilde{\mathbf{x}}_k$

$$\hat{\mathbf{x}}_k^+ = \mathbf{x}_k + \tilde{\mathbf{x}}_k^+ \qquad \hat{\mathbf{x}}_k^- = \mathbf{x}_k + \tilde{\mathbf{x}}_k^- \tag{4.73}$$

and inserting the equation for $\hat{\mathbf{x}}_k^-$ and Eq. (4.71) in Eq. (4.72), the estimation error $\tilde{\mathbf{x}}_k^+$ becomes:

$$\tilde{\mathbf{x}}_k^+ = (\mathbf{K}'_k + \mathbf{K}_k\mathbf{C}_k - \mathbf{I})\,\mathbf{x}_k + \mathbf{K}'_k\tilde{\mathbf{x}}_k^- + \mathbf{K}_k\mathbf{v}_k\ . \tag{4.74}$$

Consider the expected value ($E[\cdot]$) of Eq. (4.74). By definition $E[\mathbf{v}_k] = 0$. If $E[\tilde{\mathbf{x}}_k^-] = 0$ as well, then the estimator (Eq. (4.72)) will be unbiased for any given $\mathbf{x}_k$ if $\mathbf{K}'_k + \mathbf{K}_k\mathbf{C}_k - \mathbf{I} = 0$. Hence, substituting for $\mathbf{K}'_k$ in Eq. (4.72):

$$\hat{\mathbf{x}}_k^+ = (\mathbf{I} - \mathbf{K}_k\mathbf{C}_k)\hat{\mathbf{x}}_k^- + \mathbf{K}_k\mathbf{y}_k \tag{4.75}$$

that can be rearranged as

$$\hat{\mathbf{x}}_k^+ = \hat{\mathbf{x}}_k^- + \mathbf{K}_k(\mathbf{y}_k - \mathbf{C}_k\hat{\mathbf{x}}_k^-) \ . \tag{4.76}$$

The corresponding estimation error is derived from Eqs. (4.71), (4.73) and (4.76) as

$$\tilde{\mathbf{x}}_k^+ = (\mathbf{I} - \mathbf{K}_k\mathbf{C}_k)\tilde{\mathbf{x}}_k^- + \mathbf{K}_k\mathbf{y}_k \ . \tag{4.77}$$

The error covariance matrix $\mathbf{P}_k$ changes when new measurement information is used.

$$\mathbf{P}_k^+ = E\left[\tilde{\mathbf{x}}_k^+\tilde{\mathbf{x}}_k^{-T}\right] = (\mathbf{I} - \mathbf{K}_k\mathbf{C}_k)\mathbf{P}_k^-(\mathbf{I} - \mathbf{K}_k\mathbf{C}_k)^T + \mathbf{K}_k\mathbf{R}_k\mathbf{K}_k^T \tag{4.78}$$

where $\mathbf{P}_k^-$ and $\mathbf{P}_k^+$ are the prior and updated error covariance matrices, respectively [182].

From Eq. (4.76), the updated estimate is equal to the prior estimate corrected by the error in predicting the last measurement and the magnitude of the correction is determined by the "gain" $\mathbf{K}_k$. If the criterion for choosing $\mathbf{K}_k$ is to minimize a weighted scalar sum of the diagonal elements of the error covariance matrix $\mathbf{P}_k^+$, the cost function J_k could be

$$J_k = E\left[\tilde{\mathbf{x}}_k^+\mathbf{S}\ \tilde{\mathbf{x}}_k^+\right] \tag{4.79}$$

where $\mathbf{S}$ is a positive semidefinite matrix. If $\mathbf{S} = \mathbf{I}$, $J_k = trace\left[\mathbf{P}_k^+\right]$ which is equivalent to minimizing the length of the estimation error vector. The optimal choice of $\mathbf{K}_k$ is derived by taking the partial derivative of J_k with respect to $\mathbf{K}_k$ and equating it to zero:

$$\mathbf{K}_k = \mathbf{P}_k^-\mathbf{C}_k^T\left(\mathbf{C}_k\mathbf{P}_k^-\mathbf{C}_k^T + \mathbf{R}_k\right)^{-1} \ . \tag{4.80}$$

Substituting Eq. (4.80) in Eq. (4.78) provides a simpler expression for $\mathbf{P}_k^+$ [182]:

$$\mathbf{P}_k^+ = (\mathbf{I} - \mathbf{K}_k\mathbf{C}_k)\,\mathbf{P}_k^- \ . \tag{4.81}$$

The equations derived so far describe the state estimate and error covariance matrix behavior *across* a measurement. The extrapolation of these entities between measurements is

$$\hat{\mathbf{x}}_k^- = \mathbf{F}_{k-1}\hat{\mathbf{x}}_{k-1}^+ \tag{4.82}$$

$$\mathbf{P}_k^- = \mathbf{F}_{k-1}\mathbf{P}_{k-1}^+\mathbf{F}_{k-1}^T + \mathbf{Q}_{k-1} \ . \tag{4.83}$$

Discrete Kalman filter equations are summarized in Table 4.1.

Table 4.1. Summary of discrete Kalman filter equations

Description	Equation	Other
Process model	4.70	$\mathbf{w}_k \sim N(0, \mathbf{Q}_k)$
Measurement (output) model	4.71	$\mathbf{v}_k \sim N(0, \mathbf{R}_k)$
Initial conditions		$E[\mathbf{x}(0)] = \hat{\mathbf{x}}_0$ $E\left[(\mathbf{x}(0) - \hat{\mathbf{x}}_0)(\mathbf{x}(0) - \hat{\mathbf{x}}_0)^T\right] = \mathbf{P}_0$ $E[\mathbf{w}_k \mathbf{v}_j^T] = 0$ for all $j,\ k$
State estimate extrapolation	4.82	
Error covariance extrapolation	4.83	
State estimate update	4.76	
Error covariance update	4.81	
Kalman gain matrix	4.80	

Continuous Kalman Filter

The formulation of continuous Kalman filter parallels that of the discrete Kalman filter. For the state-space model of the unforced system

$$\dot{\mathbf{x}} = \mathbf{A}(t)\mathbf{x} + \mathbf{E}(t)\mathbf{w} \tag{4.84}$$

and an output equation similar to Eq. (4.71)

$$\mathbf{y} = \mathbf{C}(t)\mathbf{x} + \mathbf{v} \tag{4.85}$$

the propagation of error covariance becomes

$$\begin{aligned}\dot{\mathbf{P}}(t) &= \mathbf{A}(t)\mathbf{P}(t) + \mathbf{P}(t)\mathbf{A}^T(t) + \mathbf{E}(t)\mathbf{Q}(t)\mathbf{E}^T(t) \\ &\quad -\mathbf{P}(t)\mathbf{C}^T(t)\mathbf{R}^{-1}(t)\mathbf{C}(t)\mathbf{P}(t)\end{aligned} \tag{4.86}$$

where $\mathbf{AP} + \mathbf{PA}^T$ is the contribution of the unforced system without the effect of measurements, $\mathbf{EQE}^T$ accounts for the increase in uncertainty because of process noise, and $\mathbf{PC}^T\mathbf{R}^{-1}\mathbf{CP}$ accounts for the reduction in uncertainty as a result of measurements. Equation (4.86) is called the *matrix Riccati equation* and its solution is described in most advanced systems science and control books [17] and provided in most control toolbox software such as Matlab. The continuous Kalman filter equations for white measurement noise are given in Table 4.2 where the last row shows the Kalman gain matrix when the process and measurement noises are correlated. The explicit statement of time dependency is eliminated by excluding

Table 4.2. Summary of continuous Kalman filter equations

Description	Equation
Process model	$\dot{\mathbf{x}} = \mathbf{Ax} + \mathbf{Ew} \quad \mathbf{w}_k \sim N(0, \mathbf{Q})$
Output model	$\mathbf{y} = \mathbf{Cx} + \mathbf{v} \quad \mathbf{v}_k \sim N(0, \mathbf{R})$
Initial conditions	$E[\mathbf{x}(0)] = \hat{\mathbf{x}}_0 \quad \mathbf{R}^{-1}$ exists $E\left[(\mathbf{x}(0) - \hat{\mathbf{x}}_0)(\mathbf{x}(0) - \hat{\mathbf{x}}_0)^T\right] = \mathbf{P}_0$
State estimate	$\dot{\hat{\mathbf{x}}} = \mathbf{Ax} + \mathbf{K}\left(\mathbf{y} - \mathbf{C}\hat{\mathbf{x}}\right), \quad \hat{\mathbf{x}}(0) = \mathbf{x}_0$
Error covariance	$\dot{\mathbf{P}} = \mathbf{AP} + \mathbf{PA}^T + \mathbf{EQE}^T - \mathbf{PC}^T\mathbf{R}^{-1}\mathbf{CP}, \ \mathbf{P}(0) = \mathbf{P}_0$
Kalman gain matrix	$\mathbf{K} = \mathbf{PC}^T\mathbf{R}^{-1}$ when $E[\mathbf{w}(t)\mathbf{v}^T(\tau)] = 0$ $= (\mathbf{PC}^T + \mathbf{EZ})\mathbf{R}^{-1}$ when $E[\mathbf{w}(t)\mathbf{v}^T(\tau)] = \mathbf{Z}\delta(t - \tau)$

the arguments (t) from the matrices in the equations of Table 4.2 for compactness. Time dependency of system matrices **A, C**, and **E** will indicate which matrices in other equations are time dependent.

If the process is excited by deterministic inputs $\mathbf{u}$ (either a deterministic disturbance or a control signal), the procedures for computing $\mathbf{P}$ and $\mathbf{K}$ remain the same, but the estimators are modified. For the discrete time process, state equation Eq. (4.70) becomes

$$\mathbf{x}_k = \mathbf{F}_{k-1}\mathbf{x}_{k-1} + \mathbf{G}_{k-1}\mathbf{u}_{k-1} + \mathbf{w}_{k-1} \tag{4.87}$$

and the state estimator is

$$\begin{aligned} \hat{\mathbf{x}}_k^+ &= \mathbf{F}_{k-1}\hat{\mathbf{x}}_{k-1}^+ + \mathbf{G}_{k-1}\mathbf{u}_{k-1} \\ &\quad + \mathbf{K}_k\left(\mathbf{y}_k - \mathbf{C}_k\mathbf{F}_{k-1}\hat{\mathbf{x}}_{k-1}^+ - \mathbf{C}_k\mathbf{G}_{k-1}\mathbf{u}_{k-1}\right) . \end{aligned} \tag{4.88}$$

For continuous processes, state equation Eq. (4.84) becomes

$$\dot{\mathbf{x}}(t) = \mathbf{A}(t)\mathbf{x}(t) + \mathbf{B}(t)\mathbf{u}(t) + \mathbf{E}(t)\mathbf{w}(t) \tag{4.89}$$

and the corresponding state estimator is

$$\dot{\hat{\mathbf{x}}}(t) = \mathbf{A}(t)\hat{\mathbf{x}}(t) + \mathbf{B}(t)\mathbf{u}(t) + \mathbf{K}(t)\left(\mathbf{y}(t) - \mathbf{C}(t)\hat{\mathbf{x}}(t)\right) . \tag{4.90}$$

Steady State Kalman Filter

When the process and measurement dynamics are represented by linear constant coefficient equations (**A, B, E** or **F, G, H** are not functions of time) and the driving noise statistics are stationary (**Q, R** are not functions of time), the filtering process may reach a steady state. Computing the steady state value of $\mathbf{P}$ denoted by $\mathbf{P}_\infty$ and inserting it in state estimator equations reduce the computational burden for state estimation significantly. For $\dot{\hat{\mathbf{P}}} = \mathbf{0}$ the Riccati equation becomes

$$\mathbf{AP}_\infty + \mathbf{P}_\infty\mathbf{A}^T + \mathbf{EQE}^T - \mathbf{P}_\infty\mathbf{C}^T\mathbf{R}^{-1}\mathbf{CP}_\infty = 0 \ . \tag{4.91}$$

The Kalman filter gain becomes a constant:

$$\mathbf{K}_\infty = \mathbf{P}_\infty\mathbf{C}^T\mathbf{R}^{-1} \tag{4.92}$$

and the corresponding steady-state Kalman filter is

$$\dot{\hat{\mathbf{x}}}(t) = \mathbf{A}\hat{\mathbf{x}}(t) + \mathbf{Bu}(t) + \mathbf{K}_\infty\left(\mathbf{y}(t) - \mathbf{C}\hat{\mathbf{x}}(t)\right) \ . \tag{4.93}$$

Extended Kalman Filter

The linear filters were developed so far for linear process models. State estimation can be extended to nonlinear processes described by nonlinear stochastic differential equations:

$$\dot{\mathbf{x}}(t) = \mathbf{f}(\mathbf{x}(t), t) + \mathbf{w}(t) \tag{4.94}$$

where $\mathbf{f}(\cdot)$ is a nonlinear function of the state and $\mathbf{w}(t)$ is $N(0, \mathbf{Q}(t))$ noise vector. The objective is to estimate $\mathbf{x}(t)$ from sampled nonlinear measurements

$$\mathbf{y}_k = \mathbf{h}_k(\mathbf{x}(t_k)) + \mathbf{v}_k \qquad k = 1, 2, \cdots \tag{4.95}$$

where $\mathbf{h}_k$ depends on both the time index k and $\mathbf{x}(t_k)$, and $\mathbf{v}_k$ is $N(0, \mathbf{R}_k)$ noise vector. Hence the estimation is for a process with continuous dynamics and discrete-time measurements. While the general solution to this problem is challenging, a practical estimation algorithm that relies on Taylor series expansion of $\mathbf{f}$ can be formulated [182]. The filter developed is called *extended* Kalman filter (EKF) [257].

Consider the expansion of $\mathbf{f}$ about the current estimate (conditional mean) of the state vector $\bar{\mathbf{x}} = \hat{\mathbf{x}}$:

$$\mathbf{f}(\mathbf{x}, t) = \mathbf{f}(\hat{\mathbf{x}}, t) + \left.\frac{\partial \mathbf{f}}{\partial \mathbf{x}}\right|_{\mathbf{x}=\hat{\mathbf{x}}} (\mathbf{x} - \hat{\mathbf{x}}) + \cdots \tag{4.96}$$

Taking the expectation of both sides of the equation yields

$$\hat{\mathbf{f}}(\mathbf{x}, t) = \mathbf{f}(\hat{\mathbf{x}}(t), t) + \mathbf{0} + \cdots \tag{4.97}$$

and using Eq. (4.94)

$$\dot{\hat{\mathbf{x}}}(t) = \mathbf{f}(\hat{\mathbf{x}}(t)) \qquad t_{k-1} \leq t < t_k \ . \tag{4.98}$$

By using the first two terms of the expansion in Eq. (4.96) an approximate differential equation for the estimation error covariance matrix is obtained:

$$\dot{\mathbf{P}}(t) = \mathbf{F}(\hat{\mathbf{x}}(t), t)\mathbf{P}(t) + \mathbf{P}(t)\mathbf{F}^T(\hat{\mathbf{x}}(t), t) + \mathbf{Q}(t) \qquad t_{k-1} \leq t < t_k \quad (4.99)$$

where $\mathbf{F}(\hat{\mathbf{x}}(t), t)$ is a matrix whose ij th element is

$$F_{ij}(\hat{\mathbf{x}}(t), t) = \left. \frac{\partial f_i(\mathbf{x}(t), t)}{\partial x_j(t)} \right|_{\mathbf{x}(t)=\hat{\mathbf{x}}(t)} . \qquad (4.100)$$

To complete the filtering algorithm, update equations that account for new measurement information must be developed. Assume that the estimate of $\mathbf{x}(t)$ and its associated covariance matrix are propagated using Eqs. (4.98) and (4.99) and denote the solutions at time t_k by $\hat{\mathbf{x}}_k^-$ and $\mathbf{P}_k^-$. When a new measurement $\mathbf{y}_k$ is received the updated estimates are

$$\hat{\mathbf{x}}_k^+ = \hat{\mathbf{x}}_k^- + \mathbf{K}_k(\mathbf{y}_k - \mathbf{h}_k(\hat{\mathbf{x}}_k^-)) \qquad (4.101)$$

with

$$\mathbf{P}_k^+ = (\mathbf{I} - \mathbf{K}_k\mathbf{H}_k(\hat{\mathbf{x}}_k^-))\mathbf{P}_k^- . \qquad (4.102)$$

The same approach as the linear case is used to determine the optimal filter gain matrix:

$$\mathbf{K}_k = \mathbf{P}_k^-\mathbf{H}_k^T(\hat{\mathbf{x}}_k^-)\left(\mathbf{H}_k(\hat{\mathbf{x}}_k^-)\mathbf{P}_k^-\mathbf{H}_k^T(\hat{\mathbf{x}}_k^-) + \mathbf{R}_k\right)^{-1} \qquad (4.103)$$

where

$$\mathbf{H}_k(\hat{\mathbf{x}}_k^-) = \left. \frac{\partial \mathbf{h}_k(\mathbf{x}(t_k))}{\partial x(t_k)} \right|_{x(t_k)=\hat{x}_k^-} \qquad (4.104)$$

resulting from Taylor series expansion of $\mathbf{h}_k(x_k) = \mathbf{h}_k(\hat{x}_k^-) + \mathbf{H}_k(\hat{x}_k^-)(x_k - \hat{x}_k^-) + \cdots$ about $\hat{x}_k^-$.

EKF equations for a continuous process with discrete-time measurements are summarized in Table 4.3. EKFs for continuous measurements and more advanced filters for nonlinear systems are discussed in [17, 182]. Applications of EKFs to batch processes are illustrated in [72, 120].

Kalman Filters for Processes with Nonstationary Stochastic Disturbances

The literature on Kalman filters focuses mostly on deterministic systems subjected to arbitrary noise to account for modeling error (Eq. 4.70) and measurement error (Eq. 4.71). This framework implies that the true process states can never drift away for prolonged periods from their values predicted by the deterministic model equations [356]. Consequently, the Kalman filter will not contain any integration terms and will not track a

Table 4.3. Summary of extended Kalman filter equations for continuous process with discrete-time measurements

Description	Equation	Other
Process model	4.94	$\mathbf{w}(t) \sim N(0, \mathbf{Q}(t))$
Measurement model	4.95	$\mathbf{v}(t) \sim N(0, \mathbf{R}(t))$
Initial conditions		$\mathbf{x}(0) \sim N(\hat{\mathbf{x}}_0, \mathbf{P}_0)$
Assumptions		$E[\mathbf{w}(t)\mathbf{v}_j^T] = 0$, for all k and all t
State estimate propagation	4.98	
Error covariance propagation	4.99	
State estimate update	4.101	
Error covariance update	4.102	
Kalman gain matrix	4.103	

real shift in the level of the process variables caused by nonstationary disturbances such as changes in the impurity level of the feedstock. Stochastic nonstationary disturbances force the process to drift away from deterministic model predictions. The presence of a disturbance state model in addition to white noise variables $\mathbf{w}_k$ and $\mathbf{v}_k$ will provide the necessary information for tracking the trajectories of state variables. A common practice for eliminating the offset caused by nonstationary disturbances (instead of using the disturbance state model) is to increase the Kalman filter gain $\mathbf{K}_k$ either directly or indirectly by augmenting the magnitude of the state noise covariance matrix $\mathbf{Q}_{k-1}$ (Refer to Eqs. (4.80) and (4.83)). This will reduce the bias, but will also increase the sensitivity of the Kalman filter to measurement noise, just like the effect of increasing the proportional gain of a feedback controller with only proportional action. The addition of the nonstationary disturbance model will have an effect similar to integral action in feedback controllers to eliminate the offset.

Since most Kalman filter applications for processes with nonstationary disturbances are for processes with external inputs $\mathbf{u}$ and involve processes that are typically nonlinear, the incorporation of the nonstationary disturbance model will be illustrated using an EKF for processes with external inputs. The nonlinear process is described by

$$\begin{aligned} \frac{d\mathbf{x}^d}{dt} &= \mathbf{f}^d(\mathbf{x}, \mathbf{u}, t) \qquad \mathbf{x}_0 = \mathbf{x}(t=0) \\ \mathbf{y}(t) &= \mathbf{h}(\mathbf{x}, \mathbf{u}, t) \end{aligned} \tag{4.105}$$

where $\mathbf{x}$ represents the complete vector of state variables, and $\mathbf{x}^d$ is the modeled deterministic subset of $\mathbf{x}$. The complete state vector $\mathbf{x}$ consists of $\mathbf{x}^d$ and stochastic state variables $\mathbf{x}^s$ which include model parameter and disturbance states that may vary with time in some stochastic manner and may be unknown initially [300, 356]. Performing local linearization and discretization of model equations

$$\mathbf{x}_k^d = \mathbf{F}_{k-1}^d \mathbf{x}_{k-1}^d + \mathbf{F}_{k-1}^s \mathbf{x}_{k-1}^s + \mathbf{G}_{k-1} \mathbf{u}_{k-1} + \mathbf{w}_{k-1}^d \tag{4.106}$$

where $\mathbf{x}_{k-1}^d$ is the zero-mean Gaussian white noise vector with covariance $\mathbf{Q}^d$. The true dynamics of stochastic states $\mathbf{x}^s$ are usually unknown and often they are assumed to follow a simple nonstationary random walk behavior [300, 356]

$$\mathbf{x}_k^s = \mathbf{x}_{k-1}^s + \mathbf{w}_{k-1}^s \tag{4.107}$$

where $\mathbf{w}_{k-1}^s$ is a white noise vector with covariance $\mathbf{Q}^s$. Usually $\mathbf{Q}^s$ is a diagonal matrix with elements representing the change of magnitude in stochastic states (disturbances) in one sampling interval. The elements of $\mathbf{Q}^s$ are tuning parameters to give good tracking behavior. The optimal one-step-ahead prediction of the stochastic states is

$$\hat{\mathbf{x}}^s(t_k|t_{k-1}) = \mathbf{x}_{k-1}^s(t_{k-1}|t_{k-1}) \; . \tag{4.108}$$

If more information is available about the stochastic states, an ARIMA model can be constructed to represent their dynamics.

The combined linearized dynamic/stochastic model of the system is

$$\mathbf{x}_k = \mathbf{F}_{k-1} \mathbf{x}_{k-1} + \mathbf{G}_{k-1} \mathbf{u}_{k-1} + \mathbf{w}_{k-1} \tag{4.109}$$

where

$$\mathbf{x}_k = \begin{bmatrix} \mathbf{x}_k^d \\ \mathbf{x}_k^s \end{bmatrix} \qquad \mathbf{w}_k = \begin{bmatrix} \mathbf{w}_k^d \\ \mathbf{w}_k^s \end{bmatrix} \tag{4.110}$$

$$\mathbf{F}_k = \begin{bmatrix} \mathbf{F}_k^d & \mathbf{F}_k^s \\ \mathbf{0} & \mathbf{I} \end{bmatrix} \qquad \mathbf{G}_k = \begin{bmatrix} \mathbf{G}_k^d \\ \mathbf{0} \end{bmatrix} \qquad \mathbf{Q}_k = \begin{bmatrix} \mathbf{Q}^d & \mathbf{0} \\ \mathbf{0} & \mathbf{Q}^s \end{bmatrix} . \tag{4.111}$$

The measurements are represented by Eq. (4.95), which can be modified if the inputs $\mathbf{u}$ directly affect the measurements. The Kalman filter, and the recursive relations to compute the filter gain matrix $\mathbf{K}_k$, and the covariance propagation matrices $\mathbf{P}_k^-$ and $\mathbf{P}_k^+$ are given by Eqs. (4.101), (4.103), (4.102), and (4.99) or (4.83), respectively.

A challenge in this approach is the selection of covariance matrices $\mathbf{Q}_k$, $\mathbf{R}_k$, and $\mathbf{P}_0$. Process knowledge and simulation studies must be used to find an acceptable set of these tuning parameters to prevent biased and poor estimates of state variables. Knowledge of the initial state $\mathbf{x}_0$ affects the accuracy of estimates as well. If $\mathbf{x}_0$ initially unknown, the EKF can be restarted from the beginning with each new measurement using the updated estimate $\hat{\mathbf{x}}_0|k$. Convergence to the unknown $\mathbf{x}_0$ is usually achieved during the early states of the estimation, but there is a substantial increase in computational load. If feedback control is used during the batch run, rapid convergence to initially unknown disturbance and parameter states can be achieved using this reiterative Kalman filter. One approach to implement the reiterative estimation of $\mathbf{x}_0$ is to combine a recursive nonlinear parameter estimation procedure with the EKF [300].

4.3.4 Batch Modeling with Local Model Systems

Most batch processes are nonlinear to some degree and may be represented better by nonlinear models. An alternative is to develop a series of local, preferably linear models to describe parts of the batch operation, then link these models to describe the whole batch operation. These models are referred to as local models, operating-regime based models [156, 260], or linear parameter-varying models [312].

Operating-regime Based Models.

Consider the general state space representation in Eqs. (4.44-4.45)

$$\begin{aligned} \dot{\mathbf{x}} &= \mathbf{f}(\mathbf{x}, \mathbf{u}, \mathbf{w}) \\ \mathbf{y} &= \mathbf{h}(\mathbf{x}, \mathbf{u}, \mathbf{v}) \end{aligned} \tag{4.112}$$

where $\mathbf{x}, \mathbf{u}, \mathbf{y}$ are the state, input (control) and output (measurement) vectors, $\mathbf{w}$ and $\mathbf{v}$ are disturbance and measurement noise vectors, and $\mathbf{f}$ and $\mathbf{g}$ are nonlinear function systems. Assume that the batch run can be partitioned into i operating regimes that can be represented sufficiently well by *local* model structures

$$\begin{aligned} \dot{\mathbf{x}} &= \mathbf{f}_i(\mathbf{x}, \mathbf{u}, \mathbf{w}, \boldsymbol{\theta}_i) \\ \mathbf{y} &= \mathbf{h}_i(\mathbf{x}, \mathbf{u}, \mathbf{v}, \boldsymbol{\theta}_i) \end{aligned} \tag{4.113}$$

parameterized with the vector $\boldsymbol{\theta}_i$. Each local model will be valid in its particular operating regime. Denote by ϕ_i the operating point (described by some $\mathbf{x}$, $\mathbf{u}$, $\mathbf{y}$) representing a specific regime Φ_i. The whole batch run (the full range of operation) is composed of N regimes: $\{\Phi_1, \cdots, \Phi_N\} = \boldsymbol{\Phi}$. The selection of variables to characterize an operating regime will be process dependent, containing a subset of state variables, inputs, and disturbances. Assume the existence of a smooth *model validity function* ρ_i that has a value close to 1 for operating points where the model i of Eq.(4.113) is a good description of the process, and close to 0 otherwise. Define an *interpolation function* ω_i with range [0,1] that is the normalization of ρ_i:

$$\omega_i(\phi) = \frac{\rho_i(\phi)}{\sum_{j=1}^{N} \rho_j(\phi)} \tag{4.114}$$

such that $\sum_{i=1}^{N} \omega_i(\phi) = 1$. To guarantee a global model, not all local model validity functions should vanish at any operating point ϕ.

The modeling framework consists of three tasks [156]:

- **Decompose the operating range of the process** into a number of operating regimes that completely cover the whole range of operation (complete batch run). This can be achieved based on process knowledge or by using computerized decomposition tools on the basis of an informative data sequence [156].
- **Develop a local model structure** using process knowledge and data. Assign local model validity functions.
- **Identify local model parameters.** The unknown parameter sets $\theta_1, \cdots, \theta_N$ are identified. If the models are linear, many model identification methods and tools that are readily available can be used. Attention must be paid during data collection to generate data that contain significant information for all operating regimes.

Linear Parameter Varying Models

The local models approach was extended such that a *linear parameter varying* (LPV) *model* was obtained by *interpolation* between multiple local models to emulate the behavior of a nonlinear batch process [312]. First consider the nonlinear process model Eq. 4.112 without disturbances.

$$\begin{aligned} \dot{\mathbf{x}} &= \mathbf{f}(\mathbf{x}, \mathbf{u}) \\ \mathbf{y} &= \mathbf{h}(\mathbf{x}, \mathbf{u}) \ . \end{aligned} \tag{4.115}$$

Batch processes are operated to track an optimal profile obtained by off-line optimization or empirical methods using historical information from previous batches. Local models are obtained by linearizing the model at different points in time on the optimal profile. Denote the optimal profile by $(\mathbf{x}_{0,i}, \mathbf{y}_{0,i})$, $i = 1, \cdots, N$ where N represents the number of operating points. Linearize Eq. 4.115 using Taylor series expansion and neglecting terms higher than first order. The resulting model is similar to Eqs. 4.61-4.62

$$\begin{aligned} \dot{\bar{\mathbf{x}}} &= \mathbf{A}_i\bar{\mathbf{x}} + \mathbf{B}_i\bar{\mathbf{u}} \\ \bar{\mathbf{y}} &= \mathbf{C}_i\bar{\mathbf{x}} + \mathbf{D}_i\bar{\mathbf{u}} \end{aligned} \tag{4.116}$$

where the deviation variables (Eq. 4.60) are defined using $\mathbf{x}_{ss} = \mathbf{x}_{0,i}$ and $\mathbf{y}_{ss} = \mathbf{y}_{0,i}$ and the elements of the Jacobian matrices are derived by evaluating the derivatives of $\mathbf{f}$ and $\mathbf{g}$ at the corresponding operating point $(\mathbf{x}_{0,i}, \mathbf{y}_{0,i})$ such that

$$\mathbf{A}_i = \left.\frac{\partial \mathbf{f}}{\partial \mathbf{x}}\right|_{\mathbf{x}_{0,i}, \mathbf{y}_{0,i}} . \tag{4.117}$$

A linear time-varying global model is constructed by using the local models to approximate the nonlinear dynamics of a batch run. The time-varying model is obtained by interpolating between local models by using model validity functions $p_i(t)$ which are similar to the interpolation function ω_i of Eq. 4.114. Model validity functions are the estimates of the validity of various local models in different operating points and for N local models

$$\mathbf{p}(t) = [p_1(t), p_2(t), \cdots, p_N(t)] \qquad \sum_{i=1}^{N} p_i(t) = 1 . \tag{4.118}$$

The state space matrices of the global model are then parametrized in terms of $p(t)$ as a LPV model:

$$\{\mathbf{A}[\mathbf{p}(t)], \mathbf{B}[\mathbf{p}(t)], \mathbf{C}[\mathbf{p}(t)], \mathbf{D}[\mathbf{p}(t)]\} . \tag{4.119}$$

The LPV model dynamics can be constructed in terms of model validity functions as

$$\begin{aligned} \dot{\bar{\mathbf{x}}} &= \mathbf{A}[\mathbf{p}(t)]\bar{\mathbf{x}} + \mathbf{B}[\mathbf{p}(t)]\bar{\mathbf{u}} \\ \bar{\mathbf{y}} &= \mathbf{C}[\mathbf{p}(t)]\bar{\mathbf{x}} + \mathbf{D}[\mathbf{p}(t)]\bar{\mathbf{u}} \end{aligned} \tag{4.120}$$

and the model Jacobians are of the form

$$\mathbf{A}[\mathbf{p}(t)] = \sum_{i=1}^{N} p_i(t)\mathbf{A}_i . \tag{4.121}$$

The LTV model can be extended for state estimation. An adaptive technique using the Bayesian framework has been developed [40]. Consider the discrete time version of the state space model Eq. (4.116) with zero-mean Gaussian process and measurement noises $\mathbf{w}$ and $\mathbf{v}$, respectively

$$\begin{aligned} \bar{\mathbf{x}}_{k+1} &= \mathbf{F}(p_i)\bar{\mathbf{x}}_k + \mathbf{G}(p_i)\bar{\mathbf{u}}_k + \mathbf{w}_k \\ \bar{\mathbf{y}}_k &= \mathbf{C}(p_i)\bar{\mathbf{x}}_k + \mathbf{D}(p_i)\bar{\mathbf{u}}_k + \mathbf{v}_k \end{aligned} \tag{4.122}$$

with noise covariance matrices $\mathbf{Q}_k$ and $\mathbf{R}_k$ for $\mathbf{w}$ and $\mathbf{v}$, respectively. A moving horizon estimator that updates the p_is based on a past window of data of length n_e can be developed. At the start of the batch, the number of data samples is fewer than n_e and the measurement data history is given by data of different length

$$\begin{aligned} \mathbf{Y}_k &= [\mathbf{y}_k, \mathbf{y}_{k-1}, \cdots, \mathbf{y}_0] & k &\leq n_e \\ \mathbf{Y}_k &= [\mathbf{y}_k, \mathbf{y}_{k-1}, \cdots, \mathbf{y}_{k-n_e}] & k &> n_e \end{aligned} \tag{4.123}$$

Let $p(j|\mathbf{Y}_k)$ represent the probability that model j is describing the batch process based on measurements collected until time k. Bayes theorem can be used to compute the posterior probability $p(j|\mathbf{Y}_k)$ given the *prior* probability (computation of $p(j)$ before measurements at time k are used) $p(j|\mathbf{Y}_{k-1})$:

$$\begin{aligned} p(j|\mathbf{Y}_k) &= p\left(j|\mathbf{y}_k, \mathbf{Y}_{k-1}\right) = \frac{f\left(\mathbf{y}_k|j, \mathbf{Y}_{k-1}\right) p(j|\mathbf{Y}_{k-1})}{p(\mathbf{y}_k)} \\ &= \frac{f\left(\mathbf{y}_k|j, \mathbf{Y}_{k-1}\right) p(j|\mathbf{Y}_{k-1})}{\sum_i^N f\left(\mathbf{y}_k|i, \mathbf{Y}_{k-1}\right) p(i|\mathbf{Y}_{k-1})} \end{aligned} \tag{4.124}$$

where $f\left(\mathbf{y}_k|j, \mathbf{Y}_{k-1}\right)$ is the probability distribution function (PDF) of the outputs at time k computed by using model j and the measurement history $\mathbf{Y}_{k-1}$ collected until time $k-1$. A Kalman filter is designed for each model in order to evaluate the PDFs. The jth Kalman filter assumes that the jth model matches the plant and any mismatch is caused by process and measurement noises with covariance matrices $\mathbf{Q}_j$ and $\mathbf{R}_j$. The update of state variables of the jth model is

$$\hat{\bar{\mathbf{x}}}_{j,k|k-1} = \mathbf{F}_j \hat{\bar{\mathbf{x}}}_{k-1|k-1} + \mathbf{G}_j \bar{\mathbf{u}}_{k-1} \tag{4.125}$$

where $\hat{\bar{\mathbf{x}}} = \sum_{j=1}^N p_{j,k} \hat{\bar{\mathbf{x}}}_{j,k|k}$. Following the derivations of Kalman filters in Section (4.3.3), measurements at time k provide a correction to the updates

$$\hat{\bar{\mathbf{x}}}_{j,k|k} = \hat{\bar{\mathbf{x}}}_{j,k|k-1} + \mathbf{K}_j \left(\bar{\mathbf{y}}_k - \mathbf{C}_{j,k} \hat{\bar{\mathbf{x}}}_{j,k|k-1} - \mathbf{D}_{j,k} \bar{\mathbf{u}}_k\right) \tag{4.126}$$

where $\mathbf{K}_j$ denotes the Kalman filter gain for the jth model. The outputs of the jth model are

$$\bar{\mathbf{y}}_{j,k} = \mathbf{C}_j \hat{\bar{\mathbf{x}}}_{j,k|k} + \mathbf{D}_j \bar{\mathbf{u}}_k \ . \tag{4.127}$$

If the assumptions about the accuracy of the jth model and noise characteristics hold, then the model residuals $\varepsilon_{j,k} = \bar{\mathbf{y}}_k - \bar{\mathbf{y}}_{j,k}$ will have zero mean and covariance $\mathbf{\Omega}_j = \mathbf{C}_j \mathbf{P}_j \mathbf{C}_j^T + \mathbf{R}_j$. Here $\mathbf{P}_j$ is the state covariance matrix from the jth Kalman filter. The PDF is computed using [312]

$$f\left(\mathbf{y}_k|j, \mathbf{Y}_{k-1}\right) = f(\mathbf{y}_k|j) = f(\varepsilon_{j,k}) = \frac{\exp(-\frac{1}{2}\varepsilon_{j,k}\mathbf{\Omega}_j^{-1}\varepsilon_{j,k}^T)}{\left((2\pi)^N \det(\mathbf{\Omega}_{j,k})\right)^{1/2}} \ . \tag{4.128}$$

Estimates of model probabilities are then obtained by substituting recursively Eq. (4.128) into Eq. (4.124) [312]. To reduce the time required for the computations to converge to the correct probability model, the probabilities are initialized as [312]:

$$\begin{aligned} p(j|\mathbf{Y}(0)) &= \begin{cases} p_1, & j=1 \\ \frac{1-p_1}{N-1} & j>1 \end{cases} & k &\leq n_e \\ p\left(j|\mathbf{Y}(k-n_e)\right) &= \frac{1}{N} & k &> n_e \end{aligned} \tag{4.129}$$

where $p_1 \geq (1-p_1)/(N-1)$ and N is the number of local models. The relative magnitude of p_1 with respect to p_i depends on the expected magnitude of disturbances, for large disturbances p_1 is closer to p_i [312].

Local model dynamics are affected by disturbances entering the batch process, and the PDFs of various local models may become identical. The proximity of model outputs to the optimal profiles may be used to select the best local model with a moving horizon Bayesian estimator (MHBE) with time-varying tuning parameters [312]. The aim of the MHBE is to assign greater credibility to a model when plant outputs are closer to the outputs around which the model is identified. This reduces the covariance of model residuals $\mathbf{\Omega}_{i,k}$ for model i at time k. This approach is implemented by reducing the noise covariance matrices $\mathbf{Q}_{i,k}$ and $\mathbf{R}_{i,k}$ which may be used as the tuning parameters for the respective Kalman filters. Relating these covariances to deviations from optimal output trajectories

$$\begin{aligned} \mathbf{Q}_{i,k} &= \mathbf{Q}\exp\left(\sigma \parallel \mathbf{y}_k - \mathbf{y}_{0,i} \parallel\right) \\ \mathbf{R}_{i,k} &= \mathbf{R}\exp\left(\sigma \parallel \mathbf{y}_k - \mathbf{y}_{0,i} \parallel\right) \end{aligned} \tag{4.130}$$

where $\mathbf{y}_{0,i}$ is the optimal output profile for local model i and σ is a tuning parameter, $\mathbf{Q}_{i,k}$ and $\mathbf{R}_{i,k}$ of the most likely local model is reduced. The Euclidian norm in Eqs. 4.130 is defined as $\parallel \mathbf{x} \parallel^2 = \mathbf{x}^T\mathbf{x}$. Consequently, the residual covariance matrix $\mathbf{\Omega}_{i,k}$ is reduced as well and the probability of model i is increased. The parameter σ reflects the trust in the model and at higher values promotes rapid transition between models [312]. Case studies reported in [312] indicate that model predictive control of a batch reactor using LVS models provided better control than model predictive control with extended Kalman filters.

Multiple ARX Models

A different way of approximating the nonlinear process behavior is based on the use of linear ARX models for each sampling instant. Define the input ($\mathbf{u}$), output ($\mathbf{y}$) and disturbance ($\mathbf{d}$) sequences over the entire batch run as

$$\begin{aligned}\mathbf{u} &= [\mathbf{u}^T(0)\ \ \mathbf{u}^T(1)\ \ \cdots\ \ \mathbf{u}^T(N-1)]^T \\ \mathbf{y} &= [\mathbf{y}^T(1)\ \ \mathbf{y}^T(2)\ \ \cdots\ \ \mathbf{y}^T(N)]^T \\ \mathbf{d} &= [\mathbf{d}^T(1)\ \ \mathbf{d}^T(1)\ \ \cdots\ \ \mathbf{d}^T(N)]^T\end{aligned} \tag{4.131}$$

where N is the batch run data length. Given the initial condition $\mathbf{y}_{\text{ini}} = \mathbf{y}(0)$, a nonlinear model relating the outputs to inputs and disturbances is expressed as

$$\mathbf{y} = \mathcal{M}(\mathbf{u}, \mathbf{y}_{\text{ini}}, \mathbf{d}) = \mathcal{N}(\mathbf{u}, \mathbf{d}) + \mathcal{N}_{\text{ini}}(\mathbf{y}_{\text{ini}}) \ . \tag{4.132}$$

The nonlinear model $\mathcal{N}(\mathbf{u}, \mathbf{d})$ can be approximated with one linear model for each sampling instant [70]. Denote a specific or optimal output trajectory by $\mathbf{y}_0$, the batch index by k and define the output error trajectory $\mathbf{e}^y = \mathbf{y}_0 - \mathbf{y}_k$. The state equations for the output error are

$$\begin{aligned}\breve{\mathbf{e}}^y_{k+1} &= \breve{\mathbf{e}}^y_k - \mathbf{G}^y \Delta \mathbf{u}_{k+1} - \mathbf{G}^y_{\text{ini}} \Delta \mathbf{y}_{\text{ini},k+1} + \mathbf{w}^y_k \\ \mathbf{e}^y_k &= \breve{\mathbf{e}}^y_k + \mathbf{v}^y_k\end{aligned} \tag{4.133}$$

where $\Delta\mathbf{u}_{k+1} = \mathbf{u}_{k+1} - \mathbf{u}_k$, $\Delta\mathbf{y}_{\text{ini},k+1} = \mathbf{y}_{\text{ini},k+1} - \mathbf{y}_{\text{ini},k}$, $\mathbf{w}^y_k$ and $\mathbf{v}^y_k$ are zero-mean, independently and identically distributed random noise sequences with respect to k, and $\breve{\mathbf{e}}^y_k$ is the noise-free (cannot be measured) part of the error trajectory. Matrices $\mathbf{G}^y$ and $\mathbf{G}^y_{\text{ini}}$ are linear system approximations. The same modeling approach can be applied to secondary outputs $\mathbf{s}$ (outputs that are not used in control systems) and quality variables $\mathbf{q}$ and the resulting models can be combined. Define the error trajectory vector for the controlled outputs, secondary outputs and quality variables as

$$\mathbf{e}_k = \begin{bmatrix} \mathbf{e}^y_k \\ \mathbf{e}^s_k \\ \mathbf{e}^q_k \end{bmatrix}, \qquad \mathbf{e}_{\text{ini},k} = \begin{bmatrix} \Delta\mathbf{y}_{\text{ini},k} \\ \Delta\mathbf{s}_{\text{ini},k} \end{bmatrix} . \tag{4.134}$$

The resulting combined model then is [70]

$$\begin{aligned}\breve{\mathbf{e}}^{k+1} &= \breve{\mathbf{e}}_k - \mathbf{G}\Delta\mathbf{u}_{k+1} - \mathbf{G}_{\text{ini}}\mathbf{e}_{\text{ini},k+1} + \mathbf{w}_k \\ \mathbf{e}_k &= \breve{\mathbf{e}}_k + \mathbf{v}_k \ .\end{aligned} \tag{4.135}$$

To enable the use of a Kalman filter for estimation of initial conditions $\mathbf{e}_{\text{ini}}$, the initial condition can be modeled with a batchwise random walk model [70]:

$$\begin{aligned}\mathbf{e}_{\text{ini},k+1} &= \mathbf{e}_{\text{ini},k} + \boldsymbol{\nu}_{\text{ini},k} \\ \mathbf{z}_{\text{ini},k} &= \mathbf{e}_{\text{ini},k} + \boldsymbol{\xi}_{\text{ini},k}\end{aligned} \tag{4.136}$$

where $\boldsymbol{\nu}_{\mathrm{ini},k}$ is a batchwise white disturbance, $\mathbf{z}_{\mathrm{ini},k}$ is the measured value of $\mathbf{e}_{\mathrm{ini},k}$, and $\boldsymbol{\xi}_{\mathrm{ini},k}$ is a batchwise white measurement noise.

The general form of the model has a very large dimension. Structural information is used in [70, 200] to reduce the dimensions of the model.

4.4 Functional Data Analysis

Functional data are data generated by an inherent functional relationship in a process. The relationship may not be known explicitly, but its existence is assumed based on the knowledge about the process. Variations of daily weather temperature over the year, height and weight growth of a child over the years, or trajectories of process variables during a batch are functional data. The goals of functional data analysis (FDA) are to represent the data in ways that facilitate further analysis, determine patterns in data and study important sources of pattern in data, explain variations in output variables in terms of input variations, and conduct comparative studies between sets of data [495]. Hence, modeling, analysis, and diagnosis activities are conducted in a framework that is different from "analysis of large data sets" approach. Indeed, a functional observation is considered as a single datum rather than a sequence of individual observations. The focus is on the trajectory of a process variable during the batch rather than the several hundred measured values for the variable.

The FDA approach detects and removes characteristics in data by applying a linear operator that consists of weighted sums of various orders of derivatives rather than subtracting the assumed characteristics from the original data. The derivative terms in the linear operator provides physical insight such as acceleration in production of a certain biological species for a specific time period during the batch. The differential equation representation of a dynamic process is well-accepted in physical sciences and engineering, and the interpretation of these differential equations provides significant information about the characteristics of the process. This approach can also be used for nonlinear trajectories by finding the mean trajectories and centering the data with respect to the mean before implementing the principal differential analysis (PDA) method in order to eliminate most of the nonlinearity in data.

FDA starts by converting raw functional data (measurements during the batch) to a functional representation. This usually involves data smoothing since most data include measurement noise, estimating the derivatives of various orders for the data trajectories, and development of functional relations that include these derivatives. The K functional observations represented by the raw data vector as $\mathbf{x} = (x_1, x_2, \cdots, x_k, \cdots, x_K)^T$ are used to define a much smaller set of m functions that are efficient approximations of these data. Data smoothing provides the ability of possessing a certain number of derivatives for the latent function, which may not be obvious in the raw data vector. Denoting the latent function at time t_k as $z(t_k)$, $x_k = z(t_k) + \epsilon_k$ where ϵ_k is the measurement error that contributes to the roughness of the data. Derivatives should not be estimated by computing differences because of the measurement error. Differencing magnifies these errors.

The *Principal Differential Analysis* (PDA) method identifies the linear differential operator L

$$L = w_0 I + w_1 D + \cdots + w_{m-1} D^{m-1} + D^m \tag{4.137}$$

that comes as close as possible to satisfying the homogeneous linear differential equation Lx_k for each observation x_k [493]. The methodology outlined and the nomenclature used follows Ramsay and Silverman [495] where a detailed treatment of the topic is provided. The differential equation model

$$D^m x_k = -w_0 x_k - w_1 D x_k + \cdots - w_{m-1} D^{m-1} x_k \tag{4.138}$$

that satisfies the data as closely as possible is sought. Since the operator is expected to annihilate the data functions x_k as nearly as possible, Lx_k can be regarded as the residual error from the fit provided by L. A least squares approach can be used to fit the differential equation model by using the minimization of sum of squared errors (SSE) criterion.

$$SSE(L) = \sum_{k=1}^{K} \int [Lx_k(t)]^2 dt \ . \tag{4.139}$$

Here, $SSE(L)$ is minimized to determine the m weight functions in Eq. 4.137, viz., $\mathbf{w} = (w_0, w_1, \cdots, w_{m-1})$. Once the operator L is determined by estimating its $\mathbf{w}$, a set of m linearly independent basis functions ξ_j that satisfy $L\xi_j = 0$ and form the null space of L can be computed. The weights $\mathbf{w}$ can be determined by *pointwise* minimization of the sum of squared errors (SSE_P) criterion:

$$SSE_P(t) = \frac{1}{K}\sum_{k=1}^{K}(Lx_k)^2(t) = \frac{1}{K}\sum_{k=1}^{K}\Big[\sum_{j=0}^{m} w_j(t)(D^j x_k)(t)\Big]^2 \tag{4.140}$$

with $w_m(t) = 1$ for all t. The representation $(Lx_k)^2(t)$ is used to underline the pointwise nature of time-variant data [495]. Defining the $K \times m$ regressor matrix $\mathbf{\Gamma}(t)$ and the K-dimensional dependent variable vector $\lambda(t)$ as

$$\mathbf{\Gamma}(t) = [(D^j x_k)(t)]_{k=1,K;j=0,m-1} \quad \text{and} \quad \lambda(t) = [-(D^m x_k)(t)]_{k=1,K} \tag{4.141}$$

the least squares solution of Eq. (4.140) gives the weights $w_j(t)$

$$\mathbf{w} = [\mathbf{\Gamma}(t)^T \mathbf{\Gamma}(t)]^{-1} \mathbf{\Gamma}^T \lambda(t). \tag{4.142}$$

Weights $\mathbf{w}$ must be available at a fine level of detail for computing the basis functions ξ_j. The resolution of $\mathbf{w}$ depends on the smoothness of the derivatives $D^j x_k$. Pointwise computation of $\mathbf{w}$ for larger orders of m is computationally intensive. Furthermore, $\mathbf{\Gamma}(t)$ may not be of full rank. One way to circumvent these problems is to approximate $\mathbf{w}$ by a fixed set of basis functions $\phi = \phi_l, l = 1, \cdots, L$. Standard basis function families such as polynomials, Fourier series, B-spline functions or wavelets could be used. The weights $\mathbf{w}$ can be approximated as

$$w_j \approx \sum_l c_{jl} \phi_l \tag{4.143}$$

where the mL coefficients $\mathbf{c} = [c_{jl}]_{j=1,m;l=1,L}$ are stored as a column vector. The estimates $\hat{\mathbf{c}}$ are the solution of $\mathbf{Rc} = -\mathbf{s}$ resulting from the minimization of the quadratic form

$$C + \mathbf{c}^T \mathbf{R} \mathbf{c} + 2\mathbf{c}^T \mathbf{s} \tag{4.144}$$

where C is a constant independent of $\mathbf{c}$, $\mathbf{R} = [R_{ij}]_{i=0,m-1;j=0,m-1}$ and $\mathbf{s} = [s_j]_{j=0,m-1}$ with

$$R_{ij} = \frac{1}{K} \int \phi(t)\phi(t)^T \sum_{k=1}^{K} D^i x_k(t) D^j x_k(t) dt \tag{4.145}$$

$$s_j = \frac{1}{K} \int \phi(t) \sum_{k=1}^{K} D^i x_k(t) D^m x_k(t) dt \ . \tag{4.146}$$

The integrals are evaluated numerically by using traditional tools such as the trapezoidal rule.

An alternative computation of $\mathbf{w}$ can be made by attaching a penalty term to Eq. (4.139):

$$PSSE(L) = \frac{1}{K} \sum_{k=1}^{K} (Lx_k)^2(t) + \sum_{j=0}^{m-1} \theta_j \int w_j(t)^2 dt \tag{4.147}$$

where θ_j controls the roughness of the estimated weight functions. The solution can be found by using the pointwise approach which results in

$$\hat{\mathbf{w}} = [\mathbf{\Gamma}(t)^T \mathbf{\Gamma}(t) + K\mathbf{\Theta}]^{-1} \mathbf{\Gamma}^T \boldsymbol{\lambda}(t) \tag{4.148}$$

where $\mathbf{\Theta} = diag(\theta_0, \cdots, \theta_{m-1})$. Alternatively, the basis function approach can be used to compute $\mathbf{w}$.

The FDA framework can be used to model the trajectories of the process variables of a batch process. This model can be used to generate the estimates of the future portion of the trajectories for implementing SPM during the progress of the batch. The FDA based prediction provides remarkable improvement in trajectory estimation illustrated in Figures 4.7 and 4.8. In this illustration, data are generated using the simulator based on the unstructured nonlinear multivariable model of penicillin fermentation (Section 2.7.1) and data-based models are developed using multiway PCA (MPCA) and FDA frameworks. The trajectories to the end of the batch are estimated based on these models and the "data collected" (Solid curves in Figures 4.7 and 4.8) up to the present time in the current batch.

Two cases are generated for penicillin concentration profile to illustrate and compare estimation methods. Curves labelled 1-3 are based on the estimation methods described in Section 6.5.1, curve 4 is based on the PDA model and curve 5 is the PDA based estimation with EWMA-type local weights on "measured" and "estimated" data. In the first case, data generated under normal operation are used. Estimation performed starting at 150 h onward to the end of the batch run resulted in comparatively close results for all methods (Figure 4.7(a)). Mean-centered profiles are also given in Figure 4.7(b) to provide a magnified look at the predictions. Although the predictions are close in this case, PDA with EWMA-type local weightings produced the best result (Curve 5). The problem can also be cast into a framework of Kalman filter-type correction of the predicted values. The results are identical when the EWMA weight and Kalman filter gain are matched. A drift disturbance in the substrate feed rate from the start of fed-batch operation until the end of the batch is generated as the second case. Estimation is started at 180 h onward to the end of the batch (Figure 4.8). The best estimates are given by PDA with local weighting of data (Curve 5).

The FDA approach provides a framework to develop methods for data pretreatment, adjustment of data length of different batches, detection of landmarks for the beginning and ending of various stages during the batch, PCA, and estimators for final product properties. Landmark detection and data synchronization using FDA are discussed in Section 6.3.3. Furthermore, the differential equations generated can be converted to a state-space

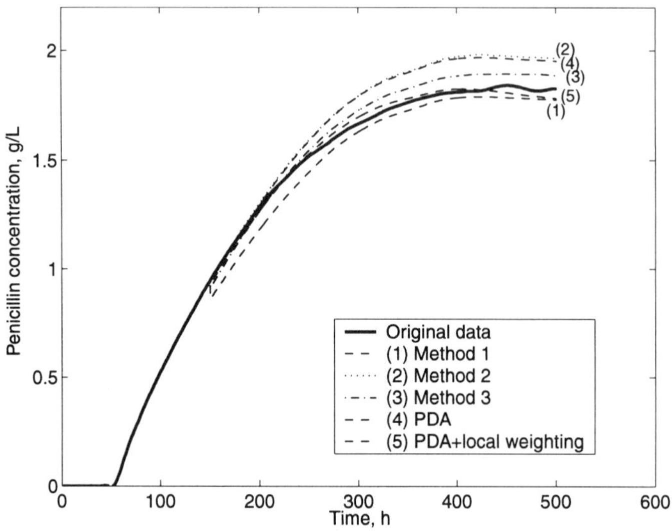

(a) Raw profiles

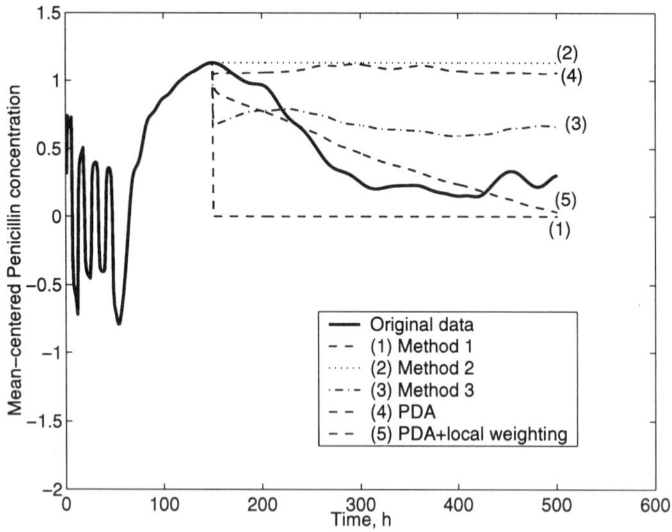

(b) Mean-centered profiles

Figure 4.7. Estimates of the penicillin concentration trajectory under normal operation.

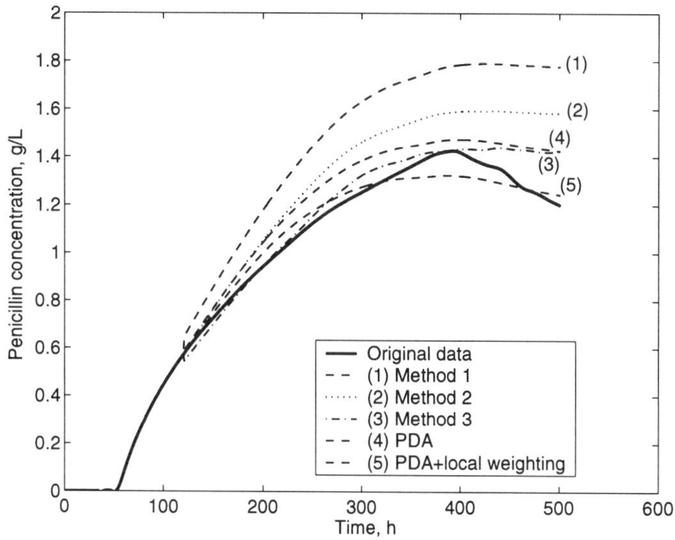

(a) Raw profiles

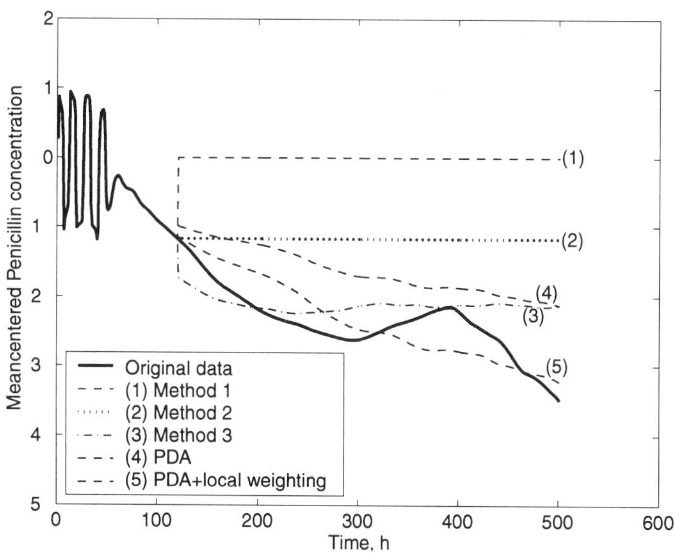

(b) Mean-centered profiles

Figure 4.8. Estimates of the penicillin concentration trajectory under disturbance.

representation, enabling the representation of the process model in a form that can be used efficiently in process control.

4.5 Multivariate Statistical Paradigms for Batch Process Modeling

Data collected from batch or fed-batch processes have a three-dimensional structure. This array is different than the two-dimensional structure (variables $\times$ time) resulting from continuous process data. The dimensions in batch data are batch runs, variables and time (Figure 4.9). Data are arranged into a three-dimensional array $(I \times J \times K)$ where I is the number of batches, J is the number of variables and the K is the number of sampling times in a given batch. Consequently, the PCA and PLS based methods discussed in Sections 4.1 and 4.2.4 must be modified to handle three-dimensional data.

4.5.1 Multiway Principal Component Analysis–MPCA

MPCA is based on PCA [661]. It is equivalent to performing ordinary PCA on a large two-dimensional matrix constructed by unfolding the three-way array. The use of MPCA for batch process monitoring was proposed in mid 1990s [433] and applied to monitor a polymerization reactor. It has been extended to nonlinear PCA [130] and wavelet decomposition based multiscale PCA techniques for analysis of batch process data [39].

Batch process data are arranged into a three-dimensional array $\underline{\mathbf{X}}$ The underbar is used to denote a three dimensional matrix. MPCA decomposes the $\underline{\mathbf{X}}$ array into a summation of the product of score vectors $\mathbf{t}_a$ and loading matrices $\mathbf{P}_a$, plus a residuals array $\underline{\mathbf{E}}$ that is minimized in a least-squares sense, as

$$\underline{\mathbf{X}} = \sum_{a=1}^{A} \mathbf{t}_a \otimes \mathbf{P}_a + \underline{\mathbf{E}} \tag{4.149}$$

where $\otimes$ is the Kronecker product ($\underline{\mathbf{X}} = \mathbf{t} \otimes \mathbf{P}$ is $\underline{X}(i,j,k) = t(i)P(j,k)$) and A is the number of principal components (PC) retained [661].

This three-way array is unfolded and scaled properly prior to MPCA (Figure 4.10). Unfolding of three-way array $\underline{\mathbf{X}}$ can be performed in six possible ways. For instance, $\underline{\mathbf{X}}$ can be unfolded to put each of its vertical slices $(I \times J)$ side by side to the right, starting with the slice corresponding to the first time interval. The resulting two-dimensional matrix has dimensions $(I \times JK)$. This particular unfolding allows one to analyze variability among the batches in $\underline{\mathbf{X}}$ by summarizing information in the data with respect to

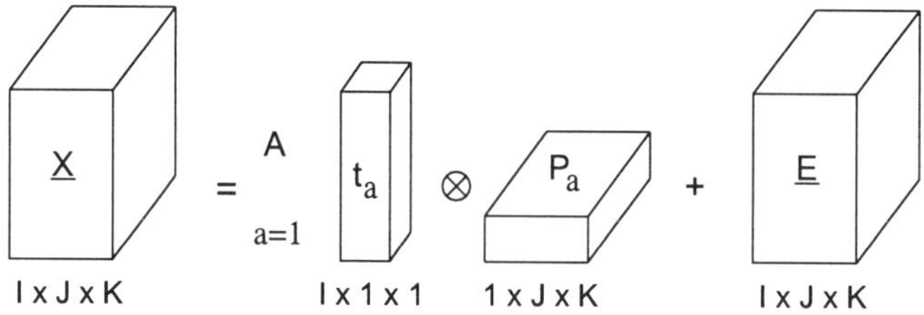

Figure 4.9. Batch process data arrangement and decomposition in three-way array [433].

variables and their time variation. A mathematically equivalent unfolding would be to take slices off the side of $\underline{\mathbf{X}}$ and place them down the time axis, which also forms a $(I \times JK)$ dimensional matrix. The latter unfolding orders the matrix with the history of each variable kept together while the former orders the matrix with all the measurements taken at the same time kept together. After mean-centering and scaling the unfolded data matrix, PCA is applied. Each of the $\mathbf{p}$, however, is really an unfolded version of the loadings matrix $\mathbf{P}_a$. After vectors $\mathbf{p}$ are obtained, $\mathbf{P}_a$ can be obtained by reversing the unfolding procedure. Similarly, the three-way array $\underline{\mathbf{E}}$ can be formed by folding the PCA residual matrix $\mathbf{E}$. For the unfolded $\mathbf{X}$:

$$\mathbf{X} = \sum_{a=1}^{A} \mathbf{t}_a \mathbf{p}_a^T + \mathbf{E} = \hat{\mathbf{X}} + \mathbf{E} \tag{4.150}$$

MPCA explains variation of measured variables about their average trajectories. Subtracting the average trajectory from each variable (accomplished by mean centering the columns of the unfolded matrix $\mathbf{X}$) removes most of the nonlinear behavior of the process (*see* Figure 4.11). Batch process models, developed based on historical data of batch runs yielding good products, using MPCA provide the foundation to develop statistical process monitoring and quality control systems in Section 6.4.

4.5.2 Multiway Partial Least Squares–MPLS

Traditional SPC methods in batch processes are usually limited to end-product quality measurements [598, 614] or to a single variable measured throughout the batch. Most batch processes operate in open loop with respect to product quality variables due to lack of on-line sensors for tracking these variables. Upon completion of the batch, off-line quality measurements are usually made in the laboratory. MPCA makes use of process

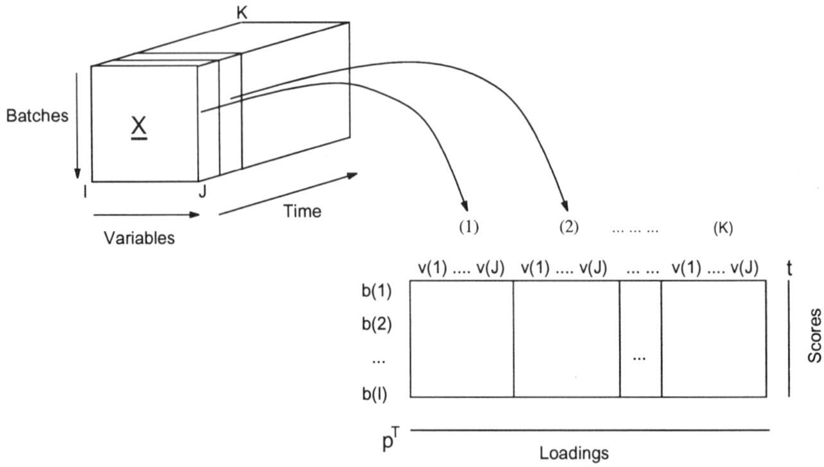

Figure 4.10. Batch data representation and unfolding process. The rows are batches and columns are the variables, v_j, sampled at each time τ_k [433].

variable trajectory measurements ($\mathbf{X}$) taken throughout the duration of the batch to help classify a batch as 'good' or 'bad'. MPLS [661] is an extension of PLS that is performed using both process data ($\mathbf{X}$) and the product quality data ($\mathbf{Y}$) to predict final product quality during the batch [434]. The unfolding process, mean-centering and scaling issues apply to MPLS technique as well (Section 4.5.1). There is also a $\mathbf{Y}$ matrix of quality variables in addition to three-way data matrix, as shown in Figure 4.10. After unfolding this three-way array into two dimensions, the algorithm explained for PLS in Section 4.2.4 is applied to this unfolded three-way array [298, 434].

For batch data, MPLS decomposes the $\mathbf{X}(I \times JK)$ and $\mathbf{Y}(I \times M)$ matrices into a summation of A score vectors $[\mathbf{t}(I \times 1), \mathbf{u}(I \times 1)]$, loading vectors $[\mathbf{p}(JK \times 1), \mathbf{q}(M \times 1)]$, weights $\mathbf{w}(JK \times 1)$ and model residual matrices $\mathbf{E}(I \times JK)$, $\mathbf{F}(I \times M)$. $\mathbf{t}$, $\mathbf{u}$, $\mathbf{p}$, $\mathbf{q}$ and $\mathbf{w}$ can be combined into $\mathbf{T}(I \times A)$, $\mathbf{U}(I \times A)$, $\mathbf{P}(JK \times A)$, $\mathbf{Q}(M \times A)$ and $\mathbf{W}(JK \times A)$ matrices to build matrix form of equations Eq. 4.19 and 4.20 in Section 4.2.4. A denotes the number of latent variables included in the MPLS model.

$$\mathbf{X} = \mathbf{T}\mathbf{P}^T + \mathbf{E}, \quad \mathbf{Y} = \mathbf{T}\mathbf{Q}^T + \mathbf{F} \tag{4.151}$$

where $\mathbf{T}$ is given by

$$\mathbf{T} = \mathbf{X}\mathbf{W}(\mathbf{P}^T\mathbf{W})^{-1}. \tag{4.152}$$

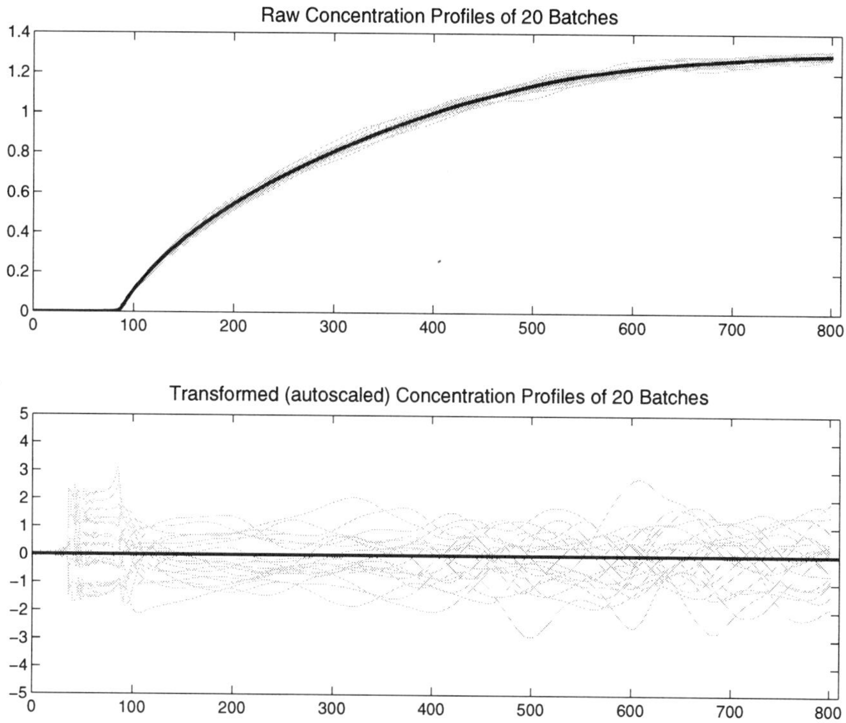

Figure 4.11. Autoscaling of a set of reference trajectories about the mean trajectory (darker line).

This decomposition summarizes and compresses the data with respect to both **X** and **Y** variables and time into low dimensional spaces that describe the process operation which is most relevant to final product quality. **T** matrix carries information about the overall variability among the batches. The **P** and **W** convey the time variation of measured variables about their mean trajectories and weights applied to each variable at each time instant within a batch giving the scores for that batch. **U** represents the inner relationship between **X** and **Y**, and summarizes the **Y** variables (quality variables) with some information coming from the **X** block (process variables). **Q** relates the variability of process measurements to final product quality [243, 434, 661].

MPLS will detect unusual operation based on large scores and classify a batch as 'good' or 'bad' as MPCA does and in addition, it will indicate if the final product qualities are not well predicted by process measurements when the residuals in the **Y** space $[SPE_Y = \sum_{c=i}^{M} \mathbf{F}(i,c)^2]$ are large. The **W**, **P**, and **Q** matrices of MPLS model bear all the structural information about how the process variables behaved and how they are related to the final quality variables. Implementation of this technique is discussed in Section 6.4.4.

4.5.3 Multiblock PLS and PCA Methods for Modeling Complex Processes

Multiblock data analysis has its origins in path analysis and path modeling in sociology and econometrics. In situations where the number of variables is very large or the process that is analyzed is large and consists of many different stages, it is logical to group variables in a meaningful way, either based on their similarity, or their origin in the system or process, and then summarize each group that is called *block*. Each block may be divided into sub-blocks according to process phases and stages (several **X**-blocks of process variables and/or **Y** blocks of quality variables). If the focus is on the process measurements space, several MPCA models can be developed out of sub-blocks, however in regression models separate projections of each block can be put together as a block and the resulting block scores are then treated as predictor and response variables on the "super level (or upper level)" of the model. The resulting models are called *hierarchical* projection models.

A version of multiblock PCA (MBPCA) called as "consensus PCA" (CPCA) was introduced by Wold et al. [662] as a method for comparing several blocks of descriptor variables (process variables) measured on the same objects (batches). A consensus direction is sought among all the blocks. One of the classical applications of CPCA is the testing of food and beverages especially wines by a number of judges (or samplers). Each judge (b is an index for judges) tastes each of the N samples and gives his/her opinion in terms of K_b variables such as sweetness, color, tannic taste, etc. A consensus matrix **T** (super score) will then contain the overall opinion of the judges about the same object while the super weight showing the relative importance of each judge in the consensus score (Figure 4.12). Wold et al. [665] also suggested a slightly different multiblock PCA algorithm called "hierarchical PCA" (HPCA). The only difference is the normalization step where in HPCA, $\mathbf{t}_b$ and $\mathbf{t}_T$ are normalized instead of $\mathbf{w}_T$ and $\mathbf{p}_b$ in CPCA, and the super weight only shows if the direction of the super score is present in the block in HPCA. In both algorithms the super score will show the direction most dominant in the consensus block **T**. However, because the block scores are normalized in HPCA, it will search the most dominant direction in these normalized scores. In CPCA, block scores will be combined in $\mathbf{T}$ as they are calculated for each block and hence the super score will just be the direction most dominant in the block scores. This difference between the two methods intensifies as one direction becomes stronger in only a single block [665].

To prevent the domination of one block due to large variance with respect to other blocks, an initial block scaling is performed by modifying autoscaling according to a function the number of variables m contained in each block (see Eq. 4.153). Typically this function is chosen to be between the square root and the fourth root of m, giving each block the total weight of between one and $\sqrt{m}$ [640, 665].

$$\mathbf{X} = [\ \mathbf{X}_1/\sqrt{m_{X1}}, \ldots, \mathbf{X}_b/\sqrt{m_{Xb}}\ ]\ . \tag{4.153}$$

Additional scaling factors can also be introduced to some particular $\mathbf{X}$ and/or $\mathbf{Y}$ blocks in hierarchical models with many blocks to scale up or down the importance of those blocks. Since larger blocks have usually a greater importance than the smaller ones, a mild weighting according to size can be assigned as [665]

$$d_b = 1 + 0.5\ \log_{10} K_b. \tag{4.154}$$

A convergence problem in these original algorithms is reported and resolved by Westerhuis et al. [640]. An adaptive version of HPCA for monitoring of batch processes has been reported by Rännar et al. [496]. The details of this advantageous technique is discussed along with case studies in Section 6.5.2.

The application of *multiway* MBPCA is also suggested for batch process data. In multiway MBPCA, blocking of the variables is done as explained above. Kosanovich et al. [291] have grouped the data from a batch polymerization reactor based on operational stages while Undey et al. [604, 605] have extended the same approach by dividing one process unit into two operational phases and included additional data for analysis from a second process unit for the case of multistage pharmaceutical wet granulation. A detailed example is given in Section 6.4.5 for this case. A nonlinear version of multiway MBPCA based on artificial neural networks is also suggested with some improvement on the sensitivity of the monitoring charts in the literature [129, 130].

When development of regression (projection based) models is aimed between multiple $\mathbf{X}$ and $\mathbf{Y}$ blocks, hierarchical PLS (HPLS)or MBPLS methods can be used. HPLS is an extension of the CPCA method. After a CPCA cycle on multiple $\mathbf{X}$ blocks, a PLS cycle is performed with the super block $\mathbf{T}$ and $\mathbf{Y}$ [640, 662, 665]. An application of HPLS was given by

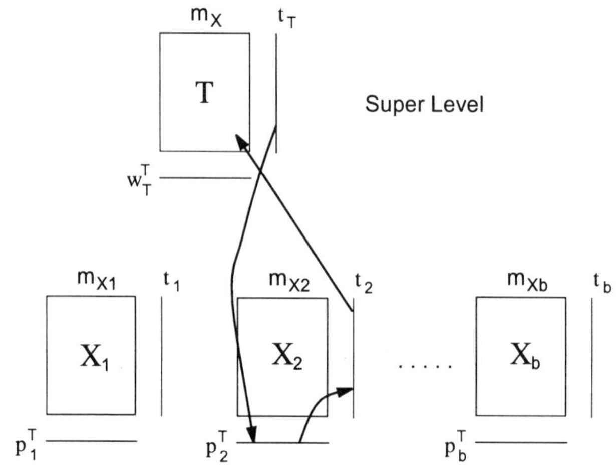

Figure 4.12. CPCA and HPCA methods [640, 665]. **X** data matrix is divided into b blocks ($\mathbf{X}_1$, $\mathbf{X}_2$,..., $\mathbf{X}_b$) with block b having m_{Xb} variables.

Wold et al. [665] the modeling process data (hundreds of variables) from a catalytic cracker.

Another PLS algorithm called multiblock PLS (MBPLS) has been introduced to deal with data blocks [629, 640, 666]. The algorithm could handle many types of pathway relationships between the blocks. It is logically specified from left to right. Left end blocks are defined as blocks that predict only while right end blocks are blocks that are predicted but do not predict. Interior blocks both predict and are predicted. The main difference between this method and HPLS is that in MBPLS, each **X** block is used in a PLS cycle with **Y** block to calculate the block scores $\mathbf{t}_b$, while in HPLS $\mathbf{t}_b$ is calculated as in CPCA. The basic methodology is illustrated in Figure 4.14 where there is a single **Y** block and two **X** blocks and the algorithm is given as

1. Start by selecting one column of **Y**, $\mathbf{y}_j$, as the starting estimate for **u**.

2. Perform part of a PLS round on each of the blocks $\mathbf{X}_1$ and $\mathbf{X}_2$ to get ($\mathbf{w}_1$,$\mathbf{t}_1$) and ($\mathbf{w}_2$,$\mathbf{t}_2$) as in Eq. 4.21 stated in PLS algorithm in Section 4.2.4.

3. Collect all the score vectors $\mathbf{t}_1$, $\mathbf{t}_2$ in the consensus matrix **T** (or composite block).

4. Make one round of PLS with $\mathbf{T}$ as $\mathbf{X}$ (Eqs. 4.21-4.23) to get a loading vector $\mathbf{v}$ and a score vector $\mathbf{t}_c$ for $\mathbf{T}$ matrix, as well as a loading vector $\mathbf{q}$ and a new score vector $\mathbf{u}$ for the $\mathbf{Y}$ matrix.

5. Return to step 2 and iterate until convergence of $\mathbf{u}$.

6. Compute the loadings $\mathbf{p}_1 = \mathbf{X}_1^T\mathbf{t}_1/\mathbf{t}_1^T\mathbf{t}_1$ and $\mathbf{p}_2 = \mathbf{X}_2^T\mathbf{t}_2/\mathbf{t}_2^T\mathbf{t}_2$ for the $\mathbf{X}_1$ and $\mathbf{X}_2$ matrices.

7. Compute the residual matrices $\mathbf{E}_1 = \mathbf{X}_1 - \mathbf{t}_1\mathbf{p}_1^T, \mathbf{E}_2 = \mathbf{X}_2 - \mathbf{t}_2\mathbf{p}_2^T, \mathbf{F} = \mathbf{Y} - \mathbf{t}_c\mathbf{q}^T$.

8. Calculate the next set of latent vectors by replacing $\mathbf{X}_1, \mathbf{X}_2$ and $\mathbf{Y}$ by their residual matrices $\mathbf{E}_1, \mathbf{E}_2$, and $\mathbf{F}$, and repeating from step 1.

This algorithm has been applied to monitoring a polymerization reactor [355] where process data are divided into blocks of data, with each block representing a section of the reactor. An increased sensitivity in multivariate charts is reported for this reactor. The reason behind sensitivity improvement is that these charts for individual blocks are assessing the magnitude of the deviations relative to normal operating conditions in that part of the process only, and not with respect to variations in all variables of the process.

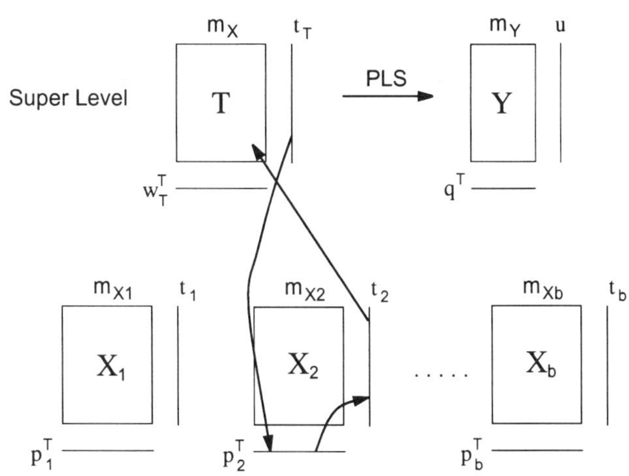

Figure 4.13. HPLS method [640, 665]. $\mathbf{X}$ data matrix is divided into b blocks ($\mathbf{X}_1$, $\mathbf{X}_2$,..., $\mathbf{X}_b$) with block b having m_{Xb} variables while only one $\mathbf{Y}$ block containing m_Y variables is present.

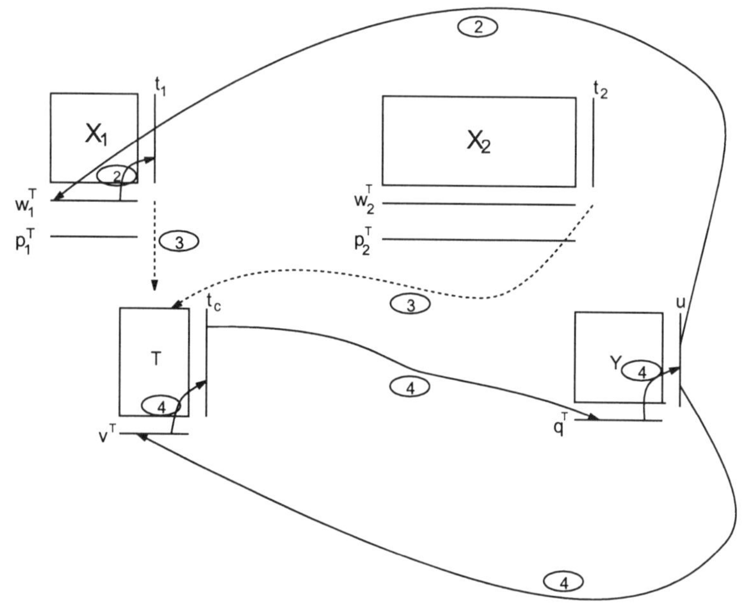

Figure 4.14. Multiblock PLS algorithm [355, 629].

Another application of the same algorithm is also reported for the case of wet granulation and tableting [638]. An improvement (with respect to ordinary PLS method) in prediction of a number of pharmaceutical tablet properties was reported.

When extra information is available (such as feed conditions, initial conditions, raw material qualities, etc.), this information should be incorporated in the *multiway* MBPLS framework. A general block interaction can be depicted for a typical batch process as shown in Figure 4.15. In this typical multiblock multiway regression case (multiblock MPLS), the blocks are the matrix containing a set of initial conditions used for each batch, $\mathbf{Z}(I \times N)$, the three-way array of measurements made on each variable at each batch, $\underline{\mathbf{X}}(I \times J \times K)$, and $\mathbf{Y}(I \times M)$ containing quality measurements made on batches. Kourti et al. have presented an implementation of this MBPLS technique [297] by dividing process data into two blocks based on different polymerization phases and also incorporating a matrix of initial conditions. An improvement in the interpretation of multivariate charts and fault detection sensitivity on individual phases are reported. The ability to relate the faults detected to initial conditions was another benefit of multiblock modeling that included relations between initial conditions and final product quality.

4.5.4 Multivariate Covariates Regression

The *principal covariates* regression proposed by de Jong and Kiers [119] can also be used to develop predictive models [554]. In this method, the relationship between the predictor block $\mathbf{X}$ and the dependent variable block $\mathbf{Y}$ is expressed by finding components scores $\mathbf{T}$ where the A column vectors $\mathbf{t}_i$, $i = 1, \ldots, A$. span the low-dimensional subspace of $\mathbf{X}$ that accounts for the maximum amount of variation in both $\mathbf{X}$ and $\mathbf{Y}$.

$$\mathbf{T} = \mathbf{XW} \qquad \mathbf{X} = \mathbf{TP}_X + \mathbf{E}_X \qquad \mathbf{Y} = \mathbf{TP}_Y + \mathbf{E}_Y \tag{4.155}$$

where $\mathbf{W}$ is a $p \times A$ matrix of component weights, $\mathbf{E}_X$ and $\mathbf{E}_Y$ contain the unique factors of $\mathbf{X}$ and $\mathbf{Y}$, respectively, and the loading matrices $\mathbf{P}_X$ $(A \times p)$ and $\mathbf{P}_Y$ $(A \times m)$ contain the regression parameters relating the variables in $\mathbf{X}$ and $\mathbf{Y}$, respectively [119]. The model is fitted to the data in the least-squares sense by maximizing the weighted average of R^2_{XT}, the percentage of variance in $\mathbf{X}$ accounted for by $\mathbf{T}$ and R^2_{YT}, the percentage of

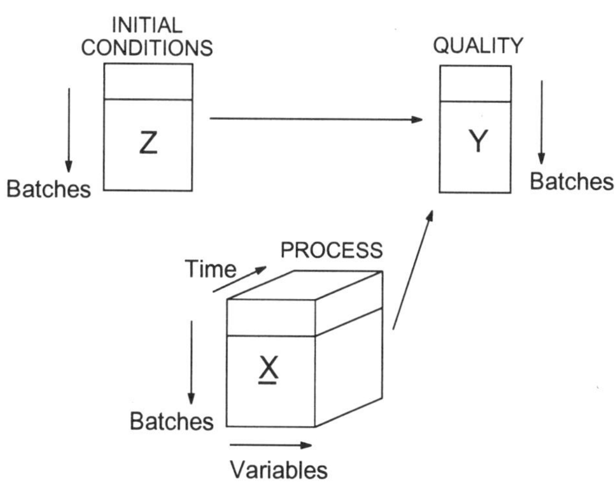

Figure 4.15. Multiway multiblock regression problem [297, 556].

variance in $\mathbf{Y}$ accounted for by $\mathbf{T}$: $\alpha \cdot R_{XT}^2 + (1-\alpha) \cdot R_{YT}^2$, with $0 < \alpha < 1$. The least-squares loss function may be written in terms of the residuals as a minimization problem:

$$\sigma(\mathbf{W}, \mathbf{P}_X, \mathbf{P}_Y) = \alpha \|\mathbf{X} - \mathbf{XWP}_X\|^2 + (1-\alpha)\|\mathbf{Y} - \mathbf{XWP}_Y\|^2 \quad (4.156)$$

with the Frobenius matrix norm $\|\cdot\|$ and constraint $\mathbf{T}^T\mathbf{T} = \mathbf{W}^T\mathbf{X}^T\mathbf{XW} = \mathbf{I}_A$, where $\mathbf{I}_A$ is an $A \times A$ identity matrix. The method can be extended to multi-block data. While de Jong and Kiers [119] consider prediction of $\mathbf{Y}_k$, $k > 1$ from a single $\mathbf{X}$, it is also possible to formulate problems where many $\mathbf{X}$ blocks are used to predict $\mathbf{Y}$.

4.5.5 Other Three-way Techniques

There are several methods for decomposing the three-way array that are more general than MPCA [179]. These methods include parallel factor analysis (PARAFAC) [553, 555] and Tucker models [179, 597].

The Tucker Model. The Tucker model decomposes the 3-dimensional data as

$$x_{ijk} = \sum_{l=1}^{L} \sum_{m=1}^{M} \sum_{n=1}^{N} a_{il} b_{jm} c_{kn} z_{lmn} + e_{ijk} \quad (4.157)$$

where a_{il} is an element of the $(I \times L)$ loading matrix of component i, b_{jm} is an element of the $(J \times M)$ loading matrix of the second component j, and c_{kn} is an element of the $(K \times N)$ loading matrix of the third component k. z_{lmn} denotes an element of the three-way core matrix $\mathbf{Z}$ and e_{ijk} is an element of the three-way residual matrix $\mathbf{E}$. The core matrix $\mathbf{Z}$ represents the magnitude and interactions between the variables [305]. $\mathbf{A}, \mathbf{B}, \mathbf{C}$ are column-wise orthonormal, and $\mathbf{A}, \mathbf{B}, \mathbf{C}$ and $\mathbf{Z}$ are chosen to minimize the sum of the squared residuals. In the Tucker model, each mode (I, J, K) may have a different number of PCs, which is defined -not chosen- by the least squares algorithm.

The PARAFAC Model. The PARAFAC model yields a trilinear model of $\underline{\mathbf{X}}$

$$x_{ijk} = \sum_{g=1}^{G} a_{ig} b_{jg} c_{kg} + e_{ijk} \quad (4.158)$$

x_{ijk} and e_{ijk} are the same as in Eq. 4.157. a_{ig}, b_{jg}, and c_{kg} are the elements of the loading matrices $\mathbf{A}$ $(I \times G)$, $\mathbf{B}$ $(J \times G)$, and $\mathbf{C}$ $(K \times G)$. $\mathbf{A}$, $\mathbf{B}$, and $\mathbf{C}$ are chosen to minimize the sum of squared residuals. Under certain conditions, the solution does not converge due to the *degeneracy* of the problem [552, 555]. *Degeneracy* refers to the fact that the loadings $\mathbf{A}$, $\mathbf{B}$, and **C** are rotation-dependent *i.e.*, there can be no multiple solutions for the calculated set of loadings. In terms of data manipulations, the PARAFAC model is a simplification of the Tucker model in two ways [555]:

1. The number of components in all three modes (I, J, K) are equal, and
2. There is no interaction between latent variables of different modes.

The use of MPCA and three-way techniques have been reported for SPM in [641, 646].

4.6 Artificial Neural Networks

Although this chapter is devoted to empirical modeling techniques for modeling linear systems, artificial neural networks (ANNs) which can be used to model both linear and nonlinear systems are discussed here as well because of their popularity. Following a short historical perspective, summarizing foundations of ANNs. Due to availability of numerous ANN software on different platforms, there is no need to construct ANN models from scratch unless a very special, custom application is aimed.

The "neural networks" have been inspired from the way the human brain works as an information-processing system in a highly complex, nonlinear and massively parallel fashion. In its most general form, the following definition of a neural network has been suggested as an adaptive machine [226]:

> *A neural network is a massively parallel distributed processor made up of simple processing units, which has a natural propensity for storing experiential knowledge and making it available for use. It resembles the brain in two respects:*
>
> 1. *Knowledge is acquired by the network from its environment through a learning process.*
> 2. *Interneuron connection strengths, known as synaptic weights, are used to store the acquired knowledge.*

ANNs have a large number of highly interconnected *processing elements* also called as *nodes* or artificial *neurons.* The first computational model of a biological neuron, namely the *binary threshold unit* whose output was either 0 or 1 depending on whether its net input exceeded a given threshold, has been proposed by McCulloch and Pitts in 1943 [377]. Their interpretation has united the studies of neurophysiology and mathematical logic. The next major development in neural networks came in 1949 when Hebb, in his book named *The Organization of Behavior* [227], explicitly proposed that the connectivity of the brain is continually changing as an organism learns differing functional tasks, and that "neural assemblies" are created by such changes. This suggested that a system of neurons, assembled in a finite state automaton, could compute any arbitrary function, given suitable values of weights between the neurons [390]. 15 years after these pioneering fundamental studies, automatically finding suitable values for those weights was introduced by Rosenblatt [520] in his work on the *perceptron* which is a function that computes a linear combination of variables and returns the sign of the result. It was proposed that this iterative learning procedure (so-called *perceptron convergence theorem*) always converged to a set of weights that produced the desired function, as long as the desired function is computable by the network [521, 643]. Interest in neural networks was gradually revived from about 1985s when Rumelhart et al. [527] popularized a much faster learning procedure called *back-propagation*, which could train a multi-layer perceptron to compute any desired function.

Other commonly used names for ANNs include *parallel distributed processors, connectionist models (or networks), self-organizing systems, neurocomputing systems, and neuromorphic systems.* ANNs can be seen as "black-box" models for which no prior knowledge about the process is needed. The goal is to develop a process model based only on the input-output data acquired from the process. There are benefits and limitations of using ANNs for empirical modeling [226, 483]:

Benefits

- *Adaptive Behavior.* ANNs have the ability to adapt, or learn, in response to their environment through training. A neural network can easily be *retrained* to deal with minor changes in the operational and/or environmental conditions. Moreover, when it is operating in a *nonstationary* environment, it can be designed to adjust its synaptic weights in real time. This is especially a valuable asset in adaptive pattern classification and adaptive control.

- *Nonlinearity.* A neural network is made of interconnections of neurons and is itself nonlinear. This special kind of nonlinearity is distributed throughout the network. The representation of nonlinear behavior by nonlinear structure is a significant property, since the inherent characteristic of most fermentations/biological processes is highly nonlinear.

- *Pattern Recognition Properties.* ANNs perform multivariable pattern recognition tasks very well. They can learn from examples (training) by constructing an *input-output mapping* for the system of interest.

In the pattern classification case an ANN can be designed to provide information about similar and unusual patterns. Training and pattern recognition must be made by using a closed set of patterns. All possible patterns to be recognized should be present in the data set.

- *Fault Tolerance.* A properly designed and implemented ANN is usually capable of robust computation. Its performance degrades gracefully under adverse operating conditions and when some of its connections are severed.

Limitations

- *Long Training Times.* When structurally complex ANNs or inappropriate optimization algorithms are used, training may take unreasonably long times.

- *Necessity of Large Amount of Training Data.* If the size of input-output data is small, ANNs may not produce reliable results. ANNs provide more accurate models and classifiers when large amounts of historical data rich in variations are available.

- *No Guarantee of Optimal Results.* Training may cause the network to be accurate in some operating zones, but inaccurate in others. While trying to minimize the error, it may get trapped in local minima.

- *No Guarantee of Complete Reliability.* This general fact about all computational techniques is particularly true for ANNs. In fault diagnosis applications, for instance, ANNs may misdiagnose some faults 1% of the time while other faults in the same domain 25% of the time. It is hard to determine a priori (when backpropagation algorithm is used) what faults will be prone to higher levels of misdiagnosis.

- *Operational Problems Associated with Implementation.* There are practical problems related to training data set selection [302, 334].

4.6.1 Structures of ANNs

ANNs have a number of elements. The basic structure of ANNs typically includes multilayered, interconnected neurons (or computational units) that nonlinearly relate input-output data. A nonlinear model of a neuron, which forms the core of the ANNs is characterized by three basic attributes (Figure 4.16):

A set of synaptic weights (or connections), describing the amount of influence a unit (a synapse or node) has on units in the next layer; a positive weight causes one unit to excite another, while a negative weight causes one unit to inhibit another. The signal x_j at the input synapse j connected to neuron k in Figure 4.16 is multiplied by weight w_{kj} (Eq. 4.159).

A linear combiner (or a summation operator) of input signals, weighted by the respective synapses of the neuron.

An activation function with limits on the amplitude of the output of a neuron. The amplitude range is usually given in a closed interval [0,1] or [-1,1]. Activation function $\varphi(\cdot)$ defines the output of a neuron in terms of the activation potential υ_k (given in Eqs. 4.160 and 4.161). Typical activation functions include the unit step change and sigmoid functions.

A neuron k can be described mathematically by the following set of equations [226]:

$$u_k = \sum_{j=0}^{m} w_{kj} x_j \tag{4.159}$$

$$\upsilon_k = u_k + b_k \tag{4.160}$$

and

$$y_k = \varphi(\upsilon_k) \tag{4.161}$$

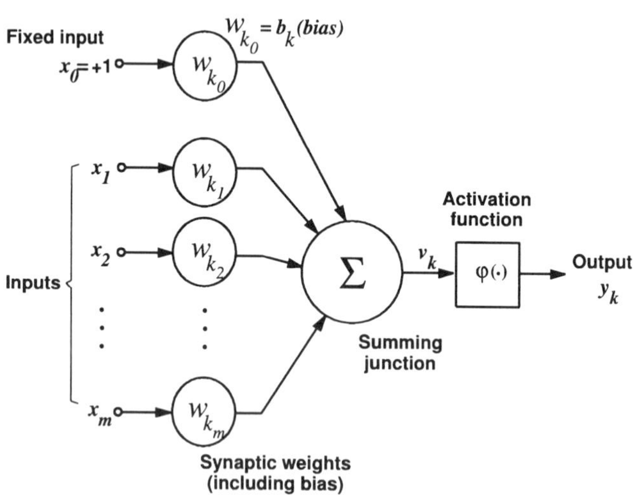

Figure 4.16. A nonlinear model of a single neuron [226].

where $x_1, x_2, \ldots, x_j, \ldots, x_m$ are the input signals; $w_{k1}, w_{k2}, \ldots, w_{kj}, .., w_{km}$ are the synaptic weights of neuron k, u_k is the linear combiner output of the input signals, b_k is the bias, v_k is the activation potential (or induced local field), $\varphi(\cdot)$ is the activation function, and y_k is the output signal of the neuron. The bias is an external parameter providing an affine transformation to the output u_k of the linear combiner.

Several activation functions are available. The four basic types illustrated in Figure 4.17 are:

1. *Threshold Function.* Also known as McCulloch-Pitts model [377]

$$\varphi(v) = \begin{cases} 1, & v \geqslant 0 \\ 0, & v < 0. \end{cases} \tag{4.162}$$

2. *Piecewise-linear Function.*

$$\varphi(v) = \begin{cases} 1, & v \geqslant +\frac{1}{2} \\ v, & +\frac{1}{2} > v > -\frac{1}{2} \\ 0, & v \leqslant -\frac{1}{2} \end{cases} \tag{4.163}$$

where the amplification factor inside the linear region of operation is assumed to be the unity.

3. *Sigmoid Function.* This s-shaped function is by far the most common form of activation function used. A typical expression is

$$\varphi(v) = \frac{1}{1 + e^{-av}} \tag{4.164}$$

where a is the slope parameter.

4. *Hyperbolic Tangent Function.* This is a form of sigmoid function but it produces values in the range $[-1, +1]$ instead of $[0, 1]$

$$\varphi(v) = \tanh(v) = \frac{e^{v} - e^{-v}}{e^{v} + e^{-v}} . \tag{4.165}$$

Processing units (neurons) are linked to each other to form a network associated with a learning algorithm. A neural network can be formed with any kind of topology (architecture). In general, three kinds of network topologies are used [226]:

Single-layer feedforward networks include input layer of source nodes that projects onto an output layer of neurons (computation nodes), but not vice versa. They are also called *feedforward* or *acyclic* networks. Since the computation takes place only on the output layer nodes, the input layer does not count as a layer (Figure 4.18(a)).

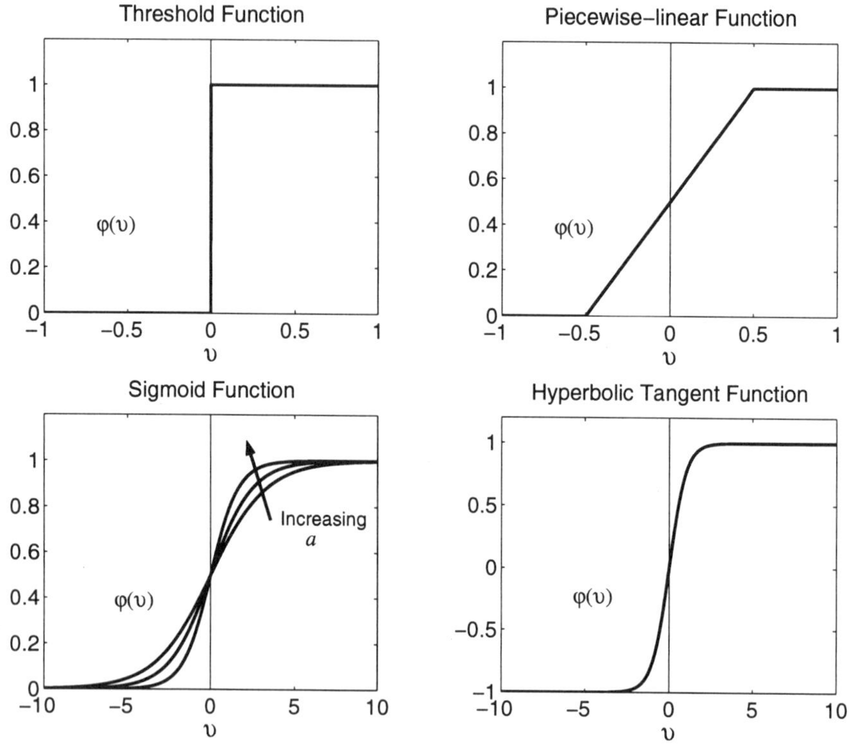

Figure 4.17. Activation functions [226].

Multilayer feedforward networks contain an input layer connected to one or more layers of hidden neurons (hidden units) and an output layer (Figure 4.18(b)). The hidden units internally transform the data representation to extract higher-order statistics. The input signals are applied to the neurons in the first hidden layer, the output signals of that layer are used as inputs to the next layer, and so on for the rest of the network. The output signals of the neurons in the output layer reflect the overall response of the network to the activation pattern supplied by the source nodes in the input layer. This type of networks are especially useful for pattern association (i.e. mapping input vectors to output vectors).

Recurrent networks differ from feedforward networks in that they have at least one *feedback* loop. An example of this type of network is given in Figure 4.18(c) which is one of the earliest recurrent networks called

Jordan network [306]. The activation values of the output units are fed back into the input layer through a set of extra units called the *state units.* Learning takes place in the connection between input and hidden units as well as hidden and output units. Recurrent networks are useful for pattern sequencing (i.e., following the sequences of the network activation over time). The presence of feedback loops has a profound impact on the learning capability of the network and on its performance [226]. Applications to chemical process modeling and identification have been reported [97, 616, 679].

Before proceeding with training the network, an appropriate network architecture should be declared. This can be done either in static or dynamic manner. Many *ad hoc* techniques for static network structure selection are based on pruning the redundant nodes by testing a range of network sizes, i.e., number of hidden nodes. However, techniques for network architecture selection for feedforward networks have been proposed [301, 335, 482, 627, 628]. Reed [499] gives a partial survey of pruning algorithms and recent advances can be found in the neural network literature [144, 404].

Having specified the network architecture, a set of input-output data is used to *train* the network, i.e. to determine appropriate values for the weights associated with each interconnection. The data are then propagated forward through the network to generate an output to be compared with the actual output. The overall procedure of training can be seen as *learning* for the network from its environment through an interactive process of adjustments applied to its weights and bias levels. A number of learning rules such as *error-correction, memory-based, Hebbian, competitive, Boltzmann* learning have been proposed [226] to define how the network weights are adjusted. Besides these rules, there are several procedures called *learning paradigms* that determine how a network relates to its environment. The *learning paradigm* refers to a model of the environment in which the network operates. There are two main classes of learning paradigms:

Learning with teacher (*supervised learning*), in which a *teacher* provides output targets for each input pattern, and corrects the network's errors explicitly. The teacher can be thought of as having knowledge of the environment (presented by the historical set of input-output data) so that the neural network is provided with desired response when a training vector is available. The desired response represents the optimum action to be performed to adjust neural network weights under the influence of the training vector and error signal. The *error signal* is the difference between the desired response (historical value)

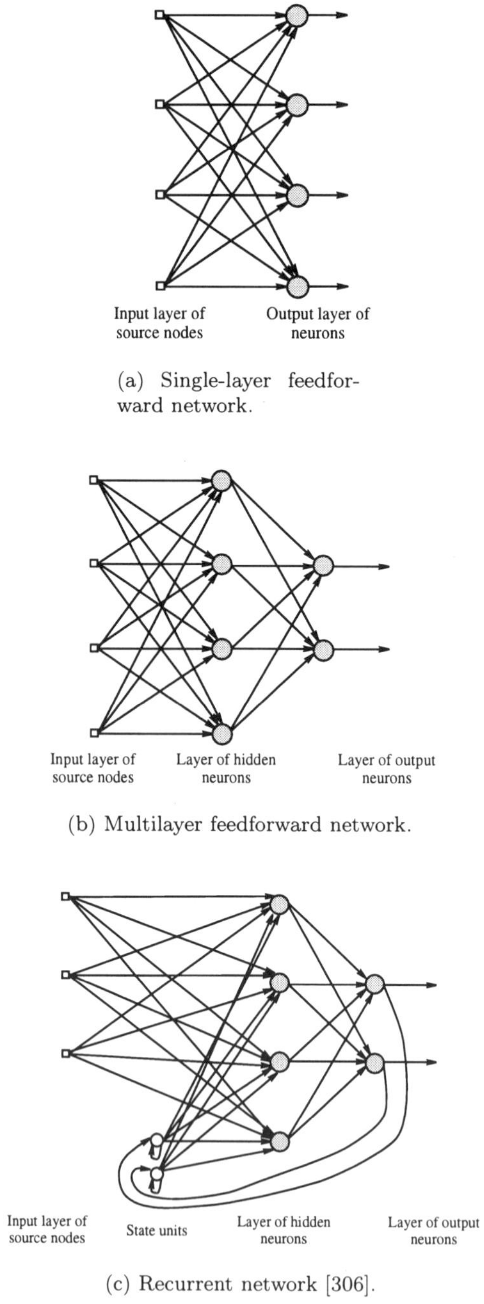

Figure 4.18. Three fundamentally different network architectures.

and the actual response (computed value) of the network. This corrective algorithm is repeated iteratively until a preset convergence criteria is reached. One of the most widely used supervised training algorithms is the *error backpropagation* or *generalized delta rule* proposed by Rumelhart and others [527, 637].

Learning without a teacher, in which there is no teacher, and the network must find the regularities in the training data by itself. This paradigm has two subgroups

1. ***Reinforcement learning/Neurodynamic programming***, where learning the relationship between inputs and outputs is performed through continued interaction with the environment to minimize a scalar index of performance. This is closely related to *Dynamic Programming* [53].
2. ***Unsupervised learning***, or *self-organized learning* where there is no external teacher or critic to oversee the learning process. Once the network is tuned to the statistical regularities of the input data, it forms internal presentations for encoding the input automatically [48, 226].

4.6.2 ANN Applications in Fermentation Industry

Application of ANN models in biochemical and fermentation industries concentrate mostly on soft sensor development for estimating infrequently measured quality variables such as biomass concentration using process variables that are frequently measured. Developing such empirical models is almost similar to developing statistical regression models. There are numerous applications for different types of fermentations in the literature. Applications include use of ANN models to estimate biomass concentration in continuous mycelial fermentation [28, 649], improve yield in penicillin fermentations [125], and develop on-line estimators for batch and cell-recycle systems [268]. Applications in general for soft sensor technology [25, 75, 98], optimization and fault diagnosis [92, 345, 587] and experimental design based on ANNs [194] have also been reported. There are also hybrid neural network-first principles models that incorporate fundamental models with ANN models for better accuracy on predictions [170, 481, 588]. One approach is to use a first principle model to explain as much variation as possible in data. The remaining significant variation is modeled by an ANN. This hybrid structure is inspired from the first principle-time series model combinations that rely on the same philosophy. In another approach, the parameters of a first principles model are estimated by ANN in a hybrid structure [481]. Another hybrid structure [588] creates series parallel

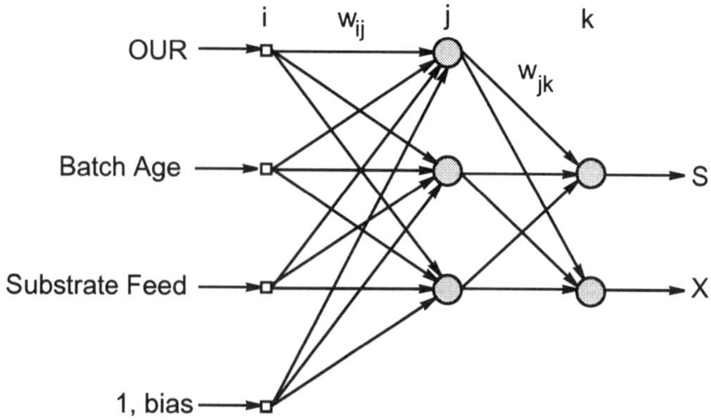

Figure 4.19. A hypothetical feedforward ANN with one hidden layer for estimating substrate and biomass concentrations in fed-batch penicillin fermentation (OUR: Oxygen uptake rate, S: Substrate conc., X: Biomass conc., w_{ij} and w_{jk} weight vectors associated with interconnected layers).

structure that combines parametric models based on fundamental process knowledge and nonparametric models based on process data.

Three-layer feedforward networks with backpropagation learning algorithm are dominantly used in the applications mentioned above. A hypothetic neural network structure is shown in Figure 4.19 for estimating biomass and substrate concentrations in penicillin fermentation utilizing frequently measured variables such as feed rates and off gas concentrations.

There are many educational and commercial software packages available for development and deployment of ANNs. Most of those packages include data preprocessing modules such as Gensym's NeurOn-Line Studio [185].

ANNs can be seen as autoassociative regression models. They resemble statistical modeling techniques in that sense. But the lack of statistical inference and robustness issues may cause problems. Care should be taken (e.g., data pre-processing, appropriate selection of input-output data set) during deployment. The advantages/disadvantages summarized in the introductory paragraph of Section 4.6 should be taken into consideration prior to deciding if ANNs are appropriate for a specific application.

4.7 Extensions of Linear Modeling Techniques to Nonlinear Model Development

Several paradigms are available for developing nonlinear dynamic input-output models of processes. These models have the capability to describe pathological dynamic behavior and to provide accurate predictions over a wider range of operating conditions compared to linear models. ANNs were introduced in the previous section. Chapter 5 presents system science methods for nonlinear model development. Various other nonlinear model development paradigms such as time series models, Volterra kernels, cascade (block-oriented) models and nonlinear PLS have been developed. Extensions of linear empirical model development techniques based on time series models and PLS are introduced in this section to expand the alternatives available to build nonlinear models. Polynomial models, threshold models, and models based on spline functions can describe various types of nonlinear behavior observed in many physical processes. Polynomial models include bilinear models, state dependent models, nonlinear autoregressive moving average models with exogenous inputs (NARMAX), nonlinear polynomial models with exponential and trigonometric functions (NPETM), canonical variate nonlinear subspace models, and multivariate adaptive regression splines (MARS). A unified nonlinear model development framework is not available, and search for the appropriate nonlinear structure is part of the model development effort. Use of a nonlinear model development paradigm which is not compatible with the types of nonlinearities that exist in data can have a significant effect on model development effort and model accuracy. Various nonlinear time series modeling paradigms from system identification and statistics literature are summarized in Section 4.7.1. A special group of nonlinear models based on the extension of PLS is presented in Section 4.7.2.

4.7.1 Nonlinear Input-Output Models in Time Series Modeling Literature

More than twenty nonlinear time series model structures have been proposed during the last four decades. They could be classified based on features such as the types of variables used in the model, and the way the model parameters appear in equations. The characteristic features of various nonlinear time series (NLTS) models are discussed in this subsection.

The three basic groups of variables used in NLTS models are:

1. Previous values of the dependent variable that yield *autoregressive* (AR) terms,

2. Sequences of independent and identically distributed (*iid*) random vectors (white noise) that provide *moving average* (MA) terms,

3. Input variables with nonrandom features that are called *external (exogenous)* (X) variables.

Volterra series models [624] do not utilize previous values of the dependent variable, while nonlinear autoregressive moving average models with exogenous variables (NARMAX) (Eqs. 4.171-4.174) use all three types of variables. Model structures are either linear or nonlinear in the parameters. Model parameter estimation task is much less computation intensive if the model parameters appear in a linear structure. This permits use of well-developed parameter estimation techniques for linear modeling paradigms. NARMAX, *bilinear* (Eq. 4.170), and *threshold* models (Eq. 4.177) are linear in the parameters, while *exponential* models are nonlinear in the parameters.

Volterra models have been utilized by Wiener [644] for the study of nonlinear systems by constructing transformations of Volterra series in which the successive terms are orthogonal. Expressing $y(t)$ as a function of current and past values of a zero mean white noise process $e(t)$

$$y(t) = H(e(t), e(t-1), \cdots) \, . \tag{4.166}$$

H can be expanded as a Taylor series about the point $\mathbf{a}$

$$\begin{aligned} y(t) \; &= \; \mu + \sum_{i=1}^{\infty} g_i e(t-i) + \sum_{i=1}^{\infty}\sum_{j=1}^{\infty} g_{ij} e(t-i)e(t-j) \\ &\quad + \sum_{i=1}^{\infty}\sum_{j=1}^{\infty}\sum_{k=1}^{\infty} g_{ijk} e(t-i)e(t-j)e(t-k) + \ldots \end{aligned} \tag{4.167}$$

where

$$\mu = H\,|_{\mathbf{a}} \qquad g_i = \left(\frac{\partial H}{\partial e(t-i)}\right)\Big|_{\mathbf{a}} \cdot \qquad g_{ij} = \left(\frac{\partial^2 H}{\partial e(t-i)\partial e(t-j)}\right)\Big|_{\mathbf{a}} \cdot \tag{4.168}$$

When input $u(t)$ and output $y(t)$ are both observable, the Volterra series can be represented in terms of the input by replacing $e(t)$ by $u(t)$. If the system is *linear*, only the first derivative term is present and the model is completely characterized by the transfer function g_i of the system. For nonlinear processes, additional terms in Eq. (4.167) must be included, and the *generalized transfer functions* concept is used [479].

Exponential models of order k have the basic form [211]

$$y(t) = \sum_{j=1}^{k} \left[\alpha_j + \beta_j \ exp(-\delta(y(t-1))^2)y(t-j)\right] + e(t) \tag{4.169}$$

where $e(t)$ is a sequence of *iid* random variables, α_j, β_j, and δ are model parameters. Since δ is in the argument of the exponential term, the model estimation problem is computationally more challenging.

Bilinear models [394] cannot describe several types of nonlinearities such as limit cycles, but they have a simple form that can describe processes where products of two variables appear in equations derived from first principles. The general form of a bilinear model is

$$y(t) + \sum_{j=1}^{p} a_j y(t-j) = \sum_{j=0}^{r} c_j e(t-j) + \sum_{i=1}^{m}\sum_{j=1}^{k} b_{ij} y(t-i)e(t-j) \tag{4.170}$$

where $c_0 = 1$ and $e(\cdot)$ represents another variable or white noise. With suitable choices of parameters, bilinear models can approximate a "well behaved" Volterra series relationship over a *finite* time interval [83].

Nonlinear Polynomial Models. An important class of nonlinear polynomial models has been proposed by Billings and his coworkers [93, 95, 336]. Depending on the presence of autoregressive (AR), moving average (MA) terms, and/or exogenous (X) variables, they are denoted by acronyms such as NAR, NARX, or NARMAX. NARMAX models consist of polynomials that include various linear and nonlinear terms combining the inputs, outputs and past errors. Once the model structure, monomials to be included in the model, has been selected, identification of parameter values (coefficients of monomials) can be formulated as a standard least squares problem. The number of candidate monomials to be included in a NARMAX model ranges from about a hundred to several thousands for moderately nonlinear systems. Determination of the model structure by stepwise regression type of techniques becomes inefficient. An algorithm that efficiently combines structure selection and parameter estimation has been proposed [290] and extended to MIMO nonlinear stochastic systems [94].

The *NARMAX model* [336] of a discrete time multivariable nonlinear stochastic system with r inputs and m outputs is

$$\begin{aligned}\mathbf{y}(t) = \; & \mathbf{f}(\mathbf{y}(t-1), \cdots, \mathbf{y}(t-n_y), \mathbf{u}(t-1), \cdots, \mathbf{u}(t-n_u), \\ & \mathbf{e}(t-1), \cdots, \mathbf{e}(t-n_e)) + \mathbf{e}(t)\end{aligned} \tag{4.171}$$

where

$$\mathbf{y}(t) = \begin{pmatrix} y_1(t) \\ \vdots \\ y_m(t) \end{pmatrix}, \quad \mathbf{u}(t) = \begin{pmatrix} u_1(t-1) \\ \vdots \\ u_r(t-1) \end{pmatrix}, \quad \mathbf{e}(t) = \begin{pmatrix} e_1(t) \\ \vdots \\ e_m(t) \end{pmatrix} \tag{4.172}$$

are the system output, input and noise, respectively, n_y, n_u, and n_e are the maximum lags in the output, input and noise, respectively, $\{e(t)\}$ is a zero mean *iid* sequence, and $\mathbf{f}(\cdot)$ is some vector valued nonlinear function.

NARMAX models can be illustrated by a NAR model

$$\begin{aligned} y_q(t) = \ & f_q(y_1(t-1), \cdots, y_1(t-n_y), \cdots, y_m(t-1), \cdots, y_m(t-n_y)) \\ & + e_q(t), \quad q = 1, m \end{aligned} \tag{4.173}$$

Writing $f_q(\cdot)$ as a polynomial of degree l yields

$$\begin{aligned} y_q(t) = \theta_0^{(q)} + \sum_{i_1=1}^{n} \theta_{i_1}^{(q)} z_{i_1}(t) + \sum_{i_1=1}^{n} \sum_{i_2=i_1}^{n} \theta_{i_1 i_2}^{(q)} z_{i_1}(t) z_{i_2}(t) + \cdots + \\ \sum_{i_1=1}^{n} \cdots \sum_{i_1=i_l-1}^{n} \theta_{i_1 \cdots i_l}^{(q)} z_{i_1}(t) \cdots z_{i_l}(t) + e_q(t), \quad q = 1, m \end{aligned} \tag{4.174}$$

where $n = m \times n_y$, $z_1(t) = y_1(t-1)$, $z_2(t) = y_1(t-2)$, $\cdots$, and $z_{mn_y}(t) = y_m(t-n_y)$. All terms composed of $z_{i_1}(t), \cdots, z_{i_l}(t)$ in Eq. (4.173) are thus provided. Hence, for each q, $1 \leq q \leq m$, Eq. 4.174 describes a linear regression model of the form

$$y_q(t) = \sum_{i=1}^{M} p_i(t)\theta_i + e(t), \quad t = 1, \cdots, N \tag{4.175}$$

where $M = \sum_{i=1}^{l} m_i$ with $m_i = m_{i-1} \cdot (n_y \cdot m + i - 1)/i$, N is the time series data length, $p_1(t) = 1$, $p_i(t)$ are monomials of degree up to l composed of various combinations of $z_1(t)$ to $z_n(t)$ ($n = m \times n_y$), $e(t)$ are the residuals, and θ_i are the unknown model parameters to be estimated.

A new methodology has been proposed for developing multivariable additive NARX (Nonlinear Autoregressive with eXogenous inputs) models based on subspace modeling concepts [122]. The model structure is similar to that of a Generalized Additive Model (GAM) and is estimated with a nonlinear Canonical Variate Analysis (CVA) algorithm called CANALS. The system is modeled by partitioning the data into two groups of variables. The first is a collection of "future" outputs, the second is a collection of past input and outputs, and "future" inputs. Then, future outputs are predicted in terms of past and present inputs and outputs. This approach is similar to linear subspace state-space modeling [316, 415, 613]. The appeal of linear and nonlinear subspace state-space modeling is the ability to develop models with error prediction for a future window of output (window length selected by user) and with a well-established procedure that minimizes trial-and-error and iterations. An illustrative example of such modeling is presented based on a simulated continuous chemical reactor that exhibits multiple steady states in the outputs for a fixed level of the input [122].

Models with a small number of monomials are usually adequate to describe the dynamic behavior of most real processes. Methods have been developed for the combined structure selection and parameter estimation problem based on Gram-Schmidt orthogonalization [94]. The selection of monomials is carried out by balancing the reduction in residuals and increase in model complexity. Criteria such as *Akaike Information Criteria* (AIC) are used to guide the termination of modeling effort. A variant of AIC is given in Eq. 4.42

$$AIC(k) = Nlog\left(\frac{1}{N}\hat{\mathbf{E}}^T\hat{\mathbf{E}}\right) + 2k \tag{4.176}$$

where $\mathbf{E} = (e(1) \quad \cdots \quad e(N))^T$ and k is the number of parameters θ in the model. AIC and similar criteria balance minimization of prediction error (residuals) and model complexity (parsimony). The addition of new monomials to the model is terminated when AIC is minimized.

A common problem with polynomial models is the explosion of the predicted variable magnitude. This may be caused by assigning large values to some predictors which are raised to high powers, or to the existence of unstable equilibrium points. This type of behavior necessitates the *censoring* of the predicted variable value. One censoring method is based on embedding the whole prediction term as the argument of a sigmoid function as is done in ANN. Consequently, the predicted value reaches a lower or upper limit as the magnitude of the argument increases. Other censoring methods rely on fixed upper and lower limits, or upper and lower limits that are linear functions of the value predicted by the uncensored polynomial model.

Threshold Models provide a nonlinear description by using different submodels in different ranges of a variable. A piecewise linear model with AR and MA terms takes the form

$$y(t) = a_0^{(j)} + \sum_{i=1}^{m} a_i^{(j)} y(t-i) + \sum_{i=0}^{(m-1)} b_i^{(j)} e(t-i) \tag{4.177}$$

where the appropriate parameter set $(a_i,\ b_i)$ is selected based on $y(t-d) \in R_j,\ j = 1, l$. Here $R_j = (r_{j-1}, r_j)$ with the linearly ordered real numbers $r_0 < r_1 < \cdots < r_l$ called the threshold parameters, and d is the delay parameter [24]. The identification of threshold models involves estimation of model parameters and selection of d and r_j. The threshold model (Eq. 4.177) can be reduced to an AR structure by setting $b_0^{(j)} = 1$ and $b_i^{(j)} = 0,\ i = 1, m-1$. External input variables can also be incorporated and the condition for selection of parameter sets may be based on the input variables. The submodels may also be nonlinear functions such as NARX and NARMAX models.

Models Based on Spline Functions. Spline functions provide a non-parametric nonlinear regression method with piecewise polynomial fitting. A spline function is a piecewise polynomial where polynomials of degree q join at the knots $K_i,\ i = 1, k$ and satisfy the continuity conditions for the function itself and for its $q-1$ derivatives [658]. Often continuity of the first and second derivatives are enough, hence cubic splines ($q = 3$) have been popular. One-sided and two-sided power univariate basis functions for representing qth order splines are

$$b_q(y - K_i) = (y - K_i)_+^q \qquad \text{and} \qquad b_q^{\pm}(y - K_i) = [\pm(y - K_i)]_+^q \tag{4.178}$$

where subscript + indicates that the term is evaluated for positive values, the basis function has a value of zero for negative values of the argument.

The **multivariate adaptive regression splines** (MARS) method [169] is an extension of the recursive partitioning method. Friedman [169] describes the evolution of the method and presents algorithms for building MARS models. An introductory level discussion with applications in chemometrics is presented by Sekulic and Kowalski [539]. Spline fitting is generalized to higher dimensions and multivariable systems by generating basis functions that are products of univariate spline functions

$$B_m^{(q)}(\mathbf{y}) = \prod_{r=1}^{R_m} [\pm(y_{v(r,m)} - K_{rm})]_+^q \tag{4.179}$$

where R_m is the maximum number of allowable variable interactions and $y_{v(r,m)}$ denote predictor variables. The final model is of the form

$$f(\mathbf{x}) = a_0 + \sum_{m=1}^{M} a_m B_M^{(q)} \tag{4.180}$$

where a_0 is the coefficient of the constant basis function B_1 and the sum is over all the basis functions $B_m^{(q)}$ produced by the selection procedure.

Basis function selection is carried out in two steps. The first step is *forward recursive partitioning* which selects candidate basis functions. The second step is *backward stepwise deletion* which removes splines that duplicate similar information. Both steps are implemented by evaluating a lack-of-fit function [169]. A recent study reports the comparison of models developed by MARS and ANN with sigmoid functions [480].

Nonlinear Polynomial Models with Exponential and Trigonometric Terms (NPETM). If process behavior follows nonlinear functions such as trigonometric, exponential, or logarithmic functions, restricting model structure to polynomials would yield a model that has a large number of terms and acceptable accuracy over a limited range of predictor variables. Basically, several monomials are included in the model in order to describe approximately the functional behavior of that specific exponential or trigonometric relation. For example,

$$y_1(t) = \theta_1 y_1(t-1)e^{\theta_2 y_2(t-1)} + y_1^2(t-2) + \theta_3 \sin(\theta_4 y_1(t-2)) \ . \quad (4.181)$$

Consequently, inclusion of such functions in the pool of candidate terms would reduce the number of terms needed in the model and improve model accuracy. If the argument of such functions includes a parameter to be estimated (parameters θ_2 and θ_4 in Eq. 4.181), the model is not linear in the parameters and the parameter estimation problem becomes more challenging. If the nature of the functional relationship is not known *a priori*, the coupled problem of model structure determination and parameter estimation may not converge unless the initial guess is somewhat close to the correct values. Physical insight to the process or knowledge about model structure based on earlier modeling efforts provide vital information for the initial guess.

NPET models are feasible when changes in operating conditions necessitate a remodeling effort starting from an existing model. Some monomials and/or parameter values need to be changed, but the exponential or trigonometric type relations that are known remain the same for such cases. NPET models should be avoided when new models are being developed and information about the functional form of the nonlinearity is not known. The large number of nonlinear function types and the existence of unknown parameters in the function argument creates a search domain that is too large for most practical applications.

Cascade Systems

Cascade structures [210, 367] are composed of serially connected static nonlinear and dynamic linear transfer function blocks. This structure is appropriate when the process has static nonlinearities. The structure is called

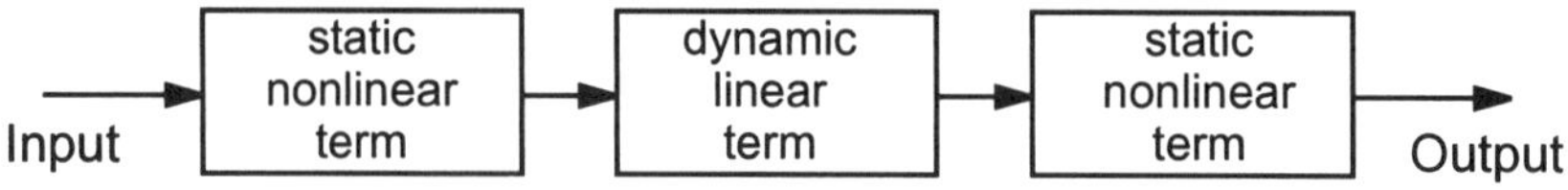

Figure 4.20. General structure of Wiener-Hammerstein cascade model.

a Hammerstein model if the static nonlinear element precedes the dynamic linear element, a Wiener model if the nonlinear element succeeds the linear element. In general, the models have polynomial terms that describe a continuous nonlinear steady state characteristic and/or a continuous function of a dynamic parameter.

Extensions of Linear Models for Describing Nonlinear Variations

Two other alternatives can be considered for developing linear models with better predictive capabilities than a traditional ARMAX model for nonlinear processes. If the nature of nonlinearity is known, a transformation of the variable can be utilized to improve the linear model. A typical example is the knowledge of the exponential relationship of temperature in reaction rate expressions. Hence, the *log* of temperature with the rate constant can be utilized instead of the actual temperature as a regressor. The second method is to build a recursive linear model. By updating model parameters frequently, mild nonlinearities can be accounted for. The rate of change of the process and the severity of the nonlinearities are critical factors for the success of this approach.

4.7.2 Nonlinear PLS Models

Linear PLS decomposes two variable blocks $\mathbf{X}$ and $\mathbf{Y}$ as $\mathbf{X} = \mathbf{T}\mathbf{P}^T$ and $\mathbf{Y} = \mathbf{U}\mathbf{Q}^T$ such that $\mathbf{X}$ is modeled effectively by $\mathbf{T}\mathbf{P}^T$ and $\mathbf{T}$ predicts $\mathbf{U}$ well by a linear model (inner relation) $\mathbf{U} = \mathbf{T}\mathbf{B}$ where $\mathbf{B}$ is a diagonal matrix. To model nonlinear relationships *between* $\mathbf{X}$ and $\mathbf{Y}$, their projections should be nonlinearly related to each other [664]. One possibility is to use a polynomial function such as

$$\mathbf{u}_a = c_{0a} + c_{1a}\mathbf{t}_a + c_{2a}\mathbf{t}_a^2 + \mathbf{h}_a \tag{4.182}$$

where a represents the model dimension, c_{0a}, c_{1a}, and c_{2a} are constants, and $\mathbf{h}_a$ is a vector of residuals. This quadratic function can be generalized to other nonlinear functions of $\mathbf{t}_a$:

$$\mathbf{u}_a = f(\mathbf{t}_a) + \mathbf{h}_a \tag{4.183}$$

where $f(\cdot)$ may be a polynomial, exponential, or logarithmic function.

Another framework for expressing a nonlinear relationship between **X** and **Y** can be based on splines [660] or smoothing functions [160]. Splines are piecewise polynomials joined at knots (denoted by z_j) with continuity constraints on the function and all its derivatives except the highest. Splines have good approximation power, high flexibility and smooth appearance as a result of continuity constraints. For example, if cubic splines are used for representing the inner relation:

$$u = b_0 + b_1 t + b_2 t^2 + b_3 t^3 + \sum_j^J b_{j+3}(t - z_j)_+^3 \tag{4.184}$$

where the J knot locations and the model coefficients b_k are the free parameters of the spline function. There are $K + J + 1$ coefficients where K is the order of the polynomial. The term $b_{j+3}(t - z_j)_+^3$ denotes a function with values 0 or $b_{j+3}(t - z_j)^3$ depending on the value of t:

$$b_{j+3}(t - z_j)_+^3 = \begin{cases} b_{j+3}(t - z_j)^3 & : \quad t > z_j \\ 0 & : \quad t < z_j \end{cases} . \tag{4.185}$$

The desirable number of knots and degrees of polynomial pieces can be estimated using cross-validation. An initial value for J can be $N/7$ or $\sqrt(N)$ for $N > 100$ where N is the number of data points. Quadratic splines can be used for data without inflection points, while cubic splines provide a general approximation for most continuous data. To prevent over-fitting data with higher-order polynomials, models of lower degree and higher number of knots should be considered for lower prediction errors and improved stability [660]. B splines that are discussed in Section 6.3.3 provide an attractive alternative to quadratic and cubic splines when the number of knots is large [121].

Other nonlinear PLS models that rely on nonlinear inner relations have been proposed [209, 581]. Nonlinear relations within X or Y can also be modeled. Simple cures would include use of known functional relationships for specific variables based on process knowledge such as exponential relationships of temperature in reactions. More sophisticated approaches are also available, including use of artificial neural networks to generate empirical nonlinear relations.

5

System Science Methods for Nonlinear Model Development

Contributing author

İnanç Birol

Department of Chemical Engineering
Northwestern University
Evanston, IL

One of the nature's greatest mysteries is the reason why she is understandable. Yet, she *is* understandable, and the language she speaks is mathematics. The history of science is full of breakthroughs when the mathematics of a certain type of behavior is understood. The power of that understanding lies in the fact that, stemming from it, one can build a model for the phenomenon, which enables the *prediction* of the outcome of experiments that are yet to be performed.

Now, it is a part of scientific folklore [195], how Lorenz [350] realized the phenomenon that was later given the name *deterministic chaos*; how it stayed unnoticed on the pages of the Journal of Atmospheric Sciences for a period of time only to be rediscovered by other scientists; and how it unfolded a new scientific approach. In fact, the existence of chaotic dynamics has been known to mathematicians since the turn of the century. The birth of the field is commonly attributed to the work of Poincaré [473]. Subsequently, the pioneering studies of Birkhoff [55], Cartwright [89], Littlewood [344], Smale [551], Kolmogorov [284] and others built the mathematical foundations of nonlinear science. Still, it was not until the wide utilization of digital computers in late-1970s for scientific studies, that the field made its impact on sciences and engineering. It has been demonstrated that chaos is relevant to problems in fields as diverse as chemistry, fluid mechanics, biology, ecology, electronics and astrophysics. Now that it has been shown to manifest itself almost anywhere scientists look, the focus is shifted from cataloging chaos, to actually learning to live with it. In this chapter, we are going to introduce basic definitions in nonlinear system theory, and present methods that use these ideas to analyze chaotic experimental time series data, and develop models.

5.1 Deterministic Systems and Chaos

Assuming that we are living in a *causal* universe, the cause of the events in the future are attributed to the events of the present and the past. Considering an isolated deterministic system that is characterized by an n-dimensional real phase space (denoted by $\mathbf{x} \in \mathcal{R}^n$), to describe this dependence, we can write a set of n differential equations

$$\frac{d\mathbf{x}}{dt} = \mathbf{f}(\mathbf{x}) \tag{5.1}$$

or a set of n difference equations

$$\mathbf{x}(i+1) = \mathbf{f}(\mathbf{x}(i)) \tag{5.2}$$

with $\mathbf{f}(\cdot)$ representing an n-dimensional vector function of $\mathbf{x}$. Under general conditions, the existence and uniqueness properties of solutions hold, and Eq (5.1) or (5.2) determines the trajectory (or orbit) of the dynamical system, given the initial conditions. Eq (5.1) defines a continuous *flow* for a continuous system, and Eq (5.2) defines a discrete *map* for a discrete system. Note that both definitions are for *autonomous systems*, meaning there is no explicit time dependence in the system equations. Even if there were explicit terms in t or i in the system equations, we could augment the system order by one, to have $x_{n+1} = t$ with $f_{n+1} = 1$ for the continuous case, and $x_{n+1} = i$ with $f_{n+1} = i + 1$. Thus, without loss of generality, we will confine our interest to autonomous systems.

Poincaré Map

Although most physical systems manifest continuous dynamics, maps arise naturally in many applications. Furthermore, even when the natural statement of a problem is in continuous time, it is often possible and sometimes desirable to transform the continuous dynamics to a map. Note, however

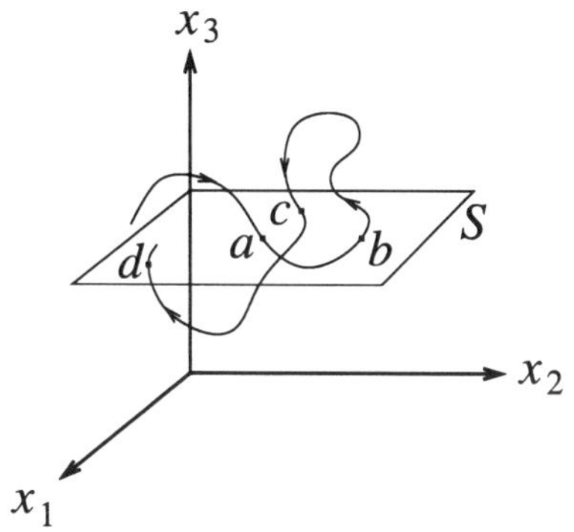

Figure 5.1. The Poincaré surface of section, S, defined by x_3 =constant, for a third order system.

that, this transformation is not necessarily a mere time discretization, but can as well be performed by the Poincaré surface of section technique, shown in Figure 5.1 for a third order system. The trajectory of the system can be visualized by a parametric plot of the states x_k, in phase space (or state space) (x_1, x_2, x_3). Thus, the trace of this parametric plot is an instance of the system dynamics, and the arrows show the direction of time.

If we choose a surface, S, in this space, say, defined by x_3 =constant, and label the points that the system trajectory crosses S, as $a, b, c, \ldots$, we can collect a two dimensional information about those crossings, given by the coordinates (x_1, x_2). If the point a corresponds to the ith crossing of the surface of section at $t = t_i$, we can define a two-dimensional vector, $\mathbf{y}(i) = (x_1(t_i), x_2(t_i))$. Given $\mathbf{y}(i)$, we can reconstruct an initial condition for the system dynamics, and solve the model equations —analytically or numerically— to find the next crossing point, b. In other words, point a uniquely determines point b for a given system dynamics. Therefore, there exists a two-dimensional map

$$\mathbf{y}(i+1) = \mathbf{g}(\mathbf{y}(i)) \tag{5.3}$$

which can be iterated to find all subsequent crossings of S. Using this Poincaré surface of section technique, we can, in general, discretize an nth order continuous flow into an $n-1$st order map, called the Poincaré map. Note that, the common time discretization via strobing is a special Poincaré map, where a periodic function of the system time, e.g., $sin(wt)$, is considered as a state variable, and the surface of section is selected, such as $sin(wt) = 0$.

Phase Volume

The way the phase space volume changes in time is an important property of systems with continuous or discrete dynamics. Select a subset of the phase space with a positive finite (hyper-) volume, and evolve the points in this subset in time. If the volume defined by the new subset is always equal to the initial volume, the dynamics under investigation belongs to a *conservative* system, such as a Hamiltonian system. If, on the other hand, that volume is changing in time, we have a *nonconservative* system. If the phase volume of the system always increases, the system will be structurally unstable, and the trajectories will diverge to infinity. Thus we cannot observe such systems for long, and they are not of much interest. The class of systems with shrinking phase volume in time are called *dissipative* systems. They are structurally stable, and the methods introduced in this chapter are directed at studying such systems.

The rate of change of the phase space volume for a continuous flow defined by Eq (5.1) is given by the trace of the tangent flow matrix, (or the Jacobian matrix evaluated along the flow),

$$r = \mathrm{Tr}\left(\frac{\partial \mathbf{f}}{\partial \mathbf{x}}\right). \tag{5.4}$$

If this rate is positive (negative), then the phase space volume grows (shrinks) in time.

Example 1 *Phase volume change of a flow*

Consider the "deterministic non-periodic flow" of Lorenz [350], given by,

$$\begin{aligned}\frac{dx_1}{dt} &= -\sigma(x_1 - x_2)\\ \frac{dx_2}{dt} &= -x_1x_3 + \rho x_1 - x_2\\ \frac{dx_3}{dt} &= x_1x_2 - \beta x_3\end{aligned} \tag{5.5}$$

with σ, ρ and β are real positive constants. The tangent flow matrix of the system can be found as

$$\frac{\partial \mathbf{f}}{\partial \mathbf{x}} = \begin{bmatrix} -\sigma & \sigma & 0 \\ -x_3 + \rho & -1 & -x_1 \\ x_2 & x_1 & -\beta \end{bmatrix} \tag{5.6}$$

which has a trace $r = -\sigma - 1 - \beta$, that is less than zero. Thus, an initial phase space volume, $V(0)$ shrinks with time as $V(t) = V(0)e^{rt}$.

A similar definition is made for the map of Eq. (5.2), using the magnitude of the determinant of the tangent flow matrix,

$$r = \left|\det\left(\frac{\partial \mathbf{f}}{\partial \mathbf{x}}\right)\right|. \tag{5.7}$$

Eq. (5.7) defines the factor by which the n-dimensional phase space volume changes. If this factor is greater (less) than one, the phase space volume grows (shrinks) at the next iteration.

Example 2 *Phase volume change of a map*

Consider the two dimensional Hénon map [231], given by

$$\begin{aligned} x_1(i+1) &= \alpha - x_1^2(i) + \beta x_2(i) \\ x_2(i+1) &= x_1(i), \end{aligned} \tag{5.8}$$

where α and β are constants. The tangent flow matrix for this system

$$\frac{\partial \mathbf{f}}{\partial \mathbf{x}} = \begin{bmatrix} -2x_2(i) & \beta \\ 1 & 0 \end{bmatrix}, \tag{5.9}$$

has a constant determinant $r = |\beta|$. The *hyper volume* defined in this phase space is in fact an *area*, since the phase space is two dimensional. If the "absolute value" of the parameter β is less than one, then the area shrinks by a factor of $|\beta|$. □

Systems with phase space contraction, such as the ones presented in the last two examples, are commonly characterized by the presence of *attractors.* The trajectories of the system originating from a specific region of the phase space are attracted to a bounded subset of the phase space, called the attractor, and that specific region that hosts all such initial conditions is called the *basin of attraction* for that attractor.

Chaos— Sensitive Dependence on Initial Conditions

In a two dimensional phase space, possible scenarios are limited for a continuous flow. Distinct trajectories cannot intersect because of the existence and uniqueness conditions. They either diverge to infinity, or converge to a limit point[1] or a limit cycle[2] However, when the phase space dimension increases to three, something fascinating happens. The trajectories can enjoy the freedom of staying confined in the phase space, without converging to a limit point or to a limit cycle. Let us illustrate this with an example.

---

[1]Limit point: a final state, where the trajectory of a dynamics converge.

[2]Limit cycle: the periodic motion displayed by the trajectory of a dynamics.

Example 3 *Driven pendulum*

Using Newton's second law, the dynamics of a damped, sinusoidally driven pendulum are expressed as

$$\frac{d^2\theta}{dt^2} + \xi\frac{d\theta}{dt} + \sin(\theta) = a\cos(\beta t) \tag{5.10}$$

where θ is the angular displacement from the vertical, ξ is the damping coefficient accounting for friction, and a and β are the forcing amplitude and forcing frequency, respectively. Note that, this is a non-autonomous second order system. Applying the definitions,

$$\begin{aligned}\frac{dw}{dt} &= -\xi w - \sin(\theta) + a\cos(\phi)\\ \frac{d\theta}{dt} &= w\\ \frac{d\phi}{dt} &= \beta\end{aligned} \tag{5.11}$$

we can transform the system into an autonomous third order system. First, consider a pendulum with no forcing ($a = 0$), which reduces the phase space dimension to two. The system will have infinitely many steady states, located at $\theta = \pm k\pi$ and $w = 0$, with $k = 0, 1, 2, \ldots$. The steady states for even k (corresponding to the lower vertical position) are stable, and those for odd k (corresponding to the upper vertical position) are unstable. If we also set $\xi = 0$ to eliminate friction, the pendulum will swing back-and-forth, or rotate in an infinite loop, determined solely by its initial conditions, as shown in Figure 5.2.a. Note that, the pendulum with no friction is a Hamiltonian system, hence it conserves the phase space volume due to Liouville theorem. If, however, we consider a finite friction, the energy of the trajectories will eventually be consumed, and the pendulum will come to a halt at one of its steady states (Figure 5.2.b). To have a better understanding of the mechanism of this, take a closer look at the trajectories near the two types of steady states. Near the stable ones, (Figure 5.3.a) the trajectories spiral down to the steady state. Near the unstable steady states, trajectories approach the (saddle) steady state from one direction, and are repelled in another direction. There are some trajectories that seem to violate the uniqueness of the solution; they approach the steady state from opposite directions to meet at the steady state, and diverge from it in opposite direction, starting from the steady state. If we consider the time aspect of the problem, the uniqueness condition is not actually violated, since it takes infinitely long to converge to the steady state, on the trajectories that end in there. Similarly, for the trajectories that emanate from the steady state, it takes infinitely long to diverge from the steady state. Another property of the trajectories that converge to this steady state is the partitioning of the phase space in the sense that trajectories on the right hand side of this trajectory cannot cross over to the left hand side of it, and vice versa. Therefore, they define the boundaries of the basins of attraction.

Next, introducing the forcing back into the system, we have a three dimensional phase space. For certain combinations of driving amplitude and frequency, we observe a rich dynamic behavior, which neither converges to a steady state, nor gets attracted by a limit cycle. Instead, the trajectories explore a finite subset of the phase space, converging to a *strange* attractor.

When the system converges to a steady state (also called a limit point), the limit set of the system in phase space is an object of zero dimension. When it converges to a limit cycle, the limit set is still an object of integer dimension (one). However, when the system exhibits a rich dynamic behavior, such as the one shown in Figure 5.2.c, the limit set is a fractal object with a non-integer dimension. We will discuss the concept of dimension in the next section in more detail.

One way of identifying chaotic behavior is using the Poincaré surface of section technique. For example, let us consider the periodically driven pendulum again, and use a surface of section on the angle of the forcing term ϕ. If we operate the system with $\xi = 0.4$, $a = 1$ and $\beta = 2/3$, it converges to a periodic trajectory which gives a single point in the Poincaré surface of section of Figure 5.4.b. If we operate it with $\xi = 0.4$, $a = 1.4$ and $\beta = 2/3$, the dynamics would be richer, and we observe a fractal object resembling the shape in the projection of the attractor on the (θ, w)-plane (Figure 5.4.d). This kind of an attractor is called a *strange attractor*.

The manifestation of chaos in the dynamics of a system is often associated with a sensitive dependence on its initial conditions. If we initialize our driven pendulum with slightly different initial conditions around its strange attractor, initially nearby trajectories diverge exponentially in time as shown by solid and dotted curves in Figure 5.5.a.

If we observe a block of initial conditions, shown in Figure 5.5.b, for 4 units of simulation time, the volume element that we started with shrinks in one direction, and is stretched in another. If we keep on observing the system, since the trajectories stay confined in a certain region of the phase space, the volume element cannot perform this shrinking and stretching without eventually folding on itself (Figure 5.5.c). This *stretch-and-fold* routine repeats itself as the dynamics further evolves. Hence, we will eventually find points that were arbitrarily close initially, separated in the phase space by a finite distance. In fact, the stretch-and-fold is the very mechanism that generates the fractal set of the strange attractor. □

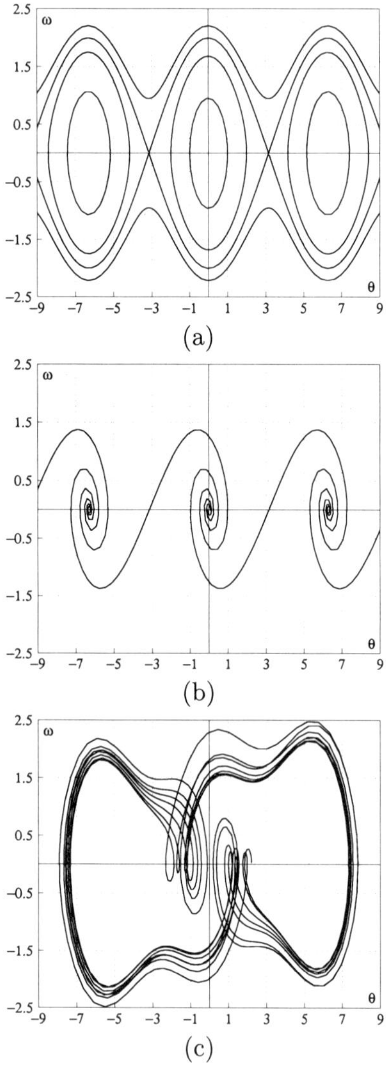

Figure 5.2. Phase space of pendulum, projected on (θ, w) plane. (a) Trajectories for no friction case with several initial conditions either oscillating or rotating around the stable steady state. (b) Trajectories for several initial conditions converge to the stable steady state, when there is friction. (c) The chaotic trajectory of the driven pendulum with friction.

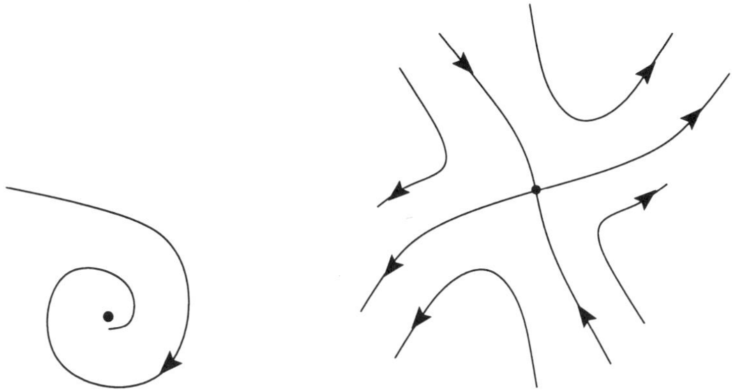

(a) (b)

Figure 5.3. Trajectories near steady states of the pendulum, $\theta = i\pi$ (a) for $i =$ even, and (b) for $i =$ odd.

There are several different measures that quantify this fingerprint of chaos. For example, the exponential divergence and convergence of nearby trajectories in different axial directions is measured by the *Lyapunov exponents* of the system. An nth order system has n Lyapunov exponents associated with the exponential rate of growth or shrinkage along its principal axes, and the set of all n Lyapunov exponents of a system is called its *Lyapunov spectrum.*

The notion of Lyapunov exponents can best be visualized by considering the experiment of putting a droplet of ink in a glass of water. The sphere described by the ink at the instant it is dropped, represents a family of nearby trajectories. In the course of time, the droplet gets deformed slowly, first into an ellipsoid, and then diffuses in the liquid. If we watch the droplet in slow motion, we can see that it is stretched in some directions, and squeezed in others. After some time, we see a folding occurring, as if to keep the droplet in the glass. In this analogy, the stretch of the droplet corresponds to a positive Lyapunov exponent, and the squeeze to a negative one. Since the phase volume is conserved in this system, i.e., total amount of ink in the glass is constant, the amount of squeezes is equal to the amount of stretches. Therefore, the sum of the Lyapunov exponents is equal to zero.

To put the idea in mathematical context, consider the autonomous continuous flow of Eq. (5.1), and observe the long-term evolution of an infinitesimal n-sphere of initial conditions. As in the ink analogy, the sphere will become an n-ellipsoid due to the locally deforming nature of the flow.

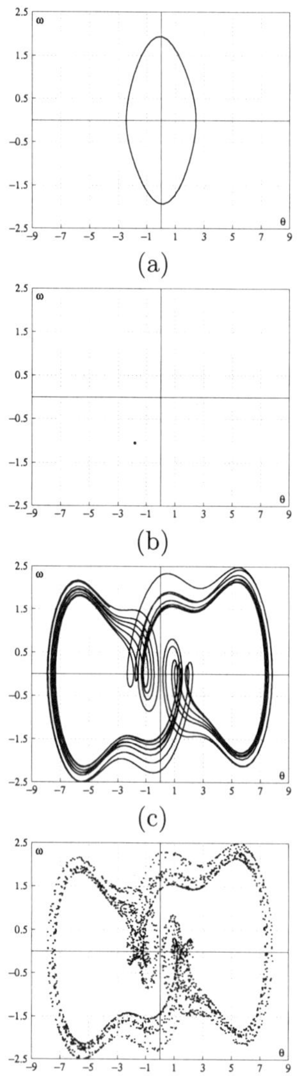

Figure 5.4. Periodically driven pendulum (a) goes to a limit cycle for $\xi = 0.4$, $a = 1$ and $\beta = 2/3$. (b) Strobing it with a frequency that is a multiple of the oscillation frequency results in a single point in the Poincaré section. (c) If we operate the system with $\xi = 0.4$, $a = 1.4$ and $\beta = 2/3$, it leads to this chaotic orbit, and (d) strobing this motion results in a fractal object in the Poincaré section.

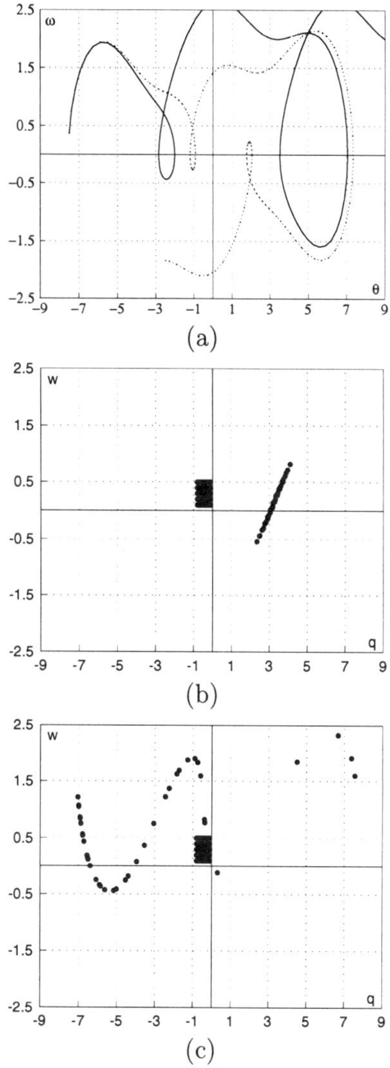

Figure 5.5. (a) Trajectories of two realizations of the periodically driven pendulum for two nearby initial conditions $(\theta_0, w_0, \phi_0) = (0.36564, -7.4964, 100)$ (solid curve) and $(\theta_0, w_0, \phi_0) = (0.36564, -7.4964, 100.1)$ (dotted curve). (b) How a rectangular block of initial points get deformed in a simulation interval of 4 units, stretching in one direction and squeezing in the other. (c) Same initial rectangular block after a simulation time of 10 units.

Using the length of the ith ellipsoidal principal axis $p_i(t)$, we can define the ith Lyapunov exponent of the system from

$$\lambda_i = \lim_{t \to \infty} \frac{1}{t} \ln \frac{p_i(t)}{p_i(0)} \tag{5.12}$$

when the limit exists. λ_i are conventionally ordered from largest to the smallest. Note that, this definition is akin to the definition of eigenvalues for linear systems, but unlike the eigenvalues, there is no unique direction associated with a given Lyapunov exponent. This is understandable, since the eigenvalue is a local definition, and, characterizes a steady state, while the Lyapunov exponent is a time average associated with a principal axis, that continuously changes orientation as it evolves.

As one classifies linear systems using their eigenvalues, Lyapunov spectra can be used to classify the asymptotic behavior of nonlinear systems. For example, for a system to be dissipative, the sum of its Lyapunov exponents should be negative. Likewise, if we have a Hamiltonian system, the sum of its Lyapunov exponents should be zero, due to the volume preserving property of such systems. A continuous dynamical system is chaotic, if it has at least one positive Lyapunov exponent.

In the investigation of chaotic systems, we have mentioned that third-order systems have a special importance. For third order dissipative systems, we can easily classify the possible spectra of attractors in four groups, based on Lyapunov exponents.

1. $(-,-,-)$: a fixed point,
2. $(0,-,-)$: a limit cycle,
3. $(0,0,-)$: a 2-torus,
4. $(+,0,-)$: a strange attractor.

Therefore, the last configuration is the only possible third-order chaotic system. However, in a continuous fourth-order dissipative system, there are three possible types of strange attractors with Lyapunov spectra $(+,0,-,-)$, $(+,0,0,-)$ and $(+,+,0,-)$. Note that, all three configurations have at least one vanishing Lyapunov exponent. In fact, it is required by the theorem of Haken [215] that the system should have at least one zero Lyapunov exponent, if the trajectory of its attractor does not have a fixed point. The last case where there are two positive Lyapunov exponents is called the *hyper chaos*.

The classical Lyapunov exponent computation method of Wolf et al. [669] is based on observing the long time evolution of the axes of an infinitesimal sphere of states. It is implemented by defining the principal

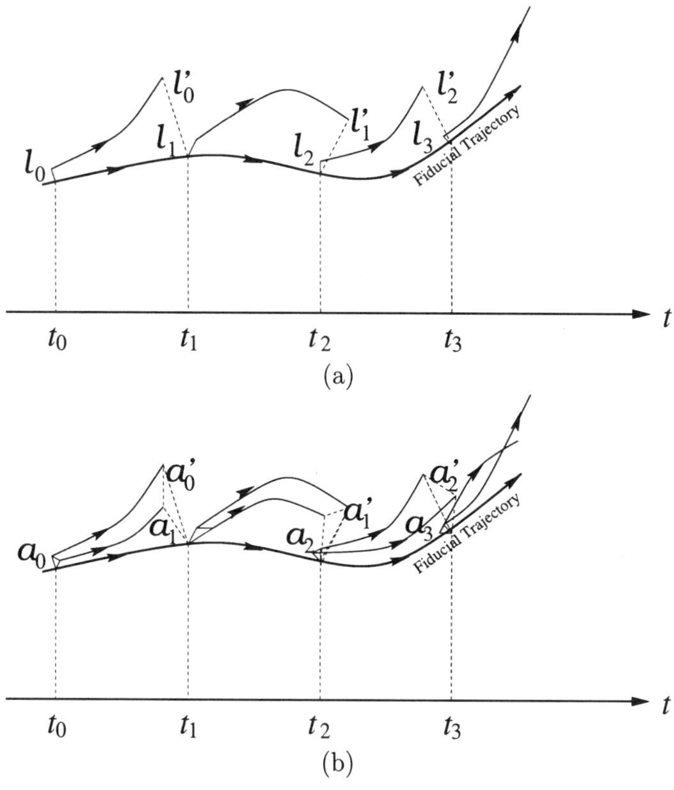

Figure 5.6. Time evolution of the fiducial trajectory and the principal axis (axes). (a) The largest Lyapunov exponent is computed from the growth of length elements. (b) The sum of the largest two Lyapunov exponents is computed from the growth of area elements.

axes, with initial conditions that are separated as small as the computer arithmetic allows, and by evolving these using the nonlinear model equations. The trajectory followed by the center of the sphere is called the *fiducial trajectory*. The principal axes are defined throughout the flow via the linearized equations of an initially orthonormal vector frame "anchored" to the fiducial trajectory. To implement the procedure, the fiducial trajectory on the attractor is integrated simultaneously with the vector tips defining n arbitrarily oriented orthonormal vectors. Eventually, each vector in the set tends to fall along the local direction of most rapid growth (or a least rapid shrink for a non-chaotic system). On the other hand, the collapse toward a common direction causes the tangent space orientation of all axis vectors to become indistinguishable. Therefore, after a certain interval, the principal axis vectors are corrected into an orthonormal set, using the Gram-Schmidt reorthonormalization. Projection of the evolved vectors onto the new orthonormal frame correctly updates the rates of growth of each of the principal axes, providing estimates of the Lyapunov exponents. Following this procedure, the rate of change of a length element, l_i, around the fiducial trajectory, as shown in Figure 5.6.a, would indicate the dominant Lyapunov exponent, with

$$\lambda_1 = \sum_{k=0}^{K} \frac{1}{t_{k+1} - t_k} \ln \frac{l'_k}{l_k}. \tag{5.13}$$

Similarly, the rate of change of an area element, as shown in Figure 5.6.b would indicate the sum of the largest two Lyapunov exponents, with

$$\lambda_1 + \lambda_2 = \sum_{k=0}^{K} \frac{1}{t_{k+1} - t_k} \ln \frac{a'_k}{a_k}. \tag{5.14}$$

The idea can be generalized to higher dimensions, considering volume elements for the largest three Lyapunov exponents, hypervolume elements for the largest four Lyapunov exponents, and so on.

The reorthonormalization procedure can further be implemented in every infinitesimal time step. This continuum limit of the procedure can be expressed by the set of differential equations

$$\begin{aligned}
\frac{d\mathbf{x}}{dt} &= \mathbf{f}(\mathbf{x}) \\
\frac{d\mathbf{e}_i}{dt} &= \left[\frac{\partial \mathbf{f}}{\partial \mathbf{x}} - \sum_{j=1}^{i} \mathbf{e}_j \mathbf{e}_j^T \frac{\partial \mathbf{f}}{\partial \mathbf{x}} - \sum_{j=1}^{i-1} \mathbf{e}_j \mathbf{e}_j^T \left(\frac{\partial \mathbf{f}}{\partial \mathbf{x}} \right)^T \right] \mathbf{e}_i \\
\frac{d\lambda_i}{dt} &= \frac{1}{t}(-\lambda_i + \mathbf{e}_i^T \frac{\partial \mathbf{f}}{\partial \mathbf{x}} \mathbf{e}_i)
\end{aligned} \tag{5.15}$$

where $\mathbf{e}_i \in \mathcal{R}^n$ with $i = 1, \ldots, n$ stand for the orthonormal basis vectors of the tangent flow around a fiducial trajectory[645]. The initial conditions of the state variables are set by the original system. Although the augmenting variables can be initialized arbitrarily, it would be a good practice to select an orthonormal set for $\mathbf{e}_i$, such as $\{\mathbf{e}_i\} = \mathbf{I}_n$, and a neutral guess for λ_i, such as $\lambda_i = 0$. In the limit $t \to \infty$, the set $\{\lambda_i\}$ will give the Lyapunov spectrum of the system.

The concept of Lyapunov exponents is illustrated for continuous flows in this presentation, but it can be carried to discrete maps as well. For instance, if the map (or a time series data for that matter) represents samplings of a continuous flow, the amount of growth or contraction associated with the ith Lyapunov exponent will be $\sigma_i = e^{\lambda_i \Delta t}$, and is called the *Lyapunov number*. Therefore, when ith Lyapunov exponent is positive (negative) the ith Lyapunov number will be greater (less)than unity.

Routes to Chaos

Unlike the continuous flows, discrete maps need not have a minimum phase space dimension to exhibit chaotic behavior. Since the values are attained at discrete instances, orbit crossings in data representations are mostly superfluous, hence do not pose the existence-uniqueness problems of continuous flows. Even a first order discrete map can produce chaotic behavior, as shown in the following example.

Example 4 *Logistic map*

The logistic map is a one dimensional nonlinear system, given by the difference equation

$$x(i+1) = \mu x(i)(1 - x(i)) \tag{5.16}$$

which was originally proposed to model population dynamics in a limited resource environment [375]. The population size, $x(i)$, at instant i is a normalized quantity. It can be easily shown that a choice of μ in the range $[0, 4]$ guarantees that, if we start with a physically meaningful population size, i.e., $x(0) \in [0, 1]$, the population size stays in $[0, 1]$.

If we simulate the system with an initial condition $x(0) = 0.1$, we will obtain the results shown in Figure 5.7 for various μ values. The system goes to a steady state for $\mu = 1$ and $\mu = 2$, but as μ is further increased to 2.9, the behavior of the convergence to a steady state is qualitatively different than the previous cases. It reaches the steady state in an oscillatory manner. For $\mu = 3.3$, the oscillations are not damped anymore, and we have periodic behavior, every other value of x_i being equal for large i. The system is said to have a *two-period* oscillation in this regime. For $\mu = 3.5$, the asymptotic behavior of the system is similar to the previous case. This time, we have a four-period oscillation though. The demonstrated increase in the period is actually common to many nonlinear systems, and is called *period doubling*. The period-doubling mechanism is a route to chaos, that has been studied extensively, since it is encountered in many dynamical systems. One interesting finding is that, period doubling may be characterized by a universal number independent of the underlying dynamics. In our example, if we label the kth period doubling value of μ with μ_k, then

$$\delta = \lim_{k\to\infty} \frac{\mu_k - \mu_{k-1}}{\mu_{k+1} - \mu_k} \tag{5.17}$$

where $\delta \simeq 4.669$ is the Feigenbaum number [150]. Naturally, the period doubling values of the parameter, μ_k, depend on the system dynamics. However, the number δ is *universal* (i.e., the same) for all one-dimensional maps. A detailed study of the link between the Feigenbaum number and the period doublings is beyond the scope of this text. (Interested reader can check the rigorous works of Lanford [313] and Collet and Eckmann [109, 110]) However, its implication for our case is important, which suggests that as $\mu = 3.569$ is approached, an infinite number of period doublings will occur.

Although different types of dynamics can be visualized by plotting the time evolution of the system for a specific parameter set (Figure 5.7), or by plotting orbits or Poincaré surface of sections in the phase space, they are far from representing the global behavior of the system for a range of parameter values. The *bifurcation diagram* provides a summary of the dynamics by plotting the essential dynamics for a range of parameter values. For example, if we plot the steady states of the system versus the system parameter, μ, we obtain the bifurcation diagram of Figure 5.8.a. The solid curve corresponds to the stable steady states, and the dashed curve to the unstable ones. The steady state loses its stability at $\mu = 3$, which is where the period doubling occurs.

Alternatively, we can plot the limiting values of $x(i)$ for the logistic map versus the system parameter, μ, to obtain the bifurcation diagram of Figure 5.8.b. This bifurcation diagram summarizes the period doubling cascade that leads to chaos. It shows that, there is a period doubling at $\mu = 3$, and another around $\mu = 3.45$, and indeed the system goes chaotic around $\mu = 3.569$, as suggested by the Feigenbaum number. Looking at this bifurcation diagram, we not only have a complete picture of how the period doubling cascade results in chaotic behavior, but we may also disclose some types of behavior that we did not know have existed, such as the *window* around $\mu = 3.83$, where the system exhibits periodic behavior. The logistic map is attracted by a three-period orbit when operated with

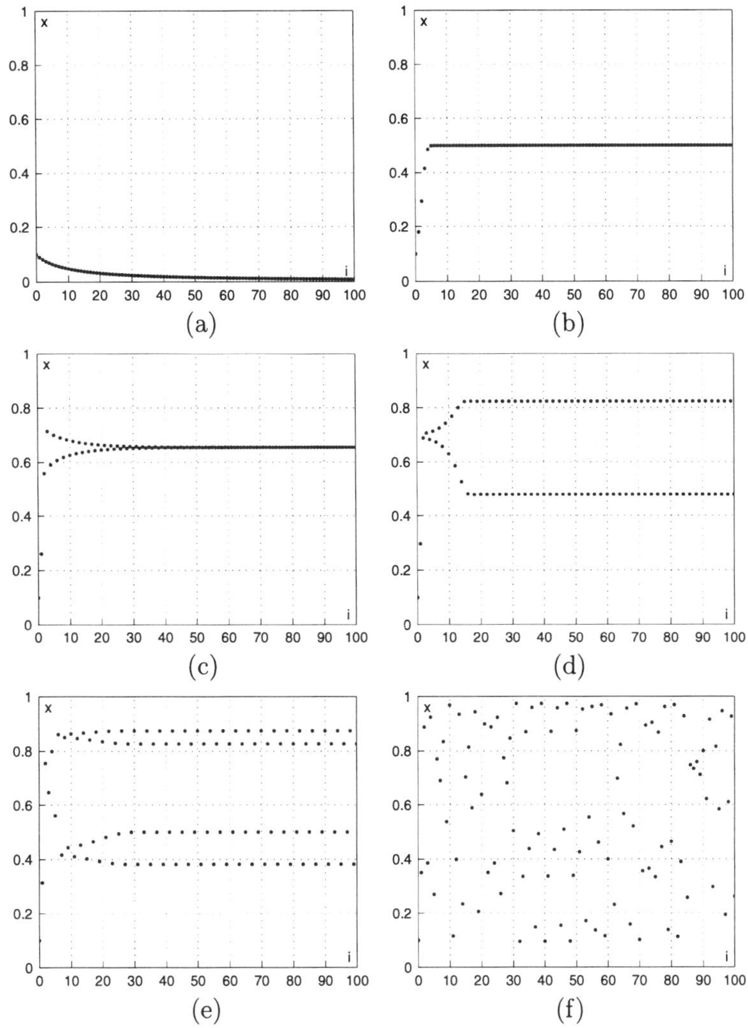

Figure 5.7. Simulation results of the logistic map for (a) $\mu = 1$, (b) $\mu = 2$, (c) $\mu = 2.9$, (d) $\mu = 3.3$, (e) $\mu = 3.5$, and (f) $\mu = 3.9$.

$\mu = 3.83$ (Figure 5.9), which is an evidence that this map has a region in the parameter space (μ for this example) that it experiences chaotic behavior.

After this periodic window, we again observe a chaotic regime, but this time the chaos is reached via the *intermittency* route. There are three documented intermittency routes [540]

1. a real eigenvalue crosses the unit circle at +1;
2. a complex conjugate pair of eigenvalues cross the unit circle;
3. a real eigenvalue crosses the unit circle at -1.

The logistic map exhibits a type 1 intermittency after the three-period window. □

Apart from the period doubling and intermittency routes, there is a third route to chaos, called the *quasiperiodicity* route, originally suggested to understand turbulence [419]. In this mechanism, the stable steady state of a dynamical system becomes unstable at a certain value of the bifurcation parameter, in a special manner. In a continuous system, with the changing bifurcation parameter, a complex conjugate pair of eigenvalues cross the imaginary axis, making the steady state unstable and creating a stable limit cycle around it. This transition of the complex conjugate eigenvalues through the imaginary axis with changing bifurcation parameter is called the *Hopf bifurcation.* After a Hopf bifurcation that makes the steady state unstable, the dynamics can have another Hopf bifurcation, this time making the stable limit cycle unstable. Right after this second bifurcation, the trajectory is confined on a torus, and the states of the system show a quasiperiodic behavior. Some further change in the bifurcation parameter may result in a third Hopf bifurcation, taking the trajectory on a three torus, which decays to a strange attractor after certain infinitesimal perturbations.

Example 5 *Autocatalytic reactions*

Consider the cubic autocatalysis

$$R + 2P \xrightarrow{k_p} 3P \tag{5.18}$$

with decay

$$P \xrightarrow{d_p} D \tag{5.19}$$

as a paradigm for population dynamics of sexually reproducing species [64, 65], with k_p and d_p representing the birth and death rates of the species P, respectively. If we let these reactions occur in two coupled identical

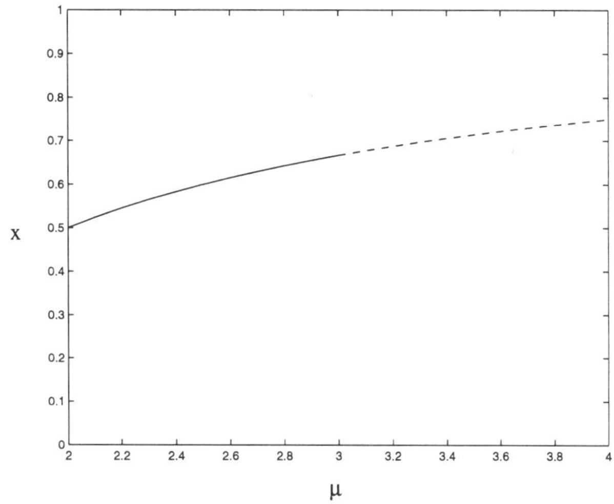

(a)

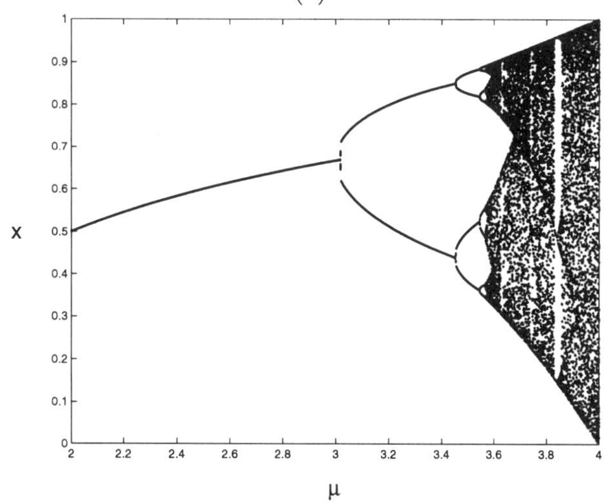

(b)

Figure 5.8. Bifurcation diagrams of the logistic map, (a) showing the stable (solid curve) and the unstable (dashed curve) steady states of the system versus the system parameter, μ, and (b) showing the limiting values of $x(i)$ versus the system parameter, μ.

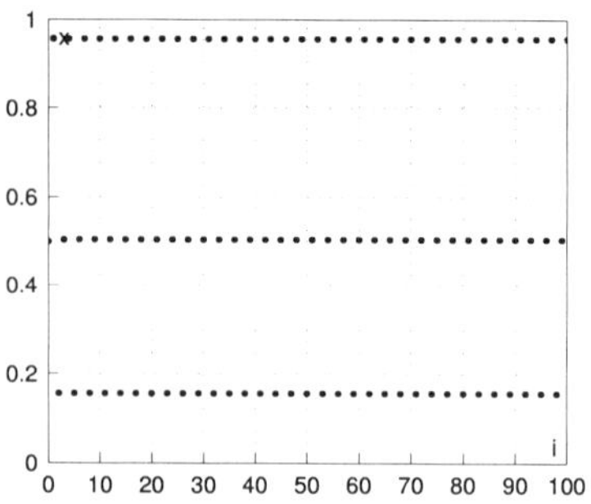

Figure 5.9. Logistic map with $\mu = 3.83$ results in a three-period orbit.

continuous stirred tank reactors fed with a volumetric flow rate f, and a coupling strength g, as shown in Figure 5.10, we can write the material balance equations as

$$\begin{aligned} \frac{dr_i}{dt} &= -k_p r_i p_i^2 + f(r_0 - r_i) + g(r_j - r_i) \\ \frac{dp_i}{dt} &= k_p r_i p_i^2 - (f + d_p)p_i + g(p_j - p_i) \end{aligned} \tag{5.20}$$

where r_i is the resource concentration in the ith tank, r_0 is the resource concentration in the feed, p_i is the concentration of species P in the ith tank, and $i = 1, 2$, $j = 1, 2$ with $i \neq j$.

Investigating a species P with $k_p = 25$ and $d_p = 0.1$ in a setup with concentration, $r_0 = 1$ and coupling strength $g = 0.002$, we can use the feed flow rate f as a bifurcation parameter to plot the bifurcation diagram of Figure 5.11. For $f = 0.006728$, we get the chaotic trajectory of Figure 5.12. □

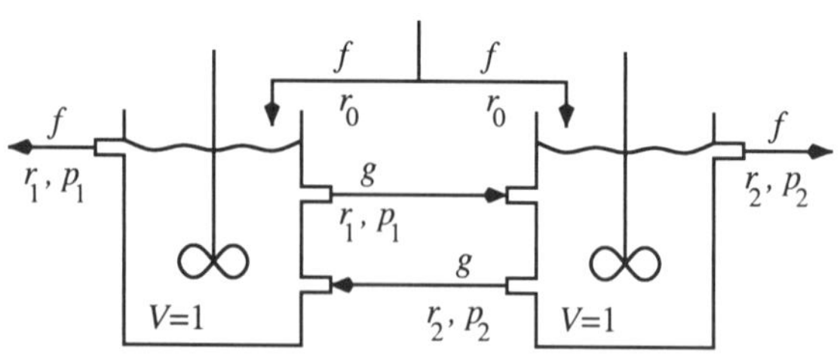

Figure 5.10. Two coupled identical CSTRs with cubic autocatalytic species P.

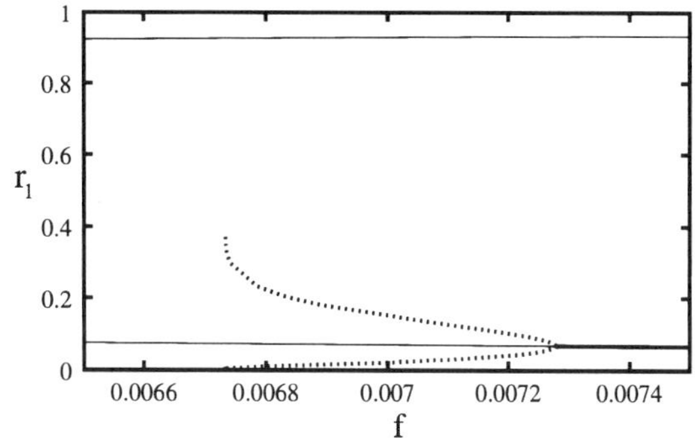

Figure 5.11. Bifurcation diagram of the autocatalytic species P, showing the stable steady states (thick curve), the unstable steady states (thin curve), and the stable limit cycle (dotted curve), using the feed flow rate, f as the bifurcation parameter.

5.2 Nonlinear Time Series Analysis

The tasks in the analysis of time series observed from nonlinear systems are not very different from those involved in the analysis of linear systems. However, the methods for the analysis are substantially different. We can classify the tasks as, (1) state-space reconstruction, (2) signal filtering, (3) system classification, and (4) model development.

5.2.1 State-Space Reconstruction

If the source of the signal were an autonomous linear system, looking at the frequency spectrum of the signal would be utmost informative, thus the Fourier domain would be an appropriate space to examine the signal. If the linear system had an explicit time component hosting some burst of high frequency events localized in time domain, then a linear transform, such as wavelet transform, would be useful. Much of the contemporary signal processing toolkits (e.g. Matlab [373]) are based on the ability to perform a linear transformation that converts a low-order ordinary differential equation to another domain, and perform matrix manipulations on the transformed algebraic representation of the measurements.

In the case of a nonlinear source, we are not likely to find any simplification from using a linear transformation, such as Fourier, since the processes that give rise to chaotic behavior are fundamentally multivariate. Consequently, we need to reconstruct the (multidimensional) state-space of

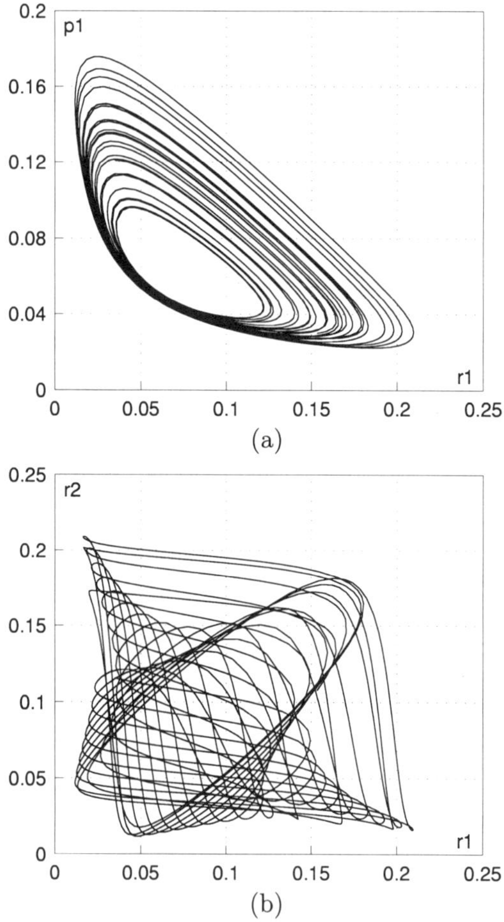

Figure 5.12. Two projections of the chaotic orbit of the autocatalysis system.

the system, as accurately as possible, using the information in the (usually scalar) time series. This reconstruction will result in vectors in an m-dimensional space that unfolds the structure the orbits follow in the multidimensional phase space. Therefore, the focus now is, how to choose the components of the m-dimensional vectors, and of course, how to determine the value of m itself.

The answer to this lies in a combination of concepts in dynamics about nonlinear systems as generators of information, and in geometry ideas about how one unfolds an attractor using coordinates established on the basis of their information content. The result of this operation will be a set of m-dimensional vectors, that replace the original scalar data we have filtered.

Although multidimensional measurements are becoming more common because of the wide availability of computer driven data acquisition systems, such measurements do not often cover all degrees of freedom of the underlying dynamics. Furthermore, scalar measurements still constitute the majority of the recorded time series data.

The most commonly used phase space reconstruction technique utilizes the so called *delay coordinates*. If we represent the measured scalar time series by $\{y_i\}$, then we can reconstruct a phase space using

$$\hat{\mathbf{x}} = [y_{i-(m-1)\tau}, y_{i-(m-2)\tau}, \ldots, y_i]^T \tag{5.21}$$

where m is called the *embedding dimension*, and τ the *time delay*. The embedding theorems of Takens [582] and Sauer et al. [535] show that, under some conditions, if the sequence $\{y_i\}$ is representative of a scalar measurement of a state, and m is selected *large enough*, the time delay coordinates provide a one-to-one image of the orbit with the underlying dynamics.

Example 6 *Time delay representation of the blood oxygen concentration signal*

Consider the blood oxygen concentration (measured by ear oximetry) data set recorded from a patient in the sleep laboratory of the Beth Israel Hospital in Boston, Massachusetts [196] (Figure 5.13.a). The data were a part of the Santa Fe Institute Time Series Prediction and Analysis Competition in 1991, and belonged to a patient with sleep apnea. The data were collected with the patient taking a few quick breaths and then stopping breathing for up to 45 seconds. If we could develop a viable low-dimensional model for the system, we could predict stoppage of breathing from the preceding data, which would be a medically significant application.

Consider a time delay representation of the blood oxygen concentration in two dimensions. If we select a time delay of 2 seconds, the autocorrelation of the data would overshadow the representation (Figure 5.13.b). Selecting a larger time delay of 25 seconds would present a more spread signal in the reconstructed phase space (Figure 5.13.c).

It is apparent that, if the frequency of measurements is higher than the dynamical fluctuations of the system, choosing a too small time delay would result in a highly correlated state variables. On the other hand, since our data set is of finite length, we cannot have a too large time delay. Thus, there should be an *optimum* way of selecting this parameter. Furthermore, if we select a low dimensional reconstruction space, the orbits would intersect, and we would not be able to untangle the dynamics. Many authors point out that it would be safe, in terms of representing the dynamics in a multidimensional phase space, if we select a *large enough* dimension, m. However, since our goal is to model the dynamics (as opposed to its representation), we should seek the smallest possible m, that would untangle the dynamics. Then again, there should be an *optimum* way of selecting m.

First, consider the choice of time delay τ. If we were to make multivariate measurements on the patient of the previous example, we would prefer measuring his heart rate, rather than his blood oxygen concentration measured from his arm. In other words, we would not like to measure closely related (or correlated, in mathematical terms) quantities. Based on this idea, some authors (c.f. [340]) propose the least time delay that minimizes the correlation between $y(t-\tau)$ and $y(t)$. Others [165] argue that, since the underlying dynamics is nonlinear, and the correlation coefficient is a linear concept, we should be selecting the time delay by monitoring the *mutual information content* of the series $y(t-\tau)$ and $y(t)$, which quantifies the amount of information gathered (in bits) about signal $y(t)$ by measuring $y(t-\tau)$. The mutual information content between these signals is defined as,

$$I_\tau = \sum_{y(t-\tau),y(t)} P_{ab}(a = y(t-\tau), b = y(t)) \log_2 \frac{P_{ab}(a = y(t-\tau), b = y(t))}{P_a(a = y(t-\tau))P_b(b = y(t))} \tag{5.22}$$

where P_{ab} is estimated from the normalized histogram of the joint distribution, and P_a and P_b are the marginal distributions for $y(t-\tau)$ and $y(t)$, respectively. Similar to the correlation coefficient guided selection of the time delay, we should select the τ where I_τ attains its first local minimum. Note that, for our example data, the choice of τ with both methods is around 25 seconds (Figure 5.14). The drift towards zero in both quantities is due to the finite data size of 17,000, with a sampling rate of 2 measurements per second. Although both methods of choosing a time delay give useful guidelines, one should always make a reality check with the data.

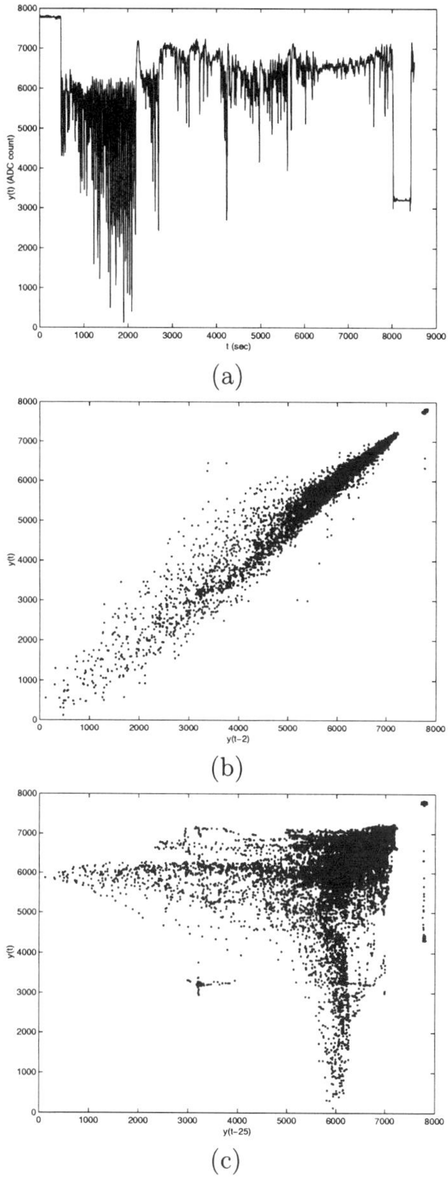

Figure 5.13. (a) Time series data of the blood oxygen concentration of a sleep apnea patient. (b) Selection of a too small time delay ($\tau = 2$) hides the information in the data. (c) Selecting a more appropriate time delay ($\tau = 25$) reveals more detail in the data.

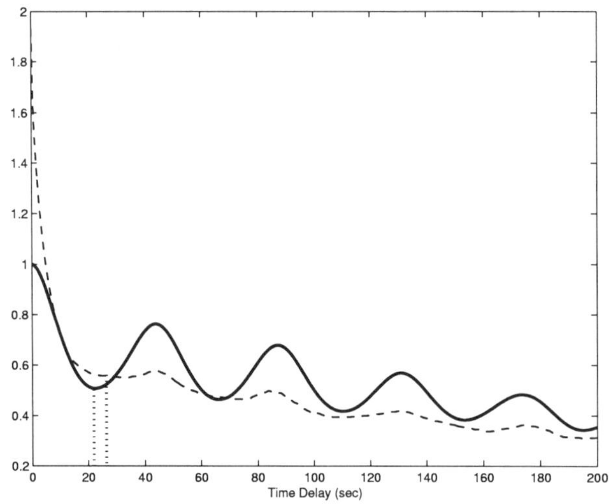

Figure 5.14. The correlation coefficient (solid line), and the mutual information content (dashed line) versus the time delay.

When analyzing chaotic signals, it is always easy to use a certain algorithm that gives out a certain result. Yet, the results should always be scrutinized before being adapted, as chaotic systems usually defy haute couture solutions. For a review of different methods to select the time delay, see [233] and the references therein. □

According to the embedding theorems of Takens [582] and Sauer et al. [535], if the attractor has a dimension d, then an embedding dimension of $m > 2d$ is sufficient to ensure that the reconstruction is a one-to-one embedding. The geometric notion of dimension can be visualized by considering the hyper-volume occupied by a hyper-cube of side r in dimension d. This volume will be proportional to r^d, and we may get a sense of dimension by measuring how the density of points in the phase space scale when we examine *small* r's. One of the methods to compute *a* dimension of the attractor is called *the box counting method* [90]. To evaluate d, we count the number of boxes necessary to cover all the points in the data set. If we evaluate this number, $N(r)$ for two *small* values of r, then we can estimate d as

$$d = \frac{\log N(r_1)/N(r_2)}{\log r_1/r_2}. \tag{5.23}$$

Here, "small" r refers to a value that is much less than the radius of the data set, yet much greater than the distance between the second nearest neighbors in the data set. The choice of the second nearest neighbors is to eliminate the impossibility of comparing with a zero distance, should we have a repeating pattern in the data set. Although it is computationally straightforward to come up with a numerical value for d using this algorithm, producing an accurate value is often of doubt. A rule of thumb is to use at least $10^{d/2}$ data points to compute d [526].

Note that the box counting dimension computed from Eq (5.23) need not be integer. In fact, it is certainly non-integer for a strange attractor, hence the name *strange*. On the other hand, the phase space dimension m is an integer, and to host the attractor, it should be grater than or equal to the box counting dimension of the attractor, d. Although we stick with the box counting dimension in our arguments, there are other definitions of (fractal) dimensions, such as correlation and Lyapunov dimensions, but selecting one or the other would not change the line of thought, as different measurements of dimension of the same strange attractor should not differ in a way to contain an integer value in the range. Thus, no matter which definition we use for the fractal dimension, we have the same necessary condition $m \geq d$, and the same sufficient condition $m > 2d$.

Using these guidelines, one may be tempted to use an embedding dimension equal to the next integer value after $2d$. In an ideal case, where there is no noise in the infinitely many data points, such a selection would be sound and safe. However, in a more realistic setup, if m is chosen too large, the noise in the data will decrease the density of points defining the attractor. In this analysis we are interested in finite dimensional deterministic systems, whereas noise is an infinite dimensional process that fills each available dimension in a reconstructed phase space. Increasing m beyond what is minimally required has the effect of unnecessarily increasing the level of contamination of data with noise [669]. A method to determine the minimal sufficient embedding dimension is called the *false nearest neighbor* method [276].

Suppose that the minimal embedding dimension for our dynamics is m_0, for which a time delay state-space reconstruction would give us a one-to-one image of the attractor in the original phase space. Having the topological properties preserved, the neighbors of a given point are mapped onto neighbors in the reconstructed space. If we try to embed the attractor in an m-dimensional space with $m < m_0$, the topological structure would no longer be preserved. Points would be projected into neighborhoods of other points to which they would not belong in higher dimensions. Such data points are called *false neighbors*. To find the minimal embedding dimension, we should require the fraction of the false neighbors to be less than a heuristic value.

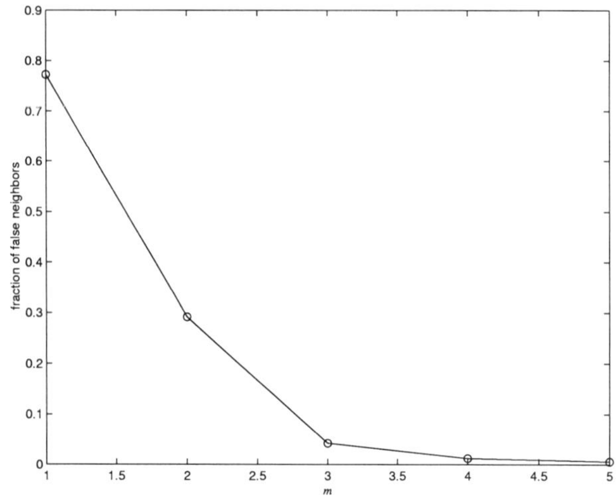

Figure 5.15. The fraction of false nearest neighbors as a function of the embedding dimension.

Example 7 *Embedding the blood oxygen concentration signal*

If we investigate the time series of the blood oxygen concentration signal of the previous example for false nearest neighbors, we can see that an embedding dimension of $m = 4$ would be enough to reconstruct the state-space (Figure 5.15). □

5.2.2 Nonlinear Noise Filtering

Every modeling effort starts with measuring some quantity, with the ultimate goal of understanding the process that generated it. Although making some measurements can be fairly easy, finding the signal out of the measurement is a task on its own. That is, we have to identify the signal which is possibly contaminated by fluctuations in the system, or by disturbances in the environment, or by the measurement procedure itself. Thus, before using the measurement for model development, it is often desirable to filter the measurement, and obtain as clear a signal as possible. In linear system theory, the process that generated the signal is assumed to send a frequency spectrum that has a finite range with sharp peaks, and the contamination is assumed to have a broadband spectrum. Then, the separation of the signal of interest from the noise becomes an exercise of distinguishing narrowband signals from broadband signals. Methods for this [172] are over fifty years old and are well developed.

In more general terms, in filtering the noise from the signal, we are separating the information-bearing signal and the interference from the environment. In the case of a narrowband signal, such as signals from a linear system, in a broadband environment, the distinction is quite straightforward. The frequency domain is the appropriate space to perform the separation, and looking at the Fourier spectrum is sufficient to differentiate the signal from noise.

Similar to the linear case, if the nonlinear process signal and the contamination are located in significantly distinct frequency bands, the Fourier techniques are still indicative. In sampling dynamic systems, if for example the Fourier spectrum of the system is bounded from above at a cut-off frequency, f_c, Shannon's sampling theorem states that, by choosing a sampling frequency, $f_s > 2f_c$, the signal can be perfectly reconstructed [172]. However, in the case of signals that come from sources that are dynamically rich, such as chaotic systems, both the signal and the contamination are typically broadband, and Fourier analysis is not of much assistance in making the separation. It is shown analytically that, the frequency spectrum of a system that follows intermittency route to chaos has a $1/f$ tail [540]. When the orbits converge to a strange attractor, which is a fractal limit set, it again has a $1/f$ tail in the frequency domain. Thus, for dynamically rich systems, no matter how high one considers the cut-off, the filtered portion of the signal will still have more information. This can be easily seen from the signal to noise ratio of a signal s, whose power content up to a frequency f_b is P, and for frequencies greater than f_b, it goes proportional to $1/f$. This ratio

$$\mathrm{SNR} = \frac{P + \alpha \int_{f_b}^{f_c} df/f}{\alpha \int_{f_c}^{\infty} df/f} \tag{5.24}$$

with α a real positive proportionality constant, vanishes for all $f_c < \infty$. Furthermore, we cannot practically consider a very large f_c, since most of the measurements are done by the aid of digital computers with finite clock frequencies. Nevertheless, we will be gathering measurements from such sources with finite sampling frequencies, and still wish to filter the data for the underlying signal. Another problem caused by finite sampling is the so called *aliasing effect.* That is, in the Fourier domain, the power contributions coming from the replicas of the original signal centered at the multiples of the sampling frequency are not negligible either.

If we can make the assumption that the signal we seek to separate is coming from a low-order system with specific geometric structure in its state space, we can make use of a deterministic system model or a Markov chain model, and seek for model parameters or transition probabilities via a time domain matching filter. The geometric structure of a system in its state space is characteristic for each chaotic process, which enables us to distinguish its signal from others. These separating techniques have a significant assumption about the nature of the process generating the signal, that is, the 'noise' we wish to separate from the 'signal' should be coming from a high-order chaotic source. Depending on the a priori information we have about the underlying system dynamics, various filtering problems can be stated.

- If we know the exact dynamics that generated the signal,

$$\mathbf{x}_{i+1} = \mathbf{f}(\mathbf{x}_i) \tag{5.25}$$

  with $\mathbf{x}_i \in \mathcal{R}^n$ (i.e., an n-dimensional real vector) and $\mathbf{f}(\cdot) : \mathcal{R}^n \to \mathcal{R}^n$ (i.e., an n-dimensional vector function that takes an n-dimensional argument), we can use this knowledge to extract the signal satisfying the dynamics. This method is referred as the regression technique.

- If we have a filtered signal from the system of interest extracted at some prior time, we can use this pivot signal to establish a statistics of the evolution on the attractor, and use it to separate the signal in the new set of measurements. This is gray box identification.

- If we know nothing about the underlying process and have just one instance of measurements, then we must start by making simplifying assumptions. Such assumptions may be that the dynamics is deterministic, and that it has a low-dimensional state space. This is black box identification.

Although as the problem bleaches out, the task of separating the signal from noise gets easier, the real life cases unfortunately favor darker shade situations. Various linear filtering and modeling techniques were discussed in Chapter 4.

To filter out noise in the time series signal, we will make use of the serial dependencies among the measurements, that cause the delay vectors to fill the available m-dimensional space in an inhomogeneous fashion. There is a rich literature on nonlinear noise reduction techniques [117, 295]. In this section we will briefly discuss one approach that exploits the geometric structure of the attractor by using local approximations.

The method is a simple local approximation that replaces the central coordinate of each embedding vector by the local average of this coordinate. The practical issues in implementing this technique are as follows [228]. If the data represents a chaotic dynamics, initial errors in the first and the last coordinates will be magnified through time. Thus, they should not be replaced by local averages. Secondly, except for oversampled data sets, it is desirable to choose a small time delay. Next, the embedding dimension, m, should be chosen higher than $2d+1$, with d being the fractal dimension of the attractor. Finally, the neighborhood should be defined by selecting a neighborhood radius r such that, r should be large enough to cover the extent of the contaminating noise, yet smaller than the typical radius of curvature of the attractor. These conditions may not always be satisfied simultaneously. As we have been stressing repeatedly for other aspects of nonlinear data analysis, the process of filtering should be carried out in several attempts, by trying different tuning parameters, associated with a careful evaluation of the results, until they look reasonably satisfactory.

The filtering algorithm is as follows:

1. Pick a small time delay, τ, a large enough odd embedding dimension, m, and an *optimum* neighborhood radius, r.

2. For each embedding vector $\hat{\mathbf{x}}$ (as defined in Eq (5.21)) calculate a filtered middle coordinate $\tilde{y}_{i-(m+1)\tau/2}$ by averaging over the neighborhood defined by r, as

$$\tilde{y}_{i-(m+1)\tau/2} = \frac{\sum_j y_{j-(m+1)\tau/2} U\left(r - \|\hat{\mathbf{x}}_i - \hat{\mathbf{x}}_j\|\right)}{\sum_j U\left(r - \|\hat{\mathbf{x}}_i - \hat{\mathbf{x}}_j\|\right)} \tag{5.26}$$

   where $U(\cdot)$ is the unit step function, and $\|\cdot\|$ is the vector norm. Note that the first and the last $(m-1)/2$ data points will not be filtered by this averaging.

3. Select the average magnitude of correction as the new neighborhood radius and the filtered data as the raw data, and go to Step 2 if deemed necessary after inspecting the filtered data. If the result looks satisfactory, or if the modified neighborhood radius drops below the distance between nonidentical nearest neighbors, iterations stop.

Example 8 *Filtering the blood oxygen concentration signal*

Again consider the time series data of Example 6 (Figure 5.16.a). Applying the local averaging noise filtering method to the signal with an embedding dimension of $m = 5$, a time delay of $\tau = 9$ sec., and a neighborhood radius of $r = 400$, we obtain the filtered signal shown in Figure 5.16.b. Note how the orbits became crisp, and how the conic shape of the attractor became visible in the filtered signal. □

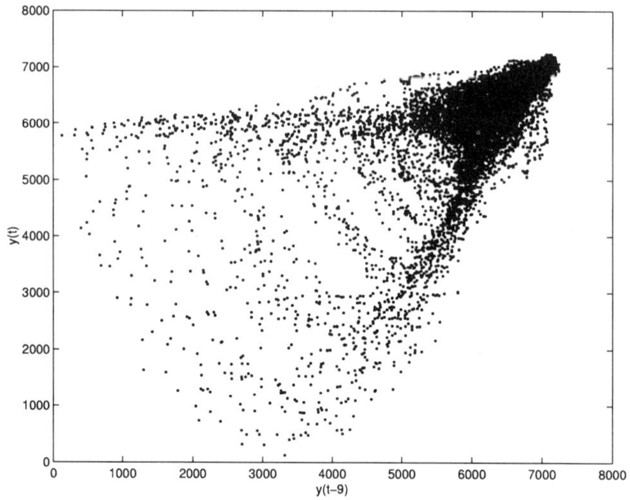

(a)

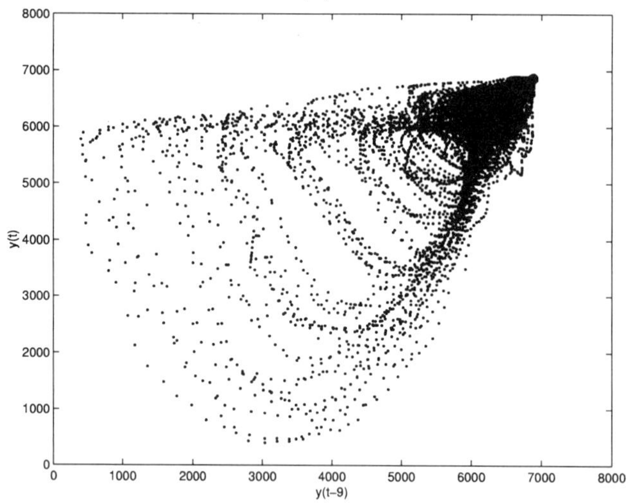

(b)

Figure 5.16. The reconstructed state space of the blood oxygen concentration signal with $\tau = 9$ sec. projected in two dimensions (a) before filtering, and (b) after filtering.

5.2.3 System Classification

With a filtered signal and the proper state space, we can investigate certain properties of the process that generated the signal. One critical question we need to address is that of classifying or identifying the source of the observations. A prerequisite for devising a suitable approximation scheme is to make predictions on the system behavior, and to solve the mathematical model if available.

In the case of linear systems, we again have the powerful tool of Fourier analysis. The locations of the sharp peaks in the frequency spectrum are characteristic of the process under investigation. If we drive the linear system with more power, the peaks will get higher, and if we observe the system starting from a different origin in time, there will be a phase shift associated with the measurements, yet in all cases, the locations of the peaks are the same. Quantities such as the locations of peaks in frequency domain are invariants of a linear system dynamics, and can be used to classify the system. A powerful example of classifying systems using frequency contents is voice recognition, where the frequency spectrum of a speech signal reveals the identity of the speaker with almost no doubt.

We have argued that frequency domain techniques are not very useful for nonlinear systems, especially when they are operated in dynamically rich regimes. Still, there are other invariants that are specific in classifying and identifying the signal source. These invariants are quantities that remain unchanged under various operations on the dynamics or the orbit. Most importantly, they remain unchanged under small perturbations in initial conditions, other than on countable specific points. Some of the invariants remain unchanged throughout the operation of the system. This guarantees that they are insensitive to initial conditions, which is apparently not true for the individual orbits. Some of them are guage invariants and stay unchanged under smooth coordinate transformations, and others are topological invariants, which are purely geometric properties of the vector field describing the dynamics. Among these invariants are the local and global Lyapunov exponents, and various fractal dimensions. Chaotic systems are notorious for the unpredictability of their orbits, and the limited predictability of chaotic systems is quantified by the local and global Lyapunov exponents of the system. Fractal dimensions associated with the source, on the other hand, reveal the topology of its attractor.

One of the hallmarks of chaotic behavior is the sensitivity of any orbit to small changes in initial condition, which is quantified by a positive Lyapunov exponent. Because of this sensitivity, it is inappropriate to compare two orbits generated by a nonlinear process directly. Generically, they will be totally uncorrelated. However, the invariants of the system will enable us to identify the source of an observation, since they are unchanged properties of the attractor of the system that has a peculiar geometry. These invariants are as useful in identifying nonlinear systems, as the Fourier spectrum is for linear systems. Therefore, system identification in nonlinear systems means establishing a set of invariants for each system of interest, and then comparing the invariants of the observation to the database of invariants.

5.3 Model Development

Now that we have identified the source of the signal, we can proceed with building a local or global model for that source, working within the coordinate system established. In linear systems, the task is relatively simple. As discussed in Section 4.7, observations y must somehow be linearly related to observations and the forcing u applied at earlier times. If time series models (Section 4.7) are used, this leads to an ARMA model of the form

$$y(i) = \sum_{k=1}^{K} a_k y(i-k) + \sum_{l=1}^{L} b_l u(i-l) \tag{5.27}$$

where the coefficients a_k and b_l are to be determined from the observations, typically using a least-squares or an information-theoretic criterion. If we take the z-transform of the Eq (5.27), we obtain the transfer function

$$H(z) = \frac{Y(z)}{U(z)} = \frac{\sum_{l=1}^{L} b_l z^l}{1 - \sum_{k=1}^{K} a_k z^k} \tag{5.28}$$

that defines the process dynamics.

Availability of a reliable model gives us the unprecedent power of predicting the outcome of hypothetical operations of the system, thus in many cases enable us to device methods of controlling it. In the case of discrete linear systems described, the choice of coefficients should be consistent with any prior knowledge of spectral peaks the system has. The denominator of the transfer function (after possible pole-zero cancellations) holds all that information, in terms of poles in the z-plane. Much of the linear signal processing literature is devoted to efficient and effective ways of choosing the coefficients in this kind of linear modeling cases of varying complexity. From the dynamical systems point of view, this kind of modeling consists of autocorrelating the signal, and time averaging the forcing function.

When chaotic dynamics is concerned, such models will not be of much use, since they cannot evolve on a strange attractor, that is, they cannot have any positive Lyapunov exponents, or equivalently, they will always have zero Kolmogorov-Sinai entropy. Nonlinear modeling of chaotic processes is based on the idea of a compact geometric attractor on which our observations evolve. The attractor of a chaotic system is a fractal object called a strange attractor. Due to its fractal nature, the orbit of a particular trajectory is folded back on itself by the nonlinear dynamics. Thus, in the neighborhood of any orbit $\mathbf{x}(i)$, other orbit points $\mathbf{x}^{(r)}(i)$ with $r = 1, \ldots, N_B$, arrive in the neighborhood at quite different times than i. One can then build various forms of interpolation functions, which account for whole neighborhoods of state space, and how they evolve from near $\mathbf{x}(i)$ to the whole set of points near $\mathbf{x}(i+1)$. The use of state space information in the modeling of the temporal evolution of the process is the key innovation in modeling chaotic systems. The general procedure would work for non-chaotic systems as well, but is likely to be less successful, because the neighborhoods are underpopulated to make reliable statistical inferences.

The implementation of this idea is to build parameterized nonlinear functions that take $\mathbf{x}(i)$ into $\mathbf{x}(i+1)$ as

$$\mathbf{x}(i+1) = \mathbf{f}(\mathbf{x}(i); \mathbf{a}) \tag{5.29}$$

and then use various criteria to determine the parameters $\mathbf{a}$. Thus building an understanding of local neighborhoods, one can build up a global nonlinear model by piecing the local models to capture much of the attractor structure.

The main departure from linear modeling techniques is to use the state space and the attractor structure dictated by the data itself, rather than to resort to some predefined algorithmic approach. It is likely that there is no algorithmic solution [511] to how to choose a model structure for chaotic systems, as the data from the dynamics dictate properties that are characteristic for the underlying structure.

If we are going to build a continuous model, we need the time derivatives of the measured quantities, which are generally not available. However, one should avoid numerical differentiation whenever possible, as it amplifies measurement noise. One remedy is to smooth the data before taking the time derivatives. The smoothing techniques usually involve least-squares fit of the data using some known functional form, e.g., a polynomial. Instead of approximating the time series data by a single (thus of high order) polynomial over the entire range of the data, it is often desirable to replace each data point by the value taken on by a (low order) least-squares polynomial relevant to a subrange of $2M+1$ points, centered, where possible, at the point for which the entry is to be modified. Thus, each smoothed value replaces a tabulated value. For example, if we consider a first order least squares fit with three points, the smoothed values, $\tilde{y}_i$, in terms of the original values, y_i, would be computed as follows:

$$\tilde{y}_{i-1} = (5y_{i-1} + 2y_i - y_{i+1})/6 \tag{5.30}$$
$$\tilde{y}_i = (y_{i-1} + y_i + y_{i+1})/3 \tag{5.31}$$
$$\tilde{y}_{i+1} = (-y_{i-1} + 2y_i + 5y_{i+1})/6 \tag{5.32}$$

If the system is sampled with a fixed frequency, $1/\Delta t$, an interpolation formula, such as Newton's, may be used, and the resulting formula is differentiated analytically. If the sampling is not done homogeneously, then Lagrange's formulae must be used. The following differentiation formulae are obtained for uniformly sampled data points by differentiating a three-point Lagrange interpolation formula

$$\dot{y}_{i-1} \approx (-3y_{i-1} + 4y_i - y_{i+1})/2\Delta t \tag{5.33}$$
$$\dot{y}_i \approx (-y_{i-1} + y_{i+1})/2\Delta t \tag{5.34}$$
$$\dot{y}_{i+1} \approx (y_{i-1} - 4y_i + 3y_{i+1})/2\Delta t\ . \tag{5.35}$$

Next, we can consider the reconstructed state space, and seek which set of coordinates give enough information about the time derivative of each coordinate. This is illustrated with an example.

Example 9 *Functional dependencies in the reconstructed state-space*

In the four dimensional reconstructed state space of the blood oxygen concentration signal (Table 5.1), we would like to investigate the mutual information contents of Table 5.2.

From an information theoretic point of view, the more state components we compare with a given time derivative, the more information we would gather. For example, the mutual information between $\dot{x}_1$ and x_1, x_2 would be greater than or equal to the mutual information between $\dot{x}_1$ and x_1. Therefore, for each $\dot{x}_k$, the last line of the Table 5.2 would be the largest entry. Of course, this would depend on the choice of time delay τ we use to reconstruct the state-space. If we plot this dependence (Figure 5.17), the mutual information contents of all time derivatives behave similarly, making a peak around $\tau = 5$ sec, and a dip around $\tau = 22$ sec. For modeling purposes, this plot suggests the use of a time delay of $\tau = 5$ sec.

For this choice of time delay, we can investigate how information are gathered about the time derivatives by filling out the mutual information table (Table 5.2) and observing the information provided by various subsets of coordinates. At this point, we again resort to our judgement about the system, and tailor the functional dependencies of the coordinates guided by our knowledge about the system and educated guess. We are interested in

Table 5.1. The mutual information contents to be computed for the four dimensional reconstructed state space of the blood oxygen concentration signal, with $k = 1, 2, 3, 4$. The mutual information content between the time derivative of each x_k and the entries in the right-hand column are computed.

$$
\begin{array}{ll}
\dot{x}_k: & x_1 \\
\dot{x}_k: & x_2 \\
\dot{x}_k: & x_3 \\
\dot{x}_k: & x_4 \\
\dot{x}_k: & x_1, x_2 \\
\dot{x}_k: & x_1, x_3 \\
\dot{x}_k: & x_1, x_4 \\
\dot{x}_k: & x_2, x_3 \\
\dot{x}_k: & x_2, x_4 \\
\dot{x}_k: & x_3, x_4 \\
\dot{x}_k: & x_1, x_2, x_3 \\
\dot{x}_k: & x_1, x_2, x_4 \\
\dot{x}_k: & x_1, x_3, x_4 \\
\dot{x}_k: & x_2, x_3, x_4 \\
\dot{x}_k: & x_1, x_2, x_3, x_4
\end{array}
$$

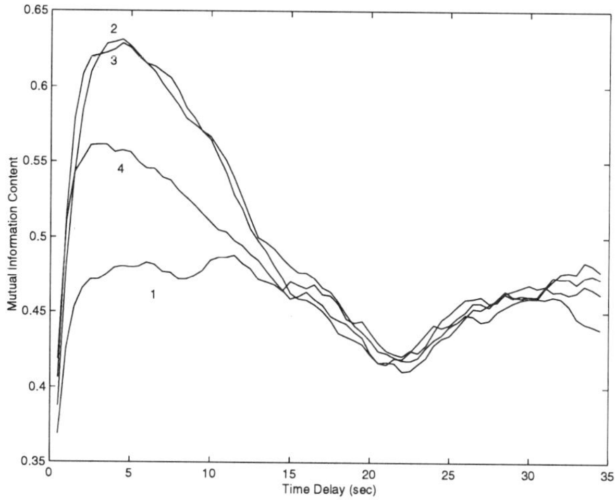

Figure 5.17. The mutual information content between $\dot{x}_k$ and x_1, x_2, x_3, x_4 versus the time delay, for $k = 1, 2, 3, 4$.

Table 5.2. The mutual information contents for the four dimensional reconstructed state space of the blood oxygen concentration signal.

	$\dot{x}_1$	$\dot{x}_2$	$\dot{x}_3$	$\dot{x}_4$
x_1	2.665e-01	3.664e-01	2.493e-01	1.250e-01
x_2	2.188e-01	2.679e-01	3.651e-01	2.476e-01
x_3	1.507e-01	2.190e-01	2.675e-01	3.627e-01
x_4	7.433e-02	1.508e-01	2.186e-01	2.665e-01
x_1, x_2	4.312e-01	5.043e-01	4.183e-01	3.214e-01
x_1, x_3	3.819e-01	5.770e-01	4.283e-01	4.098e-01
x_1, x_4	3.240e-01	5.125e-01	4.398e-01	3.486e-01
x_2, x_3	2.612e-01	4.315e-01	5.029e-01	4.152e-01
x_2, x_4	2.614e-01	3.829e-01	5.751e-01	4.260e-01
x_3, x_4	2.013e-01	2.616e-01	4.302e-01	4.999e-01
x_1, x_2, x_3	4.533e-01	5.979e-01	5.318e-01	4.431e-01
x_1, x_2, x_4	4.568e-01	5.800e-01	6.052e-01	4.752e-01
x_1, x_3, x_4	4.098e-01	6.041e-01	5.422e-01	5.347e-01
x_2, x_3, x_4	2.905e-01	4.535e-01	5.959e-01	5.287e-01
x_1, x_2, x_3, x_4	4.802e-01	6.268e-01	6.257e-01	5.562e-01

finding the simplest functional relationships (fewest variables in an equation) that describe the system with the desired accuracy. For example, if measurement noise is about 10% of a signal, we would be content by about 90% of the information we could gather about the system by considering all the coordinates. This leads to

$$\dot{x}_1 = f_1(x_1, x_2), \tag{5.36}$$

$$\dot{x}_2 = f_2(x_1, x_3), \tag{5.37}$$

$$\dot{x}_3 = f_3(x_2, x_4), \tag{5.38}$$

$$\dot{x}_4 = f_4(x_3, x_4). \tag{5.39}$$

□

The conventional usage of the methods introduced in this chapter in system modeling is to reconstruct a phase space using a scalar measurement. In the following example we will demonstrate how these concepts can be used to narrow down the phase space using multivariate measurements from a fermentation process [59].

Example 10 *Phase space reduction*

Consider the starch fermentation by recombinant *Saccharomyces cerevisiae* in a batch reactor (Figure 5.18). A series of experiments were conducted in the absence of oxygen supply by changing initial starch concentrations, and time courses of

- $\mathcal{I}$, intracellular RNA (g/L),
- $\mathcal{D}$, plasmid DNA (g/L),
- $\mathcal{X}$, biomass (g/L),
- $\mathcal{S}$, starch (g/L),
- $\mathcal{R}$, reducing sugar (g/L),
- $\mathcal{G}$, glucose (g/L),
- $\mathcal{P}$, extracellular protein (mg/L)

concentrations, and

- α, α-amylase (U/mL) and
- γ, glucoamylase (U/mL)

activities were measured using appropriate analytical techniques [57, 60]. A single run of the experiment generates snapshots of the measured quantities like the one shown in Figure 5.19. Note that measurements were made at varying time intervals, and different measurements are not necessarily synchronous. The focus is different from the previous examples in that we concentrate on the underlying non-chaotic dynamics. Although the techniques described in this chapter are applicable to chaotic and non-chaotic systems, most of the nonlinear time series analysis tools require a *dense* phase space. As we demonstrated with numerous examples, chaotic systems fulfill this requirement by populating the phase space, e.g., by strange attractors. In the present example, this was achieved by repeated experiments. Figure 5.19 shows only a single realization of the experiment. When we aggregate data from many experiments involving the same system, due to variations in the controlled conditions (e.g., initial conditions) and uncontrolled conditions (e.g., environmental conditions), the phase space will be dense, and we will be able to use the tools introduced earlier.

Next, we classify pairs of measurements as INDEPENDENT, COUPLED or REDUNDANT using the heuristic scheme described in Figure 5.20. When a measurement pair is found to be INDEPENDENT, we will conclude that the

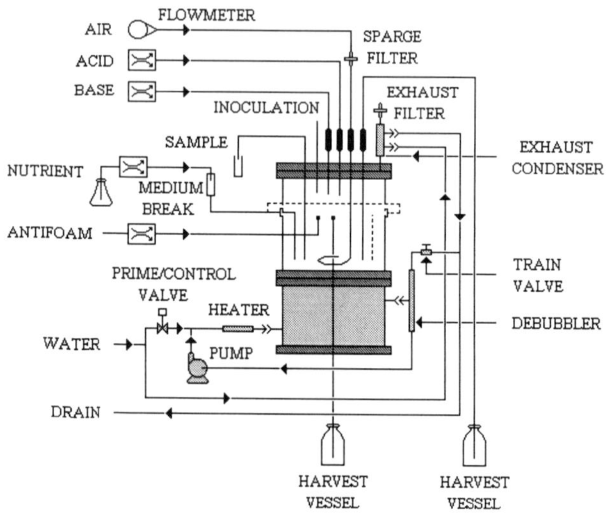

Figure 5.18. Experimental setup of the fermentation system.

measurement do not have an explicit mutual dependence in model equations. When the classification scheme indicates a COUPLED measurements, we will write those measurements in each other's model equations. If the scheme indicates REDUNDANT measurements, we will claim that making a set of measurements for one coordinate yields a remarkable amount of information about the other, hence measuring both, or considering both in a modeling attempt is not necessary.

In the classification scheme, we use the mutual information content between the measurements I normalized to data length[3], the fractal dimension of the data d and the correlation coefficient ρ. The reasoning behind this classification scheme is as follows: If two arrays of data fill the phase space they live on ($d > 1$), they are likely to be INDEPENDENT, as dependent

[3]This normalization is done by changing the base of the logarithm in Eq 5.22 from 2 to the data length N.

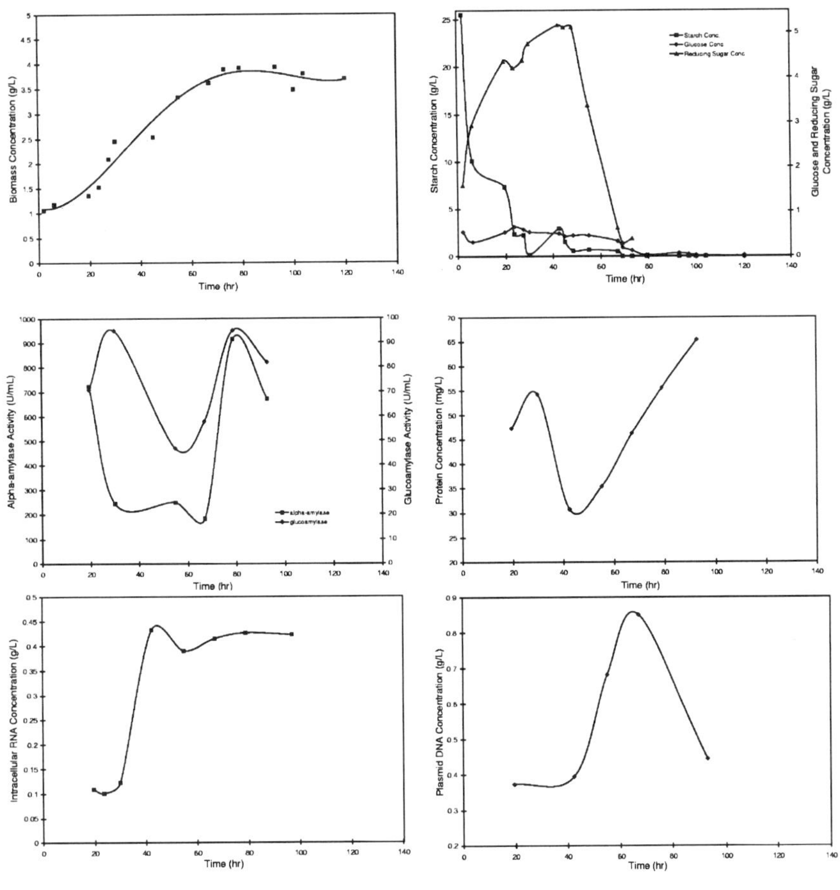

Figure 5.19. Sample experimental data.

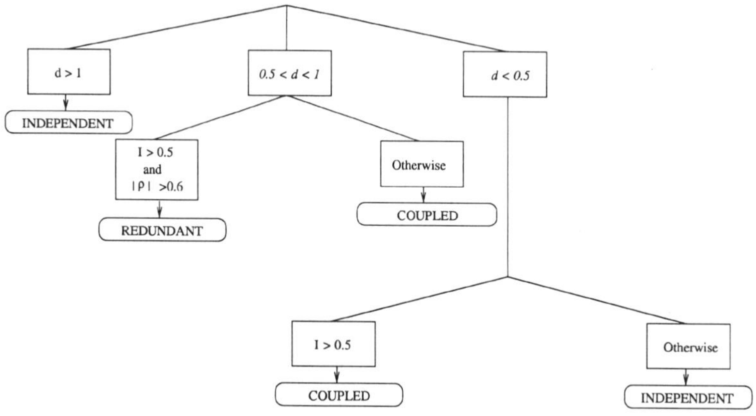

Figure 5.20. Heuristic scheme used to classify measurement pairs [58].

variables would make a loosely quilted pattern, leaving tracks on the phase space. At the other extreme, if the variables reveal a dense pattern yielding no significant information about each other, ($d < 0.5$ and $I < 0.5$), these are considered to be INDEPENDENT. If two arrays of data leave tracks on the phase space by moderately filling it and display a considerable amount of information about each other, and are highly correlated ($0.5 < d < 1$, $I > 0.5$ and $|\rho| > 0.6$), then one of the two arrays can be discarded in favor of the other, since measuring both would be REDUNDANT. For other combinations of I, d and ρ, two arrays of data will be considered to be COUPLED.

Our measurement space is 9-dimensional, and our measurement vector is composed of samples of the vector $[\mathcal{I}\ \mathcal{D}\ \mathcal{X}\ \mathcal{S}\ \mathcal{R}\ \mathcal{G}\ \mathcal{P}\ \alpha\ \gamma]^T$. When we compute the capacity dimension of this signal we find $d \simeq 2.98$. This capacity dimension yields a sufficient embedding dimension of $n = 6$. On the other hand, due to the statistical fluctuations, we may as well have a capacity dimension that is slightly above 3.0. In such a case, we should be computing an embedding dimension of $n = 7$. However, this is not the case, as the *actual* dimension of this signal must be an integer value (3 in this case), due to the assumed non-chaotic nature of the signal. Therefore, we choose $n = 6$.

This choice of embedding dimension for a 9-dimensional signal implies that at least 3 of the entries in the measurement vector should be discarded. In agreement with this finding, if we look at the eigenvalues of the covariance matrix, (see Section 4.1)

$$\begin{aligned}\{\sigma_i\} &= \{3.98\times10^4, 5.10\times10^3, 9.45\times10^1, 4.81\times10^1, 1.70\times10^0,\\ &\quad 7.18\times10^{-1}, 1.77\times10^{-2}, 2.77\times10^{-3}, 3.07\times10^{-4}\}, \qquad (5.40)\end{aligned}$$

we see that the last three eigenvalues are negligible when compared to others. If we perform a principal component analysis and select the first six dominant transformed coordinates, we will have a mean-square error less than $4.63 \times 10^{-5}\%$, which is much less than the square of the radius of error caused by the sensitivity in our measurements which is around $10^{-2}\%$.

Here, we are not after a best reduced representation, but after the most significant measured quantities in our data set. Naturally, our choice of a subset composed of the most significant measured quantities will yield a higher mean-square error in representing the data set. Nevertheless, it is desired to keep this error level as low as possible. We are to select three of the coordinates out of nine, such that the mean-square error is minimum. This gives us $\binom{9}{3} = 84$ possible ways to choose these three coordinates. To facilitate this process, consider Table 5.3 where we summarize the results of dimension (d), mutual information coefficient (I) and correlation coefficient (ρ) computations, as well as the outcome of the heuristic classification scheme (Class) on our data pairs. Note that, d, I and ρ are symmetric quantities, and the order in which they are referenced is immaterial, i.e., d, I and ρ for $\mathcal{DG}$ are the same as the d, I and ρ for $\mathcal{GD}$.

The data series pairs $\mathcal{DS}$, $\mathcal{GR}$ and $\mathcal{IX}$ show redundancies, since for all three pairs, $0.5 < d < 1$, $I > 0.5$ and $|\rho| > 0.6$. Therefore, in each pair, one coordinate can be discarded in favor of the other. Thus, we have six possible ways to choose the coordinates to be discarded. Dropping coordinates $\mathcal{D}$, $\mathcal{G}$ and $\mathcal{I}$ from the measurement vector results in a reduced covariance matrix, with eigenvalues,

$$\{\hat{\sigma}_i\} = \{3.98\times10^4, 5.10\times10^3, 9.45\times10^1, 4.81\times10^1, 1.68\times10^0, 7.12\times10^{-1}\}. \qquad (5.41)$$

Comparing the eigenvalue sets (5.40) and (5.41), we find that dropping $\mathcal{D}$, $\mathcal{G}$ and $\mathcal{I}$ gives a mean-square error of $2.41 \times 10^{-4}\%$, which is about an order of magnitude greater than that achieved by considering a principal component analysis and representing the 9-dimensional space by the first 6 coordinates. Still, this is much less than the square of our measurement sensitivity radius. We find this proximity satisfactory and reduce the phase space to $[\mathcal{X}\ \mathcal{S}\ \mathcal{R}\ \mathcal{P}\ \alpha\ \gamma]^T$. This phase space reduction will reduce any such experimental work in the future by 33% for this system.

Hereafter we concentrate on the entries below the double line in Table 5.3, where only the relations between the coordinates of the reduced phase space are considered. Looking at the mutual interactions between

Pair	d	I	ρ	Class
$\mathcal{DG}$	0.65	0.51	0.13	C
$\mathcal{DR}$	0.71	0.60	-0.12	C
$\mathcal{DS}$	0.59	0.53	-0.69	R
$\mathcal{DX}$	1.26	0.58	0.26	I
$\mathcal{D}\alpha$	0.85	0.66	0.23	C
$\mathcal{D}\gamma$	0.84	0.51	0.21	C
$\mathcal{DP}$	1.03	0.64	0.49	I
$\mathcal{GR}$	0.75	0.55	0.63	R
$\mathcal{GS}$	0.28	0.47	0.25	I
$\mathcal{GX}$	0.59	0.53	0.38	C
$\mathcal{G}\alpha$	1.11	0.51	-0.36	I
$\mathcal{G}\gamma$	0.78	0.41	-0.07	C
$\mathcal{GP}$	1.10	0.52	-0.17	I
$\mathcal{ID}$	1.40	0.62	0.13	I
$\mathcal{IG}$	0.70	0.60	0.05	C
$\mathcal{IR}$	0.79	0.62	-0.45	C
$\mathcal{IS}$	0.59	0.51	-0.34	C
$\mathcal{IX}$	0.95	0.58	0.92	R
$\mathcal{I}\alpha$	0.91	0.62	-0.08	C
$\mathcal{I}\gamma$	0.96	0.53	-0.10	C
$\mathcal{IP}$	1.33	0.63	-0.18	I
$\mathcal{R}\alpha$	1.16	0.65	-0.34	I
$\mathcal{R}\gamma$	1.00	0.48	-0.16	I
$\mathcal{RP}$	1.09	0.63	-0.09	I
$\mathcal{SR}$	0.31	0.57	0.38	C
$\mathcal{S}\alpha$	0.63	0.50	-0.53	C
$\mathcal{S}\gamma$	0.47	0.34	-0.26	I
$\mathcal{SP}$	0.77	0.50	-0.58	C
$\mathcal{XR}$	0.40	0.62	-0.07	C
$\mathcal{XS}$	0.36	0.54	-0.15	C
$\mathcal{X}\alpha$	1.01	0.61	-0.25	I
$\mathcal{X}\gamma$	0.93	0.52	-0.10	C
$\mathcal{XP}$	1.36	0.59	-0.14	I
$\alpha\mathcal{P}$	1.19	0.64	0.50	I
$\gamma\alpha$	0.87	0.53	0.46	C
$\gamma\mathcal{P}$	0.59	0.52	0.35	C

Table 5.3. Phase space dimension d, information coefficient i, and correlation coefficient ρ, corresponding to each data pair, and the result of the heuristic classification scheme. Classifications I, C and R stand for INDEPENDENT, COUPLED and REDUNDANT, respectively.

the coordinates of interest, we see that, 8 out of 15 pairs are classified as COUPLED, and 7 of them are classified as INDEPENDENT coordinates. The COUPLED pairs, $\mathcal{SR}$, $\mathcal{S}\alpha$, $\mathcal{SP}$, $\mathcal{XR}$, $\mathcal{XS}$, $\mathcal{X}\gamma$, $\gamma\alpha$ and $\gamma\mathcal{P}$ expected to appear in the equations of motion of one-another. For instance, since $\mathcal{S}$ and $\mathcal{R}$ are found to be COUPLED, the dynamics of $\mathcal{S}$ should be affected directly by the value of $\mathcal{R}$, or the dynamics of $\mathcal{R}$ should be affected directly by the value of $\mathcal{S}$, or both. To be on the safe side, we assume the "both" case. Also, a coordinate in the phase space may appear in its own dynamics, so a generic form of equations of motion should also take this into account. On the other hand, the pairs $\mathcal{R}\alpha$, $\mathcal{R}\gamma$, $\mathcal{RP}$, $\mathcal{S}\gamma$, $\mathcal{X}\alpha$, $\mathcal{XP}$ and αP are classified to be INDEPENDENT, and these pairs will not appear in the dynamics of each other.

Consequently, we propose that a suitable model for the fermentation behavior of recombinant *Saccharomyces cerevisiae* cells should be of the following form:

$$\dot{\mathcal{X}} = f_1(\mathcal{X}, \mathcal{S}, \mathcal{R}, \gamma), \tag{5.42}$$

$$\dot{\mathcal{S}} = f_2(\mathcal{X}, \mathcal{S}, \mathcal{R}, \mathcal{P}, \alpha), \tag{5.43}$$

$$\dot{\mathcal{R}} = f_3(\mathcal{X}, \mathcal{S}, \mathcal{R}), \tag{5.44}$$

$$\dot{\mathcal{P}} = f_4(\mathcal{S}, \mathcal{P}, \gamma), \tag{5.45}$$

$$\dot{\alpha} = f_5(\mathcal{S}, \alpha, \gamma), \tag{5.46}$$

$$\dot{\gamma} = f_6(\mathcal{X}, \mathcal{P}, \alpha, \gamma). \tag{5.47}$$

Writing down such a set of generic model equations for potential models reduces the computational effort of parameter estimation to about one-fourth, while increasing the reliability of the model constructed by increasing its degrees of freedom. In a modeling attempt with 100 data points, this corresponds to about a four-fold increase in reliability.

5.4 Software Resources

Several software packages have been used to generate, manipulate and present the data used in the examples presented in this chapter. Brief descriptions of various software packages used are given below:

Content [311] is an interactive program to study continuous (given as ordinary or partial differential equations) and discrete (given as iterated maps) dynamical systems. It is flexible to analyze standard and customized dynamical systems, using simulations, one- or two-parameter continuation studies and normal form analysis. It is available free of charge by anonymous ftp, from `ftp.cwi.nl/pub/CONTENT/`. The package has an online and offline documentation. A Matlab version of CONTENT is also available from http://www.math.uu.nl/people/kuznet/cm/. Platforms: several Unix flavors (including Linux) and Windows.

TISEAN [228] is a package of time series analysis programs with methods based on the theory of nonlinear deterministic dynamical systems. The name is an acronym for TIme SEries ANalysis. Software and documentation are available free of charge from http://www.mpipks-dresden.mpg.de/~tisean/. Platforms: Although it is mainly designed for a Unix based environment, its distribution is in source form (C and FORTRAN).

Maple [364] is a favorite symbolic programming environment, not restricted to dynamical systems. It has a numerical computation interface using Matlab. It is a commercial package. More information can be obtained from http://www.maplesoft.com/flash/index.html. Platforms: Unix flavors (including Linux), Windows and Macintosh.

Matlab [372] is arguably the most widely used interactive numerical programming environment in science and engineering. The name is an acronym for MATrix LABoratory. It has a symbolic computation interface using Maple. It is a commercial product. More information can be obtained from http://www.mathworks.com/. Platforms: Unix flavors (including Linux), Windows and Macintosh.

gnuplot [648] is a command-driven interactive function plotting program. It can be used to plot functions and data points in both two- and three-dimensional plots in many different formats, and will accommodate many of the needs of today's scientists for graphic data representation. gnuplot is copyrighted, but freely distributable. It can be obtained from http://www.gnuplot.info/. Platforms: Unix flavors (including Linux), VAX/VMS, OS/2, MS-DOS, Amiga, Windows, OS-9/68k, Atari, BeOS, and Macintosh.

There is a wide collection of free and commercial software packages available. Below is a list of the ones we have examined, with no particular order.

Auto [128] is a software for continuation and bifurcation problems in ordinary differential equations. Users can download the package and find documentation about it free of charge form http://indy.cs.concordia.ca auto/main.html. Platforms: Unix flavors (including Linux).

XPP [146] is a package for simulating dynamical systems that can handle differential equations, differencc equations, Volterra integral equations, discrete dynamical systems and Markov processes. The name is an acronym for X-windows Phase Plane. Data structure used by XPP is compatible with AUTO. XPP also offers a graphical user interface for AUTO. It is a free software that can be obtained from `http://www.math.pitt.edu/~bard/xpp/xpp.html`. Online documentation is also available from the same address. Platforms: Unix flavors (including Linux).

DsTool [204] is a computer program for the interactive investigation of dynamical systems. The program performs simulations of diffeomorphisms and ordinary differential equations, find equilibria and compute their one-dimensional stable and unstable manifolds. It is freely available with documentation from `http://www.cam.cornell.edu/ gucken/dstool`. Platforms: Unix flavors (including Linux).

Octave [266] is a general purpose high-level language, primarily intended for numerical computations that is mostly compatible with Matlab. Its underlying numerical solvers are currently standard Fortran ones like Lapack, Linpack, Odepack, the Blas, etc., packaged in a library of C++ classes. Users can freely download, redistribute and even modify Octave, under GNU General Public License. It is available from `http://www.octave.org/`. Platforms: Unix flavors (including Linux), Windows.

Mathematica [670] is another general purpose symbolic and numerical programming environment. It is a commercial product. More information can be obtained from `http://www.wolfram.com/`. Platforms: Unix flavors (including Linux), Windows and Macintosh.

MuPAD [147] is a system for symbolic and numeric computation, parallel mathematical programming, and mathematical visualization. The name is an acronym for Multi Processing Algebra Data tool. It is a commercial package, that is available for free for Linux. Further information and documentation is available from `http://www.mupad.de/index_uni.shtml`. Platforms: Unix flavors (free for Linux), Windows and Macintosh.

Other Resources on the Web There are many resource directories on the Web that list nonlinear dynamical systems tools. We refer interested reader to two of these.

`http://www.enm.bris.ac.uk/staff/hinke/dss/` aims to collect all available software on dynamical systems theory, and has 15 links.

`http://sal.kachinatech.com/index.shtml` is a more general directory service, called Scientific Application on Linux (SAL). Although the name suggests an exclusive Linux listing, the broad coverage of applications will also benefit the whole scientific computation community. As of March 2002, there are 3,070 entries listed in SAL.

6

Statistical Process Monitoring

Monitoring and control of batch processes are crucial tasks in a wide variety of industrial processes such as pharmaceutical processes, specialty chemicals production, polymer production and fermentation processes. Batch processes are characterized by prescribed processing of raw materials for a finite duration to convert them to products. A high degree of reproducibility is necessary to obtain successful batches. With the advent of process computers and recent developments in on-line sensors, more data have become available for evaluation. Usually, a history of the past successful and some unsuccessful batches exist. Data from successful batches characterize the normal process operation and can be used to develop empirical process models and process monitoring systems.

The goal of *statistical process monitoring* (SPM) is to detect the existence, magnitude, and time of occurrence of changes that cause a process to deviate from its desired operation. The methodology for detecting changes is based on statistical techniques that deal with the collection, classification, analysis, and interpretation of data. Traditional statistical process control (SPC) has focused on monitoring quality variables at the end of a batch and if the quality variables are outside the range of their specifications, making adjustments (hence control the process) in subsequent batches. An improvement of this approach is to monitor quality variables during the progress of the batch and make adjustments if they deviate from their expected ranges. Monitoring quality variables usually delays the detection of abnormal process operation because the appearance of the defect in the quality variable takes time. Information about quality variations is encoded in process variables. The measurement of process variables is often highly automated and more frequent, enabling speedy refinement of measurement information and inferencing about product quality. Monitoring of process variables is useful not only for assessing the status of the process, but also for controlling product quality. When process monitoring indicates abnormal process operation, *diagnosis* activities are initiated to determine the source causes of this abnormal behavior.

This chapter starts with a review of statistical monitoring techniques for a single variable system. Shewhart charts, cumulative sum (CUSUM), moving average (MA) and exponentially weighted moving average (EWMA) methods are discussed in Section 6.1. Monitoring of multivariable batch processes by using multivariate statistical process monitoring (MSPM) methods is discussed in the subsequent sections of the Chapter. Most MSPM techniques rely on empirical process models developed from process data using methods discussed in Chapter 4. Empirical models based on principal components analysis (PCA), partial least squares (PLS), functional data analysis, multiscale analysis, and artificial neural networks (ANNs) can be used for monitoring batch or continuous processes. It is usually easier to visualize these methods in terms of data from continuous processes operating around a steady state value. Consequently, the discussion in Section 6.2 focuses first on the application of these methods to generic continuous processes. Then, the determination of landmarks that separate different phases of a batch process and equalization of batch data lengths are discussed in Section 6.3. The application of multivariable statistical monitoring (MSPM) methods to batch process data is introduced in Section 6.4. The multiway PCA (MPCA) method is discussed first. Other modeling and monitoring techniques such as the multivariate covariates regression method, the multiblock MPCA and MPLS, various three-way methods, and multiscale SPM techniques based on wavelets are introduced in Section 6.4.6. On-line monitoring of batch processes during the progress of the batch is discussed in Section 6.5. Techniques based on estimation of variable trajectories, hierarchical PCA and estimation of final product quality are presented. Section 6.6 introduces a framework for monitoring successive batch runs for disturbances that evolve through several batches, leading to gradual drifts in product quality.

For the sake of simplicity and realism, it will be assumed that each measurement will be made only once (no repeated measurements) except for Section 6.1. For multivariable continuous processes, the index i will denote the variables and j the samples (measurements) with the upper limits indicated by m and n, respectively. For multivariable batch processes, traditionally i denotes the number of batches, j the number of variables, and k the number of samples, with the upper limits indicated by I, J, K, respectively. This notation will be followed in the text.

6.1 SPM Based on Univariate Techniques

Traditional statistical monitoring techniques for quality control of batch products relied on the use of univariate SPC tools on product quality variables. In this framework, each quality variable is treated as a single independent variable. The SPM techniques used for monitoring a single variable include Shewhart, cumulative sum (CUSUM), moving average (MA), and exponentially weighted moving average (EWMA) charts (Figure 6.1). For end-of-batch product quality control Shewhart and CUSUM charts are useful. MA and EWMA charts use time series data. Consequently, their use with end-of-batch product data is limited. However, they are discussed in this section for the sake of providing an overview of all popular univariate SPM techniques.

Hypothesis Testing

Often decisions have to be made about populations on the basis of sample information. A *statistical hypothesis* is an assumption or a guess about the population. It is expressed as a statement about the parameters of the probability distributions of the populations. Procedures that enable decision making whether to accept or reject a hypothesis are called *tests of hypotheses*. For example, if the equality of the mean of a variable (μ) to a value a is to be tested, the hypotheses are:

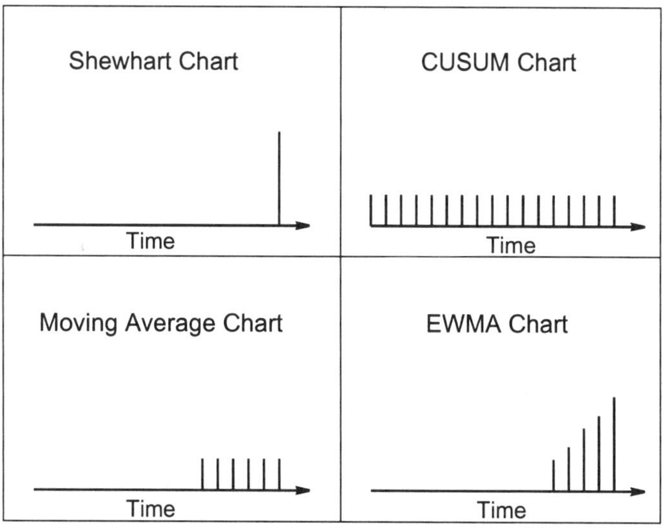

Figure 6.1. Schematic representation of univariate SPC charts.

Null hypothesis:	$\mathcal{H}_0$:	$\mu = a$
Alternate hypothesis:	$\mathcal{H}_1$:	$\mu \neq a$

Two kinds of errors may be committed when testing a hypothesis:

Type I (α) error (Producer's risk):	$P\{reject\ \mathcal{H}_0 \mid \mathcal{H}_0\ is\ true\}$
Type II (β) error (Consumer's risk):	$P\{fail\ to\ reject\ \mathcal{H}_0 \mid \mathcal{H}_0\ is\ false\}$

First α is selected to compute the confidence limit for testing the hypothesis then a test procedure is designed to obtain a small value for β, if possible. β is a function of sample size and is reduced as sample size increases. Figure 6.1 represents this hypothesis testing graphically.

6.1.1 Shewhart Control Charts

Shewhart charts indicate that a *special (assignable) cause* of variation is present when the sample data point plotted is outside the control limits. A graphical *test of hypothesis* is performed by plotting the sample mean, and the range or standard deviation and comparing them against their control limits. A Shewhart chart is designed by specifying the *centerline* (C), the

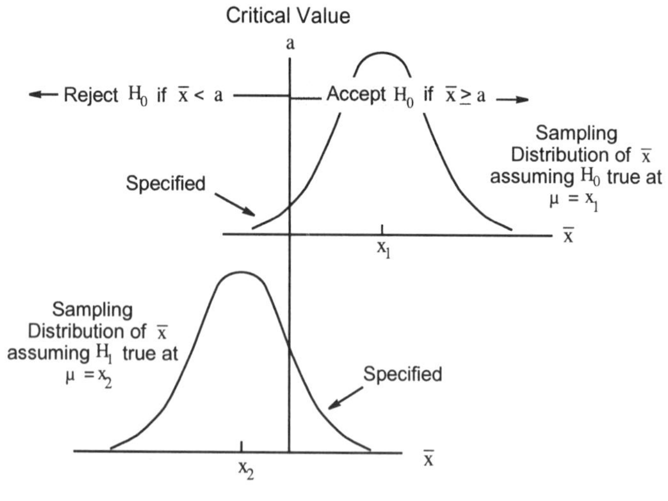

Figure 6.2. Type I and type II errors.

upper control limit (UCL) and the *lower control limit* (LCL).

Two Shewhart charts (sample mean and standard deviation or the range) are plotted simultaneously. Sample means are inspected in order to assess *between samples* variation (process variability over time). Traditionally, this is done by plotting the Shewhart mean chart ($\bar{x}$ chart, $\bar{x}$ represents average (mean) x). However, one has to make sure that there is no significant change in *within sample* variation which may give an erroneous impression of changes in *between samples* variation. The mean values at times $t-2$ and $t-1$ in Figure 6.3 look similar but within sample variation at time $t-1$ is significantly different than that of the sample at time $t-2$. Hence, it is misleading to state that between sample variation is negligible and the process level is constant. Within sample variations of samples at times $t-2$ and t are similar, consequently, the difference in variation between samples is meaningful. The *Range chart* (R chart) or

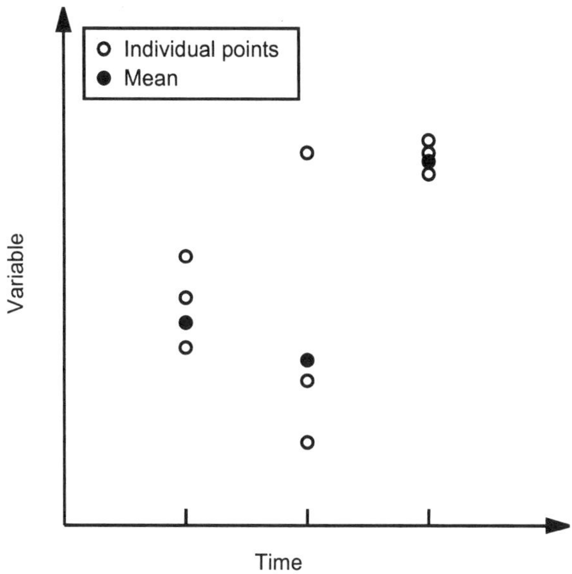

Figure 6.3. A dot diagram of individual observations of a variable.

standard deviation chart (S chart) monitors **within sample** process variation or spread (process variability at a given time). The $\bar{x}$ chart must be used along with a spread chart. The process spread must be in-control for proper interpretation of the $\bar{x}$ chart.

Usually several observations of the same variable at a specific time are used (Figure 6.3). If only one observation is available, individual values can be used to develop the x chart (rather than the $\bar{x}$ chart and the range chart is developed by using the "moving range" concept discussed in Subsection 6.1.3.

The assumptions of Shewhart charts are:

- The distribution of the data is approximately Normal.
- The sample group sizes are equal.
- All sample groups are weighted equally.
- The observations are independent.

Describing Variation

The **location** or central tendency of a variable is described by its mean, median, or mode. The **spread** or scatter of a variable is described by its range or standard deviation. For small sample sizes ($n < 6$, n=number of samples), the range chart or the standard deviation chart can be used. For larger sample sizes, the efficiency of computing the variance from the range is reduced drastically. Hence, the standard deviation charts should be used when $n > 10$.

One or more observations may be made at each sampling instant. The collection of all observations at a specific sampling time is called *a sample*. The convention on summation and representation of mean values is

$$\bar{x}_{i\cdot} = \frac{1}{n}\sum_{j=1}^{n} x_{ij} \ , \qquad \bar{x}_{\cdot\cdot} = \frac{1}{mn}\sum_{i=1}^{m}\sum_{j=1}^{n} x_{ij} \tag{6.1}$$

where m is the number of samples (groups) and n is the number of observations in a sample (sample size). The subscripts . indicate the index used in averaging. When there is no ambiguity, average values are denoted in the book using only $\bar{x}$ and $\bar{\bar{x}}$. For variables that have a Normal distribution:

Statistic	Population (size N)	Sample (size n)
Mean	$\mu = \frac{1}{N}\sum_{i=1}^{N} x_i$	$\bar{x} = \frac{1}{n}\sum_{i=1}^{n} x_i$
Variance	$\sigma^2 = \frac{1}{N}\sum_{i=1}^{N}(x_i - \mu)^2$	$S^2 = \frac{1}{n-1}\sum_{i=1}^{n}(x_i - \bar{x})^2$
Range	$R_i = max(x_i) - min(x_i) \quad i = 1, \cdots, N \ or \ n$	

Selection of Control Limits

Three parameters affect the control limit selection:

i. the estimate of average level of the variable,

ii. the variable spread expressed as range or standard deviation, and

iii. a constant based on the probability of `Type I error`, α.

The "3σ" (σ denoting the standard deviation of the variable) control limits are the most popular control limits. The constant 3 yields a `Type I` error probability of 0.00135 on each side ($\alpha = 0.0027$). The control limits expressed as a function of population standard deviation σ are:

$$\texttt{UCL} = \texttt{Target} + 3\sigma, \qquad \texttt{LCL} = \texttt{Target} - 3\sigma \tag{6.2}$$

The $\bar{x}$ chart considers *only the current data value* in assessing the status of the process. Run rules have been developed to include historical information such as trends in data. The run rules sensitize the chart, but they also increase the false alarm probability. The warning limits are useful in developing additional rules (*run rules*) in order to increase the sensitivity of Shewhart charts. The warning limits are established at "2-sigma" level, which corresponds to $\alpha/2$=0.02275. Hence,

$$\texttt{UWL} = \texttt{Target} + 2\sigma \qquad \texttt{LWL} = \texttt{Target} - 2\sigma \tag{6.3}$$

The Mean and Range Charts

Development of the $\bar{x}$ and R charts starts with the R chart. Since the control limits of the $\bar{x}$ chart depends on process variability, its limits are not meaningful before R is in-control.

The Range Chart

Range is the difference between the maximum and minimum observations in a sample.

$$R_i = x_{\max,i} - x_{\min,i} \qquad \bar{R} = \frac{1}{m}\sum_{i=1}^{m} R_i \tag{6.4}$$

The random variable R/σ is called the *relative range.* The parameters of its distribution depend on sample size n, with the mean being d_2. An estimate of σ (the estimates are denoted by a $\hat{}$) can be computed from the range data by using

$$\hat{\sigma} = \frac{\bar{R}}{d_2} \tag{6.5}$$

d_2 values as a function of n					
n	2	3	4	5	6
$d_2 =$	1.128	1.683	2.059	2.326	2.534

The standard deviation of R is estimated by using the standard deviation of R/σ, d_3:

$$\hat{\sigma}_R = d_3\sigma = d_3\frac{\bar{R}}{d_2} \tag{6.6}$$

The control limits of the R chart are

$$\text{UCL, LCL} = \bar{R} \pm 3d_3\frac{\bar{R}}{d_2} \tag{6.7}$$

Defining

$$D_3 = 1 - 3\frac{d_3}{d_2} \qquad \text{and} \qquad D_4 = 1 + 3\frac{d_3}{d_2} \tag{6.8}$$

the control limits become

$$\text{UCL} = \bar{R}D_4 \qquad \text{and} \qquad \text{LCL} = \bar{R}D_3 \tag{6.9}$$

which are tabulated for various values of n and are available in many SPC references and in the Table of Control Chart Constants in the Appendix.

The $\bar{x}$ chart

The estimator for the mean process level (centerline) is $\bar{\bar{x}}$. Since the estimate of *the standard deviation of the mean process level* σ is $\frac{\bar{R}}{d_2}$,

$$\frac{\sigma}{\sqrt{n}} = \frac{\bar{R}}{d_2\sqrt{n}} \tag{6.10}$$

The control limits for an $\bar{x}$ chart based on R are

$$\text{UCL, LCL} = \bar{\bar{x}} \pm A_2\bar{R}, \quad A_2 = \frac{3}{d_2\sqrt{n}}\ . \tag{6.11}$$

Example Consider the following data set where three measurements have been collected at each sampling time in Table 6.1. The first twenty samples are used to develop the monitoring charts and the last five samples are monitored by using these charts.

Data used in the development of the SPM charts by computing the mean and standard deviation and calculating the control limits are also plotted to check if any of these samples are out of control. If not, the charts are used as developed. If there are any out of control points, special causes for such behavior are investigated. If such causes are found, the corresponding data are excluded from the data set used for chart development and the chart limits are computed again. Since there are no data out of control for the first 20 samples, the charts are used as developed for monitoring the five "new" samples.

Table 6.1. A sample data set

No	Measurements			Mean	Range	St Dev
1	19.70	16.90	23.20	19.93	6.30	2.58
2	19.60	17.60	20.50	19.23	2.90	1.21
3	18.50	19.70	20.80	19.67	2.30	0.94
4	20.10	18.90	19.90	19.63	1.20	0.52
5	22.70	21.40	18.20	20.77	4.50	1.89
6	16.80	17.20	17.70	17.23	0.90	0.45
7	19.40	21.60	17.60	19.53	4.00	1.64
8	19.10	20.70	20.70	20.17	1.60	0.75
9	17.40	22.70	18.20	19.43	5.30	2.33
10	23.70	22.50	17.70	21.30	6.00	2.59
11	19.90	19.70	21.20	20.27	1.50	0.66
12	20.80	19.60	18.90	19.77	1.90	0.78
13	20.00	18.60	18.90	19.17	1.40	0.60
14	18.80	19.70	17.80	18.77	1.90	0.78
15	17.30	16.90	18.20	17.47	1.30	0.54
16	17.20	19.10	18.80	18.37	1.90	0.83
17	17.40	16.90	19.30	17.87	2.40	1.03
18	20.10	18.60	19.50	19.40	1.50	0.62
19	20.60	23.10	21.40	21.70	2.50	1.04
20	17.40	22.10	20.50	20.00	4.70	1.95
21	20.82	16.64	19.19	18.88	4.18	1.72
22	24.22	21.18	22.44	22.61	3.04	1.24
23	24.54	26.89	17.34	22.92	9.55	4.06
24	18.93	18.50	17.38	18.27	1.55	0.65
25	20.68	18.09	20.35	19.70	2.59	1.15

The overall mean, range, and standard deviation are 19.48, 2.80 and 1.18, respectively. The mean and range charts are developed by using the overall mean and range values in Eqs. 6.10 and 6.11. The resulting Shewhart charts are displayed in Figure 6.4. The mean of sample 22 is out of control while the range chart is in control, indicating a significant shift in level. Both the mean and range are out of control at sample 23, indicating significant change in both level and spread of the sample.

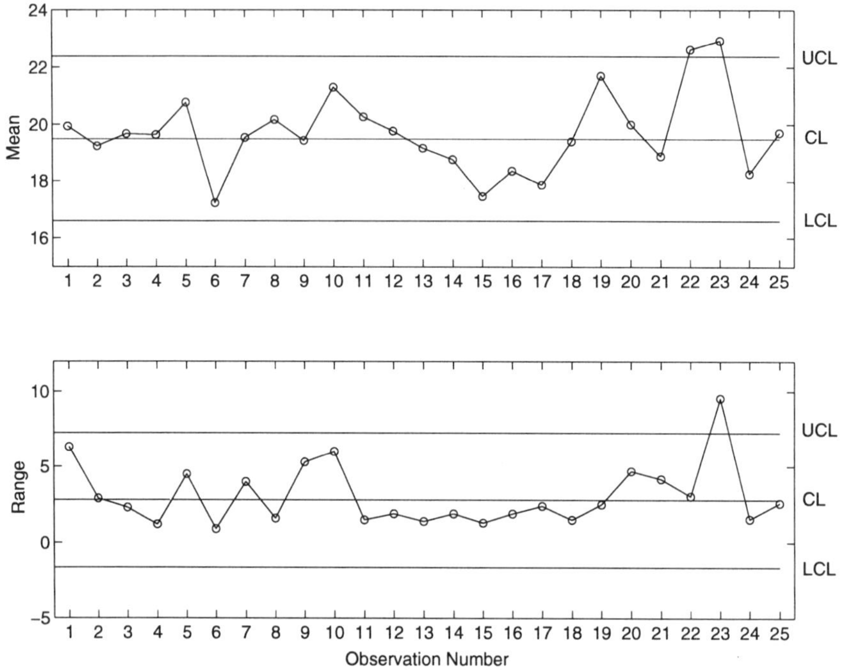

Figure 6.4. Shewhart chart for mean ($CL = \bar{\bar{x}}$) and range ($CL = \bar{R}$).

The Mean and Standard Deviation Charts

The S chart is preferable for monitoring variation when the sample size is large or varying from sample to sample. Although S^2 is an unbiased estimate of σ^2, the sample standard deviation S is not an unbiased estimator of σ. For a variable with a normal distribution, S estimates $c_4\sigma$, where c_4 is a parameter that depends on the sample size n. The standard deviation of S is $\sigma\sqrt{1-c_4^2}$. When σ is to be estimated from past data,

$$\bar{S} = \frac{1}{m}\sum_{i=1}^{m} S_i \tag{6.12}$$

and $\bar{S}/c_4$ is an unbiased estimator of σ. The exact values for c_4 are given in the Table of Control Chart Constants in the Appendix. An approximate relation based on sample size n is

$$c_4 \simeq \frac{4(n-1)}{4n-3} \tag{6.13}$$

The S Chart

The control limits of the S chart are

$$\text{UCL, LCL } = \bar{S} \pm 3\frac{\bar{S}}{c_4}\sqrt{1-c_4^2} \tag{6.14}$$

Defining the constants

$$B_3 = 1 - \frac{3}{c_4}\sqrt{1-c_4^2} \quad \text{and} \quad B_4 = 1 + \frac{3}{c_4}\sqrt{1-c_4^2} \tag{6.15}$$

the limits of the S chart are expressed as

$$\text{UCL} = B_4\bar{S} \qquad \text{and} \qquad \text{LCL } = B_3\bar{S} \tag{6.16}$$

The $\bar{x}$ Chart

When $\hat{\sigma} = \bar{S}/c_4$, the control limits for the $\bar{x}$ chart are

$$\text{UCL, LCL } = \bar{\bar{x}} \pm \frac{3}{c_4\sqrt{n}}\bar{S} \tag{6.17}$$

Defining the constant $A_3 = \frac{3}{c_4\sqrt{n}}$ the limits of the $\bar{x}$ chart become

$$\text{UCL } = \bar{\bar{x}} + A_3\bar{S} \qquad \text{and} \qquad \text{LCL } = \bar{\bar{x}} - A_3\bar{S} \tag{6.18}$$

Example The mean and standard deviation charts are developed by using the overall mean and standard deviation values in Eqs. 6.16 and 6.18. The resulting Shewhart charts are displayed in Figure 6.5. The means of samples 22 and 23 are out-of-control, while the standard deviation chart is out-of-control for sample 23, providing similar results as $\bar{x}$ and R charts.

Interpretation of $\bar{x}$ Charts

The $\bar{x}$ charts must be used along with a spread chart. The process spread must be in-control for proper interpretation of the $\bar{x}$ chart.

The $\bar{x}$ chart considers only the current data value in assessing the status of the process. In order to include historical information such as trends in data, **run rules** have been developed. The run rules sensitize the chart, but they also increase the false alarm probability. If k run rules are used simultaneously and rule i has a Type I error probability of α_i, the overall Type I error probability α_{total} is

$$\alpha_{total} = 1 - \prod_{i=1}^{k}(1-\alpha_i) \tag{6.19}$$

If 3 rules are used simultaneously and $\alpha_i = 0.05$, then $\alpha = 0.143$. For $\alpha_i = 0.01$, $\alpha = 0.0297$.

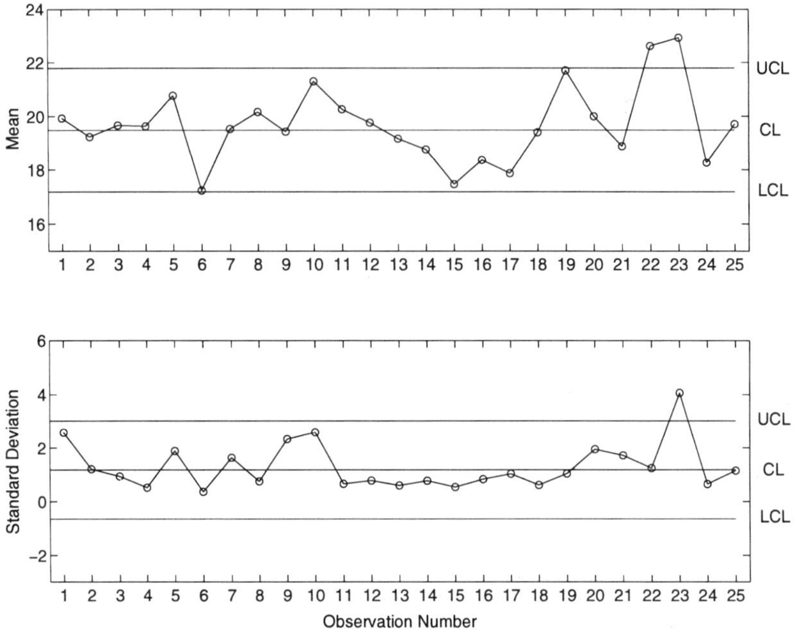

Figure 6.5. Shewhart chart for mean ($CL = \bar{\bar{x}}$) and standard deviation ($CL = \bar{S}$).

The Run Rules

Run rules, also known as Western Electric Rules [111], enable decision making based on trends in data. A process is declared out of control if any one or more of the run rules are met. Some of the criteria used as run rules are:

- One point outside the control limits.
- Two of three consecutive points outside the 2-sigma warning limits but still inside the control limits.
- Four of five consecutive points outside the 1-sigma limits.
- Eight consecutive points on one side of the centerline.
- Eight consecutive points forming a *run* up or a run down.
- A nonrandom or unusual pattern in the data.

Patterns in data could be any systematic behavior such as shifts in process level, cyclical (periodic) behavior, stratification (points clustering around the centerline), trends, or drifts.

Average Run Length (ARL)

The ARL is average number of samples (or sample averages) plotted in order to get an indication that the process is out-of-control. ARL can be used to compare the efficacy of various SPC charts and methods. ARL(0) is the *in-control ARL*, the ARL to generate an out-of-control signal even though in reality the process remains in control. The ARL to detect a shift in the mean of magnitude $k\sigma$ is represented by ARL(σ) where k is a constant and σ is the standard deviation of the variable. A good chart must have a high ARL(0) (for example ARL(0)=400 indicates that there is one false alarm on the average out of 400 successive samples plotted) and a low ARL(σ) (bad news is displayed as soon as possible).

For a Shewhart chart, the ARL is calculated from

$$\text{ARL} = E[R] = \frac{1}{p} \tag{6.20}$$

where p is the probability that a sample exceeds the control limits, R is the run length and $E[\cdot]$ denotes the expected value. For an $\bar{x}$ chart with 3σ limits, the probability that a point will be outside the control limits even though the process is in control is $p = 0.0027$. Consequently, the ARL(0) is ARL $= 1/p = 1/0.0027 = 370$. For other types of charts such as CUSUM, it is difficult or impossible to derive ARL(0) values based on theoretical arguments. Instead, the magnitude of the level change to be detected is selected and Monte Carlo simulations are run to compute the run lengths, their averages and variances.

6.1.2 Cumulative Sum (CUSUM) Charts

The cumulative sum (CUSUM) chart incorporates all the information in a data sequence to highlight changes in the process average level. The values to be plotted on the chart are computed by subtracting the overall mean μ_0 from the data and then accumulating the differences. For a sample size $n \geq 1$, denote the average of the jth sample $\bar{x}_j$. The quantity

$$S_i = \sum_{j=1}^{i} (\bar{x}_j - \mu_0) \tag{6.21}$$

is plotted against the sample number i. CUSUM charts are more effective than Shewhart charts in detecting *small process shifts*, since they combine information from several samples. CUSUM charts are effective with samples of size 1. The CUSUM values can be computed *recursively*

$$S_i = (x_i - \mu_0) + S_{i-1} . \tag{6.22}$$

If the process is in-control at the target value μ_0, the CUSUM S_i should meander randomly in the vicinity of 0. If the process mean is shifted, an *upward or downward trend* will develop in the plot. Visual inspection of changes of slope indicates the sample number (and consequently the time) of the process shift. Even when the mean is on target, the CUSUM S_i may wander far from the zero line and give the appearance of a signal of change in the mean. Control limits in the form of a V-mask were employed when CUSUM charts were first proposed in order to decide that a statistically significant change in slope has occurred and the trend of the CUSUM plot is different than that of a random walk. CUSUM plots generated by a computer became more popular in recent years and the V-mask has been replaced by upper and lower confidence limits of one-sided CUSUM charts.

One-Sided CUSUM charts are developed by plotting

$$S_i = \sum_{j=1}^{i} [\bar{x}_j - (\mu_0 + K)] \tag{6.23}$$

where K is the *reference value* to detect an increase in the mean level. If S_i becomes negative for $\mu_1 > \mu_0$, it is reset to zero. When S_i exceeds the decision interval H, a statistically significant increase in the mean level is declared. Values for K and H can be computed from the relations:

$$K = \frac{\Delta}{2}, \qquad H = \frac{d\Delta}{2} . \tag{6.24}$$

Given the α and β probabilities, the size of the shift in the mean to be detected (Δ), and the standard deviation of the average value of the variable x ($\sigma_{\bar{x}}$), the parameters in Equation 6.24 are:

$$\delta = \frac{\Delta}{\sigma_{\bar{x}}} \quad \text{and} \quad d = \left(\frac{2}{\delta^2}\right) \ln\left(\frac{1-\beta}{\alpha}\right) . \tag{6.25}$$

A *two-sided CUSUM* can be generated by running two one-sided CUSUM charts simultaneously with the upper and lower reference values. The recursive formulae for *high* and *low side* shifts that include resetting to zero are

$$\begin{aligned} S_H(i) &= max\ \left[0, \bar{x}_i - (\mu_0 + K) + S_H(i-1)\right] \\ S_L(i) &= max\ \left[0, (\mu_0 - K) - \bar{x}_i + S_L(i-1)\right] \end{aligned} \tag{6.26}$$

respectively. The starting values are usually set to zero, $S_H(0) = S_L(0) = 0$. When $S_H(i)$ or $S_L(i)$ exceeds the *decision interval* H, the process is out-of-control. Average Run Length (ARL) based methods are usually utilized to find the chart parameter values H and K. The rule of thumb for ARL(Δ) for detecting a shift of magnitude Δ in the mean when $\Delta \neq 0$ and $\Delta > K$ is

$$\text{ARL}(\Delta) = 1 + \frac{H}{\Delta - K} . \tag{6.27}$$

Two-sided CUSUM sometimes is called as the **tabular CUSUM**. Whereas the monitoring results are usually given as tabulated form, it is useful to present graphical display for tabular CUSUM. These charts are generally called as **CUSUM status charts** [400].

Example Develop the CUSUM chart to detect a shift in the mean of magnitude $\delta = \Delta/\sigma_{\bar{x}} = 2$, with $\alpha = 0.01$ and $\beta = 0.05$. Using Eq. 6.25, $d = 2.28$. H and K are computed from Eq. 6.24 as $H = 2.62$ and $K = 1.15$, and the one-sided CUSUM charts are based on Eq. 6.26. The resulting charts where the first twenty samples are used to develop the charts are shown in Figure 6.6. Since the observation 23 exceeds the decision interval H at which $S_H > H = 2.62$, we would conclude that the process is out of control at that point.

6.1.3 Moving Average Control Charts for Individual Measurements

Individual data (sample size n=1) are common in many process industries. Continuous streams of data are more common for continuous processes. MA charts may be used for monitoring successive batches by using end of batch quality measurements. MA charts may also be used for data with small variation, collected during a batch run. Instantaneous variations in a batch run may be small, but the process varies over time. Selecting a group of successive measurements close together in time will include mostly variation due to measurement and sampling error. In such situations, statistical monitoring of the process can be achieved by using *moving-average* (MA) *charts.* In MA charts, averages of the consecutive data groups of size a are plotted. The control limit computations are based on averages and standard deviation values computed from moving ranges. Since each MA point has $(a - 1)$ common data points, the successive *MAs are highly autocorrelated.* This autocorrelation is ignored in the usual construction of these charts. The MA control charts should not be used with strongly autocorrelated data. The MA charts detect small drifts efficiently (better than $\bar{x}$ chart) and they can be used when the original data do not have Normal distribution.

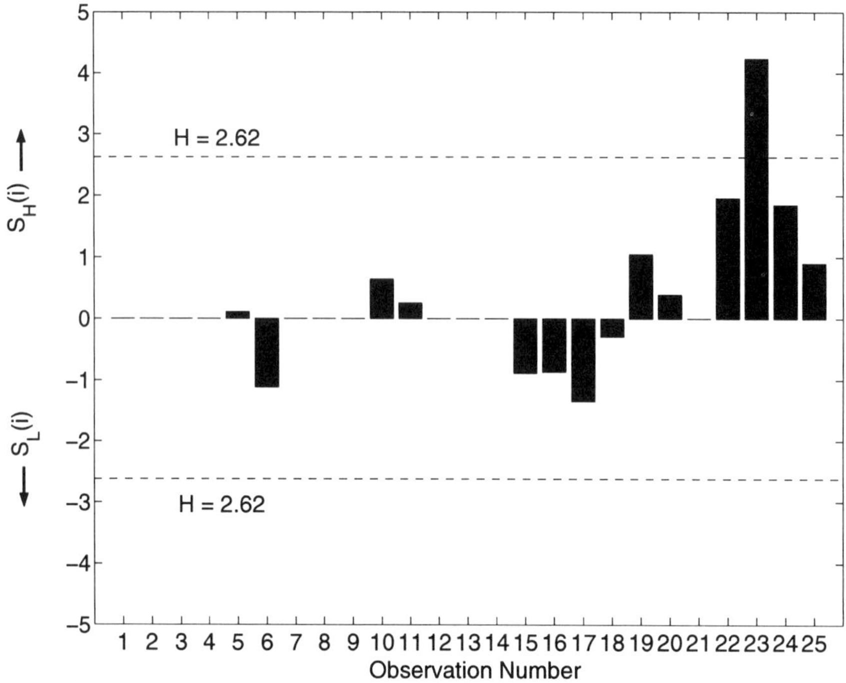

Figure 6.6. The CUSUM status chart.

The disadvantages of the MA charts are slow response to sudden shifts in level and the generation of autocorrelation in computed values.

Estimation of S for individual measurements

Three approaches can be used for estimating S:

1. If a *rational blocking* of data exists, compute an estimate of S based on it. It is advisable to compare this estimate with the estimates obtained by using the other methods to check for discrepancies.

2. *The overall S estimate.* Use all the data together to calculate an overall standard deviation. This estimate of S will be inflated by the *between-sample* variation. Thus, it is an upper bound for $\hat{S}$. If there are changes in process level, compute S for each segment separately, then combine them by using

$$S_w = \sqrt{\frac{\sum_{i=1}^{k}(n_i - 1)S_i^2}{\sum_{i=1}^{k}(n_i - 1)}} \tag{6.28}$$

where k is the number of segments with different process levels and n_i is the number of observations in each sample.

3. *Estimation of S by moving-ranges of "a" successive data points.* Use differences of successive observations as if they were ranges of n observations. A plot of S for group size a versus a will indicate if there is between-sample variation. If the plot is flat, the between-sample variation is insignificant. This approach should not be used if there is a trend in data. If there are missing observations, all groups containing them should be excluded from computations.

The procedure for estimating S by moving-ranges is:

1. Calculate moving-ranges of size a, $a = 2, 3, \cdots$, using 25 to 100 observations.

$$MR_t = \mid max(x_i) - min(x_i) \mid \ , \ \ i = (t - a + 1), t. \tag{6.29}$$

2. Calculate the mean of the ranges for each a

3. Divide the result of Step 2 by d_2 (for each a).

4. Tabulate and plot results for all a.

Process Level Monitoring by Moving-Average Charts

In a moving-average (MA) chart, the averages of consecutive groups of size a are computed and plotted. The control limit computations are based on these averages. Several original data points at the start and end of the chart are excluded, since there is not enough data to compute the moving-average at these times. MA charts detect small drifts efficiently (better than $\bar{x}$ chart). However, they respond slowly to sudden shifts in level and the MA generates autocorrelation in computed values.

Procedure For Chart Development

The procedure is outlined for m samples of size n. For individual measurements, let $n = 1$.

1. Compute the sample averages $\bar{x}_i$, $i = 1, m$.

2. Compute the moving average M_t of span a at time t as

$$M_t = \frac{\bar{x}_t + \bar{x}_{t-1} + \cdots + \bar{x}_{t-a+1}}{a} \tag{6.30}$$

3. Compute the variance of M_t

$$V(M_t) = \frac{1}{a^2} \sum_{i=t-a+1}^{t} V(\bar{x}_i) = \frac{\sigma^2}{na} \tag{6.31}$$

Hence, $\sigma = \bar{S}/c_4\sqrt{an}$ or $\sigma = \overline{MR}/d_2\sqrt{n}$, using $\overline{MR}$ for $\bar{R}$.

4. Compute the control limits with the centerline at $\bar{\bar{x}}$:

$$UCL, LCL = \bar{\bar{x}} \pm \frac{3\bar{S}}{c_4\sqrt{na}} \quad \text{or} \quad = \bar{\bar{x}} \pm \frac{3\overline{MR}}{d_2\sqrt{n}} \tag{6.32}$$

In general, the span a and the magnitude of the shift to be detected are inversely related.

Spread Monitoring by Moving-Range Charts

In a moving-range chart, the range of two consecutive sample groups of size a are computed and plotted. For $a \geq 2$,

$$MR_t = \mid max(x_i) - min(x_i) \mid , \quad i = (t-a+1), t \tag{6.33}$$

The computation procedure is:

1. Select the range size a. Often $a = 2$.

2. Obtain estimates of $\overline{MR}$ and $\sigma = \overline{MR}/d_2$ by using the moving-ranges MR_t of length a. For a total of m samples:

$$\overline{MR} = \frac{1}{m-a+1} \sum_{t=1}^{m-a+1} MR_t \tag{6.34}$$

3. Compute the control limits with the centerline at $\overline{MR}$:

$$LCL = D_3\overline{MR}, \qquad UCL = D_4\overline{MR} \tag{6.35}$$

Recall that $\sigma_R = d_3\bar{R}/d_2$, and d_2 and d_3 depend on a.

6.1.4 Exponentially Weighted Moving-Average Chart

The exponentially weighted moving-average (EWMA) z_i is defined as

$$z_i = w\bar{x}_i + (1-w)z_{i-1} \tag{6.36}$$

where $0 < w \leq 1$ is a constant weight, $\bar{x}_i$ is the mean of sample i of size n, and the starting value at $i = 1$ is $z_0 = \bar{\bar{x}}$. EWMA attaches a higher weight to more recent data and has a fading memory where old data are discarded from the average. Since the EWMA is a weighted average of several consecutive observations, it is insensitive to nonnormality in the distribution of the data. It is a very useful chart for plotting individual observations ($n = 1$). If $\bar{x}_i$ are independent random variables with variance σ^2/n, the variance of z_i is

$$\sigma_{z,i}^2 = \frac{\sigma^2}{n}\left(\frac{w}{2-w}\right)\left[1-(1-w)^{2i}\right] \tag{6.37}$$

The last term (in brackets) in Eq. 6.37 quickly approaches 1 as i increases and the variance reaches a limiting value. Often the asymptotic expression for the variance is used for computing the control limits. The weight constant w determines the memory of EWMA, the rate of decay of past sample information. For $w = 1$, the chart becomes a Shewhart chart. As $w \to 0$ EWMA approaches a CUSUM. A good value for most cases is in the range $0.2 \leq w \leq 0.3$. A more appropriate value of w for a specific application can be computed by considering the ARL for detecting a specific magnitude of level shift or by searching w which minimizes the prediction error for a historical data set by an iterative least squares procedure. 50 or more observations should be utilized in such procedures. EWMA is also known as *geometric moving average*, *exponential smoothing*, or *first order pole filter*.

Upper and the lower control limits are calculated as

$$\begin{aligned} UCL_i &= \mu_0 + 3\sigma_{z_i} \\ CL &= \mu_0 \\ LCL_i &= \mu_0 - 3\sigma_{z_i}. \end{aligned}$$

Example Develop an EWMA chart to detect a shift in the mean by using the first column of the example data set in Figure 6.6 and $w = 0.25$. Compute the variance of z by using asymptotic version of Eq. 6.37 and the values of z_i from Eq. 6.36. The resulting charts where the first twenty samples are used to develop the charts are shown in Figure 6.7. From the EWMA control chart (Figure 6.7) signal for observation 23, we conclude that the process is out of control at that point.

6.2 SPM of Continuous Processes with Multivariate Statistical Techniques

In traditional quality control of multivariable processes, a number of quality variables are monitored using Shewhart charts [542]. But because of inter-

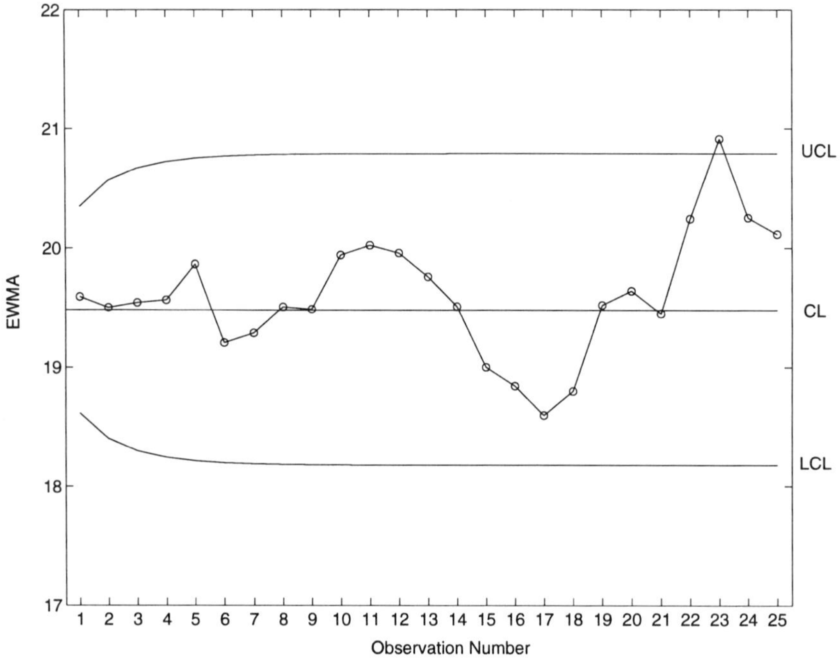

Figure 6.7. EWMA chart of observations.

actions among the variables that cause crosscorrelation, autocorrelation and colinearity, monitoring one variable at a time approach may become misleading and time consuming if the number of variables to be monitored is high. The potential for erroneous interpretations is illustrated in Figure 6.8 such that univariate charts of two quality variables (x_1 and x_2) are constructed separately and depicted as a biplot by aligning one chart perpendicular to the other. The control limits of the two individual Shewhart charts (99 % upper (UCL) and lower (LCL) confidence limits) are now shown as a rectangle. All of the observations are inside the limits. indicating an in-control situation and consequently acceptable product quality. The ellipse represents the control limits for the in-control *multivariable* process behavior with 99 % confidence. When a customer complains about low-quality product for the batch corresponding to the sample indicated by $\otimes$ in Figure 6.8, Shewhart charts do not indicate poor product quality but the multivariate limit does. Furthermore, if there are any samples outside the upper left or lower right corners of the Shewhart confidence region, but inside the ellipse, the consumer would not report them as poor quality products. The situation can be explained by inspecting the multivariate plot

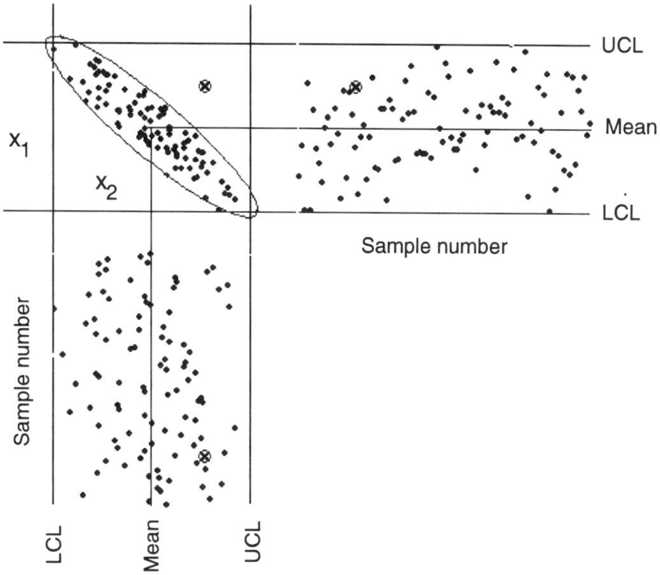

Figure 6.8. Univariate charts and biplot of two variables.

of these variables. Given the joint-confidence region defined by the ellipse, any observation that falls out of this region is considered as out-of-control.

Traditional univariate techniques based on a single variable have been reviewed in the previous section. Despite their misleading nature, univariate charts are still used in industry for monitoring multivariable processes. Several multivariate extensions of Shewhart, CUSUM and EWMA have been proposed in the literature [351, 352, 594, 672]. The multivariate perspective helps one to unveil hidden relations that reside in process data and reach correct conclusions about product quality. There is significant motivation to develop a multivariable statistical process monitoring (SPM) framework to detect the existence, magnitude, and time of occurrence of changes that cause the process to deviate from its desired operation.

Biplots are useful when only a few variables are monitored. When the process has a large number of variables, monitoring tools based on projection techniques are more effective. These techniques rely on principal components analysis (PCA) and partial least squares (PLS) introduced in Sections 4.1 and 4.2.4.

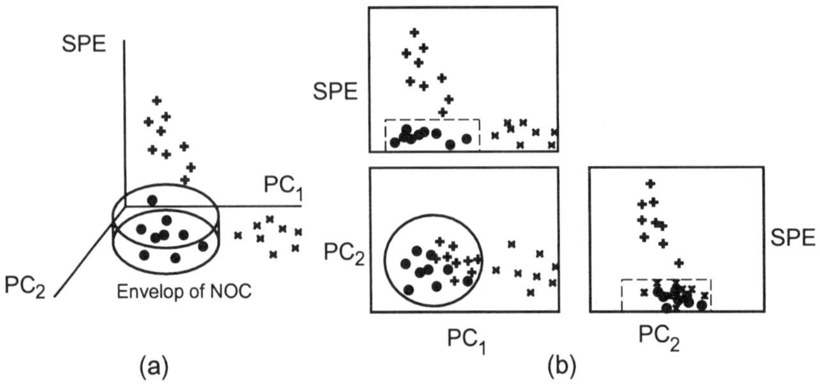

Figure 6.9. The multivariate monitoring space. (a) Three dimensional representation, (b) Two dimensional representation.

6.2.1 SPM of Continuous Processes with PCA

SPM with PCA can be implemented by graphical and numerical tools. Two types of statistics, the statistical distance T^2 and the principal components (PC) model residuals $(\mathbf{I} - \mathbf{PP^T})\mathbf{X}$ or squared prediction error (SPE) must be monitored. If a few PCs can describe the data, biplots of PC scores can be used as easy to interpret visual aids. Such biplots can be generated by projecting the data in Figure 6.9 to two dimensional surfaces PC_1-PC_2, PC_1-Error and PC_2-Error. Data representing normal operation (NO) and various faults are clustered in different regions, providing the opportunity to diagnose source causes as well [304].

Inspection of many biplots becomes inefficient and difficult to interpret when a large number of PCs are needed to describe the process. Monitoring charts based on squared residuals (SPE) and T^2 become more useful. By appending the confidence interval (UCL) to such plots, a multivariate SPM chart as easy to interpret as a Shewhart chart is obtained.

PCA techniques have been used to monitor an LDPE reactor operation [297], high speed polyester film production [635], Tennessee Eastman simulated process [488] and sheet forming processes [508]. Multiscale PCA by using wavelet decomposition has been proposed [38].

6.2.2 SPM of Continuous Processes with PLS

Modern process data acquisition systems generate large amounts of process data, such as temperatures and flow rates. Measurements of process out-

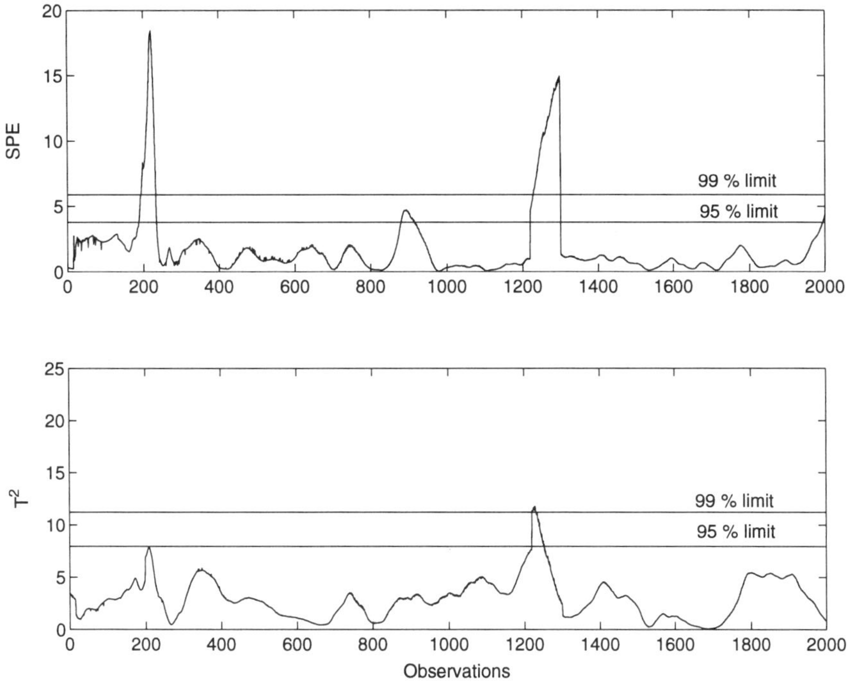

Figure 6.10. SPE and T^2 charts for continuous process monitoring based on PCA.

puts that describe product quality are collected less frequently since these measurements are expensive and time consuming. Although it is possible to measure some quality variables on-line by means of sophisticated devices, measurements are generally made off-line in the quality control laboratory. Process data contain important information about both the quality of the product and the performance of the process operation. PLS models can be used in two ways:

Quality monitoring. The correlation between the process variables and the quality variables can be determined through the PLS model. This statistical model provides information for estimating product quality from process data.

Statistical process control. PLS model can also be used to quickly detect process upsets and unexpected behavior. When an assignable cause is detected, necessary actions can be taken to prevent any damage to process performance and/or product quality.

The traditional statistical modeling methods such as Multiple Linear Regression (MLR) fail to handle process data that are correlated and collinear. PLS, as a projection method, offers a suitable solution for modeling such data.

The first step in the development of a PLS model is to determine the variables that will be considered as process variables **X** and as indicator of product quality **Y**. This selection is dependent on the measurements available and the objectives of monitoring. The reference set used to develop the multivariate monitoring chart will determine the variations considered to be part of normal operation and ideally includes all variations leading to desired process performance. If the reference set variation is too small, the procedure will cause frequent alarms, and if it is too large the sensitivity of the monitoring scheme to the abnormal operation will be poor. The normal operating data are collected from past successful process history. The reference data set selected should include the range of process variables that yield desired product quality. If the PLS model is developed for monitoring certain process conditions, the reference data set should include data collected under these conditions. Data for various batch runs are then stacked together to form the reference set that represents normal behavior of the process.

Since PLS technique is sensitive to outliers and scaling, outliers should be removed and data should be scaled prior to modeling. After data pretreatment, another decision to be made is the determination of the number of latent variables (PLS dimensions) to be retained in the model. Cumulative prediction sum of squares (CUMPRESS) vs number of latent variables or prediction sum of squares (PRESS) vs number of latent variables plots are used for this purpose. It is usually enough to consider the first few PLS

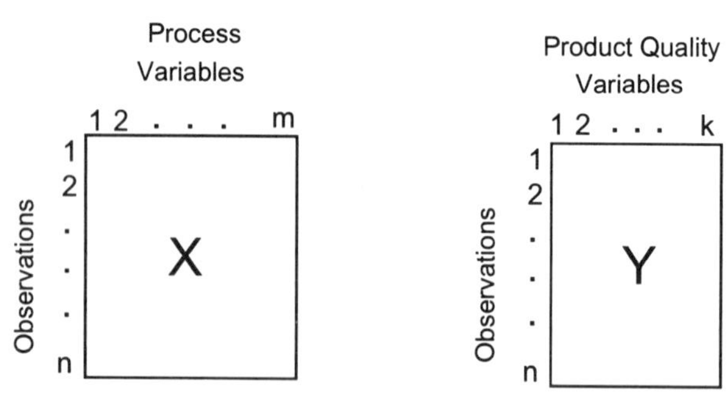

Figure 6.11. Data arrangement in PLS for continuous SPM.

dimensions for monitoring while for prediction more PLS dimensions are needed in order to improve the precision of thc predictions.

Once the PLS model is built, squared prediction error (SPE) can be calculated for either the **X** or the **Y** block model (Eqs. 6.38 and 6.39)

$$SPE_{\mathbf{X},i} = \sum_{j=1}^{m} (x_{ij} - \hat{x}_{ij})^2 \tag{6.38}$$

$$SPE_{\mathbf{Y},i} = \sum_{j=1}^{k} (y_{ij} - \hat{y}_{ij})^2 \tag{6.39}$$

where $\hat{x}$ and $\hat{y}$ are predicted observations in **X** and **Y** using the PLS model, respectively, i and j denote observations and variables in **X** or **Y**, respectively.

$\hat{x}$ and $\hat{y}$ in Eqs. 6.38 and 6.39 are calculated for new observations as follows:

$$t_{a,\text{new}} = \sum_{j=1}^{m} x_{\text{new},j} w_{a,j} \tag{6.40}$$

$$\hat{x}_{\text{new},j} = \sum_{a=1}^{A} t_{a,\text{new}} p_{a,j}^T \tag{6.41}$$

$$\hat{y}_{\text{new},j} = \hat{x}_{\text{new},j} b \tag{6.42}$$

where $w_{a,j}$ denotes the weights, $p_{a,j}$ the loadings for **X** block (process variables) of the PLS model, $t_{a,\text{new}}$ the scores of new observations and b the vector of regression coefficients.

Multivariate control charts based on squared prediction errors (SPE_X and SPE_Y), biplots of the scores ($\mathbf{t}_a$ vs $\mathbf{t}_{a+1}$) and the Hotelling's statistic (T^2) are constructed with the control limits. The control limits at significance level $\alpha/2$ for a new independent t score under the assumption of normality at any time interval are

$$\pm t_{n-1,\alpha/2} s_{\text{est}} (1 + 1/n)^{1/2} \tag{6.43}$$

where n, s_{est} are the number of observations and the estimated standard deviation of the score sample at the chosen time interval and $t_{n-1,\alpha/2}$ is the critical value of the t-student test with $n-1$ degrees of freedom at significance level $\alpha/2$ [214, 435]. The Hotelling's statistic (T^2) for a new independent **t** vector is calculated as [594]

$$T^2 = \mathbf{t}_{\text{new}}^T \mathbf{S}^{-1} \mathbf{t}_{\text{new}} \quad \sim \frac{A(n^2-1)}{n(n-A)} F_{A,n-A} \tag{6.44}$$

where **S** is the estimated covariance matrix of PLS model scores, A the number of latent variables retained in the model and $F_{A,n-A}$ the F-distribution value. The control limits on SPE charts can be calculated by an approximation of the χ^2 distribution given as $\text{SPE}_\alpha = g\chi^2_{h\alpha}$ [76]. This equation is well approximated as [148, 255, 435]

$$SPE_\alpha \cong gh\left[1 - \frac{2}{9h} + z_\alpha\left(\frac{2}{9h}\right)^{1/2}\right]^3 \tag{6.45}$$

where g is a weighting factor and h degrees of freedom for the χ^2 distribution. These can be approximated as $g = v/(2m)$ and $h = 2m^2/v$, where v is the variance and m the mean of the SPE values from the PLS model. All of the aforementioned calculations are illustrated in the following example.

Example. Consider a continuous fermentation process where monitoring will depend on how well the process is performing based on product quality. Assume that ten process variables such as aeration rate and substrate feed rate are used for the **X** block, and one quality variable, product concentration for **Y** block. As the first step of the PLS modelling the outliers are removed and both blocks are scaled appropriately (autoscaling is used for this case). A PLS model is built to relate ten process variables with one quality variable. A data window of 100 observations are taken as in-control operation. In order to decide the number of latent variables to be retained in the model, PRESS and CUMPRESS values are calculated based on cross-validation (Figure 6.12).

Only the first two latent variables are used in the monitoring procedure since they explained 88.10% variation in **Y** (Table 6.2) and the decrease in the CUMPRESS value by adding the third latent variable is small (only an additional 0.83% of the variance of **Y**). A step decrease (30% off the set point) into substrate feed rate was introduced after 100*th* observation until 150*th* observation (Figure 6.12). It is desired to detect this change based on its effects on the quality variable (product concentration). Both

Table 6.2. Percent variance captured by PLS model

	X-block		Y-block	
LV no.	This LV	Total	This LV	Total
1	27.19	27.19	75.42	75.42
2	10.98	38.17	12.67	88.10

SPE (Figure 6.12(c)) and T^2 (Figure 6.12(e)) charts for **X** block have detected this change on time. Biplot of the latent variables also shows an excursion from the in-control region defined by ellipses and the score values come back to the in-control region after the change is over (Figure 6.12(b)). SPE of **Y** block shows an out-of-control situation as well (Figure 6.12(f)). Although the disturbance is over after 150*th* observation (Figure 6.12(c)-6.12(e)), product quality seems to deteriorate because the prediction capability of PLS model becomes poor after 150th observation (Figure 6.12(d)) suggesting a change in the quality space which is different than the one reflected by PLS model.

6.3 Data Length Equalization and Determination of Phase Landmarks in Batch Fermentation

Most batch processes, including many fermentation processes, pass through several phases based on complex physiological phenomena during the progress of the batch (Figure 6.13). In this book, we used the term "stage" to refer to different process operations such as fermentation and separation, and the term "phase" to refer to distinct episodes in time during the progress of the batch where qualitatively different activities take place. Since batch fermentation time varies from batch to batch due to complex physiological behavior and operational changes, the data sets for different batches will have different lengths and shifted phase changing points or process landmarks. These shifts can affect monitoring activities and generate false alarms. Consequently, alignment of landmarks is necessary for comparing similar events. Multivariate analysis requires the data to be stacked in a matrix (or in a three-way array) prior to empirical modelling. Several techniques have been suggested for batch data length synchronization and equalization [270, 296, 418, 522, 641]. Cutting batch data lengths to length of the variable with the shortest data sequence is not recommended because of significant information loss generated by discarding data. When the time between the shortest batch and the longest batch is large, or the process in question is very sensitive to small changes in operational or environmental conditions, robust and generic methods are needed to synchronize and equalize data lengths.

Two problems will be addressed in this section: equalization of batch data lengths, and detection and alignment of phase change landmarks. Three methods are discussed for equalizing batch data lengths:

- Indicator Variable Technique

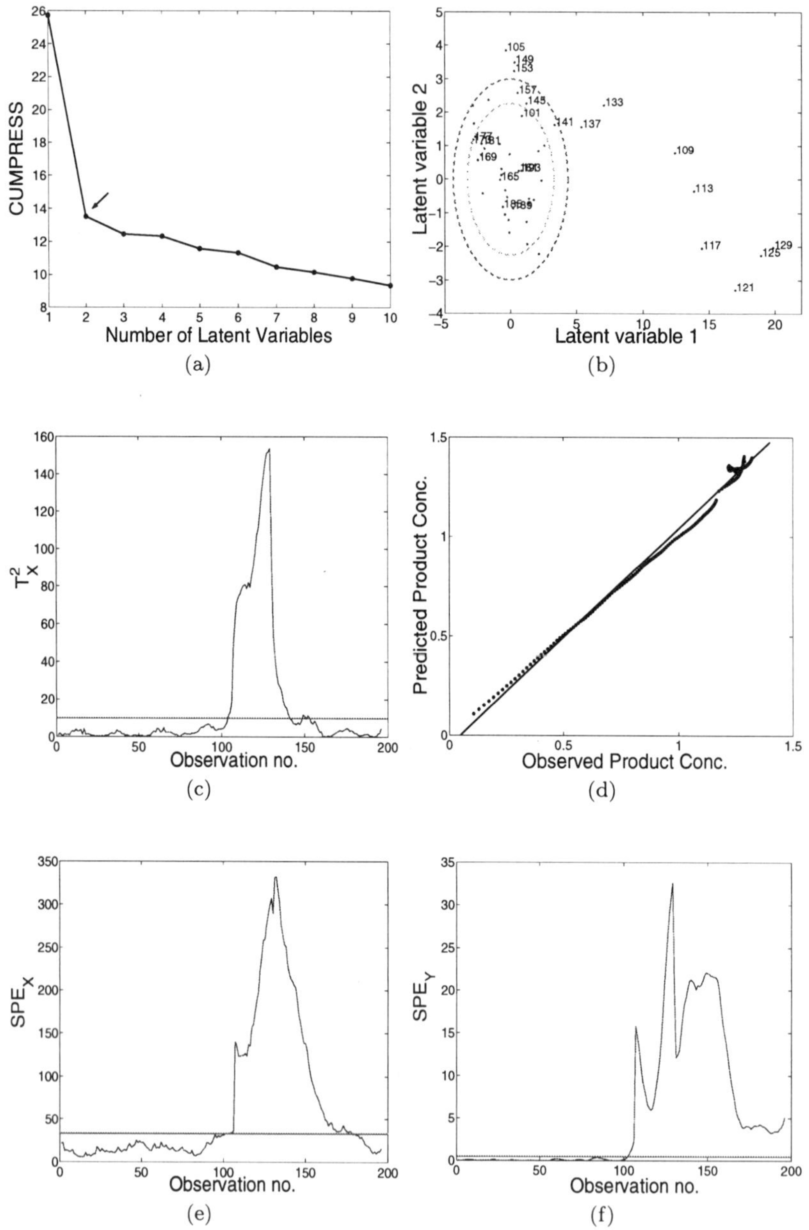

Figure 6.12. SPM charts based on PLS model for monitoring a faulty case (step decrease in substrate feed rate).

- Dynamic Time Warping (DTW)
- Curve registration

The simple popular technique based on an *indicator variable* is discussed in Section 6.3.1. Then, the *dynamic time warping* method is presented in Section 6.3.2. Finally, time warping by functional data analysis (also called *curve registration*) is discussed in Section 6.3.3 and comparative examples are provided.

6.3.1 Indicator Variable Technique

This technique is based on selecting a process variable to indicate the progress of the batch instead of time. Each new observation is taken relative to the progress of this variable. The indicator variable should be smooth, continuous, monotonic and spanning the range of all other process variables within the batch data set. Linear interpolation techniques are used to transform batch-time dimension into indicator variable dimension. This variable should be chosen appropriately such that it also shows the *maturity* or *percent completion* of each batch. This variable can be for example percent conversion or percent of a component fed to the fermenter. For monitoring new batches, data are collected from all process variables

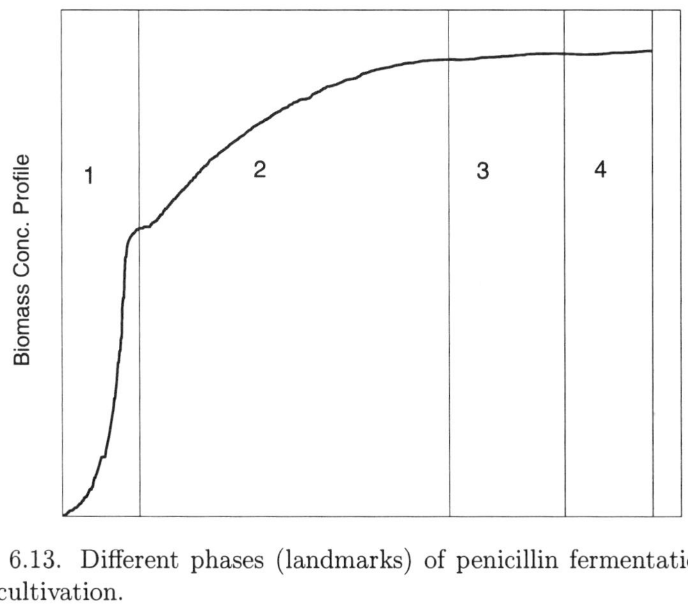

Figure 6.13. Different phases (landmarks) of penicillin fermentation in a batch cultivation.

at specified time intervals and then adjusted with respect to the indicator variable. In this technique, a measure of the *maturity* or *percent completion* of any batch is provided by the percentage of its final value that has been attained by the indicator variable at the current time. Several successful applications of this approach can be found in the literature, mostly for batch/semi-batch polymerization processes, reaction extent or percent of component fed being the indicator variables [296, 418]. An application for fermentation processes has been also given in the literature [522].

Choosing an indicator variable in batch fermentation processes depends on the process operation and characteristics. If the process is a batch fermentation, the choice of this variable is simpler than processes with batch and fed-batch phases. For batch fermentations, there may be several variables, which can serve as indicator variables such as substrate concentration, product concentration or product yield. In the fed-batch case, in addition to the aforementioned variables, percent substrate fed is also an indicator variable. This percentage is calculated by fixing the total amount of substrate added into the fermenter based on some performance criteria. This end point (total amount of substrate fed), which is eventually reached in all batches, defines a maturity point. For more complex operations such as batch operation followed by fed-batch operation, which is very common for non-growth associated products such as antibiotics, different approaches to choosing indicator variables can be considered. Batch and fed-batch phases of the operation can be treated separately so that appropriate indicator variables can be determined for individual phases. Implementation of this two-phase operation is illustrated in the following example.

Example. Assume that data are available from 5 runs of a batch followed by fed-batch penicillin fermentation. Potential process variables are shown in Figure 6.14 for all batches before data pretreatment. Based on simulation studies, data were collected using 0.2 h of sampling interval on each variable for each batch resulting in total batch lengths varying between 403.8 h (2019 observations) and 433.6 h (2168 observations). When these variables are assessed for use as an indicator variable, none of them seem appropriate. Most of these variables contain discontinuities because of the two operating regions (batch and fed-batch) and some of them are not smooth or monotonically increasing/decreasing. Since none of the variables can be chosen as an indicator variable that spans the whole duration of fermentation, a different approach is suggested. The solution is to look for different indicator variables for each operating region. In order to achieve this mixed approach fermentation data are analyzed. For the first operating (batch operation) region, substrate concentration in the fermenter can be

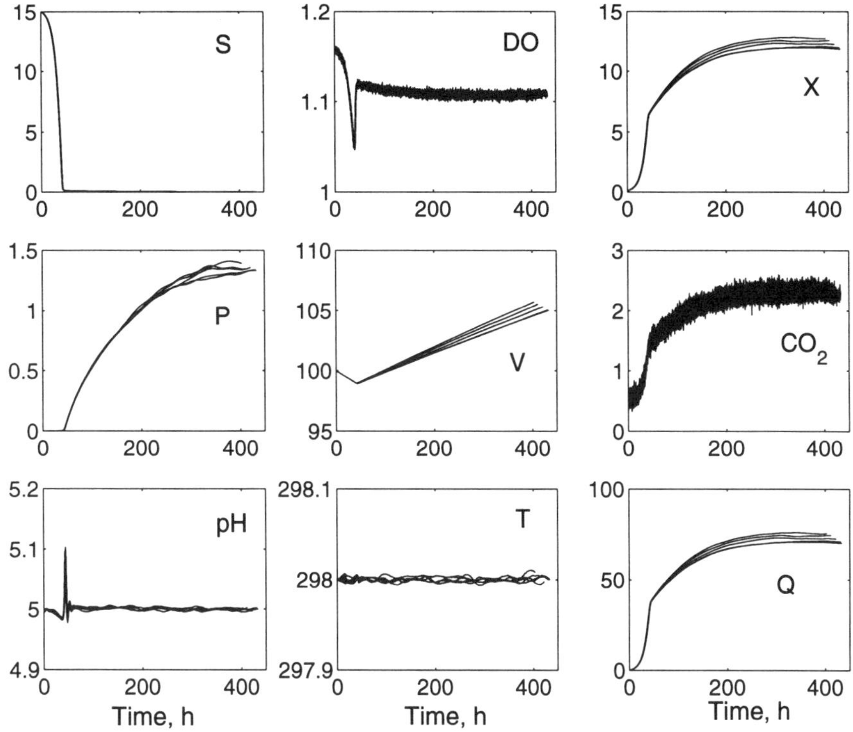

Figure 6.14. Output variables for the five batches. S: Substrate conc., DO: Dissolved oxygen conc., X: Biomass conc., P: Penicillin conc., V: Culture volume, CO_2: CO_2 conc., T: Temperature in the fermenter and Q: Generated heat [62, 603].

considered as a good candidate since it can be started from the same initial value and terminated at the same final value for each batch. The initial and final substrate concentrations are fixed to 15 g/L and 0.4 g/L, respectively to implement this idea. Instead of reporting data as a function of time for these batches, data are reported on each variable for each batch at every decrease of 0.5 g/L in substrate concentration using linear interpolation.

Choosing substrate concentration decrease (substrate consumption) as an indicator variable for the batch operation provides another advantage, it defines the end of batch operation or in other words the switching point to fed-batch operation. Since the operating conditions are slightly different and there are some random changes in microbial phenomena, the switching point is reached at different times for each batch resulting in different number of observations (Figure 6.15). While the number of observations is varying between 210 and 218 before equalization, after implementing the indicator variable technique there are only 147 observations taken from each batch.

Substrate concentration cannot be used for the fed-batch operation region since substrate is added continuously to promote penicillin production and biomass maintenance such that substrate concentration approximately constant until the end of the run. In the second region (fed-batch), amount of substrate fed to the fermenter can be considered as an indicator variable since it somehow defines the end point of the fermentation. Figure 6.16 is used to decide the approximate amount of substrate needed to reach the desired final penicillin concentrations under normal operating conditions. Analysis shows approximately 16 L of substrate would be necessary for each batch. To calculate this amount, fermentations were carried out fairly long enough (*approx.* 600 h) to see a decline in penicillin concentration. A mean trajectory is then calculated and its maximum is used to determine the total amount of substrate to be added. Batch/fed-batch switching times, maximum and final penicillin concentrations for each batch including mean trajectory values are given in Table 6.3. The following calculations then become straightforward

$$\bar{P}_{\text{final}} = 1.3525\ g/L\ ,\ \bar{t}_{\text{switch}} = 42\ h,\ \bar{t}_{\text{final}} = 420\ h,$$

$$V_{\text{total substrate added}} = \sum_i F_i \Delta t_i \tag{6.47}$$

where $\bar{P}_{\text{final}}$ denotes final penicillin concentration, $\bar{t}_{\text{switch}}$ beginning of the fed-batch period, $\bar{t}_{\text{final}}$ end of fermentation for the mean trajectory, F_i, instantaneous value of the substrate feed rate and Δt_i, instantaneous sampling interval (which is 0.2 h in the example). When the maximum of penicillin concentration of the mean trajectory is used by assuming that the values $\bar{P}_{\text{max}} = 1.3664\ g/L$, $\bar{t}_{\text{switch}} = 42.0\ h$, $\bar{t}_{\text{final}} = 446.2\ h$ and $\bar{F}_i = 0.0394\ L/h$ are predetermined by analyzing the mean trajectory for penicillin concentration, the approximate total amount of substrate is calculated as $\bar{F}_i \times (\bar{t}_{\text{final}} - \bar{t}_{\text{switch}}) = 0.0394 \times (446.2 - 42.0) = 15.9055\ L$. This amount can be calculated more accurately by using Eq. 6.47 resulting a closer value to 16 L. Although 16 L is not the exact outcome of the calculations, it was chosen to round off the number and introduce a little safety margin (using a little more substrate than the required minimum). The resulting final penicillin concentrations do not deviate substantially from their maximum values (Table 6.3) verifying this choice. The result of equalization is shown in Figure 6.17 for several variables based on the calculations above.

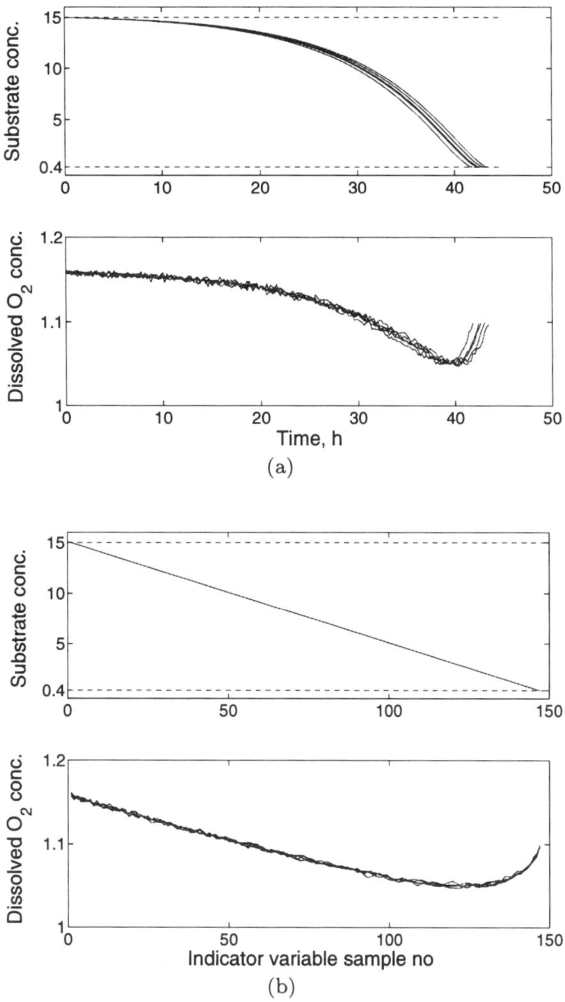

Figure 6.15. Substrate and dissolved oxygen concentrations of five batches (a) before and (b) after equalization.

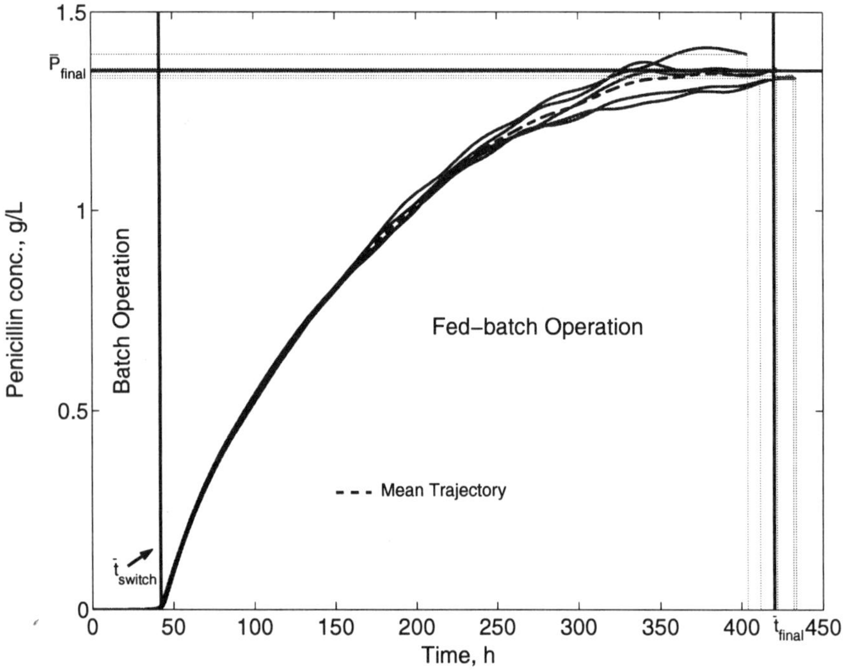

Figure 6.16. Penicillin concentration profiles of five batches.

Table 6.3. Critical values on penicillin concentration profiles

	Batch Operation		Fed-batch Operation			
Batch No	t_{switch}	No. of obs.	t_{final}	No. of obs.	P_{final}	P_{max}
1	43.2	216	432.0	1944	1.3392	1.3392
2	43.6	218	421.8	1891	1.3574	1.3574
3	42.8	214	411.6	1844	1.3470	1.3736
4	42.0	210	403.8	1809	1.3935	1.4101
5	42.6	213	433.6	1955	1.3328	1.3332
P_{mean}	42.0	210	420.0	1890	1.3525	1.3664

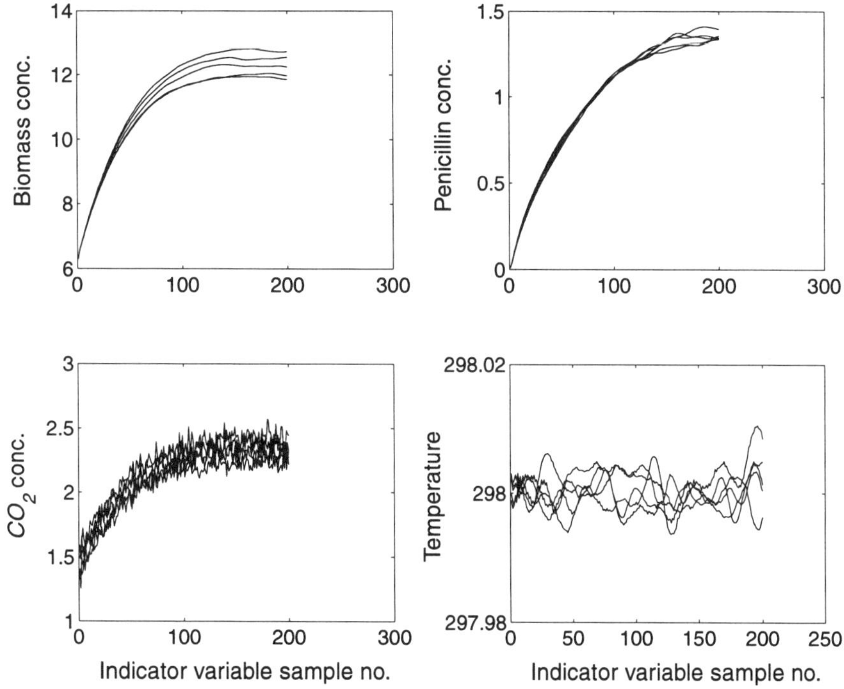

Figure 6.17. Equalized batch lengths of five batches based on indicator variable technique in fed-batch operation.

6.3.2 Dynamic Time Warping

When an appropriate indicator variable does not exist, other techniques can be implemented to synchronize data and equalize batch lengths. Dynamic Time Warping (DTW) technique is one of them. It has its origins in speech recognition. Unsynchronized feature vectors (trajectories) are a common problem in speech recognition [123, 406, 446, 547] since the same word can be uttered in varying intensities and durations by different speakers. Speech recognition systems should have the ability to interpret words independent of speakers [484]. This is analogous to batch process trajectory synchronization problem since similar events that take place in each batch run are required to be matched. DTW is a flexible, deterministic, pattern matching scheme which works with pairs of patterns. The time-varying features within the data are brought into line (synchronized) by time normalization. This process is known as "time warping" since the data patterns are locally translated, compressed, and expanded until similar features in the patterns between reference data and the new data are matched resulting in the same data length as the reference data [252, 533]. Basic description of DTW and different algorithms for implementing it have been reported [252, 406, 533, 485].

One of the pioneering implementations of DTW to bioprocesses was suggested by Gollmer and Posten [197] on the detection of important process events including the onset of new phases during fermentations. They have provided a univariate scheme of DTW for recognition of phases in batch cultivation of *S. cerevisiae* and detection of faults in fed-batch *E. coli* cultivations. Another application of DTW (by Kassidas et al. [270]) has focused on batch trajectory synchronization/equalization. They have provided a multivariate DTW framework for both off-line and on-line time alignment and discussed a case study based on polymerization reactor data.

To introduce the DTW theory, consider two sets of multivariate observations, reference set R, with dimensions $j \times P$, and test set T, with dimensions $i \times P$ (Eq. 6.48). These sets can be formed from any multivariate observations of fermentation processes (or batch processes in general) where j and i denote the number of observations in R and T, respectively, and P the number of measured variables in both sets as $p = 1, 2, \ldots, P$.

$$\begin{aligned} \mathbf{R}(j,p): &\quad Reference\ Set,\ j = 1,2,\ldots,M \\ \mathbf{T}(i,p): &\quad Test\ Set,\ i = 1,2,\ldots,N \end{aligned}$$

$$\begin{aligned} \mathbf{R} &= r_{1p}, r_{2p}, \ldots, r_{jp}, \ldots, r_{MP} \\ \mathbf{T} &= t_{1p}, t_{2p}, \ldots, t_{ip}, \ldots, t_{NP} \end{aligned} \tag{6.48}$$

Data lengths N and M will not be equal most of the time because the operating time is usually adjusted by the operators to get the desired product quality and yield in response to variations in input properties for each batch run and the randomness caused by complex physiological phenomena inherent in biochemical reactions. This problem could be overcome using linear time alignment and normalization based on linear interpolation or extrapolation techniques. Let i and j be the time indices of the observations in $\mathbf{T}$ and $\mathbf{R}$ sets, respectively. In linear time normalization, the dissimilarity between $\mathbf{T}$ and $\mathbf{R}$ for any variable trajectory is simply defined as

$$d(T,R) = \sum_{i=1}^{N} d(i,j) \tag{6.49}$$

where i and j satisfy [1]

$$i = \frac{M}{N} j. \tag{6.50}$$

[1] Since the indices i and j are integers, some round-off rule is implied in Eq.6.50 [484].

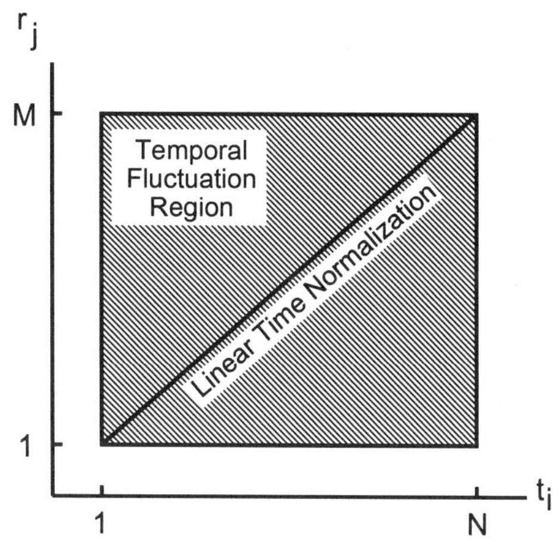

Figure 6.18. Linear time alignment of two trajectories with different durations [484].

Note that the dissimilarity measure $d(t_i, r_j)$ between $\mathbf{T}$ and $\mathbf{R}$ is denoted as $d(i, j)$ for simplicity of notation in Eq. 6.49. Hence, the distortion measure assessment will take place along the diagonal straight line of the rectangular (t_i, r_j) plane shown in Figure 6.18. Linear time normalization implicitly assumes that the temporal trajectory variations are proportional to the duration of the batch (or the number of samples made on each variable). However, since the timing differences between the two batches will be local and not global, a more general time alignment and normalization scheme would be appealing, including the use of nonlinear warping functions that relate the indices of the variables in two trajectory sets to a common "normal" time axis k. *Time warping* has been developed to deal with these issues by using the principles of *dynamic programming* (that is why it is called as dynamic time warping) [252, 485, 533].

The objective in time warping is to match the elements of each pattern (trajectory in our case) $\mathbf{T}$ and $\mathbf{R}$ so as to minimize the discrepancy in each pair of samples. Similar events will be aligned when this is achieved using a nonlinear warping function. This function will shift some feature vectors in time, compress and/or expand others to obtain minimum distances. For each vector pair in $\mathbf{T}$ and $\mathbf{R}$, DTW is performed on a $M \times N$ grid under a number of constraints. An example of pattern matching between two vectors is shown in Figure 6.19.

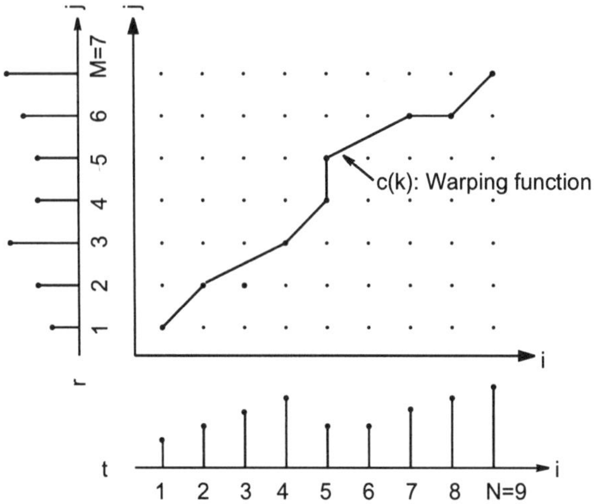

Figure 6.19. Nonlinear time warping process details. The point at (5,4) aligns $t(5)$ with $r(4)$.

Let the warping function be

$$C = c(1), c(2), \ldots, c(k), \ldots, c(K) \; ; \qquad \max(N, M) \leqslant K \leqslant N + M \quad (6.51)$$

where each c is a pair of indices to the trajectory elements being matched (position in the grid):

$$c(k) = [i(k), j(k)]. \quad (6.52)$$

Thus, C can be considered as a model of the time axis fluctuation in a given pattern sequence. For each $c(k)$ we have a cost function or a local distance measure which reflects the discrepancy between the paired samples. There are many definitions of local distance calculation such as Euclidian distance, Mahalanobis distance and Chebyshev norm [252, 441, 485, 533]. A typical local distance is the squared difference between the samples

$$d[c(k)] = d[i(k), j(k)] = (t_{i(k)} - r_{i(k)})^2. \quad (6.53)$$

The most commonly used local distance for multivariate data is the weighted quadratic distance

$$d[c(k)] = [\mathbf{T}(i(k), p) - \mathbf{R}(j(k), p)]\mathbf{W}_{\mathrm{p}}[\mathbf{T}(i(k), p) - \mathbf{R}(j(k), p)]^T \quad (6.54)$$

where $\mathbf{W}_{\mathrm{p}}$ is a positive definite weight matrix that shows the relative importance of each variable p.

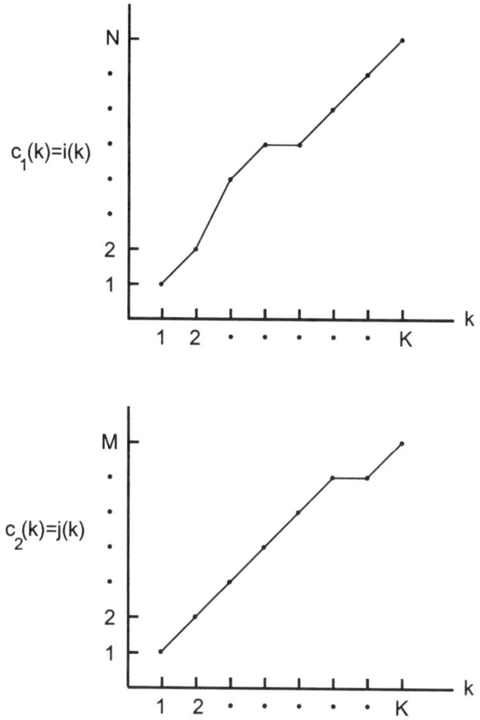

Figure 6.20. An example of time warping of two features vectors (made on the same object); time warping functions c_1 and c_2 map the individual time indices i and j, respectively, to the common time index k [484].

The warping function C is required to minimize the overall cost function for any path [406, 446, 484, 533]

$$D(C) = D[i(k), j(k)] = \frac{\sum_{k=1}^{K} d[i(k), j(k)]w(k)}{N(w)} \tag{6.55}$$

where $D[i(k), j(k)] = D(t, r)$ is a normalized total distance between the two trajectories along the path of length K, $w(k)$ is a weighting function for the local distances and $N(w)$ is a normalization factor which is a function of the weighting function. Now, the problem is reduced into an optimization problem that can be written as

$$D^*(C) = \frac{1}{N(w)} \min_{C}[D(t, r)] \tag{6.56}$$

where $D^*(C)$ denotes the minimum normalized total distance and C^* the optimal path. This optimization problem can be efficiently solved by using *dynamic programming* techniques [49, 53]. Dynamic programming is a well known optimization technique used extensively in operations research for solving sequential decision problems. The decision rules about determining the next point (location) to be visited following a current point i is called "policy". Dynamic programming determines the policy that leads to the minimum cost, moving from point 1 to point i based on the *Principle of Optimality* defined by Bellman [49] as

> *An optimal policy has the property that, whatever the initial state and decision are, the remaining decisions must constitute an optimal policy with regard to the state resulting from the first decision.*

For time normalization problem this principle can be recast as follows [270, 406, 484]:

1. A globally optimal path is also locally optimal. If C^* is determined as the optimal path, any (i,j) point on C^* is also optimal.

2. The optimal path to the grid point (i,j) only depends on the values of the previous grid points.

Before delving into the integration/formulation of dynamic programming and time warping, constraints on the warping function must be discussed.

Warping Function Constraints

There are some restrictions imposed on the warping function. Since it represents a model of the time axis fluctuation in feature patterns (batch trajectories), it should properly approximate the properties of the actual time axis. Typical essential DTW constraints include:

1. *Endpoint Constraints (Boundary Conditions).*
   For batch process trajectories, this condition is easy to visualize:

$$\begin{aligned} \textit{Beginning point} &: \quad i(1)=1,\ j(1)=1 \ \textit{or}\ c(1)=(1,1) \\ \textit{Ending point} &: \quad i(K)=N,\ j(K)=M \ \textit{or}\ c(K)=(N,M) \end{aligned} \tag{6.57}$$

2. *Monotonicity Conditions.*
   The temporal order of the measurements collected on each variable is of crucial importance to their physical meaning. Hence, imposing a reasonable monotonicity constraint (monotonically nondecreasing sequence requirement) to maintain the temporal order while performing

DTW is necessary:

$$\begin{aligned} i(k+1) &\geq i(k) \\ j(k+1) &\geq j(k) \end{aligned} \tag{6.58}$$

As shown in Figure 6.20, any path on which $D(t, r)$ is calculated will not have a negative slope. Basically, this constraint prevents the possibility of reverse warping along the time axis that is physically meaningless.

3. *Local Continuity Constraints.*
There are two reasons to impose local continuity constraints:

   (*i*) To guarantee that excessive compression or expansion of the time scale is avoided (neither too steep nor too gentle a gradient of fluctuations should be allowed).

   (*ii*) To compare events in their natural temporal order while keeping any potential loss of information to a minimum [406, 484, 533].

This is a very important constraint since it defines the set of potential preceding points (predecessors) in the grid. Obviously, it is possible to specify many sets of such local constraints. In order to visualize the concept, consider the following local transition rule between two consecutive points on the path c, an example suggested by Sakoe and Chiba [533],

$$i(k) - i(k-1) \leq 1 \text{ and } j(k) - j(k-1) \leq 1. \tag{6.59}$$

Inequalities in Eq.6.59 impose that the warping function c should not skip any points in both vectors. This can also be formulated as

$$c(k-1) = \begin{cases} [i(k), j(k-1)] \\ [i(k-1), j(k-1)] \\ [i(k-1), j(k)] \end{cases} \tag{6.60}$$

If (i, j) is the kth path point in the grid shown in Figure 6.21, then the previous path point, $c(k-1)$ can only be chosen from a set of preceding points (Eq. 6.60). In this simple example, $[i(k), j(k)]$ can only be reached by either from $[i(k), j(k-1)]$ or $[i(k-1), j(k-1)]$ or $[i(k-1), j(k)]$. This is also known as "no slope constraint" case. Obviously, the control of the slope would be of importance for the correct alignment. Sakoe and Chiba [533] have proposed a slope constraint on the warping function using a slope intensity measure $P = q/p$

(Figure 6.22). The intensity measures ensure that, if the warping function $c(k)$ moves forward in the horizontal or vertical direction p consecutive times, then it is not allowed to proceed further in the same direction before moving at least q times in the diagonal direction (Figure 6.22). The larger the P value the more rigidly the warping function will be restricted. If the slope intensity (P) is too severe, DTW would not work effectively. If it is too lax, then the discrimination between the trajectories will be degraded. Consequently, it is necessary to set a proper value. Sakoe and Chiba [533] have reported that they have observed the best results with $P = 1$ although that depends on the system under investigation. This constraint also helps reduce search paths through the grid while maintaining a reasonable distortion in time axis. A number of slope intensities has been proposed [252, 406, 533]. Local continuity can also be expressed in terms of incremental path changes [406, 484]. A path can be defined as a sequence of moves associated with a pair of coordinate increments as

$$\mathcal{P} \rightarrow (t_1, r_1)(t_2, r_2) \ldots (t_K, r_K) \tag{6.61}$$

where the indices refer to normalized time increments (or location of the points on the warping function). For illustration purposes, consider the slope intensity $P = 1$ case of Sakoe and Chiba [533] that has also been studied by Myers et al. [406] (Figure 6.23). According to this description, for instance, the path in Figure 6.19 contains the following transition sequence:

$$\mathcal{P} \rightarrow (1,1)(1,0)(1,1)(1,1)(0,1)(2,1)(1,0)(1,1). \tag{6.62}$$

Most common local transitions are given in Figure 6.24.

4. *Slope Weighting*.
Another constraint that can be included to optimal path search defined in Eq. 6.55 (distance calculation) is the *weighting function*,

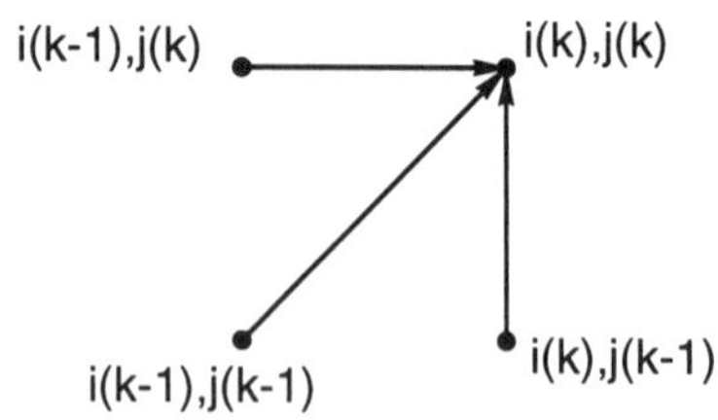

Figure 6.21. Sakoe-Chiba local transition constraint with no constraint on slope [533].

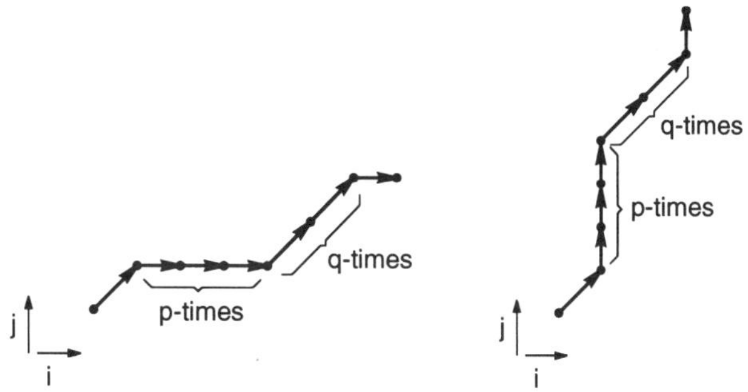

(a) Minimum slope (b) Maximum slope

Figure 6.22. Sakoe-Chiba slope constraints [533].

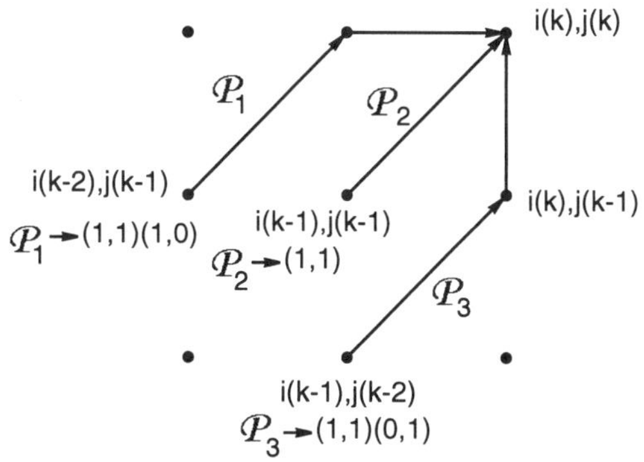

Figure 6.23. An example of local continuity constraint expressed in terms of coordinate increments (Sakoe-Chiba local transition constraint with slope intensity of 1) [406, 484, 533].

$w(k)$. This function depends only on the local path and controls the contribution of each local time distortion $d[i(k), j(k)]$.

Based on the local continuity constraint used, many slope weighting functions are possible. Sakoe and Chiba [533] have proposed the following four types of slope weighting functions and their effects on

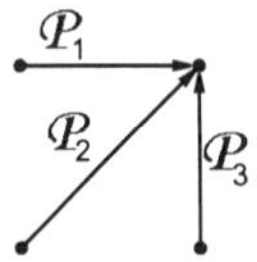

(a) Type I

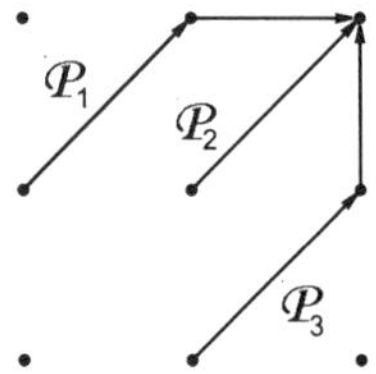

(b) Type II

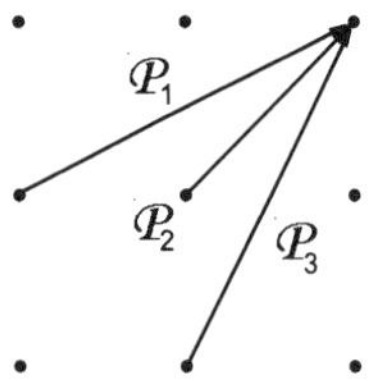

(c) Type III

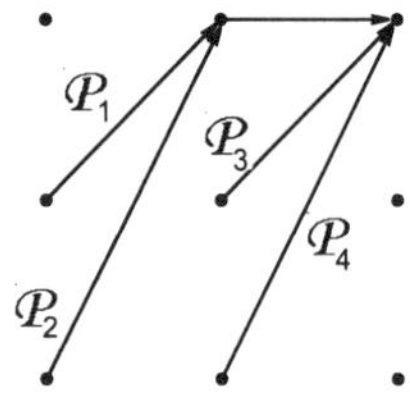

(d) Type IV

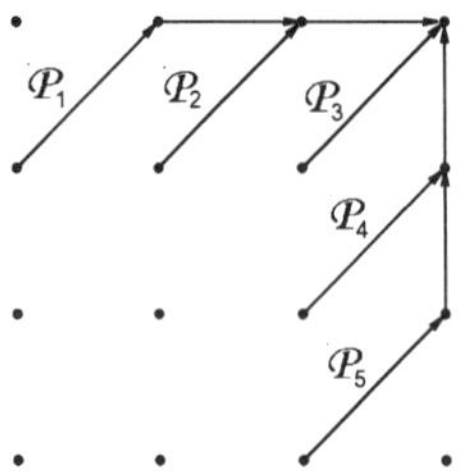

(e) Type V

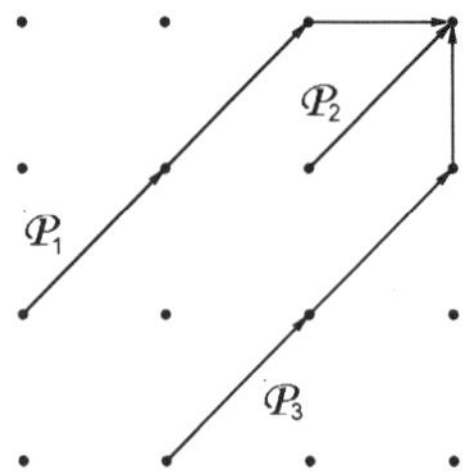

(f) Type VI

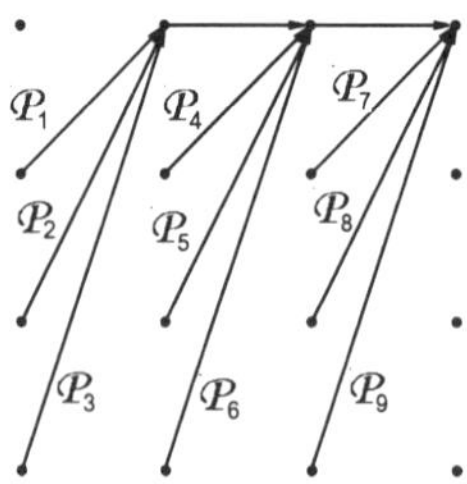

(g) Type VII

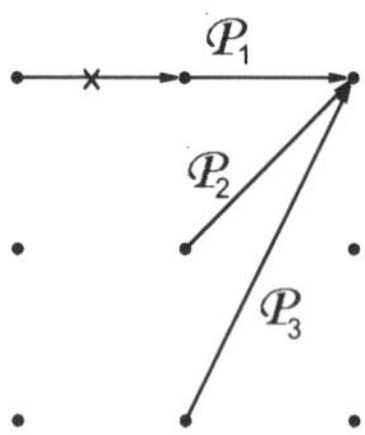

(h) Type Itakura

Figure 6.24. Local continuity constraints studied by Myers et al. [406, 533].

DTW performance were extensively studied by Myers et al. [406]:

$$\text{Type (a)}: \; w(k) = \min[i(k) - i(k-1), j(k) - j(k-1)] \tag{6.63}$$

$$\text{Type (b)}: \; w(k) = \max[i(k) - i(k-1), j(k) - j(k-1)] \tag{6.64}$$

$$\text{Type (c)}: \; w(k) = i(k) - i(k-1) \tag{6.65}$$

$$\text{Type (d)}: \; w(k) = i(k) - i(k-1) + j(k) - j(k-1) \tag{6.66}$$

where it is assumed that $i(0) = j(0) = 0$ for initialization. Figure 6.25 illustrates the effects of weighting functions on Type III local continuity constraints [406]. The numbers refer to particular weighting coefficient (calculated by the relevant formula) associated with each local path. Note that, since the increase in distortion will cause a decrease in the likelihood of proper matching, larger weightings will lead to less preferable paths. For instance, in Figure 6.25(b), Type (b) weighting will promote diagonal moves in the search grid.

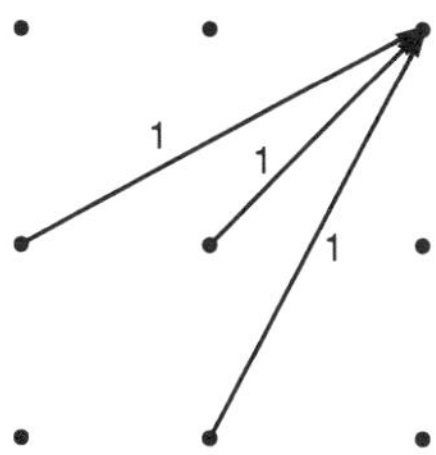

(a) w(k)=min(i(k)-i(k-1),j(k)-j(k-1))

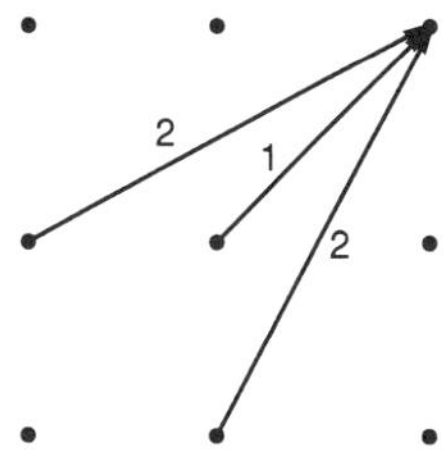

(b) w(k)=max(i(k)-i(k-1),j(k)-j(k-1))

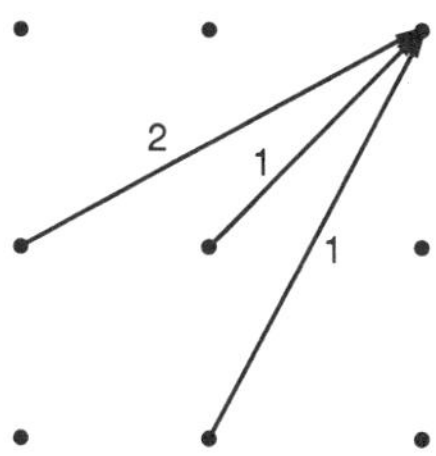

(c) w(k)=i(k)-i(k-1)

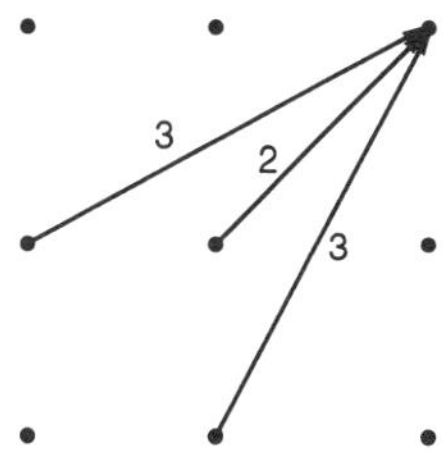

(d) w(k)=i(k)-i(k-1)+j(k)-j(k-1)

Figure 6.25. Sakoe-Chiba slope weightings for Type III local continuity constraint studied by Myers et al. [406, 533].

When these four types of weighting functions are applied to different types of local continuity constraints, sometimes 0 weight can be assigned to certain local paths. When this happens, weightings of inconsistent paths are smoothed by redistributing the existing weight equally on each move. This situation is illustrated using Sakoe and Chiba's Type II local constraint on Figure 6.26. Similar smoothing can be applied to different local continuity constraints given in Figure 6.24 if needed.

The calculation of accumulated distance in Eq. 6.55 requires an overall normalization to provide an average path distortion independent of the lengths of the two patterns being synchronized. The normalization factor $N(w)$ is a function of the slope weighting type chosen such that

$$N(w) = \sum_{k=1}^{K} w(k). \tag{6.67}$$

For instance, when Type (c) and Type (d) slope weighting constraints are used, the overall normalization factors would be

$$w(k)^{(c)} = \sum_{k=1}^{K} [i(k) - i(k-1)] = i(K) - i(0) = N \tag{6.68}$$

and

$$\begin{aligned} w(k)^{(d)} &= \sum_{k=1}^{K} [i(k) - i(k-1) + j(k) - j(k-1)] \\ &= i(K) - i(0) + j(K) - j(0) = N + M \end{aligned} \tag{6.69}$$

respectively, independent of the length of the warping functions [484]. Type (a) and type (b) slope weighting constraints will produce normalization factors that are strong functions of the actual paths, hence Type (c) and (d) are preferred in most of the applications.

5. *Global Path Constraints.*
Since local continuity constraints are imposed on the warping function, certain portions of the i, j grid (search space in mapping) are excluded from the region where the optimal warping path can be located. For each type of local constraints the allowable regions that determine the maximum and minimum amounts of expansion (or com-

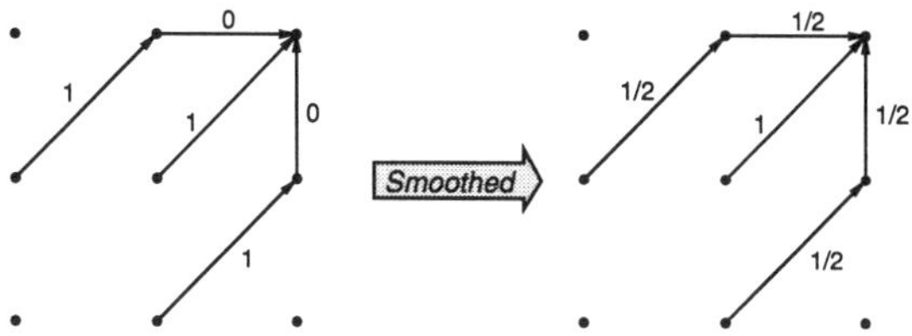

(a) w(k)=min(i(k)-i(k-1),j(k)-j(k-1))

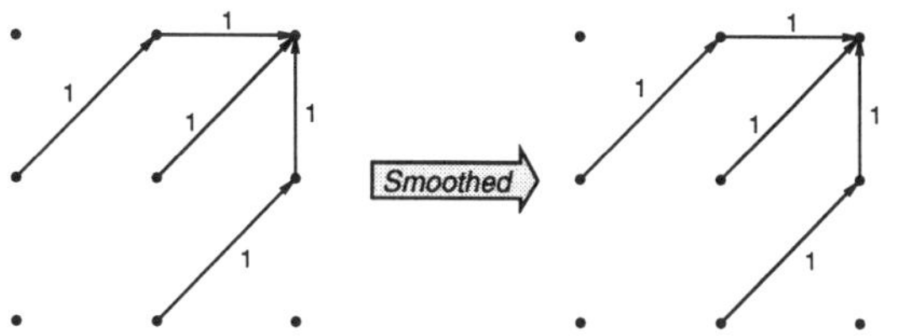

(b) w(k)=max(i(k)-i(k-1),j(k)-j(k-1))

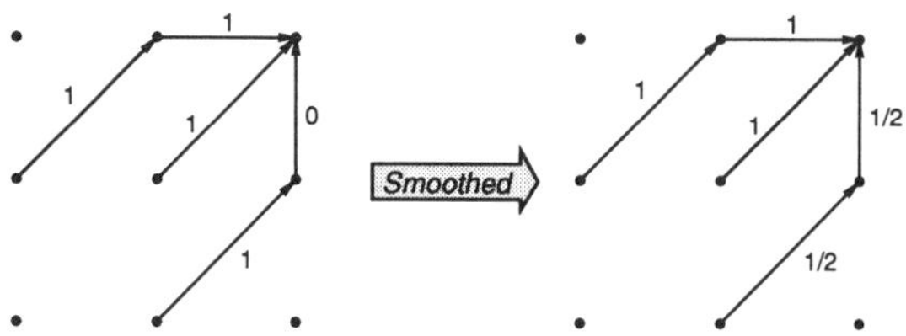

(c) w(k)=i(k)-i(k-1)

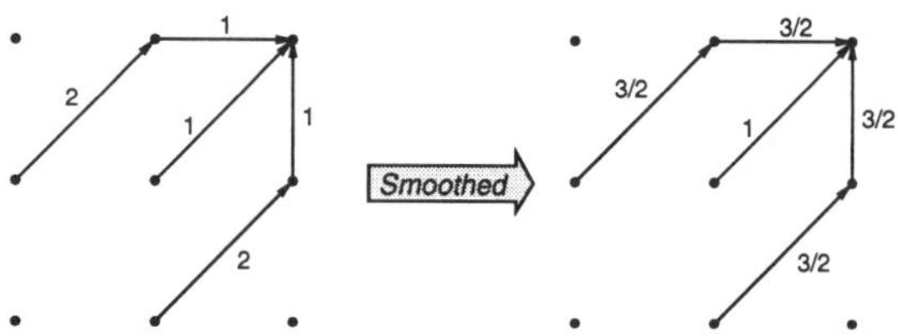

(d) w(k)=i(k)-i(k-1)+j(k)-j(k-1)

Figure 6.26. Uniformly redistributed (smoothed) slope weightings for Type II local constraint [406, 484, 533].

pression) can be defined using the two parameters $Q_{\max}$ and $Q_{\min}$:

$$Q_{\max} = \max_{\ell} \left[\sum_{k=1}^{K_\ell} p_k^{(\ell)} \Big/ \sum_{k=1}^{K_\ell} q_k^{(\ell)} \right] \tag{6.70}$$

$$Q_{\min} = \min_{\ell} \left[\sum_{k=1}^{K_\ell} p_k^{(\ell)} \Big/ \sum_{k=1}^{K_\ell} q_k^{(\ell)} \right] \tag{6.71}$$

where ℓ denotes the index of the allowable path $\mathcal{P}_\ell$ in the constraint set, and K_ℓ is the total number of moves in $\mathcal{P}_\ell$. The monotonicity conditions in Eq. 6.58 hold as

$$p_k^{(\ell)}, q_k^{(\ell)} \geq 0 \quad \text{for all } k. \tag{6.72}$$

For instance, in Sakoe-Chiba local continuity constraints given in Figure 6.23, $\ell = 1, 2, 3$, and $K_\ell = 2, 1, 2$, respectively for $\mathcal{P}_1, \mathcal{P}_2, \mathcal{P}_3$ resulting in $Q_{\max} = 2$ and $Q_{\min} = 1/2$. Normally, $Q_{\max} = 1/Q_{\min}$. The values of $Q_{\max}$ and $Q_{\min}$ for different types of local continuity constraints are given in Table 6.4.

The boundaries of the allowable regions (global path constraints) can be defined using the values of $Q_{\max}$ and $Q_{\min}$ as

$$1 + \frac{[i(k) - 1]}{Q_{\max}} \leq j(k) \leq 1 + Q_{\max}[i(k) - 1] \tag{6.73}$$

$$M + Q_{\max}[i(k) - N] \leq j(k) \leq M + \frac{[i(k) - 1]}{Q_{\max}}. \tag{6.74}$$

Eq. 6.73 defines the range of the points that can be reached using a legal path based on a local constraint from the beginning point $(1, 1)$. Likewise, Eq. 6.74 specifies the range of points that have a legal path to the ending point (N, M) defined by Itakura [252] (Itakura constraints). Figure 6.27 shows the effects of the global constraints on the optimal search region defined by the parallelogram (Itakura constraints) in the (N, M) grid. An additional global path constraint has been proposed by Sakoe and Chiba [533] as

$$|i(k) - j(k)| \leq K_0 \tag{6.75}$$

where K_0 denotes the maximum allowable absolute temporal difference between the two variable trajectories at any given sampling instance. This constraint further decreases the search range as well as the potential misalignments by trimming off the edges of the parallelogram in the grid.

Table 6.4. Allowable local path specifications and associated Q_{max} and Q_{min} values for different types of local continuity constraints given in Figure 6.24 [484]

Type	Allowable paths	Q_{max}	Q_{min}
I	$\mathcal{P}_1 \to (1,0)$ $\mathcal{P}_2 \to (1,1)$ $\mathcal{P}_3 \to (0,1)$	∞	0
II	$\mathcal{P}_1 \to (1,1)(1,0)$ $\mathcal{P}_2 \to (1,1)$ $\mathcal{P}_3 \to (1,1)(0,1)$	2	1/2
III	$\mathcal{P}_1 \to (2,1)$ $\mathcal{P}_2 \to (1,1)$ $\mathcal{P}_3 \to (1,2)$	2	1/2
IV	$\mathcal{P}_1 \to (1,1)(1,0)$ $\mathcal{P}_2 \to (1,2)(1,0)$ $\mathcal{P}_3 \to (1,1)$ $\mathcal{P}_4 \to (1,2)$	2	1/2
V	$\mathcal{P}_1 \to (1,1)(1,0)(1,0)$ $\mathcal{P}_2 \to (1,1)(1,0)$ $\mathcal{P}_3 \to (1,1)$ $\mathcal{P}_4 \to (1,1)(0,1)$ $\mathcal{P}_5 \to (1,1)(0,1)(0,1)$	3	1/3
VI	$\mathcal{P}_1 \to (1,1)(1,1)(1,0)$ $\mathcal{P}_2 \to (1,1)$ $\mathcal{P}_3 \to (1,1)(1,1)(0,1)$	3/2	2/3
VII	$\mathcal{P}_1 \to (1,1)(1,0)(1,0)$ $\mathcal{P}_6 \to (1,3)(1,0)$ $\mathcal{P}_2 \to (1,2)(1,0)(1,0)$ $\mathcal{P}_7 \to (1,1)$ $\mathcal{P}_3 \to (1,1)(1,0)(1,0)$ $\mathcal{P}_8 \to (1,2)$ $\mathcal{P}_4 \to (1,3)(1,0)(1,0)$ $\mathcal{P}_9 \to (1,3)$ $\mathcal{P}_5 \to (1,2)(1,0)$	3	1/3
Itakura	$\mathcal{P}_1 \to (1,0)$ $\mathcal{P}_2 \to (1,1)$ (Note: consecutive (1,0)(1,0) $\mathcal{P}_3 \to (1,2)$ not allowed)	2	1/2

Different slope constraints result in different forms of dynamic pattern matching that can be classified as either *symmetric* or *asymmetric* algorithms. The two most commonly used ones are Types (c) and (d)

$$\text{Type (c), } \textit{asymmetric}: \; w(k) = i(k) - i(k-1), \; i(0) = 0 \tag{6.76}$$

$$\text{Type (d), } \textit{symmetric}: \; w(k) = i(k) - i(k-1) + j(k) - j(k-1), \quad i(0) = j(0) = 0 \tag{6.77}$$

resulting in asymmetric and symmetric forms, respectively. In the asymmetric form, time normalization is realized by transforming the time axis of a feature vector onto that of the other, resulting in the expansion or compression of the second feature vector. The asymmetric algorithm will map the time index of the test trajectory (**T**) that is placed on the horizontal axis onto the time index of the reference trajectory (**R**) that is placed on the vertical axis. As a result of such matching, the optimal path will contain as many points as the test set does (which is N in our example)

$$c(i) = [i, j(i)]. \tag{6.78}$$

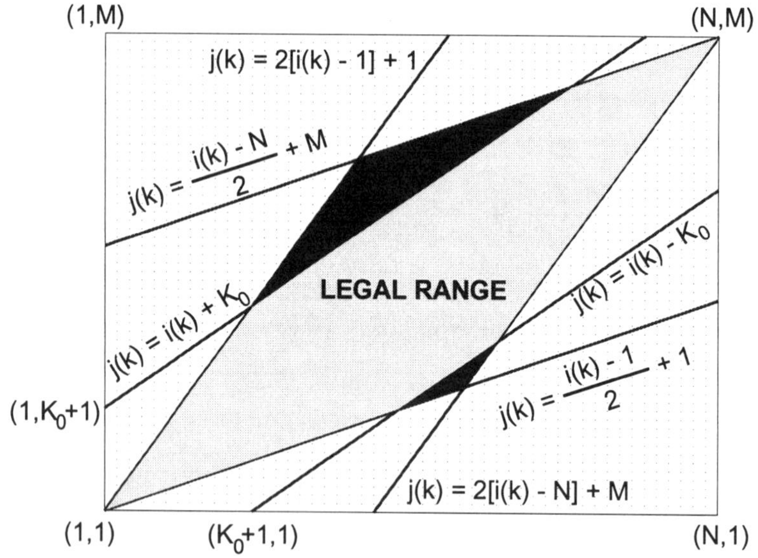

Figure 6.27. Global path constraints on the search grid ($N \times M$) [484] for $Q_{\max} = 2$ and $K_0 = 2|N - M|$ given that ($K_0 \geq |i(K) - j(K)|$).

Consequently, the optimal path will pass through each point on t but may skip some on r.

A symmetric algorithm, however, will transform both time axes onto a temporarily defined common axis with a common index, k. In this case, the optimal path will go through all the points in both trajectory sets. Furthermore, for the same reference trajectory set, the number of points in the optimal path will be different for each new test trajectory set. Although each test trajectory individually will be synchronized with the reference trajectory, they will not be synchronized with each other, resulting in a set of individually synchronized batch trajectories with unequal batch lengths. If an asymmetric DTW is used, it will skip some points in the test trajectory but will produce synchronized (with reference and each other) batch trajectories having equal length with the reference set. Depending on the choice of the reference set, some of the inconsistent features in $\mathbf{T}$ that may cause false alarms in statistical process monitoring will be left out. In order to compromise between the two extremes, solutions that are presented in the following sections have been suggested [270].

Dynamic Programming Solution

Since the search space for optimal path (minimum accumulated distance) can be reduced by imposing local and global constraints, dynamic programming can be used to develop the DTW algorithm. In addition to their contribution on more realistic pattern matching, local and global constraints reduce the search space so that the computational requirements will be lower. Dynamic programming will be used to find a move sequence on the constrained $i \times j$ grid under the given local continuity constraints (local transitions) and the boundary conditions (usually fixed end-points) to minimize the following accumulated distance starting from point $(1,1)$ to (i,j):

$$D_A(i,j) = \min_{C} \sum_{k=1}^{K} d[i(k), j(k)]w(k). \tag{6.79}$$

When point (N, M) is reached, the time normalized distance is calculated to evaluate the performance of the DTW algorithm under the chosen conditions.

Dynamic programming proceeds in phases, the first phase is the forward algorithm and second is the backward tracking (optimal path reconstruction). Assume that the test pattern is placed on the horizontal axis and the reference pattern on the vertical axis. After initialization, for each increment made on $i-axis$ (abscissa), all possible points (i.e., all points within the allowed region) along the $j-axis$ (ordinate) are considered based on the local continuity constraints chosen. For each point, a number of possible predecessors and associated accumulated costs are determined and stored to construct the optimal path at the end such that

$$D(C_k) = d[c(k)] + \min_{\text{legal c(k-1)}} [D(C_{k-1})] \tag{6.80}$$

where $D(C_k) = D_A(i,j)$. Eq. 6.80 defines the accumulated distance that is comprised of the cost of particular point [i(k),j(k)] itself and the cheapest cost path associated with it. The second term in Eq. 6.80 requires a decision on the predecessor. In Table 6.5, dynamic programming recursion equations are summarized for different local continuity constraints when Type (d) slope weighting is used. Note that slope weightings of the paths in Table 6.5 are smoothed according to Figure 6.26. To illustrate the progress of dynamic programming procedure, assume that Type (d) slope weighting (symmetric) in Eq. 6.66 and Type III local continuity constraint (Figures 6.24(c) and 6.25(d)) are used (Figure 6.24).

During the forward phase, transition cost to the accumulated distance at point (i,j) can be found by solving the following simple minimization problem (dynamic programming recursion equation)

$$D_A(i,j) = \min \begin{bmatrix} D_A(i-2,j-1) + 2d(i,j) \\ D_A(i-1,j-1) + 3d(i,j) \\ D_A(i-1,j-2) + 2d(i,j) \end{bmatrix} . \tag{6.81}$$

The local continuity constraints chosen above mean that the point (i,j) can only be reached by either points $(i-2,j-1)$, or $(i-1,j-1)$ or $(i-1,j-2)$ as shown in Figure 6.24(c). To initialize the iterative process, $D_A(1,1)$ can be assigned to $2d(1,1)$. The forward phase finishes when point $[i(K),j(K)]$ is reached and the minimum normalized distance, $D^*(C)$, in Eq.6.56 is computed (note that $N(w) = i(K) + j(K)$ for Type (d) slope weighting).

At this point the second phase, which is the reconstruction of the optimal path, starts. Starting from point $[i(K),j(K)]$ (say $i(K) = N$ and $j(K) = M$) in the search grid and using the stored information on optimal transitions, first the predecessor of point (N,M) is located, then the predecessor of the latter is identified. This is repeated until point $(1,1)$ is reached. At the end of the second phase, the optimal path is reconstructed and as a consequence pattern indices are matched accordingly.

Example Consider two artificial signals to illustrate the DTW algorithm. r is a (1×60) reference signal and t is a (1×50) test signal (Figure 6.28(a)). Boundary conditions are assumed as

$$r(0) = t(0) = 0 \quad \textit{and} \quad j(M) = 60, \; i(N) = 50. \tag{6.82}$$

Table 6.5. Accumulated distance formulations used for some of the local constraints for Type (d) slope weighting case [406, 484]

Type	Accumulated Distance Function
I	$\min \begin{bmatrix} D_A(i-1,j)+d(i,j) \\ D_A(i-1,j-1)+2d(i,j) \\ D_A(i,j-1)+d(i,j) \end{bmatrix}$
II	$\min \begin{bmatrix} D_A(i-2,j-1)+\frac{3}{2}[d(i-1,j)+d(i,j)] \\ D_A(i-1,j-1)+2d(i,j) \\ D_A(i-1,j-2)+\frac{3}{2}[d(i,j-1)+d(i,j)] \end{bmatrix}$
III	$\min \begin{bmatrix} D_A(i-2,j-1)+3d(i,j) \\ D_A(i-1,j-1)+2d(i,j) \\ D_A(i-1,j-2)+3d(i,j) \end{bmatrix}$
IV	$\min \begin{bmatrix} D_A(i-2,j-1)+2d(i-1,j)+2d(i,j) \\ D_A(i-2,j-2)+2d(i-1,j)+2d(i,j) \\ D_A(i-1,j-1)+2d(i,j) \\ D_A(i-1,j-2)+3d(i,j) \end{bmatrix}$
V	$\min \begin{bmatrix} D_A(i-3,j-1)+\frac{3}{2}[d(i-2,j)+d(i-1,j)+d(i,j)] \\ D_A(i-2,j-1)+\frac{3}{2}[d(i-1,j)+d(i,j)] \\ D_A(i-1,j-1)+2d(i,j) \\ D_A(i-1,j-2)+\frac{3}{2}[d(i,j-1)+d(i,j)] \\ D_A(i-1,j-3)+\frac{3}{2}[d(i,j-2)+d(i,j-1)+d(i,j)] \end{bmatrix}$
Itakura	$\min \begin{bmatrix} D_A(i-1,j)+d(i,j) \\ D_A(i-1,j-1)+2d(i,j) \\ D_A(i-1,j-2)+3d(i,j) \end{bmatrix}$

Type V local continuity constraint was used along with a Type (d) slope weighting function to produce the results in Figure 6.29 (*see* Figure 6.24(e) and Table 6.4 for definitions). For simplicity, Sakoe and Chiba band global path constraint was applied as shown by the shaded region in Figure 6.29. The resulting synchronized signals are now comparable since the similar features were aligned by DTW (Figure 6.28(b)). The signal magnitudes at the end points were different, and DTW has preserved this difference while adjusting the time scale.

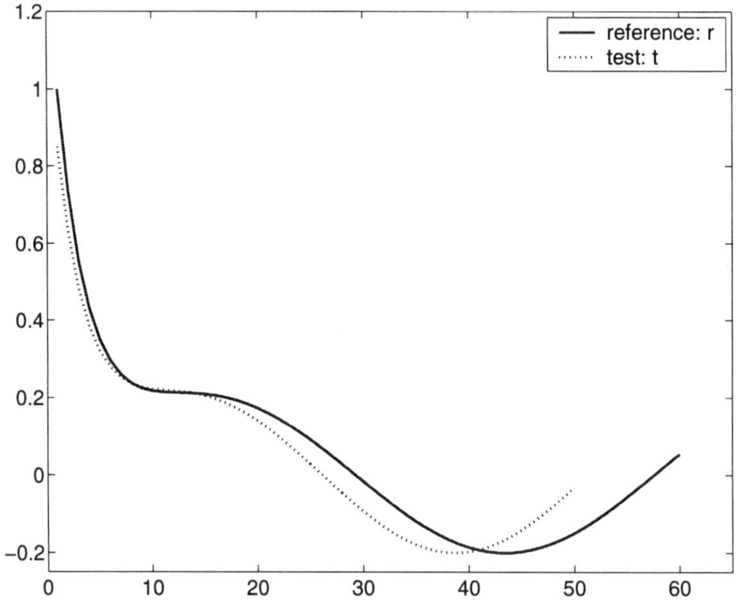

(a) Reference: r and Test: t signals before DTW

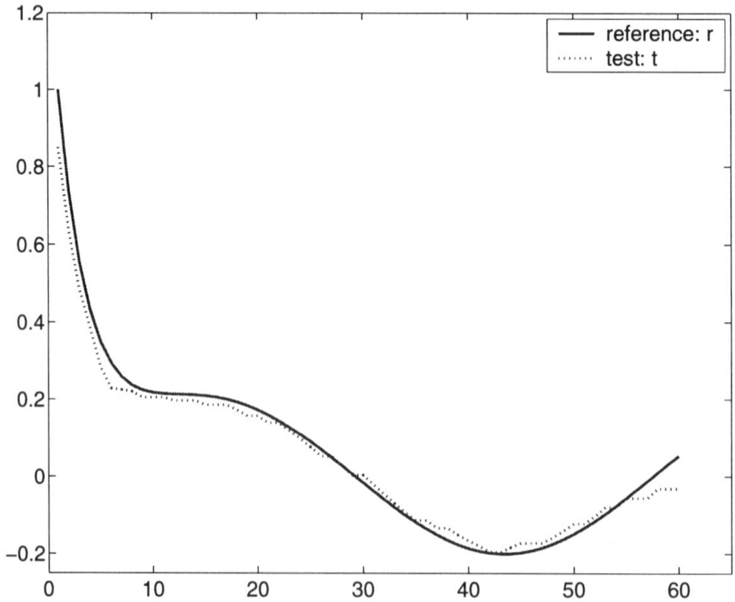

(b) Reference: r and Test: t signals after DTW

Figure 6.28. A univariate application of DTW.

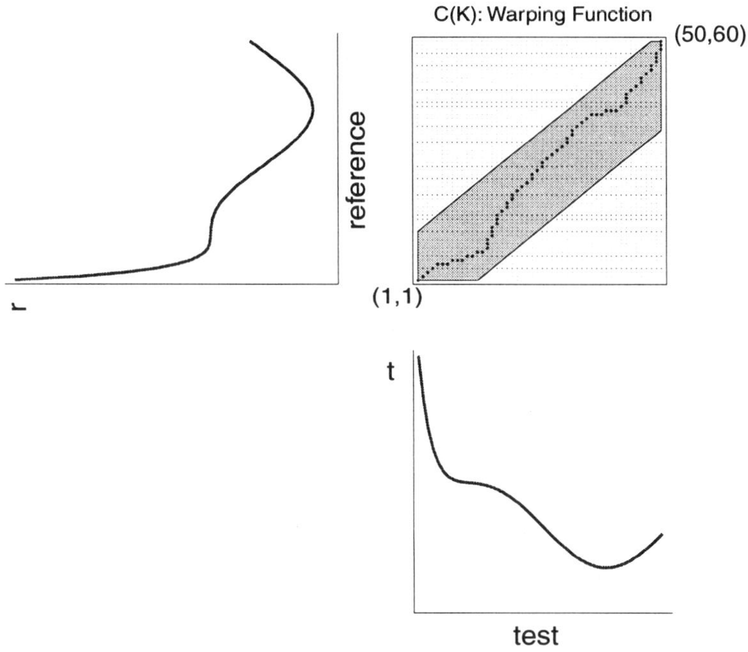

Figure 6.29. Optimal path determined by dynamic time warping function.

Multivariate Iterative DTW Framework for Equalization and Synchronization of Batch Trajectories

To implement DTW technique to equalize/synchronize a set of trajectories, some modifications need to be performed. As mentioned earlier for the univariate case, DTW works with pairs of patterns, and for each pattern only two data points (on the reference and the test set) are analyzed at each step of the algorithm. However, in the multivariate case, there are two vectors of multivariate observations to be considered. Thus, the relative importance of the variables should be taken into consideration by giving more weight to some variables such that the synchronization is conducted more on those variables. The underlying idea is to utilize variable trajectories that contain more structural information about the progress of the process. Monotonically increasing or decreasing, smooth variables will be given more weight since they resemble the time axis behavior.

It is assumed that an appropriate scaling was performed prior to implementation of DTW. Scaling is an important issue since DTW is a distance-based technique. One can calculate the means and standard deviations for each variable in each batch trajectory set, take the average of those and use to autoscale the set of trajectories to a common y-axis scale which was used in the following example in this book. The average mean and standard deviation should be stored for rescaling and scaling of future batches. The iterative synchronization procedure that will be presented here is an adaptation of Kassidas and his co-workers' approach [270].

Consider $\mathbf{T}_\ell$, $\ell = 1, \ldots, L$ set of trajectories of normal operation. Each trajectory set in these batches will contain unequal data lengths as well as unsynchronized trajectories. Each $\mathbf{T}_\ell$, is a matrix of $N_\ell \times P$ where N_ℓ is the number of observations and P is the number of variables, as given in Eq. 6.48. It is also assumed that a reference batch trajectory set $\mathbf{R}_f$ $(M \times P)$ has been defined. The objective is to synchronize each $\mathbf{T}_\ell$ with $\mathbf{R}_f$.

After scaling and choosing one of the batches as reference batch run, the next step becomes deciding on which DTW algorithm to implement. If a symmetric algorithm is used, the resulting trajectories will be of equal length that is greater than the length before synchronization since a symmetric algorithm projects the time indices of both test and reference trajectories onto a common time scale. After each round of synchronization for each batch, the resulting batch lengths will be different even though each test batch will be synchronized with the reference batch individually but not with each other. However, if one chooses to implement an asymmetric algorithm, the optimal path will go through each point on the reference batch run but could skip some points on the test set. The resulting trajectories will be of equal length with the reference set and each test set will be synchronized with each other. Since the synchronized trajectories may not contain all the data points that were in the original trajectories before synchronization, some inconsistent features of the test trajectories may be left out. A combined algorithm (a symmetric DTW followed by an asymmetric DTW) has been proposed by Kassidas et al. to compromise between the two extremes [270].

According to their proposition, conventional symmetric DTW is first applied for each batch trajectory set. The resulting expanded trajectories are then exposed to an asymmetric synchronization step. If more than one point of $\mathbf{T}$ is aligned with one point of $\mathbf{R}_f$, they suggested to take the average of these points of $\mathbf{T}$ and align this average point with the particular point of $\mathbf{R}_f$ [270]. As shown in Figure 6.30 for one variable, both t_i and t_{i+1} are aligned with the same r_j of the reference trajectory. In this case, the following averaged point is aligned with r_j instead

$$\frac{t_i + t_{i+1}}{2} \rightarrow r_j. \tag{6.83}$$

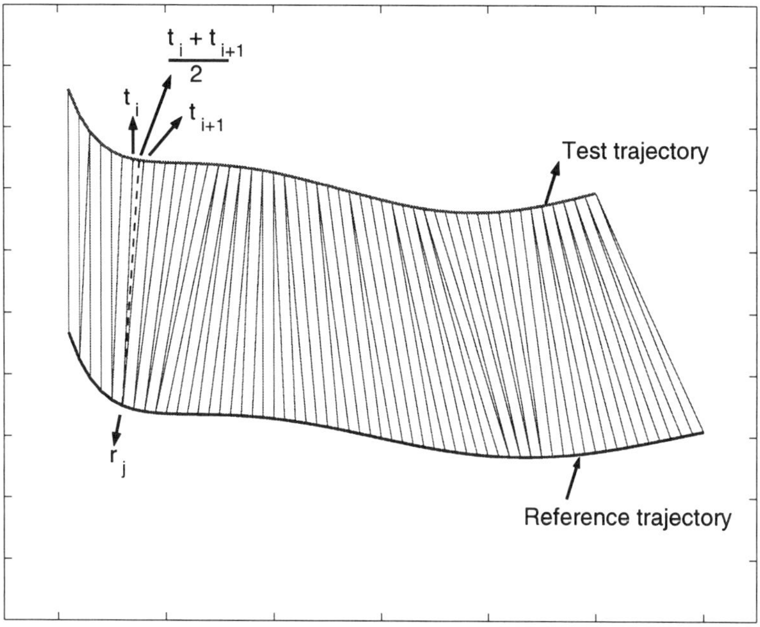

Figure 6.30. Combined symmetric/asymmetric DTW matching.

This calculation can be extended to multivariate case by using the vector notation as $(\mathbf{T_i} + \mathbf{T_{i+1}})/2 \rightarrow \mathbf{R_j}$. The resulting trajectory set will contain the same number of observations although some of them are averaged. Before applying DTW a number of calculations are required for initialization as shown in Figure 6.31. The DTW/synchronization procedure is performed for a specified number of iterations. In each iteration, the weight matrix $\mathbf{W}$ is updated to give more importance to variables bearing more consistent dynamic behavior (monotonically increasing/decreasing, smooth, non-noisy, etc.). The weight matrix can also be assigned values based on process knowledge, but when there is uncertainty or no such knowledge, automatic updating of $\mathbf{W}$ based on calculations that will be explained below becomes advantageous.

1. Select one of the batches (trajectory set) as reference batch $\mathbf{R_f} = \mathbf{T}_\ell$, (with r_f the length of the reference batch)

2. Define the maximum number of iterations

3. Apply the DTW procedure outlined previously to synchronize/equalize the test batches with the reference batch resulting in $\tilde{\mathbf{T}}_\ell, \ell = 1, \ldots, L$,

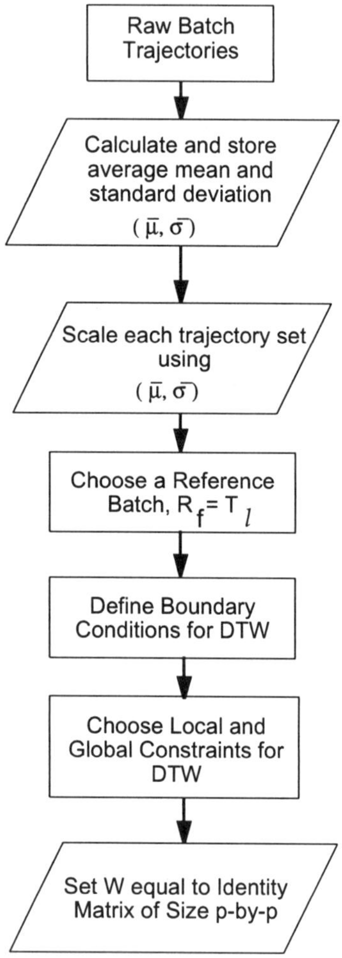

Figure 6.31. Initialization steps preceding multivariate DTW implementation.

synchronized trajectory sets having the common duration r_f (i.e. each $\tilde{\mathbf{T}}_\ell$ will be a matrix of size $(r_f \times P)$ containing the synchronized variable trajectories).

4. Compute the average trajectory for each variable

$$\bar{\mathbf{T}} = \sum_{\ell=1}^{L} \tilde{\mathbf{T}}_\ell / L \tag{6.84}$$

5. Compute the sum of squared deviations from $\bar{\mathbf{T}}$ for each variable

$$\mathbf{W}(p,p) = \left[\sum_{\ell=1}^{L} \sum_{m=1}^{r_f} \left[\tilde{\mathbf{T}}_\ell(m,p) - \bar{\mathbf{T}}(m,p) \right]^2 \right]^{-1} \tag{6.85}$$

6. Normalize the weight matrix $\mathbf{W}$ so that the sum of the weights is equal to the number of variables, p

$$\mathbf{W} = \mathbf{W} \left(\frac{P}{\left[\sum_{p=1}^{P} \mathbf{W}(p,p) \right]} \right) \tag{6.86}$$

7. For the first couple of iterations (two or three would be sufficient) use the same chosen reference trajectory $\mathbf{R}_f$. Set $\mathbf{R}_f = \bar{\mathbf{T}}$ for the subsequent iterations.

8. Go to step 3 for the next iteration.

The change in the weight matrix can also be monitored to see the convergence and to determine consistent variables.

Example Consider a set of unequal/unsynchronized trajectories of 13 variables collected from 55 normal operation runs of fed-batch penicillin fermentation ($L = 55$, $P = 13$) at 0.2 h of sampling interval. The durations of each batch trajectory set varies from 418.4 h (2092 observations) to 499.8 h (2499 observations). Batch number 36 of length 445.6 h (2228 observations) is chosen arbitrarily and it was found to be appropriate since its length is close to the median length of 451 h (2255 observations). As a result, the number of batches that will be expanded and the number of batches to be compressed will be close. Prior to DTW implementation, the necessary initial steps explained in Figure 6.31 were taken. Each variable was scaled using averaged means and standard deviations to remove the effects of different engineering units. A low-pass filter was also used to remove white noise from especially off-gas measurements such as CO_2 concentration. Figure 6.32 shows profiles of substrate, biomass, and penicillin concentrations

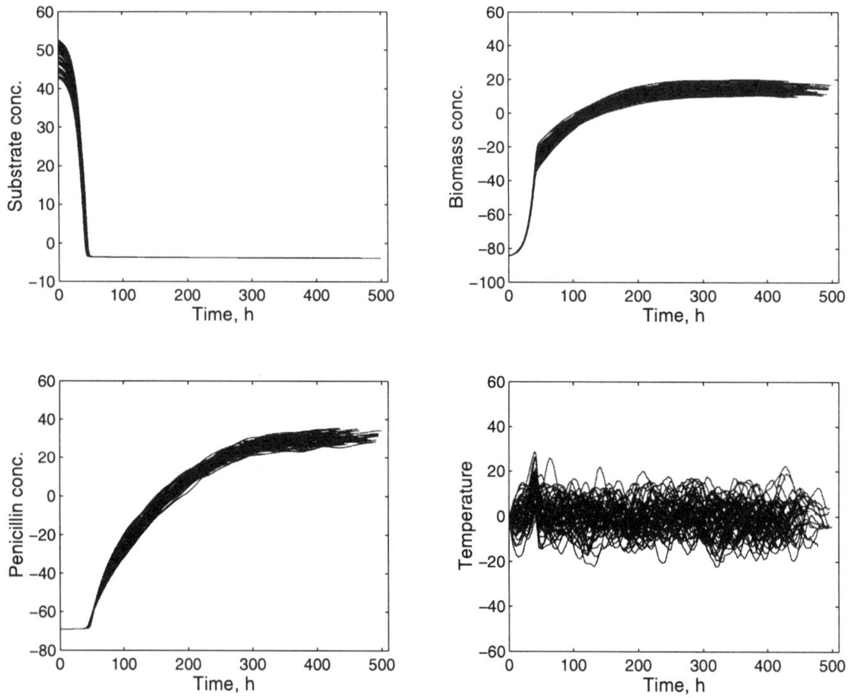

Figure 6.32. Four variables (out of 13) of 55 good performance batches before synchronization/equalization.

and temperature profiles before DTW. Fixed boundary conditions were chosen for DTW as $i(1) = j(1) = 1$ and $i(K) = N, j(K) = 2228$. Type II symmetric local continuity constraint (Figure 6.24(b)) with smoothed path weightings and Itakura global constraint together with Sakoe and Chiba band constraint (with $Q_{max} = 2$ and $K_0 = 30$) were also used. DTW synchronization procedure explained earlier was applied for a maximum of five iterations. Batch 36 was used as a reference batch in the first two iterations. Because of their monotonically decreasing, smooth behavior, biomass concentration and heat generated were assigned the highest weights throughout the calculations. The percent change in the weight matrix **W** was given to reflect this fact in Figure 6.33. The resulting profiles of the four variables (out of 13) are presented in Figure 6.34. All of the trajectories are synchronized and equalized to a common length of 445.6 h (2228 observations). □

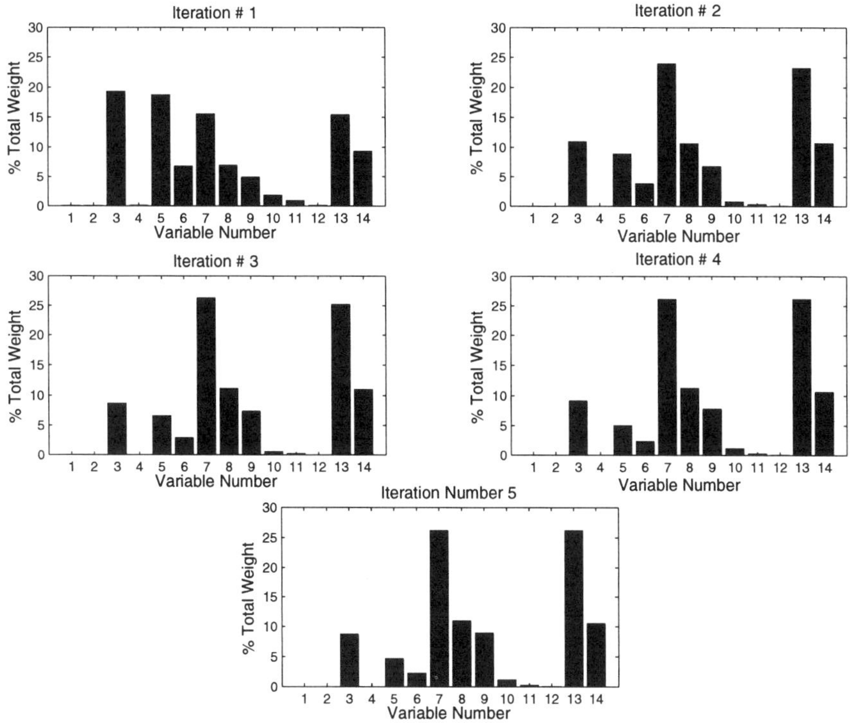

Figure 6.33. Percent weight change for the variables after five iterations.

6.3.3 Curve Registration

The curve registration problem casts the landmark alignment problem in functional data analysis (FDA) framework [495]. FDA involves the estimation of mth order linear differential operators $L = w_0 I + w_1 D + \ldots + w_{m-1} D^{m-1} + D^m$ where $Lx = 0$, and D^m denotes the mth derivative, and the weights w are functions of the independent variable t. Let N functions x_i be defined on closed real interval $[0, T_0]$ and $h_i(t)$ be a transform of t for case i with domain $[0, T_0]$. The time of events must remain in the same order regardless of the time scale. Therefore, the time warping functions h_i must be such that $h_i(t_1) > h_i(t_2)$ for $t_1 > t_2$. Define $y(t)$ to be a fixed (reference) function defined over $[0, T_0]$ to act as a template (for example a reference batch trajectory) for individual curves x_i such that after registration, the features of x_i will be aligned with the features of y. In discrete values y_i, $k = 1, \ldots, K$,

$$y_i = x_i \left[h_i(t_k) \right] + \epsilon_{ik} \tag{6.87}$$

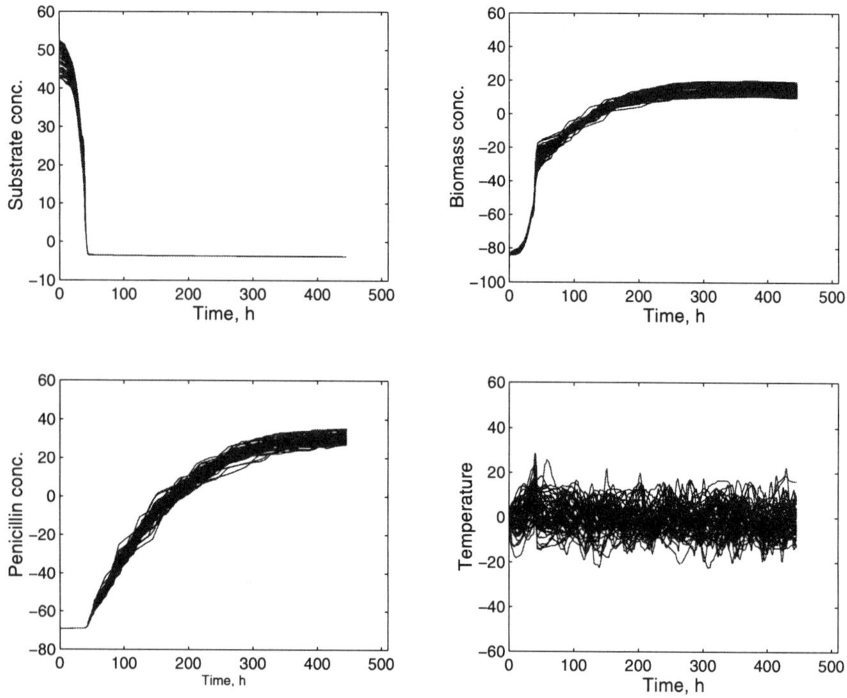

Figure 6.34. Four variables (out of 13) of 55 good performance batches after synchronization.

where ϵ_{ik} is a small residual relative to x_i and roughly centered about 0 [495]. The curve registration task is to determine the time warping functions h_i so that the de-warped trajectories $x_i[h_i(t_k)]$ can be interpreted more accurately. The h_i can be determined by using a smooth monotone transformation family consisting of functions that are strictly increasing (monotone) and have an integrable second derivative [494]:

$$D^2h = qDh. \tag{6.88}$$

A strictly increasing function has a nonzero derivative and consequently the weight function $q = D^2h/Dh$ or the curvature of h. The differential equation 6.88 has the general solution

$$\begin{aligned} h(t) &= C_0 + C_1 \int_0^t \left[exp \int_0^u q(v)dv \right] du \\ &= C_0 + C_1\{D^{-1}\ expD^{-1}q\}(t) \end{aligned} \tag{6.89}$$

where D^{-1} is the integration operator and C_0 and C_1 are arbitrary constants [495]. Imposing the constraints $h(0) = 0$ and $h(T_0) = T_i$, $C_0 = 0$ and $C_1 = T_i/[\{D^{-1}exp(D^{-1}q)\}(T_0)]$. Hence, h depends on q. The time warping function h can be estimated by minimizing a measure of the fit Υ_η of $x_i[h_i(t_j)]$ to y. A penalty term in Υ_η based on q permits the adjustment of the smoothness of h_i [495]. To estimate the warping function h_i, one minimizes

$$\Upsilon_\eta(y, x|h) = \sum_{j=0}^{m} \int \alpha_j(t) \left\| D^j y(t) - D^j x\left[h(t)\right] \right\|_j^2 dt + \eta \int q^2(t) dt \tag{6.90}$$

where $\alpha_j(t)$'s are weight functions and

$$\begin{aligned} \left\| D^j y(t) - D^j x\left[h(t)\right] \right\|_j^2 &= \\ (D^j y(t) - D^j x\left[h(t)\right])^T \mathbf{W}_j (D^j y(t) - D^j x\left[h(t)\right]) \end{aligned} \tag{6.91}$$

and $\mathbf{W}_j$ are weight matrices [495]. Weight matrices $\mathbf{W}_j$ allow for general weighting of the elements and weight functions $\alpha_j(t)$ permit unequal weighting of the fit to a certain target over time [495]. Parameter η adjusts the penalty on the degree of smoothness and q is expressed as a linear combination of B-spline bases

$$q(u) = \sum_{k=0}^{K} c_k B_k(u). \tag{6.92}$$

B-splines are used in this study as the polynomial basis for performing curve registration because calculating the coefficients of the polynomial is well defined. When estimating the solution to transforming particular waveforms into the B-spline domain, the required number of calculations increases linearly with the number of data points [494]. The derivative of Υ_η with respect to the B-spline coefficient vector $\mathbf{c}$ is

$$\begin{aligned} \frac{\partial \Upsilon_\eta(y, x|h)}{\partial \mathbf{c}} &= -2 \sum_{j=0}^{m} \int \alpha_j(t) \frac{\partial h(t)}{\partial \mathbf{c}} \left[\frac{\partial D^j x(h)}{\partial h} \right]^T \mathbf{W}_j \\ &\times (D^j y(t) - D^j x[h(t)]) dt + \eta \int \left(\frac{\partial q(t)}{\partial \mathbf{c}} \right)^2 dt. \end{aligned} \tag{6.93}$$

The derivative $[\partial D^j x(h)/\partial h]$ must be estimated with a smoothing technique to ensure monotonic increase [495].

One can use either an a priori local time stamp indicator or an optimization procedure for determining the landmarks in a reference set. The challenge of implementing multivariate landmarking is that landmarks are different (in placement and number) for different process variables. Critical issues are the selection of the process variable(s) for determining the landmarks, the number of landmarks and their locations to define clearly the progress of the batch.

One alternative is to select a process variable based on process knowledge and implement landmarking by using the trajectory of that variable. Another alternative is to use an iterative approach which will reconcile the identification of process landmarks with respect to particular trajectory landmarks:

1. Find the landmarks of the most important variable trajectories Lm_1. Align all other variable trajectories with respect to the landmarks Lm_1.

2. Calculate the principal components of the aligned set of process variables. Determine the landmarks of the first principal component Lm_{PCA}.

3. Realign the process trajectories with respect to Lm_{PCA}.

4. Recalculate the principal components of the realigned set of process variables. Determine the landmarks of the first principal component $Lm_{PCA\text{new}}$.

5. Determine if $Lm_{PCA\text{new}}$ are reasonably close to Lm_{PCA}. If so, the process landmarks are defined by $Lm_{PCA\text{new}}$. If not, return to Step 3.

The outcome of this procedure may be interpreted using several alternatives. Once $Lm_{PCA\text{new}}$ has converged, one may proceed with statistical analysis using the data warped with respect to $Lm_{PCA\text{new}}$. As an alternative, only the data identified as "most significant" (either by user or principal components) may be warped with respect to $Lm_{PCA\text{new}}$, and other process data may be warped with respect to its own optimal landmarks.

When landmarking a test trajectory with respect to a reference trajectory, two distinct cases may be considered. The first case is a simple idealized situation where all the landmarks are delayed (or advanced) by a constant time τ and is called *uniform landmark case*. The second is the *mixed landmark case* that represents a general framework where some landmarks of the new batch are delayed and others are advanced with respect to the landmarks of the reference trajectory, yielding a more challenging landmark detection problem. Furthermore, the time shifts of the landmarks

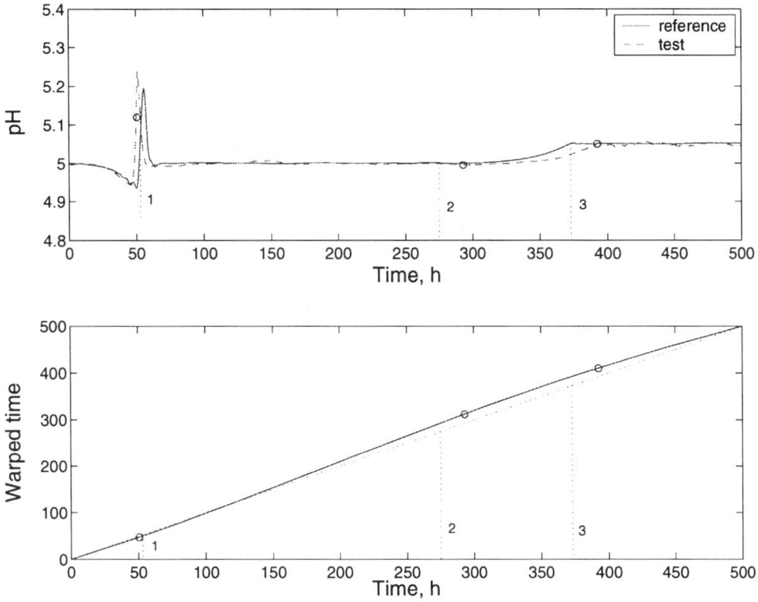

Figure 6.35. On-line landmarking: mixed case.

will vary, preventing the use of an assumption of a constant time shift τ between calculated and reference landmarks.

For illustration, consider the example in Figure 6.35, where the solid curve represents the mean pH trajectory calculated from an aligned reference set and the dashed curve in the upper figure represents a different test trajectory. The estimated landmarks (denoted by hollow circles) are mixed in leading or lagging the reference landmarks (indicated by dotted vertical lines and numbers) and they have varying time shift magnitudes. The first estimated landmark is advanced with respect to the first mean-value landmark, the second and third estimated landmarks are delayed with respect to their mean values. The advance between the first estimated and mean-value landmark is not equal to the delay between the second and the third estimated and their mean-value counterparts. When the test trajectory is warped with respect to the reference trajectory (as illustrated in the lower portion of Figure 6.35), the warped time-values (the solid line) is a curved pattern that crosses the center dashed-curve that represents linear mapping. This pattern suggests that when warping the test pattern (to align similar events to the reference pattern), the test trajectory will be stretched and compressed, respectively. In this example, the warping pattern lies slightly below the center line until the first landmark, so test values before it will be stretched to align the estimated landmark. After the first landmark, the warped curve lies above the center dashed-line, indicating that the values after the first landmark location need to be compressed to align the second and third landmarks with respect to the mean-value landmarks. This makes intuitive sense, because an estimated landmark that is advanced before a mean-value landmark must shift the data in a direction that will align similar process events.

The on-line optimization procedure sequentially searches for the optimal location of the landmarks of the test data with respect to the reference mean landmarks. The following procedure is given for the mixed landmarks case and can be modified to implement in an adaptive hierarchical PCA framework for online SPM:

1. Initialize estimated landmarks vector L_i.
2. For $i = 1, \ldots, m$ (m, number of landmarks in the reference set).
3. Collect values of test trajectory that contains landmark information up to time K. Choose time K_i for ith landmark so that it will span the reference landmarks range as

$$K_i \geq \arg\max_K(\boldsymbol{\ell_i}).$$

4. Set up search space for ith landmark. This consists of a vector of possible landmarks range for the ith landmark of reference set ($\boldsymbol{\ell_m}(1\times n_i)$ and a $n_i \times K_i$ matrix of n_i replicates of the test curve to be warped with respect to the possible landmarks' space.
   (a) Align n_i test curves by landmark registration. If $i \geq 2$ use previously estimated and stored landmarks in L_i during registration.
   (b) Calculate the sum of squared errors (SSE) between each of the n_i test curves and the reference trajectory (calculate reference trajectory from previously aligned reference set).
   (c) For $j = 1, \ldots, n_i$ (number of test curves)
   When the minimum SSE is found, select the i_jth landmark as the ith landmark for the test trajectory.
   (d) End j-loop.
   (e) Store estimated ith landmark in vector L_i.
5. End i-loop.
6. Warp test trajectory (and the rest of the trajectories in the new batch) with respect to the m estimated landmarks.

Example. The implementation of time alignment by landmark registration differs in postmortem and online analysis. When there is a database of historical batches of different lengths that alignment is to be performed, all of the multivariate batch data are re-sampled to a common number (e.g. the median length) of data points by interpolation first, then the registration technique is implemented. However, in the case of online registration with landmark detection, unknown batch length (process termination time or campaign run length is not known) of the new batch represents some difficulties in implementation. To represent online landmark detection in real-time, it is assumed that a fixed batch termination point is determined. Although all of the batches are to come to a completion at the same final time point, a temporal variation in important landmark locations is still present. The critical implementation issues of the alignment technique are presented in an orderly fashion in this example.

Determination of the landmark locations in reference batches

The regularization technique for mixed case is implemented to simulated fed-batch penicillin fermentation data of 40 batches sampled at 0.5 h on 16 variables for 500 h resulting in 1000 measurements. The decision should be made using engineering judgment on choosing appropriate variable trajectories that may contain physiological information about the location of the process landmarks. Figure 6.36 shows some of the process variable trajectories as well as two of the manipulated variables (base and acid flow rates) of a reference batch run under nominal operating conditions. Note that, the concentration of hydrogen ion (pH) is associated with biomass growth [61] as explained in Section 2.7.1, hence, becoming a good indicator for tracking physiological activities.

It is inferred that there are three process landmarks separating four distinct phases in fed-batch penicillin fermentations based on expert knowledge about penicillin fermentation. The first phase (lag phase and pre-exponential phase) corresponds to batch operation where a lag exists on the inception of penicillin production while cells are consuming substrates to grow. The landmark for the first phase can be found easily in any of the trajectories as shown in Figure 6.36. Exponential cell growth along with the start of penicillin production is observed in the second phase where glucose feed is initiated. The location of the second landmark is not apparent on each trajectory. Normally it corresponds to the time when biomass concentration begins to level off. In the vicinity of that point, hydrogen ion concentration starts to decrease and consequently the need for base addition (F_{base}) is reduced as shown in Figure 6.36. Hence, base flow rate is an excellent candidate for determining the location of the second landmark that indicates the beginning of the third phase (stationary phase). A

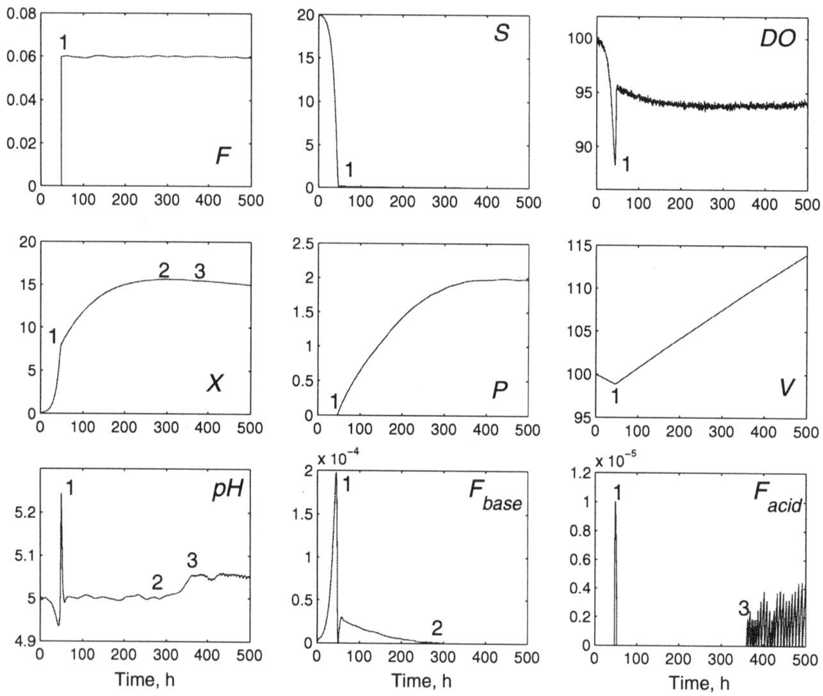

Figure 6.36. Physiologically important variable trajectories (F: glucose feed rate, S: glucose concentration, DO: percent oxygen saturation, X: biomass concentration, P: penicillin concentration, V: culture volume, F_{base}: base flow rate and F_{acid}: acid flow rate).

similar line of thought can be followed for detecting the temporal location of the third landmark which is the start of the death phase towards harvesting the fermentation. At this phase, biomass concentration begins to decline resulting in the decrease in hydrogen ion concentration level. Note that, a set point gap is defined for acid flow rate controller action to avoid excessive acid addition during the simulations resulting in a small increase at pH right after the third landmark [61]. Therefore, the instant when acid addition takes place after stationary phase can be used to determine the location of the third landmark and the beginning of the death phase.

Once the decision about the choice of the variables that contain landmark information is made, these variables (biomass concentration (X), base (F_{base}) and acid (F_{acid}) flow rates in Figure 6.36) are investigated in each batch of the reference set and landmark locations are stored. In this example, reference landmark locations matrix $\boldsymbol{\ell}_m$ is of size (3×40). Note

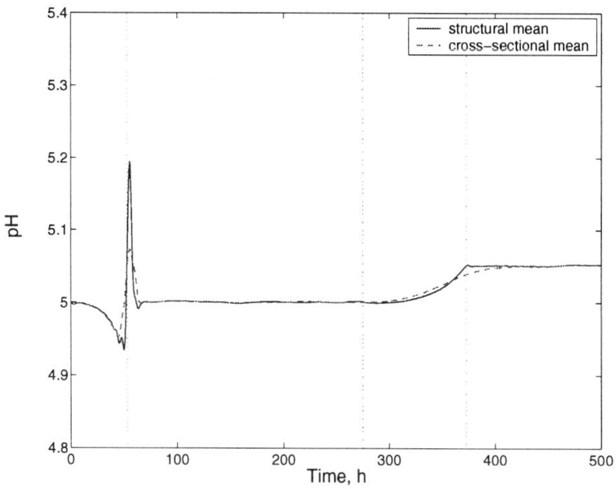

Figure 6.37. Comparison of average pH profiles. Solid curve is obtained after alignment by landmark registration.

that, instead of using one variable to determine process landmarks a combination of three variables is used in this example. The landmark location information from the inflection point of biomass concentration, the instant when base flow reaches zero after about 100 h and the instant when acid flow controller takes action after base flow decreases to zero are gathered in a landmark location vector, respectively.

Alignment of variable profiles in reference batches using landmark registration

Given the locations of landmarks in the reference batches, variable profiles are 'registered' so that similar events will be aligned with respect to mean landmarks. Mean landmark locations can be calculated from $\boldsymbol{\ell}_m$ and used for the alignment to be performed around the average landmark locations. This is an arbitrary choice. A vector of landmark locations that belong to a reference batch could also be chosen. The aforementioned curve registration technique is implemented to all variable trajectories by using the mean landmarks vector and the matrix $\boldsymbol{\ell}_m$ of landmarks in reference batches.

Comparison of cross-sectional mean trajectories prior to alignment and structural means after alignment illustrates the affects of trajectory alignment. The average pH profile (dashed curve) that is the cross-sectional mean of the reference profiles before alignment (Figure 6.37), resembles the reference profiles shown in Figure 6.38(a) but differ in the amplitudes

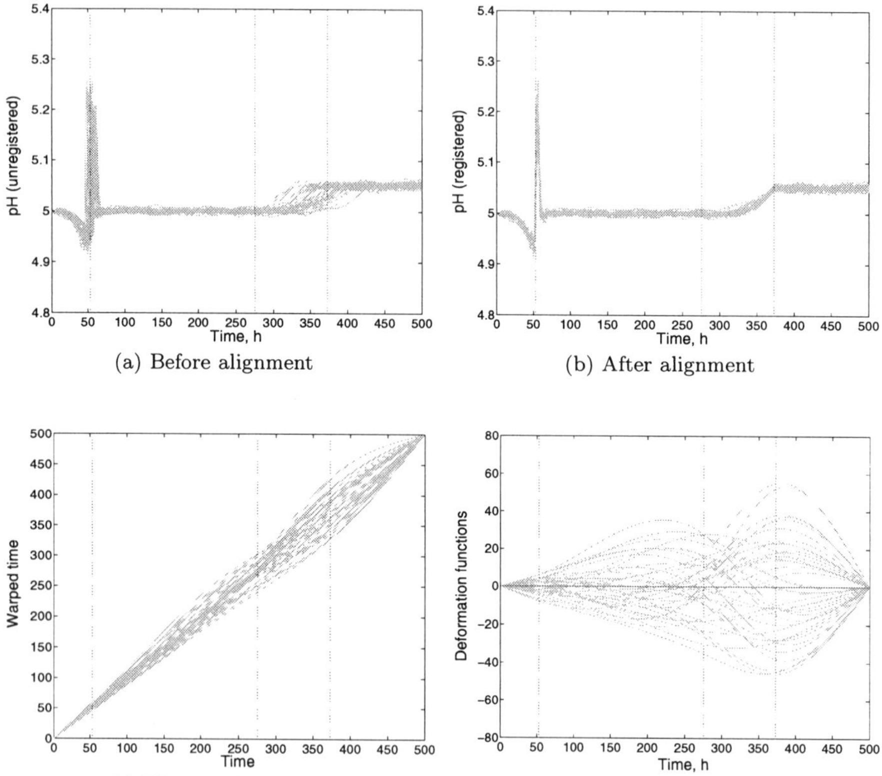

(a) Before alignment

(b) After alignment

(c) Warping functions

(d) Deformation functions

Figure 6.38. Curve registration results on pH profiles.

and temporal occurrences of landmarks. The average profile obtained after alignment by curve registration resembles reference curves structurally in both landmark locations and amplitudes of the local peaks and valleys (solid curve). Aligned pH curves are presented in Figure 6.38(b) along with corresponding mean landmark locations (vertical dotted lines). Each curve is aligned with respect to its particular landmarks allowing for comparison of similar events in the statistical analysis, hence leading to the development of more consistent MSPM frameworks. Figures 6.38(c) and 6.38(d) show warping and resulting deformation functions used during alignment of variable profiles in each batch. Deformation functions are obtained from the difference between warped and actual times, and represent a measure of time distortion required for nonlinear feature matching. The rest of the

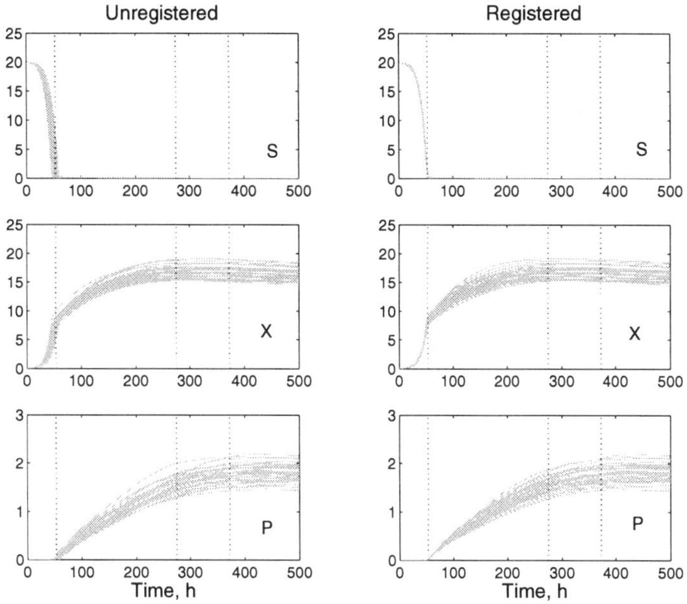

Figure 6.39. Landmark registration results for some of the concentration profiles (S: glucose concentration, X: biomass concentration, P: penicillin concentration). Dashed lines represent the locations of mean landmarks.

profiles in the reference batch set are aligned similarly so that the same set of nonlinear warping functions (Figure 6.38(c)) are used to align rest of the variable profiles. Since the same physiological events affect most variables such as concentration profiles, their landmark locations overlap in time (Figure 6.39).

Alignment of variable profiles online in real-time of a new batch using landmark registration

After aligning the reference batch profiles, the necessary information is available for implementing the alignment procedure for a new batch online in real-time. The iterative online landmark estimation procedure described earlier is used in this example. The necessary information from reference batch alignment includes reference landmark locations matrix $\boldsymbol{\ell}_m$ and its mean vector, and the aligned reference set to calculate average profiles. Since a combined landmark location vector is used from different process variables, the corresponding reference profile is used as a comparative template while implementing the online procedure. For instance, the inflection point in biomass concentration profile determines the location of the first

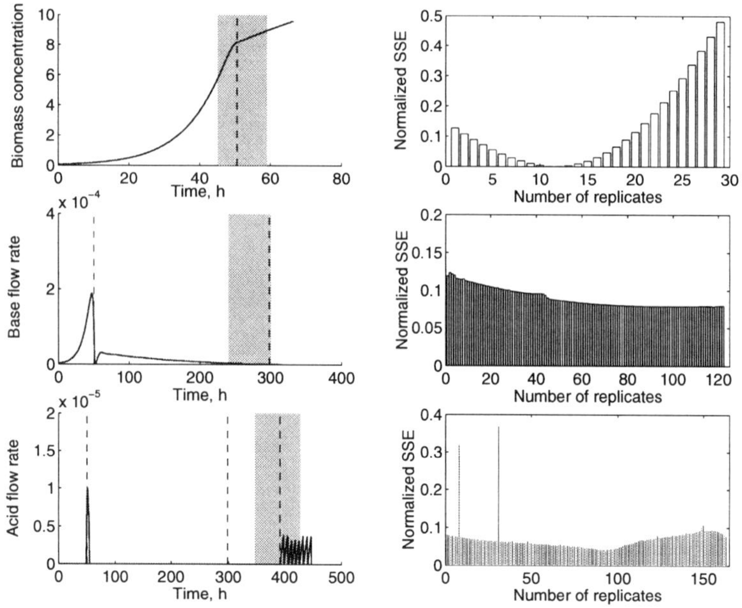

Figure 6.40. Online landmark estimation results. Dashed lines represent estimated landmark locations.

landmark in this example. Therefore, the mean biomass concentration profile that is calculated from the aligned reference set is used while searching for an estimated first landmark. Another decision should be made about the landmark search space. For practical purposes, it is chosen as the range of reference landmark locations. The search space is depicted as a band for each landmark in Figure 6.40. The plots on right side of Figure 6.40 represent sum of squared errors (SSE) between the mean and replicate profiles with potential landmarks in search space. A minimum of the SSE indicates the landmark point in the search space that gives the closest match between new batch profile and the template profile. Once the locations of the landmarks are estimated, the rest of the profiles are aligned with respect to estimated landmark(s). The results in Figure 6.40 show that the iterative technique successfully detects landmark locations. □

6.4 Multivariable Batch Processes

Batch processes often exhibit some batch-to-batch variation. Variations in charging the production recipe, differences in types and levels of impurities in raw materials, shift changes of operators, and disturbances during the progress of the batch are some of the reasons for this behavior.

Monitoring the trajectories of the process variables provides four different types of monitoring and detection activities:

- *End of batch quality control*: This is similar to the traditional quality control approach. The ability to merge information from quality variables and process variable trajectories by multivariate statistical tools enables accurate decision-making. Since process variable trajectories are available immediately at the conclusion of the batch, product quality can be inferred from them without any time delay.
- *Analysis of process variable trajectories after the conclusion of the batch*: This "postmortem" analysis of the batch progress can indicate major deviations in process variable trajectories and enable plant personnel to find out significant changes that have occurred, trace the source causes of disturbances and prevent the repetition of abnormal behavior in future batches. It can also point out different phases of production during the batch, providing additional insight about the process. Since the analysis is carried out after the conclusion of the batch, it cannot be used to improve the product of that batch.
- *Real-time on-line batch process monitoring*: The ultimate goal in batch process monitoring is to monitor the batch during its progress. This provides information about the progress of the batch while the physical, biological and chemical changes are taking place, enabling the observation of deviations from desired trajectories, implementation of interventions to eliminate the effects of disturbances, and decision to abort the batch if saving it is too costly or impossible.
- *Real-time on-line quality control*: This is the most challenging and important problem. During the progress of the batch, frequently measured process variables can be used to estimate end of batch product quality. This will provide an opportunity to foresee if there is a tendency towards the inferior product quality and take necessary actions to prevent final product quality deterioration before it is too late.

This section focuses on the first two types of activities, the off-line SPM and quality control. On-line SPM and quality control are discussed in

Section 6.5. Section 6.4.1 focuses on reference databases describing normal process operation and introduces the penicillin fermentation data used in many examples. Section 6.4.2 represents various multivariate charts for SPM. SPM of completed batches by MPCA is discussed in Section 6.4.3 and MPLS based SPM is presented in Section 6.4.4. Use of multiway/multiblock techniques for monitoring multistage/multiphase processes is discussed in Section 6.4.5. The integration of wavelet decompositions and MPCA is presented in Section 6.4.6.

6.4.1 Reference Database of Normal Process Operation

Developing empirical models as well as multivariate control charts for MSPM require a reference database comprised of past successful batches run under normal operating conditions (NOC). The historical database containing only the common cause variation will provide a reference distribution against which future batches can be compared. Selection of the reference batch records set out of a historical database depends on the objective of the monitoring paradigm that will be implemented. MPCA-based modeling is suitable if only the process variables are of interest. MPLS model will allow inclusion of final quality variables in the monitoring scheme. Initial choice of the potential NOC reference set may contain outlying batches. These batches will be found and removed at the initial round of either MPCA or MPLS modeling. As described in the earlier sections, there will be a temporal variation, in addition to amplitude variation, in process trajectories for each batch resulting in unequal/unsynchronized data. Prior to model development, it is crucial to apply one of the three equalization/synchronization techniques proposed earlier in Sections 6.3.1 (IVT), 6.3.2 (DTW) and 6.3.3 (Curve Registration). Equalized/synchronized data form a three-way array. After transforming the data by unfolding this array into a matrix and by subtracting the mean trajectory set from each batch trajectory set to remove most of the nonlinearity, MPCA and/or MPLS models can be built to investigate if the choice of the reference set is suitable for use in SPM of new batches. Once that decision is made, multivariate control chart limits are constructed according to the formulations given in Section 6.4.2. Development of a multivariate statistical process monitoring scheme will be given by means of a case study based on the unstructured mathematical model of fed-batch penicillin fermentation introduced in Section 2.7.1. A reference set of NOC generated by this simulator (Figures 6.41 and 6.42) will be used where applicable throughout the examples representing different monitoring techniques such as MPCA and MPLS along with the construction of multivariate control charts. Only for the multiblock MPCA technique, an

Table 6.6. Process variables measured throughout the batches

No.	Process Variables
1	Aeration rate, L/h
2	Agitation power, W
3	Glucose feed rate, L/h
4	Glucose feed temperature, K
5	Glucose concentration, g/L
6	Dissolved oxygen concentration, $mmole/L$
7	Biomass concentration, g/L
8	Penicillin concentration, g/L
9	Culture volume, L
10	Carbon dioxide concentration, g/L
11	pH
12	Fermenter temperature, K
13	Generated heat, $kcal$
14	Cooling water flow rate, L/h

example is chosen from pharmaceutical granules production case (Section 6.4.5).

Example. A set of data is produced using the simulator of fed-batch penicillin production (based on the unstructured model discussed in Section 2.7.1) under normal operating conditions. The values of the initial conditions and set points of input variables are slightly varied for each batch, resulting in unequal and unsynchronized batch trajectories that are typical in most experimental cases. Batch lengths varied between 375 h and 390 h. One of the batches that has a batch length of 382 h, close to the median batch length is chosen as a reference batch. Data of the other batches are equalized based on multivariate DTW algorithm discussed in Section 6.3.2. Type II symmetric local continuity constraint (Figure 6.24(b)) with smoothed path weightings, Itakura global constraint, and Sakoe and Chiba band constraint (with $Q_{\max} = 2$ and $K_0 = 50$) are used for data synchronization. Multivariate DTW synchronization procedure was applied for a maximum of five iterations (Figure 6.43).

The reference set is comprised of 42 batches containing 14 variables (sampled at 0.5 h). A three-way array of size $41 \times 14 \times 764$ is formed based on this initial analysis. The variables are listed in Table 6.6. Although on-line real-time measurement availability of some of the product related variables such as *biomass* and *penicillin concentrations* is somewhat limited in reality, it is assumed that these can be measured along with frequently measurable variables such as *feed rates* and *temperature.* If the sampling rates are different, an estimator such as Kalman filter can be used to estimate these variables from measured values of frequently measured variables. A number of product quality variables are also recorded at the end of the batch (Table 6.7 and Figure 6.41).

Three additional batches were simulated to illustrate detection, diagnosis and prediction capabilities of the MPCA and MPLS models used for both end-of-batch and on-line SPM. Fault scenarios are chosen such that they resemble the ones generally encountered in industry. First fault is a 10% step decrease in agitator power input about its set point during early in the second phase of the fermentation between 70 and 90 hrs (between the 140th and 180th samples). The second fault is a small drift in the glucose feed rate right after start of feeding in fed-batch operation. In the latter case, the abnormal operation develops slowly and none of the individual measurements reveal it clearly when their univariate charts are examined. The third fault is the same as the second fault, only the slope of the drift is higher. The first faulty batch is of length 375 h (750 samples), the second is 380 h (760 samples) and the third batch of length 382 h (764 samples). Figure 6.44 shows the trajectories of these faulty batches along with a normal batch trajectory set.

6.4.2 Multivariate Charts for SPM

The following multivariate charts are used as visual aids for interpreting multivariate statistics calculated based on empirical models. Each chart can be constructed to monitor batches or performance of one batch during its evolution.

Score biplots or 3D plots are used to detect any departure from the in-control region defined by the confidence limits calculated from the ref-

Table 6.7. Quality variables measured after the completion of batches

No.	Quality Variables
1	Final penicillin concentration, g/L
2	Overall productivity, g/h
3	Terminal yield of penicillin on biomass
4	Terminal yield of penicillin on glucose
5	Amount of penicillin produced, g

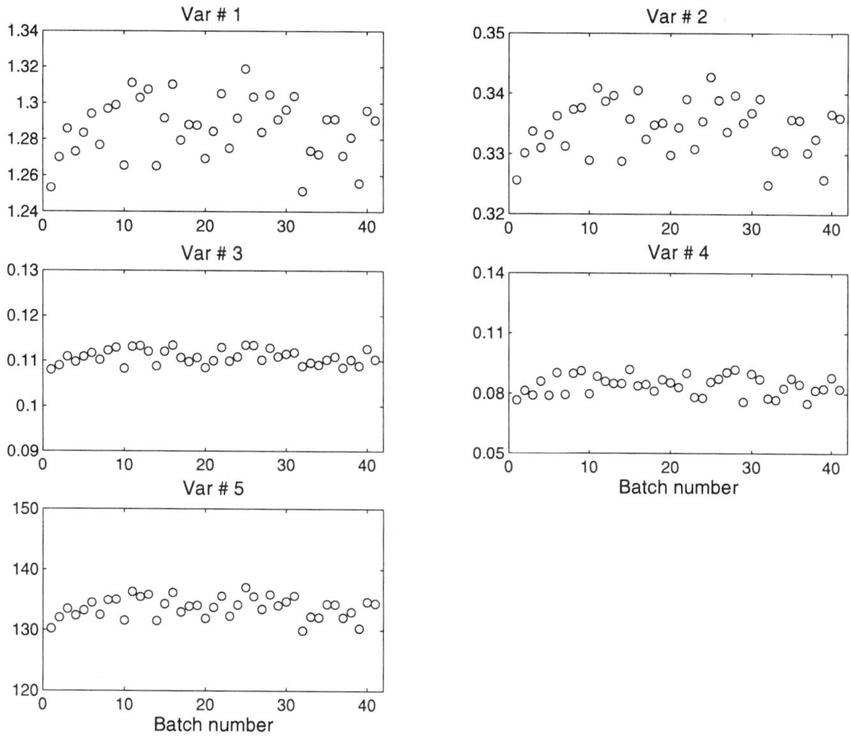

Figure 6.41. Quality variables measured at the end-of-batches for the reference set (listed in Table 6.7).

erence set. The score plots provide a summary of process performance from one batch to the next. The control limits for new independent **t** scores under the assumption of normality at significance level α at any time interval k is given by [214]

$$\pm t_{n-1,\alpha/2} s_{\text{ref}} (1 + 1/n)^{1/2} \tag{6.95}$$

where n and s_{ref} are the number of observations and the estimated standard deviation of the **t**-score sample at a given time interval k (mean is always 0) and $t_{n-1,\alpha/2}$ is the critical value of the Studentized variable with $n-1$ degrees of freedom at significance level $\alpha/2$ [435]. The axis lengths of the confidence ellipsoids in the direction of ath principal component are given by [262]

$$\pm [\mathbf{S}(a,a) F_{A,I-A,\alpha} A(I^2 - 1)/(I(I-A))]^{1/2} \tag{6.96}$$

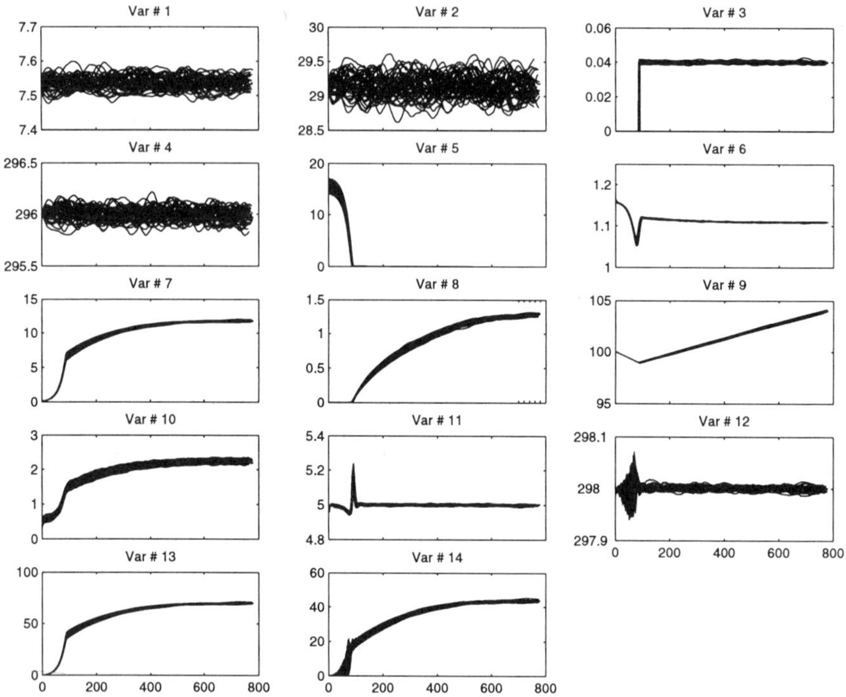

Figure 6.42. Unequal/unsynchronized raw batch trajectories of the reference set.

Table 6.8. Faulty batch information and data sets

Variable name	Fault definition	Introduction time
Agitator power input (Variable 2) Data: $\mathbf{X_1}(750 \times 14)$	10% step decrease	70 - 90 hrs (140th and 180th samples)
Glucose feed rate (Variable 3) Data: $\mathbf{X_2}(761 \times 14)$	Small downward drift	Beginning of fed-batch operation
Glucose feed rate (Variable 3) Data: $\mathbf{X_3}(764 \times 14)$	Large downward drift	Beginning of fed-batch operation

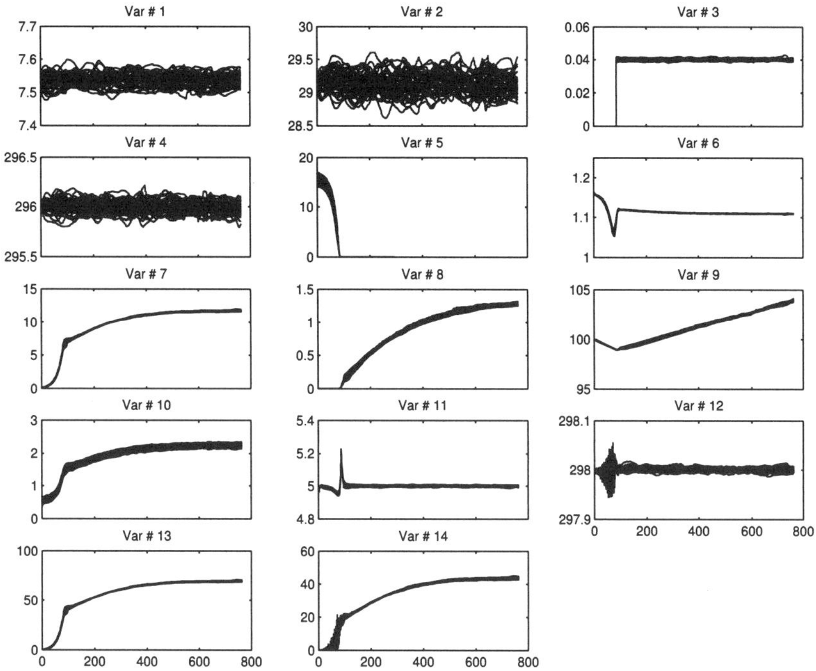

Figure 6.43. Equalized/synchronized batch trajectories of the reference set using DTW procedure in Figure 6.31.

where $\mathbf{S}$ is the estimated covariance matrix of scores and $F_{A,I-A,\alpha}$ is the F-distribution value with A and $I - A$ degrees of freedom in α significance level, I is the number of batches in the reference set, A is the number of PCs retained in the model.

Hotelling's T^2 plot detects the small shifts and deviations from normal operation defined by the model. Statistical limits on the D-statistic are computed by assuming that the data follow a multivariate Normal distribution [254, 253]. D statistics (T^2 can be written as the D-statistic) for end-of-batch SPM for batch i are

$$D_i = \frac{\mathbf{t}_a^T \mathbf{S}^{-1} \mathbf{t}_a I}{(I-1)^2} \qquad \sim B_{A/2,(I-A-1)/2} \tag{6.97}$$

where $\mathbf{t}_a$ is a vector of A scores [254] and $\mathbf{S}$ is the $(A \times A)$ estimated covariance matrix, which is diagonal due to the orthogonality of the $\mathbf{t}$ scores [594]. The statistics aforementioned in Eq. 6.97 is called

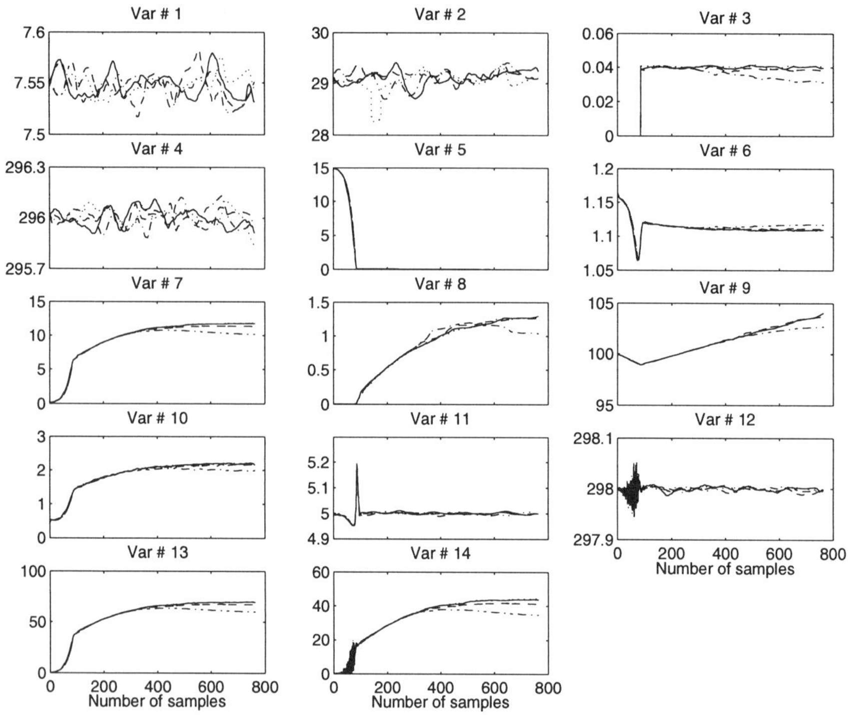

Figure 6.44. Trajectories of process variables of normal (solid curve), a batch with 10% step decrease (Variable 2) (dotted curve), with a small drift (Variable 3) (dashed curve) and with a large drift (Variable 3) (dashed-dotted curve).

Hotelling's T^2 [244] and follows the beta distribution. It can also be calculated for each batch as [254]

$$D_i = \sum_{a=1}^{A} \frac{t_{ia}^2}{\lambda_a} = \sum_{a=1}^{A} \frac{t_{ia}^2}{s_a^2} \tag{6.98}$$

where the PCA scores $\mathbf{t}$ in dimension a have variance λ_a (or estimated variance s_a^2 from the scores of the reference set), which is the a-th largest eigenvalue of the scores covariance matrix $\mathbf{S}$. If tables for the beta distribution are not readily available, this distribution can be approximated using Eq. 6.99 [594].

$$B_{A/2,(I-A-1)/2,\alpha} = \frac{(A/(I-A-1))F_{A,I-A-1,\alpha}}{1+(A/(I-A-1))F_{A,I-A-1,\alpha}} \tag{6.99}$$

T^2 values are calculated throughout the batch. As soon as the batch is complete Eq. 6.100 is applied for each observation at time interval k [435].

$$D_{ik} = \mathbf{t}_{iAk}^T (\mathbf{S})^{-1} \mathbf{t}_{iAk} \tag{6.100}$$

T^2 values for each time interval k for a new batch can also be calculated similar to Eq. 6.98 as

$$D_k = \sum_{a=1}^{A} \frac{t_{ak}^2}{\lambda_a} = \sum_{a=1}^{A} \frac{t_{ak}^2}{s_{ak}^2}. \tag{6.101}$$

D_k values follow F-distribution [594]

$$D_k \backsim \frac{A(I^2-1)}{I(I-A)} F_{A,I-A} \tag{6.102}$$

where A denotes the number of PCs and I the number of batches in the reference set.

Squared Prediction Error (SPE) plot shows large variations and deviations from the normal operation that are not defined by the model. The ith elements of the **t**-score vectors correspond to the ith batch with respect to the other batches in the database over the entire history of the batch. The **P** loadings matrices summarize the time variation of the measured variables about their average trajectories. If a new batch is good and consistent with the normal batches (used to develop MPCA model), its scores should fall within the normal range and the 'sum of the squared residuals' (Q-statistic) should be small. The Q statistic for end-of-batch SPM for batch i is written as

$$Q_i = \mathbf{e}_i \mathbf{e}_i^T = \sum_{c=1}^{KJ} \mathbf{E}(i,c)^2 \tag{6.103}$$

where $\mathbf{e}_i$ is the ith row of $\mathbf{E}$, I is the number of batches in the reference set, A is the number of PCs retained in the model, and $\mathbf{t}_a$ is a vector of A scores [254].

Statistical limits on the Q-statistic are computed by assuming that the data have a multivariate normal distribution [253, 254]. The control limits for Q-statistic are given by Jackson and Mudholkar [255] based on Box's [76] formulation (Eq. 6.104) for quadratic forms with significance level of α given in Eqs. 6.104 and 6.105 as

$$Q_\alpha = g\chi_{h,\alpha}^2 \tag{6.104}$$

$$Q_\alpha = \theta_1[1 - \theta_2 h_0(1-h_0)/\theta_1^2 + z_\alpha(2\theta_2 h_0^2)^{1/2}/\theta_1]^{1/h_0} \qquad (6.105)$$

where χ_h^2 is the chi-squared variable with h degrees of freedom and z is the standard normal variable corresponding to the upper $(1-\alpha)$ percentile (z_α has the same sign as h_0). θ values are calculated using unused eigenvalues of the covariance matrix of observations (eigenvalues that are not retained in the model) as [655]

$$\theta_i = \sum_{j=k+1}^{n} \lambda_j^i, \text{ for } i = 1, 2, \text{ and } 3. \qquad (6.106)$$

The other parameters are

$$g = \theta_2/\theta_1, \quad h = \theta_1^2/\theta_2$$
$$h_0 = 1 - 2\theta_1\theta_3/3\theta_2^2. \qquad (6.107)$$

θ_i's can be estimated from the estimated covariance matrix of residuals (residual matrix used in Eq. 6.103) for use in Eq. 6.105 to develop control limits on Q for comparing residuals on batches. Since the covariance matrices $\mathbf{E}^T\mathbf{E}$ $(JK \times JK)$ and $\mathbf{E}\mathbf{E}^T$ $(I \times I)$ have the same non-zero eigenvalues [435], $\mathbf{E}\mathbf{E}^T$ can be used in estimating θ_i's due to its smaller size for covariance estimation as

$$\mathbf{V} = \frac{\mathbf{E}\mathbf{E}^T}{I-1}, \quad \theta_i = \text{trace}(\mathbf{V}^i), \text{ for } i = 1, 2, \text{ and } 3. \qquad (6.108)$$

A simplified approximation for Q-limits has also been suggested in [148] by rewriting Box's equation (Eq. 6.104) by setting $\theta_2^2 \approx \theta_1\theta_3$

$$Q_\alpha \cong gh[1 - 2/9h + z_\alpha(2/9h)^{1/2}]^3. \qquad (6.109)$$

Eq. 6.105 can be used together with Eq. 6.108 to calculate control limits for sum of squared residuals when comparing batches (Q_i in Eq. 6.103).

In order to calculate SPE values throughout the batch as soon as the batch is complete, Eq. 6.110 is used for each observation at measurement time k [435]

$$SPE_{ik} = \sum_{j=1}^{J}(x_{ijk} - \hat{x}_{ijk})^2 = \sum_{j=1}^{J}(e_{ijk})^2. \qquad (6.110)$$

Calculated SPE values for each time k using Eq. 6.110 follow χ^2 (chi-squared) distribution (Eq. 6.104, [76]). This distribution can be well approximated at each time interval using Box's equation in Eq. 6.104 (or its modified version in Eq. 6.109). This approximation of moments is preferred because it is computationally faster than using traces of powers of the residual covariance matrix of size $(J \times J)$ at each time interval. Parameters g and h can be approximated by matching moments of the $g\chi_h^2$ distribution [435]

$$g = \frac{v}{2m}, \quad h = \frac{2m^2}{v} \tag{6.111}$$

where m and v are the estimated mean and variance of the SPE at a particular time interval k, respectively. It was reported that these matching moments were susceptible to error in the presence of outliers in the data or when the number of observations was small. Outliers should be eliminated as discussed in Section 3.4.2.

Contribution plots are used for fault diagnostics. Both T^2 and SPE charts produce an out-of-control signal when a fault occurs but they do not provide any information about the cause. Variable contributions to T^2 and SPE values indicate which variable(s) are responsible for the deviation from normal operation. T^2 statistic is used to monitor the systematic variation and SPE statistic is used to monitor the residual variation. Hence, in the case of a process disturbance, either of these statistics will exceed the control limits. If only the T^2 statistic is out of control, the model of the process is still valid but the contributions of each process variable to this statistic should be investigated to find a cause for the deviation from normal operation. If SPE is out of control, a new event is found in the data, that is not described by the process model. Contributions of each variable to SPE will unveil the responsible variable(s) to that deviation.

Contribution plots are discussed in more detail as a fault diagnosis tool in Section 8.1.

Explained variance, loadings and weights plots highlight the variabilities of batch profiles. The explained variance is calculated by comparing the real process data with the MPCA model estimates. This can be calculated as a function of batch number, time, or variable number. The value of explained variance becomes higher if the model accounts for more variability in the data and for the correlation that exists among the variables. Variance plots over time can be used as an indicator of the phenomenological/operational changes that occur during the process evolution [291]. This measure can be computed as

$$SS\ explained,\ \% = \frac{\hat{\sigma}^2}{\sigma^2} \times 100 \tag{6.112}$$

where SS stands for 'sum of squares', σ^2 and $\hat{\sigma}^2$ are the true and estimated sum of squares, respectively.

Loadings also represent variability across the entire data set. Although the loadings look like contributions, a practical difference occurs when some of the contributions of the process variables have values much smaller than their corresponding loadings and vice versa.

In the case of MPLS-based empirical modeling, variable contributions to weights ($\mathbf{W}$) carry valuable information since these weights summarize information about the relationship between $\mathbf{X}$ and $\mathbf{Y}$ blocks. There are several ways of present this infirmation as charts. The overall effect of all of the process variables on quality variables over the course of process can be plotted, or this can be performed for a specific period of the process to reflect the change of the effect of the predictor block ($\mathbf{X}$). Recently, Wold et al. [145] suggested yet another statistic as they coined the term *Variable Influence on Projection (VIP)* using the following formula

$$VIP_j = \left[\sum_{a=1}^{A} \left(w_{aj}^2 (SSY_{a-1} - SSY_a) \right) \frac{J}{(SSY_0 - SSY_A)} \right]^{1/2} \tag{6.113}$$

where J denotes the number of variables in $\mathbf{X}$-block, A the number of latent variables retained in the model, w_{aj} the weight on ath component on jth variable, SSY_0 the initial sum of squares on $\mathbf{Y}$-block, and SSY_A the sum of squares after A latent variables on $\mathbf{Y}$-block. While this equation holds for continuous process data, a small modification is needed for batch process data since in the case of $I \times JK$ data arrangement, there are JK variables. One possible modification is to calculate the mean of each j variable to obtain an overall view or this can also be done for a period of the process. The squared sum of all VIP's is equal to the number of variables in $\mathbf{X}$-block (that is J for continuous process data and JK for batch process data). VIP terms on each variable can be compared and the terms with large VIP (larger than 1) are the most relevant to explaining $\mathbf{Y}$-block. An example is given in Section 6.4.4 for the overall VIP case.

6.4.3 Multiway PCA–based SPM for Postmortem Analysis

In this section, the use and implementation of MPCA-based modeling (Section 4.5.1) are discussed for a postmortem analysis of finished batch runs to discriminate between the 'good' and the 'bad' batches. This analysis can be used to improve operation policies and discover major sources of variability among batches. MPCA can also be implemented on-line (Section 6.5). In either case, an MPCA model based on a reference set (representing normal operating conditions) selected from a historical batch database is developed.

When a batch is complete, measurements on the process variables made at each sampling instant produce a matrix of $\mathbf{X}_{\text{new}}$ $(K \times J)$. This matrix is unfolded and scaled to give $\mathbf{x}_{\text{new}}$ $(1 \times KJ)$, using the same parameters for scaling the reference batches during the model development phase. This new batch vector is tested for any unusual behavior by predicting $\mathbf{t}$ scores and residuals via the use of $\mathbf{P}$ loading matrices (Eq. 6.114) that contain most of the structural information about the deviations of variables from their average trajectories under normal operation:

$$\mathbf{t}_{\text{new}} = \mathbf{x}_{\text{new}}\mathbf{P}, \qquad \mathbf{e}_{\text{new}} = \mathbf{x}_{\text{new}} - \sum_{a=1}^{A} t_{\text{new},a}\mathbf{p}_a \tag{6.114}$$

where $\mathbf{t}_{\text{new}}$ denotes the scores of the new batch calculated by using $\mathbf{P}$ $(JK \times A)$ loadings from the MPCA model with A PCs. If the scores of the new batch are close to the origin and its residuals are small, this indicates that its operation is also similar to that of reference batches representing normal operation. The sum of squared residuals Q for the new batch over all the time periods can be calculated as $Q = \mathbf{e}^T\mathbf{e} = \sum_{\ell=1}^{KJ} e(\ell)^2$ for a quick comparison with Q values of reference batches. D statistic (Eq. 6.97) can also be used to get an overall view. These statistics give only summary information about the new batch with respect to the behavior of the reference set, they do not present instantaneous changes that might have occurred during the progress of the batch. It is a common practice to use on-line MPCA algorithms to obtain temporal SPE and T^2 values. These charts are introduced in Section 6.5.1. However, T^2 and cumulative score plots are used along with the variable contributions in this example to find out the variable(s) responsible for deviation from NO. T^2 is computed for each sampling instant using Eq. 6.101. Scores are calculated for each sampling instance and summed until the end of the batch to reach the final score value. Limits on individual scores are given in Eq. 6.95.

The MPCA model can be utilized to classify a completed batch as 'good' or 'bad'. Besides providing information on the similarity of a newly finished batch with batches in the reference set, MPCA model is also used to assess the progress during a run of a finished batch. Temporal scores evolution plots, SPE and T^2 charts, are generally used along with contribution plots to further investigate a finished batch.

Example. MPCA-based SPM framework is illustrated for a simulated data set of fed-batch penicillin production presented in Section 6.4.1. Two main

Table 6.9. Percent variance captured by MPCA model

PC no.	X-block	
	This PC	Cumulative
1	16.15	16.15
2	10.34	26.49
3	7.89	34.38
4	5.96	40.34

steps of this framework are *model development stage* using a historical reference batch database that defines normal operation and *process monitoring stage* that uses the model developed for monitoring a new batch.

MPCA model development stage: MPCA model is developed from a data set of equalized/synchronized (Figures 6.42 and 6.43), unfolded and scaled 41 good batches. Each batch contains 14 variables 764 measurements, resulting in a three-way array of size $\underline{\mathbf{X}}(41 \times 14 \times 764.)$ After unfolding by preserving the batch direction (I), the unfolded array becomes $\mathbf{X}(41 \times 10696)$. MPCA is performed on the unfolded array $\mathbf{X}$ with four principal components, resulting in scores matrix $\mathbf{T}$ of size (41×4) and loadings matrix $\mathbf{P}$ of size (10696×4). The variability of the $\mathbf{X}$ block explained by MPCA model is summarized in Table 6.9. While 4 PCs explain only 40 percent of the variation in data, the resulting MPCA model is good enough for performing various SPM tasks. Additional PCs can be included to improve model accuracy, paying attention not to include variation due mostly to noise in the model.

MPCA model statistics are summarized in a number of multivariate charts in Figure 6.45. All of the control limits are developed based on the formulations summarized in Section 6.4.2. Score biplots (with 95% and 99% confidence ellipsoids defined in Eq. 6.96) in Figures 6.45(a)-6.45(b), T^2 and Q (sum of squares of residuals) charts in Figures 6.45(c)-6.45(d), respectively, with their 95% and 99% control limits revealing that none of the 41 batches present any unexpected behavior. It can also be concluded that all of the batches are operated similarly and the scatters of the score biplots in Figures 6.45(a)-6.45(b) defines the normal operational region in the reduced space. The percent of the cumulative sum of squares explained by the MPCA model is also shown (Figures 6.45(e) and 6.45(f)) with respect to time and variables. Figure 6.45(e) summarizes the cumulative explained variance by each principal component over the course of batch

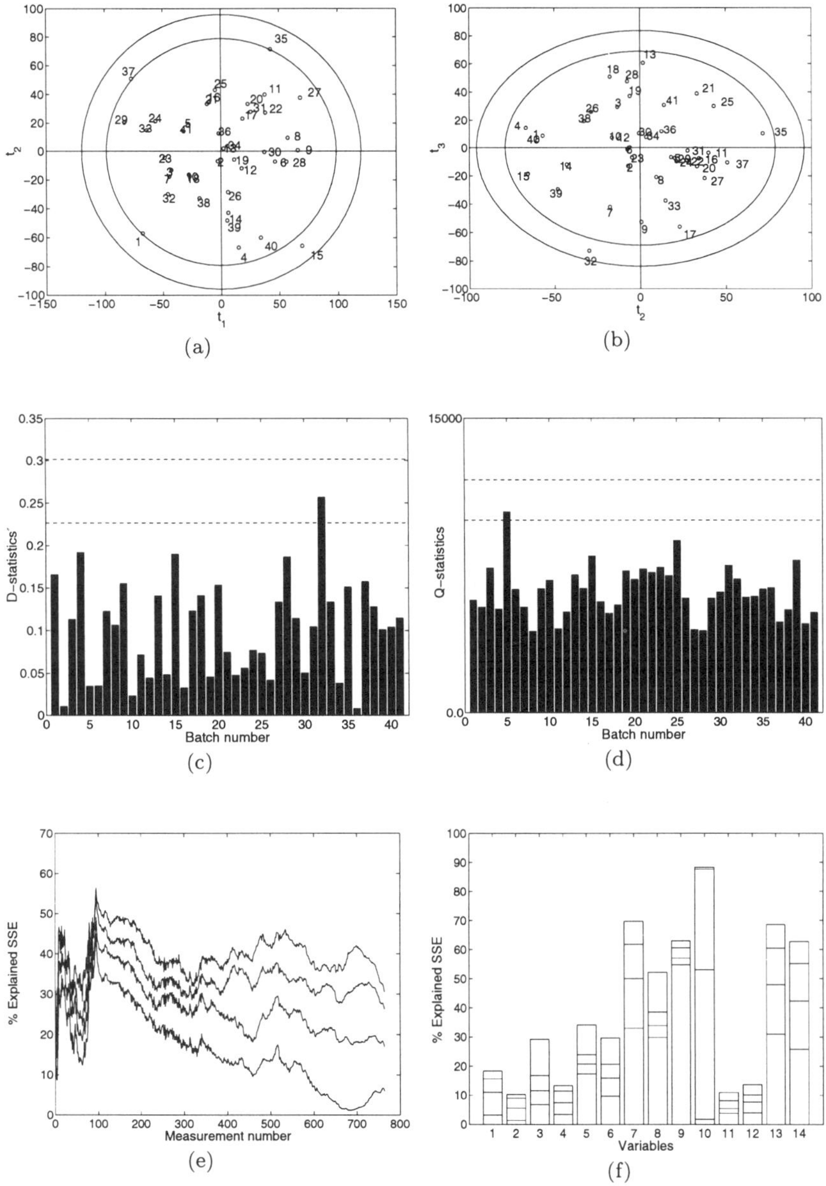

Figure 6.45. MPCA model (with four principal components) statistics.

evolution. The lowest line in both figures represents the percent explained by the first PC, the next line above shows the percent explained by the first two PCs together, and so on. Several operational and physiological phases throughout the fed-batch penicillin fermentation are detected from this plot. Comparing the relative increases in the four curves that indicate the cumulative variation explained, the first PC explains most of the variability in first phase that corresponds to the batch operation (switching from batch to fed-batch at around the measurement 85), while the second PC explains variability in the second phase (fed-batch operation/exponential growth phase). This is a common observation in MPCA because the correlation of the process variables in each phase changes over the progress of a batch. Figure 6.45(f) shows that the dominant variables in the first principal component are 5, 7, 8, 9, 13 and 14. These variables contain physiological change information and their profiles look similar (see Figure 6.31). Variable 10 and others are explained mostly by the second and third components. The first principal component explains most of the batch operation phase and exponential growth phase in fed-batch operation where most of the process dynamics take place (in the associated variables 5, 7, 8, 9, 13 and 14). The second and additional principal components capture variability mostly in the fed-batch operation where 10 (carbon dioxide evolution) is dominant. Figure 6.45(e) indicates a decrease in explained variance during the period of approximately 40th and 60th measurements for all of the 4 PCs that precedes switching to fed-batch operation, because the variability of process variables is low in this period. To increase phase-based explained variability, multiple model approaches are also suggested [130, 291, 605]. An example is given in Section 6.4.5.

Process monitoring stage: The MPCA model developed here is used to monitor finished batches to classify them as 'good' or 'bad' and also investigate past batch evolution, and detect and diagnose abnormalities. A batch scenario including a small downward drift fault is simulated (Section 6.4.1, Figure 6.44 and Table 6.8). New batch data are processed with MPCA model using Eq. 6.114 after proper equalization/synchronization, unfolding and scaling. The same set of multivariate SPM charts are plotted (Figure 6.46). Score biplots in Figures 6.46(a) and 6.46(b) detect that the new batch (batch number 42) is operated differently since its scores fall outside of the NO region defined by MPCA model scores. Both D and Q statistics also indicate an out-of-control batch. Now that the batch is classified as out-of-control, the time of the occurrence of the deviation and the variables that have contributed to increasing the values of the statistics can be determined. The aforementioned temporal T^2 chart based on cumulative scores and individual score plots can be used here. The T^2 value goes out-of-control as shown in Figure 6.47(a), the same out-of-

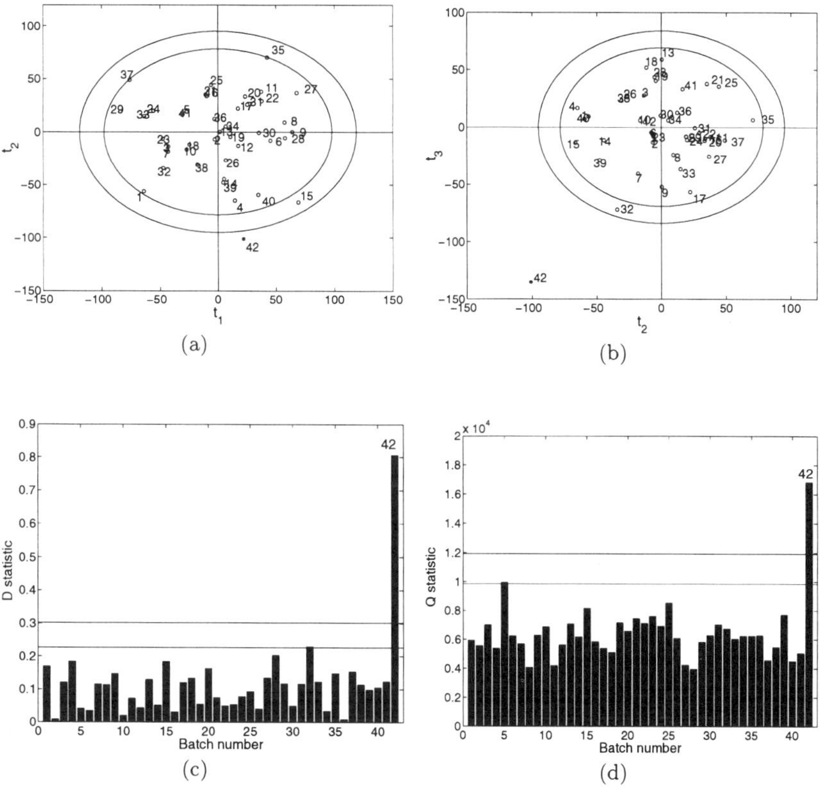

Figure 6.46. End-of-batch monitoring results of a faulty batch.

control situation is also observed with score plots (Figures 6.47(b)-6.47(c)). The first out-of-control signal is given by the PC_3 chart around the 445th measurement. When variable contributions are calculated, the responsible variables are identified. Variables 3, 5 and 8 have the highest contributions which make sense since the fault was introduced into variable 3 (glucose feed rate), which affects variables 5 and 8, glucose and penicillin concentrations, respectively.

6.4.4 Multiway PLS–based SPM for Postmortem Analysis

MPLS [661] is an extension of PLS that is performed using both process data (**X**) and the product quality data (**Y**) to predict final product quality

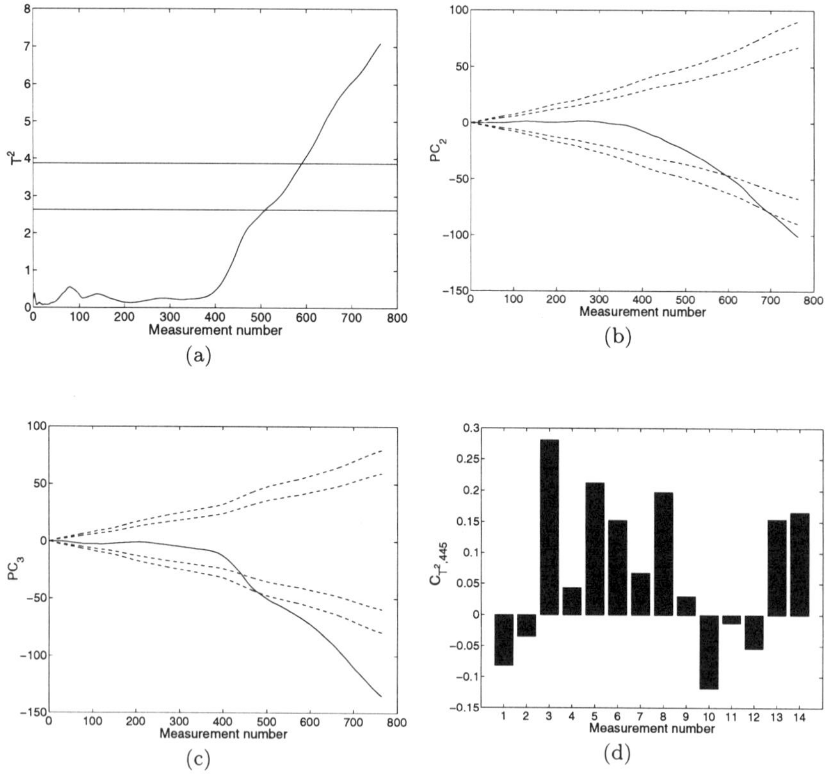

Figure 6.47. End-of-batch fault detection and diagnosis for a faulty batch.

during and/or at the end of the batch [298, 434, 663]. When a batch is finished, a block of recorded process variables $\mathbf{X}_{\text{new}}$ $(K \times J)$ and a vector of quality measurements $\mathbf{y}_{\text{new}}$ $(1 \times M)$ that are usually measured with a delay due to quality analysis, are obtained. $\mathbf{X}_{\text{new}}$ $(K \times J)$ is unfolded to $\mathbf{x}_{\text{new}}$ $(1 \times KJ)$ and both $\mathbf{x}_{\text{new}}$ and $\mathbf{y}_{\text{new}}$ are scaled similarly as the reference batch set scaling factors. Then, they are processed with MPLS model loadings and weights that contain structural information on the behavior of NOC set as

$$\hat{\mathbf{t}}_{\text{new}} = \mathbf{x}_{\text{new}} \mathbf{W} \left(\mathbf{P}^T \mathbf{W} \right)^{-1}, \qquad \mathbf{e}_{\text{new}} = \mathbf{x}_{\text{new}} - \hat{\mathbf{t}}_{\text{new}} \mathbf{P}^T \tag{6.115}$$

$$\hat{\mathbf{y}}_{\text{new}} = \hat{\mathbf{t}}_{\text{new}} \mathbf{Q}^T, \qquad \mathbf{f}_{\text{new}} = \mathbf{y}_{\text{new}} - \hat{\mathbf{y}}_{\text{new}} \tag{6.116}$$

where $\hat{\mathbf{t}}_{\text{new}}$ $(1 \times A)$ denotes the predicted $\mathbf{t}$-scores, $\hat{\mathbf{y}}_{\text{new}}$ $(1 \times M)$ the predicted quality variables, and $\mathbf{e}$ and $\mathbf{f}$ the residuals.

Example. MPLS-based SPM framework is also illustrated using simulated fed-batch penicillin production data set presented in Section 6.4.1. Similar to MPCA framework in the previous section (Section 6.4.3), the MPLS framework has two main steps: *model development stage* out of a historical reference batch data base that defines normal operation and *process monitoring and quality prediction stage* that uses model developed. The latter stage includes prediction of the product quality, which is the main difference between MPCA and MPLS based SPM frameworks. Note that in this MPLS framework, quality prediction is made at the end of batch while waiting to receive quality analysis laboratory results. It is also possible to implement MPLS on-line while predicting the final product quality as batch progresses. This version is discussed in detail in Section 6.5.1.

MPLS model development stage: MPLS model is developed from the data set of equalized/synchronized (Figures 6.42 and 6.43), unfolded and scaled 38 good batches (each containing 14 variables 764 measurements resulting in a three-way array of size $\underline{\mathbf{X}}(38 \times 14 \times 764)$. After unfolding by preserving the batch direction (I), the unfolded array becomes $\mathbf{X}(38 \times 10696)$). Three batches in the original 41 batches of data are excluded from the reference set due to their high variation. In addition to $\mathbf{X}$ block, a $\mathbf{Y}$ (38×5) block comprised of 5 quality variables measured at the end of each batch also exists (Table 6.7 and Figure 6.41). MPLS is performed between the unfolded and scaled $\mathbf{X}$ and $\mathbf{Y}$ with four latent variables resulting in scores $\mathbf{T}$ (38×4), $\mathbf{U}$ (38×4), weights $\mathbf{W}$ (10696×4) and $\mathbf{Q}$ (5×38) and loadings matrices $\mathbf{P}$ (10696×4). Explained variability on both $\mathbf{X}$ and $\mathbf{Y}$ blocks by MPLS model is summarized in Table 6.10. 38.39 % of $\mathbf{X}$ explains 97.44 % of $\mathbf{Y}$ with 4 latent variable MPLS model. Cumulative percentage of sum of squares explained by 4 latent variables on each y in $\mathbf{Y}$ block is also tabulated in Table 6.11. MPLS model statistics are summarized in a number of multivariate charts in Figure 6.49. All control limits are developed based on the formulations summarized in Section 6.4.2.

NO region is defined by the ellipsoids in Figures 6.49(a) and 6.49(b) by the MPLS model. Naturally all of the reference batches fall into these regions. Note that Figure 6.49(a) defines process measurements while Figure 6.49(b) defining final quality variables. All of the batches also are inside the control limits in sum of squared residuals as shown in Figures 6.49(e) and 6.49(f) in both process and quality spaces. Hence, MPLS model can be used to discriminate between the acceptable and 'poor' batches at the end-of-the batch. It is evident from the biplots of inner relations of the MPLS model (Figures 6.49(c) and 6.49(d)) that there is a correlation between process and product variables and this relation is linear because most of the

Table 6.10. Percent variance captured by MPLS model

LV no.	X-block		Y-block	
	This LV	Cumulative	This LV	Cumulative
1	16.11	16.11	57.58	57.58
2	9.55	25.66	26.26	83.84
3	5.84	31.50	11.78	95.62
4	6.89	38.39	1.82	97.44

nonlinearity is removed by subtracting the mean trajectories from reference set trajectories prior to analysis. Another statistic that is calculated from MPLS model is the *Variable Influence on Projection (VIP)* to investigate the effects of important process variables (predictors) on quality variables (predictees). As the formulation and interpretation details provided in Section 6.4.2, process variables that have contributions larger than 1 can be considered to exert more influence on quality as far as MPLS projection is concerned. Figure 6.48 summarizes the mean values of VIP set, i.e., over the entire course of batch run and according to these plots variables 5, 7, 8, 9, 13 and 14 are found to be important, which is meaningful since these variables carry physiological information and hence are expected to be effective on the quality.

Process monitoring and quality prediction stage: Developed MPLS model is used to monitor finished batches to classify them as 'good' or 'poor' based on how well they follow similar trajectories to achieve 'good' quality product. The same fault scenario with a small downward drift on glucose feed (see Figure 6.44 and Table 6.8) in MPCA based monitoring is used to illustrate end-of-batch MPLS framework. MPLS model is also used to predict product quality as soon as the batch finishes providing information ahead

Table 6.11. Cumulative percent variance captured by MPLS model on each quality variable

LV no.	**X**	**Y**	y_1	y_2	y_3	y_4	y_5
1	16.11	57.58	48.08	51.24	45.42	91.89	51.24
2	25.66	83.84	85.62	87.23	62.18	96.89	87.23
3	31.50	95.62	95.64	95.98	92.38	98.08	95.98
4	38.39	97.44	97.41	97.23	97.20	98.10	97.23

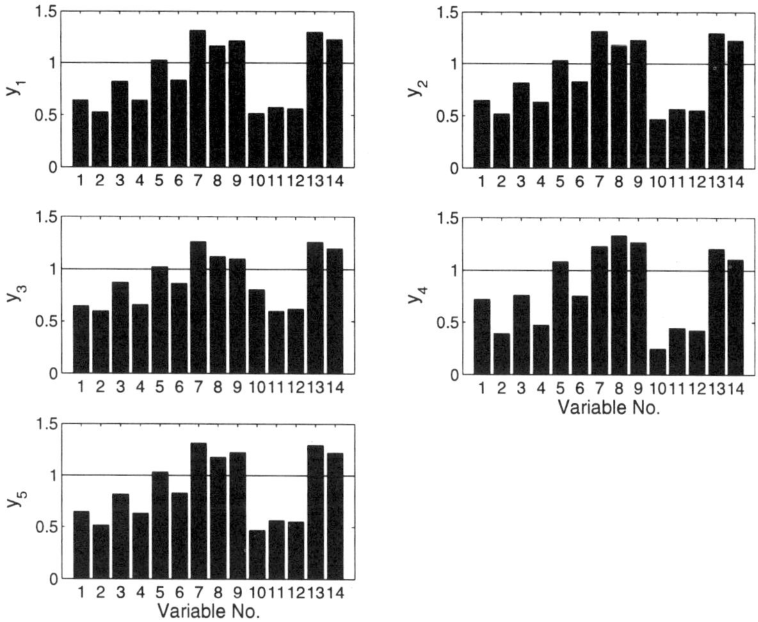

Figure 6.48. MPLS model VIP statistics. Important variables on projection are variables 5 (glucose concentration), 7 (biomass concentration), 8 (penicillin concentration), 9 (culture volume), 13 (generated heat) and 14 (cooling water flow rate).

of time for initial fast assessment before the real **Y** is available. New batch data are processed with MPLS model at the end of the batch as shown in Eqs. 6.115 and 6.116 after proper equalization/synchronization, unfolding and scaling, resulting in multivariate SPM charts (Figures 6.50 and 6.51) for detection and diagnosis. Figure 6.50 summarizes several statistics to compare new batch with the reference batches. Figures 6.50(a) and 6.50(b) indicate that there is a dissimilarity between the new batch and the NO batches in both process and quality spaces. Scores of the new batch in both spaces fall outside of the in-control regions defining NO in figures 6.50(c) and 6.50(d). These charts suggest that an unusual event occurred in new batch and should be investigated further. To find out when the process wnet out-of-control and which variables were responsible SPE_X chart and a variety of contribution plots are used (Figure 6.51). SPE_X chart of process space in Figure 6.51(a) reveals a deviation from NO and process goes out-of-control around 570th observation. The overall variable contributions to SPE_X in Figure 6.51(b) over the course of batch run indicate that variable 9

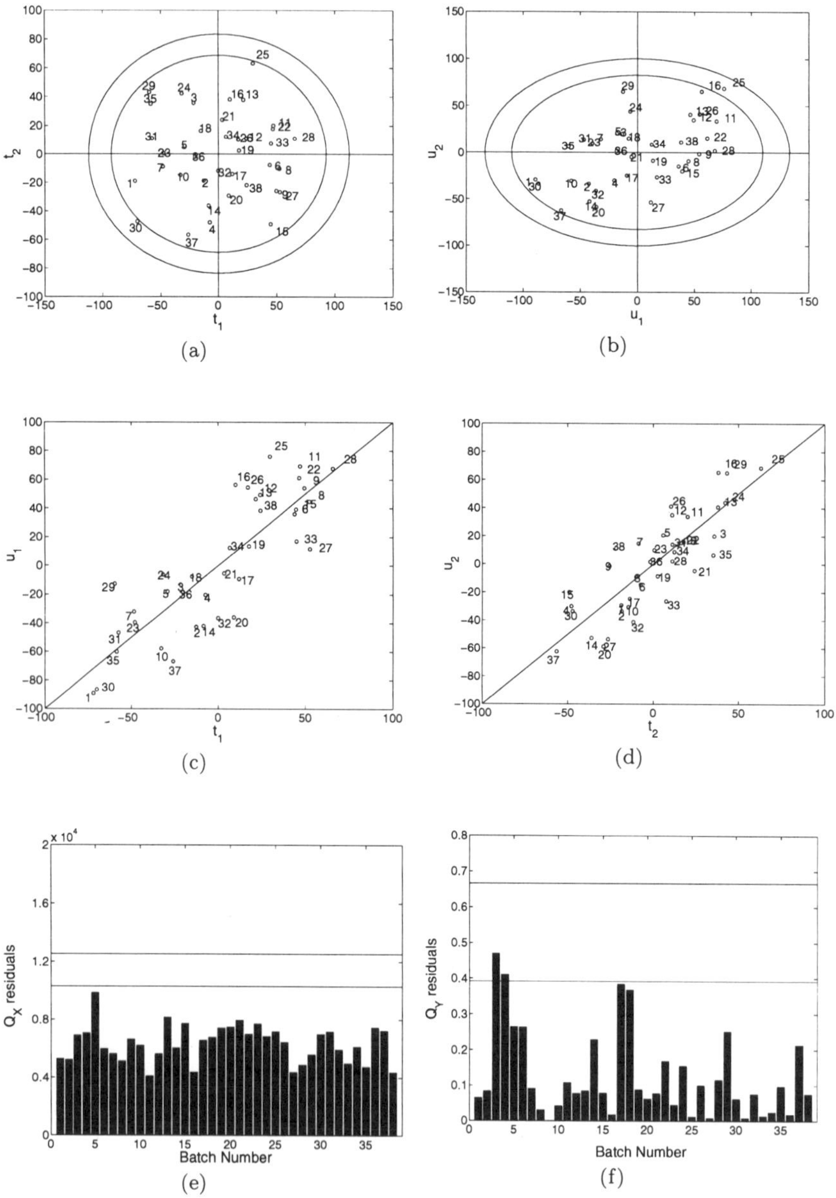

Figure 6.49. MPLS model (with four latent variables) statistics.

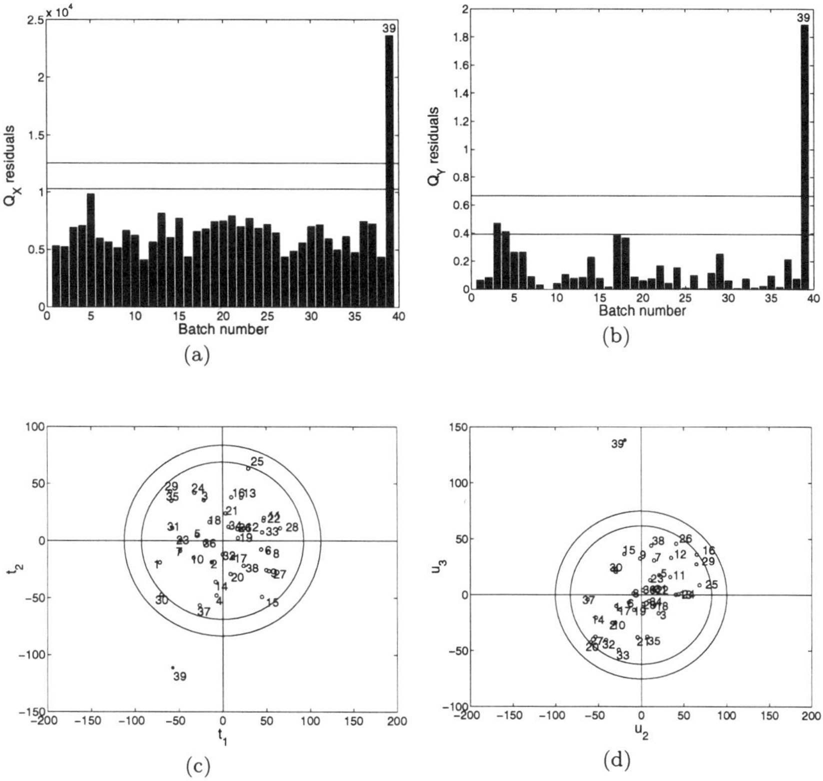

Figure 6.50. MPLS-based end-of-batch monitoring results.

(culture volume) has changed unexpectedly, hence the deviation. Variable contributions to SPE_X for a specified time interval can also be calculated to zoom the interval when out-of-control situation is observed. Figure 6.51(d) shows average variable contributions to SPE_X between 570th and 690th measurements. Variables 3, 6 and 9 are found having the highest contributions to deviation for that interval of out-of-control. A further analysis can be performed by calculating contributions to process variable weights. Since weights (**W**) bear information about the relationship between process and product variables, variable contributions to weights will reveal variable(s) that are responsible to out-of-control situation with respect to product quality. These contributions can be calculated similar to SPE_X contributions. Figure 6.51(c) shows overall absolute variable contributions to the weights over the course of the batch run. Variables 3, 6, 7, 10, 13

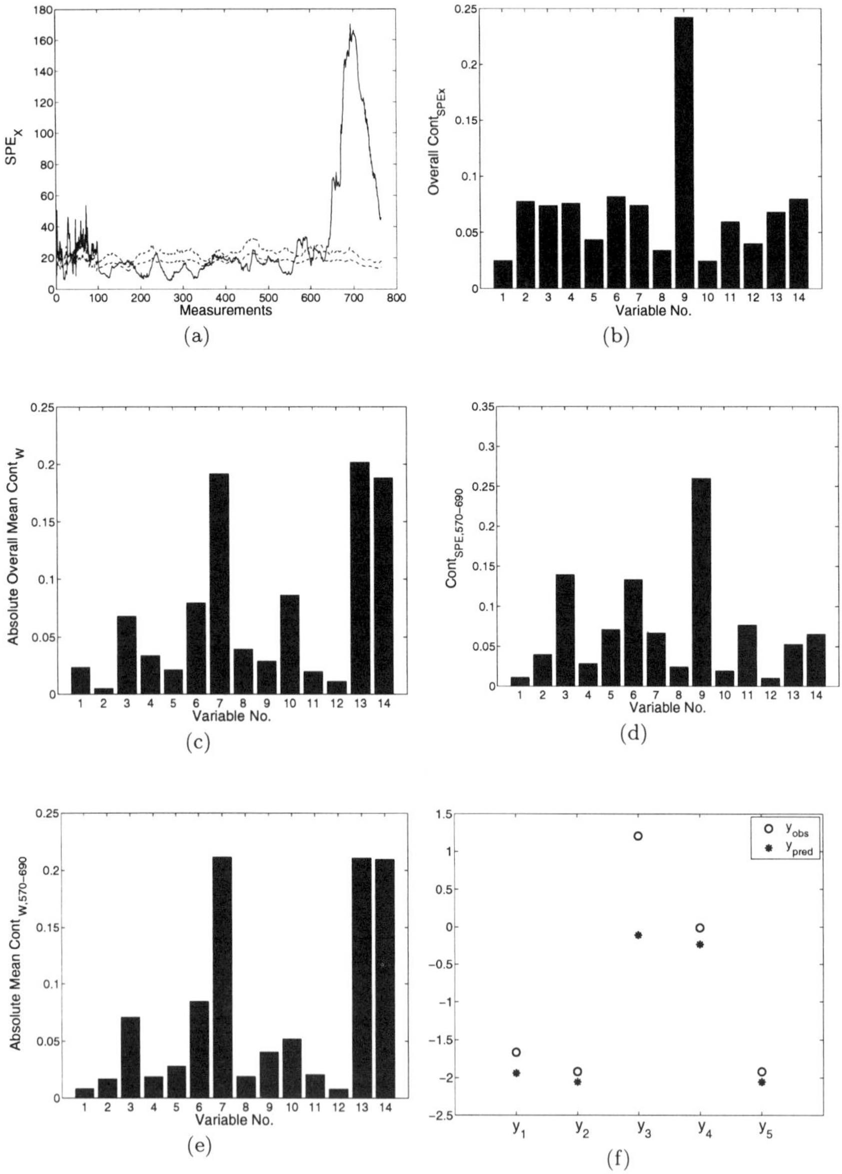

Figure 6.51. MPLS-based end-of-batch monitoring results.

and 14 have the high contributions compared to other variables. These contributions are also calculated between 570th and 690th measurements where the out-of-control occurs. Variables 3, 6, 7, 13 and 14 are found to be significant in that case (Figure 6.51(e)). Since the original disturbance was introduced into the variable 3 (glucose feed rate) as a small downward drift, its effect on the other structurally important process variables such as dissolved oxygen concentration (variable 6) and biomass concentration (variable 7) becomes more apparent as the process progresses in the presence of that disturbance. Obviously, culture volume is expected to be directly affected by this disturbance as it is found with SPE_X contribution plots. The weight contributions highlight the effect of this change on the process variables that are effective in the quality space. Variables 13 and 14 (heat generated and cooling water flow rate, respectively) being highly correlated with biomass concentration (variable7) also show high contribution. Since MPLS model can be used to predict end-of-batch quality as well, model predictions are compared with actual measurements in Figure 6.51(f). Quality variable 3 is predicted somewhat poorly. This is due to model order, and if the focus is on the prediction, this can be improved by increasing the number of latent variables retained in the MPLS model. End-of-batch quality can be predicted from the start of a new batch, this case is illustrated in Section 6.5.1.

6.4.5 Multiway Multiblock Methods

Many chemical processes consist of several distinct processing units. Data from various processing 'stages' carried in processing units and 'phases' for operational or phenomenological regions in single units provide the information about the progress of the batch. As the number of units and phases increases, the complexity of the monitoring problem also increases. The techniques presented in Section 4.5.3 are useful for monitoring these type of processes with some modifications in both data pretreatment and monitoring techniques.

Example. A pharmaceutical granule production process by wet granulation technique following a fluidized-bed drying operation was chosen as a test case in this study. Process variables were broken up into blocks that correspond to specific processing stages. The choice of blocks depends on engineering judgment and the objectives of the study. In this study, blocks are related to particular processing units. Furthermore, because of the different operating regimes occurring in each unit, it is convenient to split the data from a stage into phases (Figure 6.52). This way, the predictive and diagnostic capabilities of the multivariate statistical models can be improved to provide more accurate inferences about the whole process. These

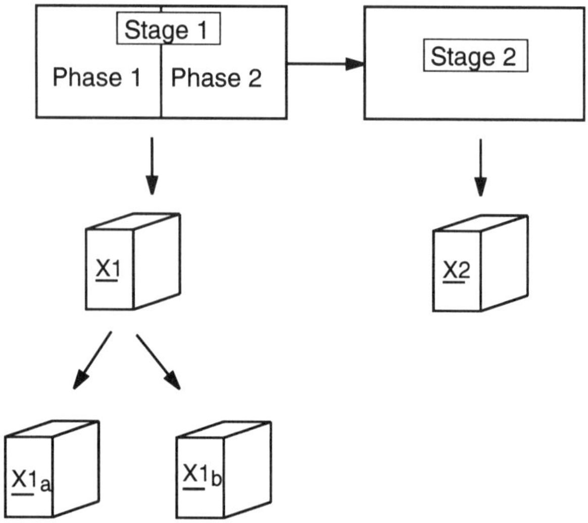

Figure 6.52. Data arrangement and blocking of variables for process stages and phases.

models can be used to eliminate major disturbances in future batches, and therefore to help adjust control limits for more consistent production, which is crucial for pharmaceutical production processes [604].

Stage 1 is the wet granulation of the fine powder mix of active ingredient(s) and other pharmaceutical excipients. The objective of the granulation is to increase the particle size by agglomerating this fine powder by adding a binder solution under continuous mixing. Particle size increase promotes higher bioavailability of the drug. At this stage, the amount of binder used and its addition rate are effective on the particle size increase. The amount of binder solution and its addition rate are predefined based on experimental design studies. We have assumed a fixed total amount of binder solution in the simulation studies. Binder addition rate, impeller speed, and power consumption are taken as measured process variables at this stage. Stage 1 is operated in two phases: *phase 1*, dry mixing for a fixed time interval, and *phase 2*, binder addition while mixing (Fig. 6.53). Since there are small fluctuations in binder flow rate at each batch, the final time of the second phase is variable, producing unequal batch length for stage 1. These differences should be eliminated prior to multivariate statistical modeling. To equalize data lengths in phase

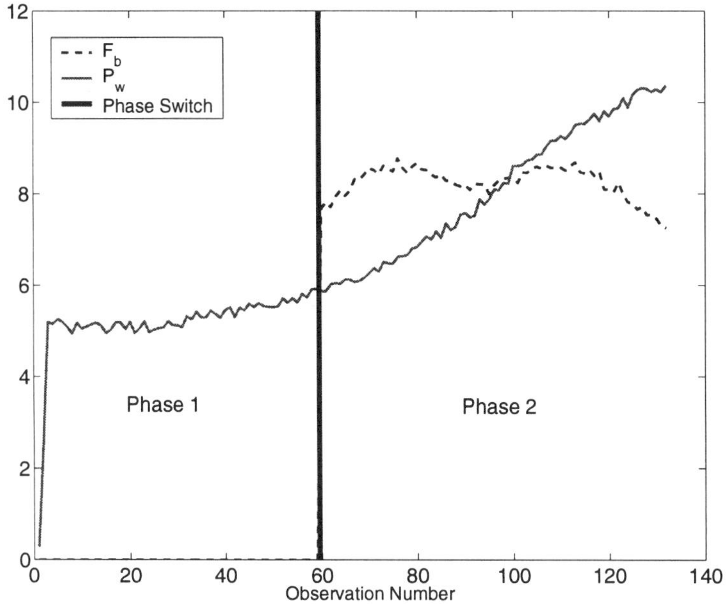

Figure 6.53. Phase structure of the first stage. F_b: Binder addition rate, P_w: Agitator power consumption.

2, we have used percent binder solution added into the granulator as an indicator variable and sampled each variable at every 1% increase in this variable, resulting in 100 observations for each batch at this phase. These equalized data are appended to the fixed data of the first phase, resulting in a total number of 160 observations for stage 1 (Figs. 6.54a and 6.54b).

Stage 2 is the drying stage where a fluid bed dryer is used. The wet granulates are dried using hot airflow to decrease their moisture content. The increase in product temperature is measured as an indicator of drying. Airflow rate, inflow air temperature, drying rate, and product moisture are also measured. Product temperature is found to be appropriate as an indicator variable for this stage, and measurements on each variable are interpolated on every 0.5 ^oC increase in product temperature, resulting in 63 observations (Figs. 6.54c and 6.54d).

MPCA model development stage for data blocks: There are two operational phases at the first stage of the granule production process. The first phase contains dry mixing for a fixed time, and the second phase involves

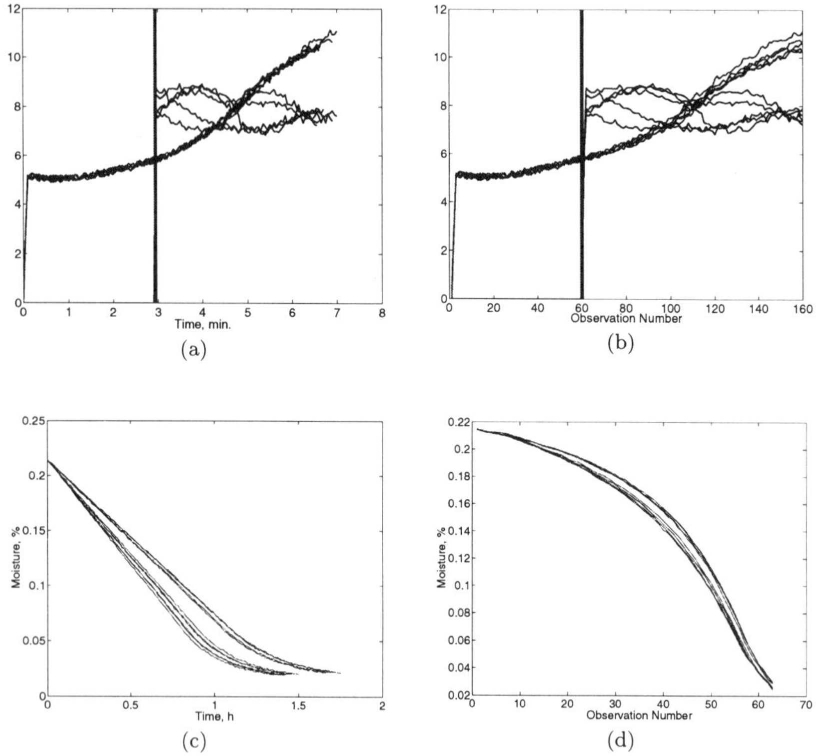

Figure 6.54. Equalized process variable trajectories of both stages using indicator variable technique.

wet massing. Since the total amount of binder to be added is fixed, the exact completion of the second phase is reached when all of the binder solution is consumed. Data from the first phase are collected based on the fixed operation time, resulting in the $\mathbf{X}_{ij1k1}$ unfolded matrix. Data arrangement in the second phase is based on a fixed indicator variable (percent binder addition), resulting in $\mathbf{X}_{ij2k2}$. The index pairs $j1$, $k1$ and $j2$, $k2$ denote variables and observation numbers of each phase, respectively. The overall performance can also be investigated by appending these matrices to form an augmented matrix

$$X_{ijk} = [X_{ij1k1} \quad X_{ij2k2}],$$

where $j = j1 + j2$ and $k = k1 + k2$.

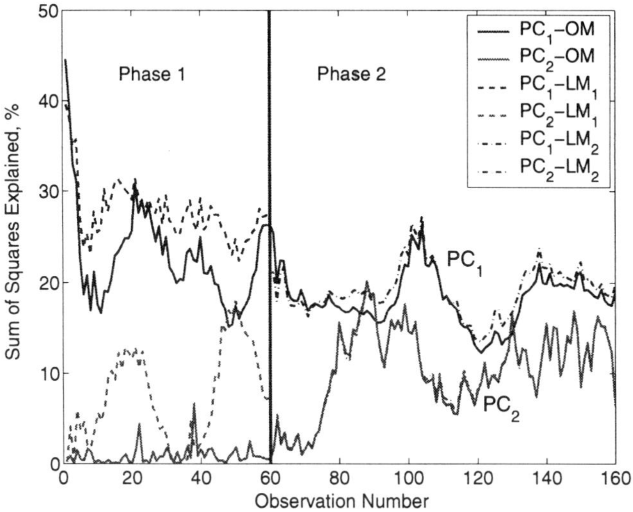

Figure 6.55. Comparison of local and overall models performances on explained variance.

Two local models for phase 1 and phase 2, and one overall model, are developed using MPCA technique and compared. Variance plots can be useful for comparing different models and investigating the changing variance structure in overall data. The variance explained is higher (45.67% more) for the local model of phase 1 than the overall model, whereas variances explained are much closer but still higher (4.22% more) for the local model of phase 2 (Fig. 6.55). This is expected, since the same event occurs in the second phase (Fig. 6.53). Local models explain more information (17.98% more for the whole process) based on computations of sum of squared errors and data lengths in each phase.

Process monitoring stage: A new batch with a small drift in impeller speed introduced at 0.5 *min* (10th observation in phase 1 of stage 1) was monitored after its completion. Note that the fault starts early in the first phase. Both SPE and T^2 plots for local and overall models in Figure 6.56 indicated that there is a deviation from the NO. Since overall model performance is not high in the first phase, the early departure is caught later with monitoring based on the overall model than with the local model (Figure 6.56b), and many false alarms are observed in the SPE plot (Figure 6.56a). The advantages of using the local model for phase 1 are:

1. The false alarms observed with the overall model for phase 1 in the SPE plot are eliminated.

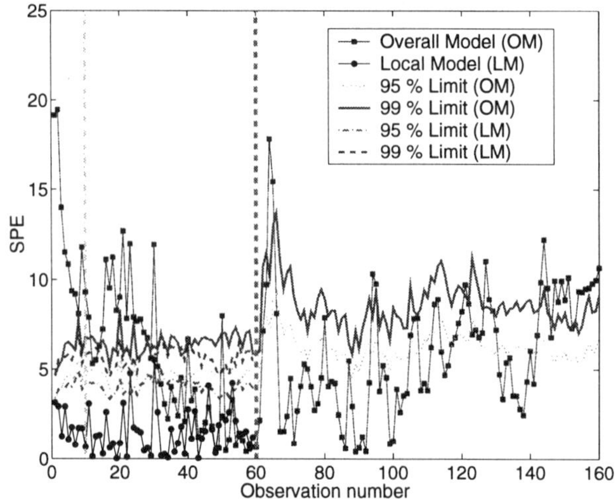

(a)

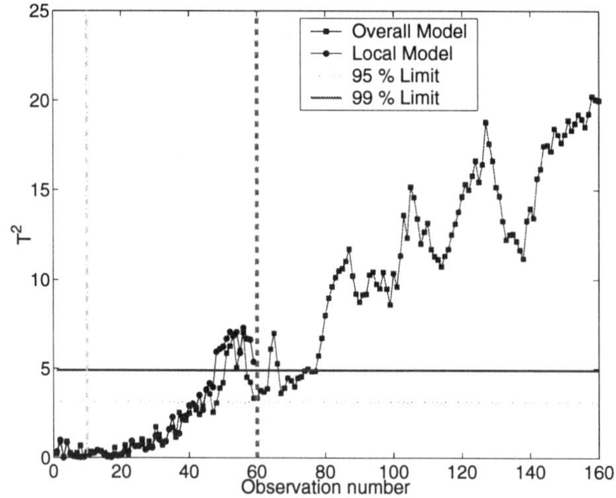

(b)

Figure 6.56. Effects of overall and local modeling on process performance monitoring.

2. Departures from the NO are detected earlier (by three observations) than with the overall model, and more consecutive out-of-control

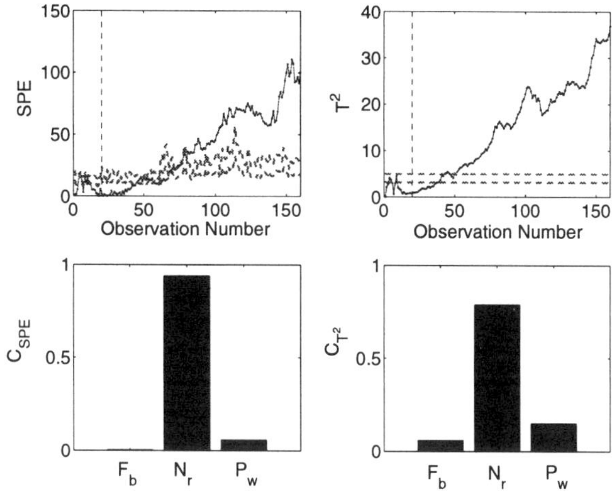

(a) Stage 1. F_b: Binder addition rate, N_r: Impeller speed, P_w:Power consumption.

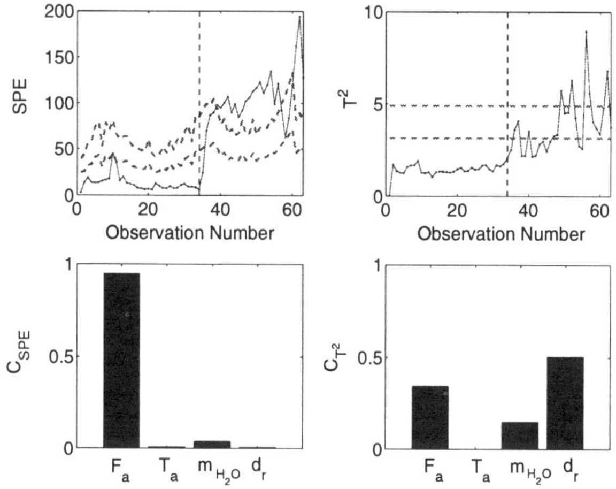

(b) Stage 2. F_a: Inflow air rate, T_a: Inflow air temperature, m_{H_2O}: Product moisture, %, d_r: Drying rate.

Figure 6.57. AHPCA based on-line monitoring of process stages under faulty operation.

points (run length of 12 observations) are observed in the T^2 plot with the local model for phase 1 than the number the overall model produces (six observations),

The case illustrates that local models provide the capability to detect earlier small trends and departures from NO that will be propagated to the next phase and eventually cause significant deviation, thus allowing process operators to improve their operations.

Online monitoring was performed in each processing stage based on adaptive hierarchical PCA (AHPCA). For this multistage process, AHPCA is limited to stages due to interstage discontinuity. To overcome this problem, different AHPCA models are developed for each stage. Different weightings can also be applied to better account for changing phase structure.

To illustrate online monitoring, a case is generated where a small drift in impeller speed is introduced (dashed line) in the first stage and a step increase (dashed line) in inflow air rate in the second stage. Each AHPCA model successfully detected and diagnosed the problem online in each stage for the overall process (Fig. 6.57).

6.4.6 Multiscale SPM Techniques Based on Wavelets

Multiscale MPCA (MSMPCA) is a combination of MPCA with wavelet decomposition. Traditional MPCA is applied at a single time scale by representing data with the same time-frequency localization at all locations. MPCA may not be suitable for processes which include measurements with different sampling rates and measurements whose power spectrum changes with time.

Another important factor is that an MPCA model will still include imbedded random noise although the random noise is reduced by selecting only the significant components. This random noise may cause failure in detecting small deviations from normal operating conditions of process. In order to improve the performance of MPCA, the random noise should be extracted from the signal in an enhanced manner. A possible solution to this shortcoming of MPCA is to apply wavelet transformation to the signal before developing MPCA model. The role of wavelet decomposition here is similar to that of filtering the signal to separate the errors. Examples of this pretreatment by filtering data such as exponential smoothing and mean filtering can be found in literature [586].

The algorithm for MSMPCA is

Model development:

1. Use a historical data set of the past batches

2. Choose a wavelet function

3. Select the number of scales

4. Apply 1-D wavelet decomposition to each variable trajectory in historical data which are the unfolded, mean-centered and scaled version of a three-way array

5. Develop MPCA models for the coefficients at each scale for the past batches

6. Reconstruct the models in a recursive manner at each scale to form the model for all scales together

Monitoring:

7. Apply the same 1-D wavelet decomposition on each variable trajectory of the batch to be monitored

8. Identify the scales that violate the detection limits as important scales

9. Reconstruct the new data by including only the important scales

10. Check the state of the process by comparing the reconstructed data with detection limits

The data set representing normal operation is decomposed to wavelet coefficients for each variable trajectory. MPCA models are developed at each scale. The overall MPCA model for all scales is obtained by reconstructing the decomposed reference data. Wavelet decomposition is applied to new batch data using the same wavelet function. For each scale, T^2 and SPE values of the new batch are compared with control limits computed based on reference data. The scales that violate the detection limits are considered as important scales for describing the critical events in current data. Inverse wavelet transform is applied recursively to the important scales to reconstruct the signal. The new batch is considered to be out-of-control if T^2 and/or SPE values of the reconstructed signal violate the control limits.

Selecting the number of scales of the wavelet decomposition is important. The optimum scale number gives maximum separation between the stochastic and deterministic components of the signal. If the scale number is chosen too small, the signal will still have noise. On the other hand, if the scale number is too large, the coarser scales will have too few a data to form an accurate model. The number of scales should be determined according to

Table 6.12. Coefficient data sets from wavelet decomposition of $\mathbf{X}$

Wavelet coefficients data sets	Data set size
Approximation coefficients at scale $m = 3$, $\mathbf{A_3}$	(40×1337)
Detail coefficients at scale $m = 3$, $\mathbf{D_3}$	(40×1337)
Detail coefficients at scale $m = 2$, $\mathbf{D_2}$	(40×2674)
Detail coefficients at scale $m = 1$, $\mathbf{D_1}$	(40×5348)

the dimension of data used. For selection of scales the following formula can be used:

$$\ell = \log_2 n - 5 \tag{6.118}$$

where ℓ is the number of scales and n is the number of observations.

Example. MSMPCA based SPM framework is illustrated for a simulated data set of fed-batch penicillin production presented in Section 6.4.1. Two main steps of this framework are *model development stage* using a historical reference batch database that defines normal operation and *process monitoring stage* that uses the model developed for monitoring of a new batch.

MSMPCA model development stage: A reference data set of equalized/synchronized (Figures 6.42 and 6.43), unfolded and scaled 40 good batches (each batch contains 14 variables 764 measurements resulting in a three-way array of size $\underline{\mathbf{X}}(40 \times 14 \times 764)$ is used. After unfolding by preserving the batch direction (I), the unfolded array becomes $\mathbf{X}(40 \times 10696)$). Each variable trajectory in $\mathbf{X}$ is decomposed into its approximation and detail coefficients in three scales using Daubechies 1 wavelet family that is chosen arbitrarily. Although Eq. 6.118 suggests four scales, the decomposition level of three is found sufficient in this case. Since the original signals can be reconstructed from their approximation coefficients at coarsest level and detail coefficients at each level, those coefficients are stored for MPCA model development (Table 6.12). Then, MPCA models with five PCs are developed at each scale and MV control limits are calculated.

Process monitoring stage: MPCA models developed at each scale are used to monitor a new batch. A faulty batch with a small step decrease on glucose feed between measurements 160 and 200 is mean-centered and scaled similarly to the reference set and 1-D wavelet decomposition is performed on variable trajectories using Daubechies 1 wavelets. This three-level decomposition is illustrated for penicillin concentration profile (variable 8 in the data set, $x_8 = a_0$) in Figure 6.58. Note that, the effect of the step change on this variable becomes more visible as one goes to coarser

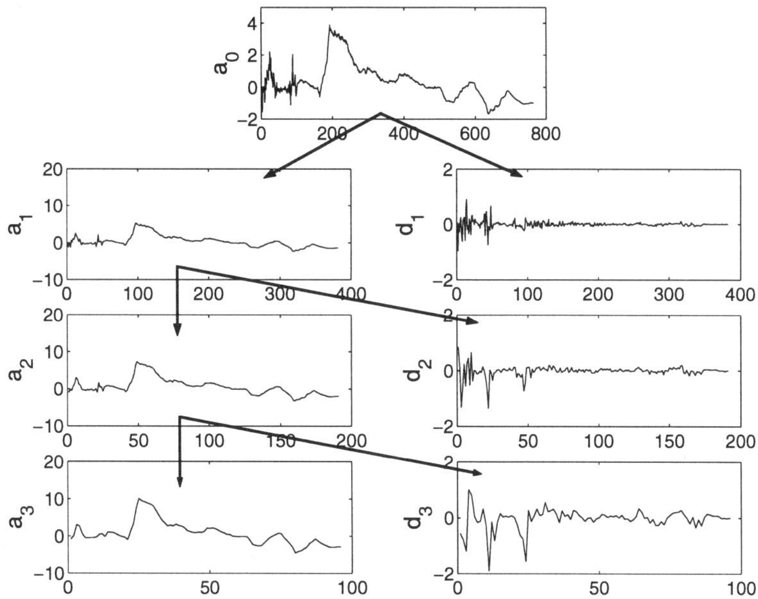

Figure 6.58. Decomposition of penicillin concentration profile of a faulty batch.

scales. Starting and end points of the fault are more apparent in the detail coefficient (d_3) of the third scale since detail coefficients are sensitive only to changes and this sensitivity increases in coarser scales. SPE on each scale is calculated based on MPCA models on scales. An augmented version of SPE values at all scales is presented in Figure 6.59. The 99% control limit is violated at scale $m = 3$ on both its approximation and detail coefficients. There are also some violation at scale two but no violation is detected at the first scale hence this scale is eliminated. Fault detection performances of conventional MPCA and MSMPCA are also compared in Figure 6.60. The lower portion of this figure represents SPE of the approximation coefficients. The first out-of-control signal is detected at point 162 and returning to NO is detected at point 208 at that scale on SPE whereas conventional MPCA detects first out-of-control signal at 165th measurement and returning point to NO at 213th measurement. In addition, MSMPCA-based SPE contains no false alarms but conventional MPCA has 16 false alarms after the process returns to NO. The advantage of MSMPCA stems from combined used of PCA and wavelet decomposition. The relationship between the variables is decorrelated by MPCA and the relationship between the stochastic measurements is decorrelated by the wavelet decomposition. MV

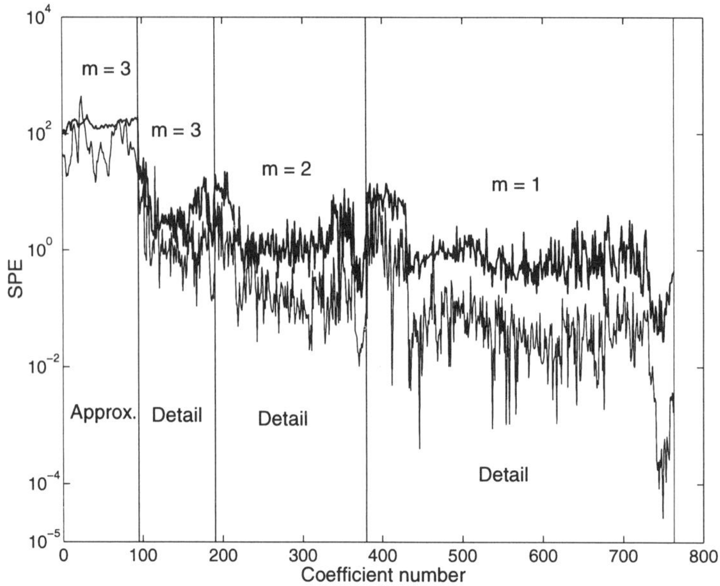

Figure 6.59. SPE on different scales of the decomposed faulty batch data. Darker line represents 99% control limit.

charts on important scales can be conveniently used on both detection and diagnosis of abnormal operation, particularly, in the case of small shifts in variable set points. Figure 6.61 represents detection and diagnosis of abnormal operation. Responsible variable(s) are detected by SPE charts and diagnosed correctly by contribution plots (averaged contributions during the period when the process is out-of-control) for this fault case as glucose feed rate (variable 3), glucose (variable 5) and penicillin concentrations (variable 8) in the fermenter. □

Methodology of on-line MSMPCA: Wavelet decomposition by nature is not suitable for on-line monitoring because future measured data is needed to calculate the current wavelet coefficient which introduces a time delay in the computation. This delay can be prevented by making the wavelet filter causal by implementing special wavelet filters named boundary corrected filters at the edges. Another source for time delay in wavelet decomposition is the dyadic downsampling. The signal is decomposed into wavelet coefficients only if it is of dyadic length. For example if the signal has three data points, the decomposition is not executed until the fourth data point is added to the signal. In order to eliminate these disadvantages, an algorithm that includes decomposition of data in a moving window has been

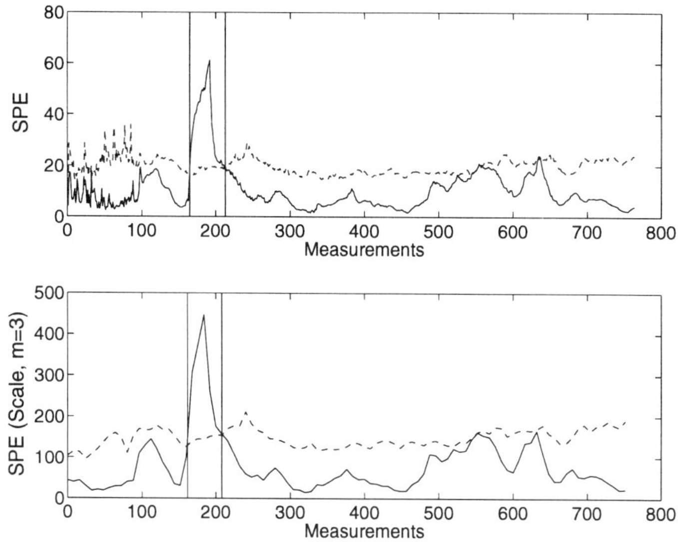

Figure 6.60. Comparison of fault detection performances. Original data (upper figure) and approximation coefficients of thirds scale (lower figure). Dashed line represents 99% control limit in both figures.

proposed [437]. This algorithm can be summarized as:

1. Decompose data in a window of dyadic length
2. Reconstruct the signal after applying thresholding to the wavelet coefficients
3. Retain only the last point of the reconstructed signal
4. Shift the window in time when new data are available and keep the window length constant

Initial window length selection is critical. If the window length is chosen too small, there may not be enough data points to decompose in coarser scales. On the other hand, if the initial window length is too large, by the time process data reaches the chosen window length, the process might have already gone out of control. The window length is kept constant to reduce the computational burden. Another difference compared to off-line algorithm is the adjustment of detection limits for each scale as for the relation:

$$C_\ell = 100 - \frac{1}{\ell + 1}(100 - C), \tag{6.119}$$

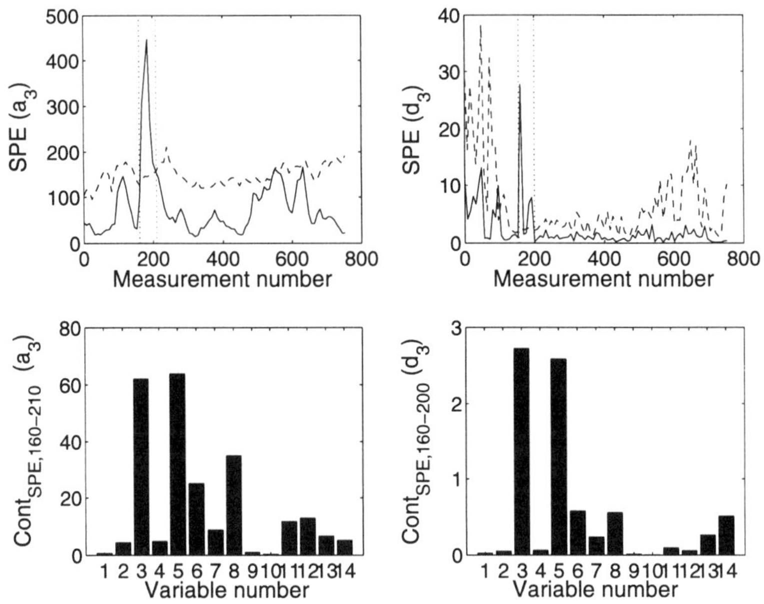

Figure 6.61. Fault detection and diagnosis by MSMPCA. Dashed line represents 99% control limit on SPE charts.

where C is the desired overall limit, C_ℓ is the adjusted confidence limit at each scale in percent, and ℓ is the number of scales of decomposition [38].

The constraints of dyadic downsampling can be eliminated by using a moving window, implementing a computational strategy similar to moving average. The increase in computational burden is a disadvantage of this approach.

6.5 On-line Monitoring of Batch/Fed-Batch Fermentation Processes

Real-time SPM during the progress of the batch can be as simple as monitoring the trajectory of each process variable and comparing it against an ideal reference trajectory. The premise for this approach is that if all variables behave as expected, the product properties will be as desired. A few control loops can be used to regulate some critical process variables. There are several problems with this approach:

1. Slight changes in many variables may seem too small for each variable, but their collective effect may be significant

2. Variations in impurity levels or other initial conditions may affect the variable trajectories, but these deviations from the reference trajectories may not cause significant product quality degradation

3. The duration of each batch may be different, causing difficulties in comparing the trajectories of the current batch to reference trajectories.

The remedies proposed fall into four groups:

1. Use the MSPM tools with variable trajectories that are combinations of real data (up to the present time in the batch) and estimates of the future portion of the trajectories to the end of the batch

2. Use hierarchical PCA that relies only on trajectory information from the beginning of the batch to the current time

3. Use MPCA or MPLS that is performed on an unfolded three-way batch data array by preserving variable direction

4. Use estimators for predicting the final product quality and base batch monitoring on this estimate.

These four approaches are discussed in the following Sections.

6.5.1 MSPM Using Estimates of Trajectories

The problem that is encountered when applying MPCA and MPLS techniques for on-line statistical process and product quality monitoring is that the $\mathbf{x}_{\text{new}}$ vector in Eqs. 6.114 and 6.115 is not complete until the end of the batch run. At time interval k, the matrix $\mathbf{X}_{\text{new}}$ has only its first k rows complete and all the future observations [$(K-k)$ rows] are missing. Several approaches have been proposed to overcome this problem for MPCA and MPLS–based on-line monitoring [433, 434, 435].

MPCA–based on-line monitoring. The future portions of variable trajectories are estimated by making various assumptions [433]. The on-line evolution of a new batch is monitored in the reduced space defined by the PCs of the MPCA model.

The incompleteness of the $\mathbf{X}_{\text{new}}$ $(K \times J)$ matrix (or $\mathbf{x}_{\text{new}}$ $(1 \times KJ)$ vector after unfolding and scaling) during the batch creates a problem for on-line monitoring. The loadings of the reference data set cannot be used with incomplete data because the vector dimensions do not match. Three approaches are suggested to fill in the missing values in $\mathbf{X}_{\text{new}}$ [433, 435].

Method 1, assumes that future observations are in perfect accordance with their mean trajectories.

Method 2, assumes that future values of disturbances remain constant at their current values over the remaining batch period.

Method 3, treats unknown future observations as missing values from the batch in MPCA model. Hence, PCs of the reference batches can be used for prediction.

All three assumptions introduce arbitrariness in the estimates of variable trajectories (Figure 6.62). Deciding which approach to use depends on the inherent characteristics of the process being monitored and information about disturbances. If process measurements do not contain discontinuities or early deviations, the third approach may be used after some data have been collected. If it is known that the disturbances in a given process are persistent, it is reported that the second approach works well [435]. When no prior knowledge exist about the process, the first estimation technique may be used.

As the new vector of variable measurements is obtained at each time k, the future portions of the trajectories are estimated for use in regular MPCA-based SPM framework as

$$\hat{\mathbf{t}}_{\text{new,k}} = \mathbf{x}_{\text{new}}^{\text{est}}\mathbf{P} \quad , \quad \mathbf{e}_{\text{new},k} = \mathbf{x}_{\text{new}}^{\text{est}} - \sum_{a=1}^{A} \hat{t}_{\text{new},ak}\mathbf{p}_a \tag{6.120}$$

where $\mathbf{x}_{\text{new}}^{\text{est}}$ denotes the *full* variable measurements vector $(1 \times KJ)$ that is estimated at each k onwards to the end of the batch run, $\hat{\mathbf{t}}_{\text{new},k}$ $(1 \times A)$, the predicted scores at sampling time k from the $\mathbf{P}$ loadings, and $\mathbf{e}_{\text{new},k}$ $(1 \times KJ)$ the residuals vector at time k. To construct the control limits for on-line monitoring of new batches, each reference batch is passed through the on-line monitoring algorithm above, as if they are new batches, and their predicted scores ($\hat{\mathbf{t}}_{\text{new},k}$) and squared prediction errors (SPE_k) are stored at each sampling interval k.

Example. MPCA-based on-line SPM framework is illustrated using the same simulated data set of fed-batch penicillin production presented in Section 6.4.1. The large downward drift fault in glucose feed rate is used as a case study (Figure 6.44 and data set $\mathbf{X_3}$ (764×14) in Table 6.8). The *model development stage* and the MPCA model developed are the same as in Section 6.4.3, with the exception that the construction of control limits is performed by passing each batch data in the reference set through the estimation-based on-line SPM procedure. The *process monitoring stage*

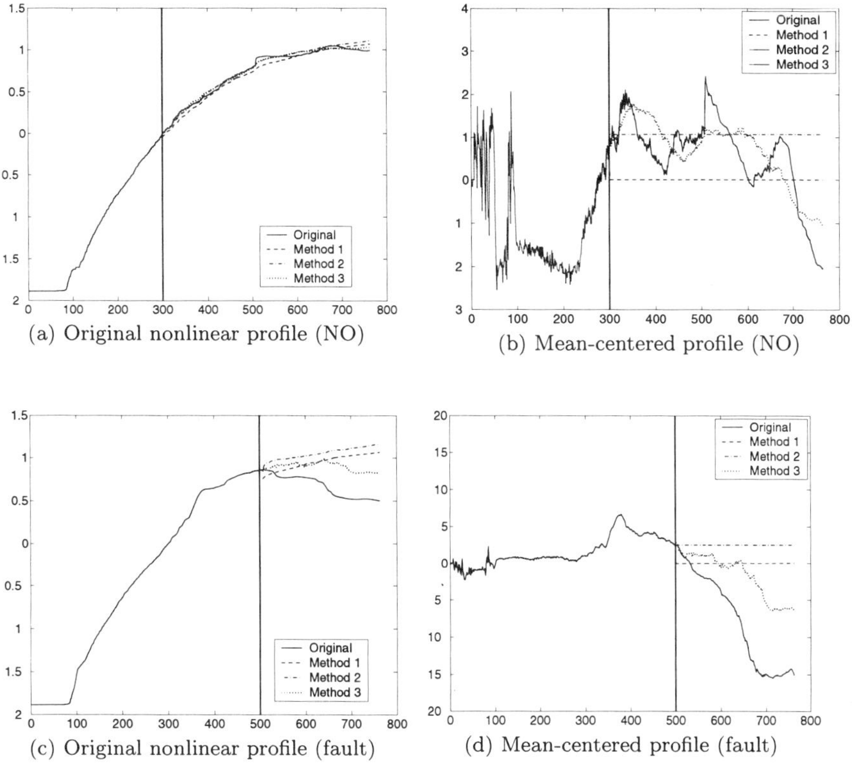

(a) Original nonlinear profile (NO)

(b) Mean-centered profile (NO)

(c) Original nonlinear profile (fault)

(d) Mean-centered profile (fault)

Figure 6.62. Methods of estimation of future values in MPCA-based on-line SPM (Autoscaled penicillin concentration profile).

depends on the estimation method used. All three methods are implemented in this example. Greater difference caused by the data estimation method used is observed in the T^2 chart in Figure 6.63(a). The out-of-control signal is first detected by the second technique (the future values of disturbances remain constant at their current values over the remaining batch period) at the 325th measurement in T^2 chart. SPE chart detected the fault around the 305th measurement in all of the techniques. Variable contributions to SPE and T^2 and scores biplots are presented for Method 2. Contribution plots revealed the variables responsible for the deviation from NO when out-of-control state is detected. Variables 3 and 5 in SPE contributions (Figure 6.63(d)) at 305th measurement and variable 3 and 5 (and 7, 13, 14 to a lesser extent) in T^2 contribution plot (Figure 6.63(c)) at 325th measurement are identified as responsible for the out-of-control situation. Variable 3 (glucose feed rate) is the main problematic variable affecting the other variables gradually. Variable 5 (glucose concentration in the fermenter) is the first variable directly affected by the drift in variable 3. Since T^2 detects the out-of-control state later, the effect of the drift develops significantly on variables such as 7 (biomass concentration in the fermenter), 13 (heat generated), and 14 (cooling water flow rate) that are signaled by the T^2 contribution plot (Figure 6.63(c)). Scores biplots also show a clear deviation from NO region defined by confidence ellipses of the reference model (Figures 6.63(e) and 6.63(f)). □

MPLS–based on-line monitoring and estimation of final product quality. Although the three estimation methods presented above can be used to deal with missing future portions of the trajectories when implementing MPLS on-line, another approach that uses the ability of PLS to handle missing values is also proposed [434]. Measurements available up to time interval k are projected onto the reduced space defined by the $\mathbf{W}$ and $\mathbf{P}$ matrices of the MPLS model in a sequential manner as for all of the A latent variables

$$\hat{\mathbf{t}}(1,a)_{\text{new},k} = \mathbf{x}_{\text{new},k} \frac{\mathbf{W}(1:kJ,a)}{\mathbf{W}(1:kJ,a)^T \mathbf{W}(1:kJ,a)} \tag{6.121}$$

$$\mathbf{x}_{\text{new},k} = \mathbf{x}_{\text{new},k} - \hat{\mathbf{t}}(1,a)_{\text{new},k} \mathbf{P}(1:kJ,a)^T \tag{6.122}$$

where $(1:kJ,a)$ indicates the elements of the ath column from the first row up to the kJth row. The missing values are predicted by restricting them to be consistent with the values already observed, and with the correlation structure that exists between the process variables as defined by the MPLS model. It is reported that this approach gives $\mathbf{t}$-scores very close to their final values as $\mathbf{X}_{\text{new}}$ is getting filled with measured data (k increases) and it works well after 10 % of the batch evolution is completed [433, 434, 435].

When a new variable measurements vector is obtained and k is incremented, scores $\hat{\mathbf{t}}(1,a)_{\text{new},k}$ can be estimated and used in MPLS (Eqs. 6.115 and 6.116). There are no residuals $\mathbf{f}$ on quality variables space during on-line monitoring since the actual values of the quality variables will be known only at the end of the batch. Each batch in the reference database is passed through the on-line MPLS algorithm as if they were new batches to construct control limits. Since MPLS provides predictions for the *final* product qualities at each sampling interval, the confidence intervals for those can also be developed [434]. The confidence intervals at significance level α for an individual predicted final quality variable $\hat{y}$ are given as [434]

$$\hat{y} \pm t_{I-A-1,\ \alpha/2} (MSE)^{1/2} \left(1 + \hat{\mathbf{t}}(\mathbf{T}^T\mathbf{T})^{-1}\hat{\mathbf{t}}^T\right)^{1/2} \tag{6.123}$$

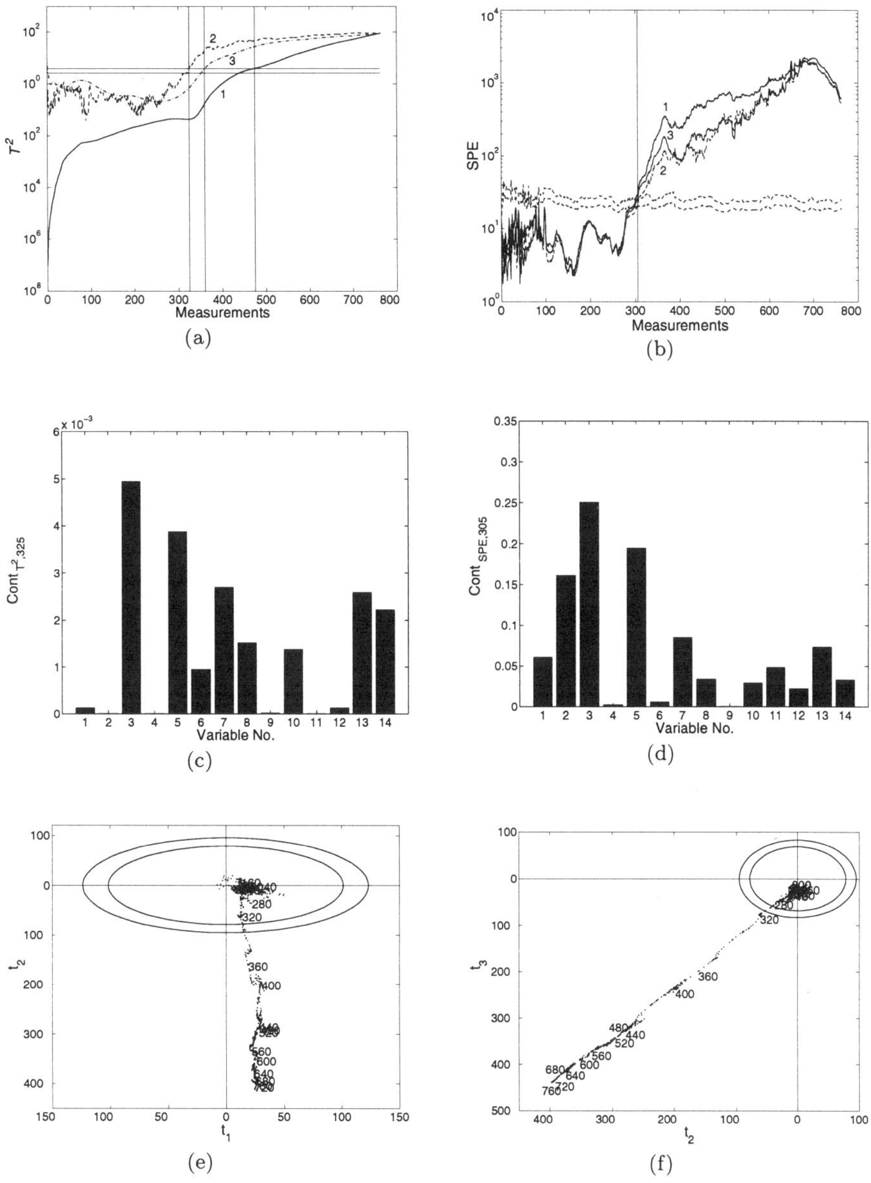

Figure 6.63. MPCA-based on-line SPM results of a faulty batch. In (a) and (b) Method 1 (Solid curve), Method 2 (Dashed curve), and Method 3 (Dash-dotted curve). (c)-(d) Variable contributions to T^2 and SPE at 325th and 305th measurements, respectively. Score biplots based on Method 2 (e) 1st vs 2nd PC and (f) 2nd vs 3rd PC.

where $\mathbf{T}$ is the scores matrix, $t_{I-A-1,\ \alpha/2}$ is the critical value of the Studentized variable with $I-A-1$ degrees of freedom at significance level $\alpha/2$ and mean squared errors on prediction (MSE) are given as

$$SSE = (\mathbf{y} - \hat{\mathbf{y}})^T(\mathbf{y} - \hat{\mathbf{y}}), \qquad MSE = SSE/(I - A - 1). \qquad (6.124)$$

In these equations, I refers to number of batches, A to number of latent variables retained in the MPLS model, and SSE to sum of squared errors in prediction.

Example. To illustrate on-line implementation of MPLS for monitoring and prediction of end product quality, the same reference set and MPLS model are used as in Section 6.4.4. All batches in the reference set are passed through the on-line algorithm to construct multivariate statistical control limits. MV charts for an in-control batch are shown in Figure 6.64. T^2, SPE, first LV and second LV charts indicate that the process is operating as expected. Figure 6.65 presents predictive capability of the model. The solid curves indicate the end-of-batch values estimated at the corresponding measurement times. The dashed curves are the 95% and 99% control limits on end-of-batch estimates. End-of-batch values of all five quality variables are predicted reasonably while the batch is in progress. The third fault scenario with a significant downward drift on substrate feed rate is used to illustrate MPLS based on-line SPM. The first out-of-control signal is generated by the SPE chart at the 305th measurement (Figure 6.66(a)), followed by the second LV plot at the 355th measurement (Figure 6.68(c)), the T^2 chart at the 385th measurement(Figure 6.66(c)) and finally by the first LV plot at the 590th measurement (Figure 6.68(a)). Contribution plots are also plotted when out-of-control status is detected on these charts. Variable contributions to SPE in Figure 6.66(b) reveal the root cause of the deviation that is variable 3 (glucose feed rate). Second highest contribution in this plot is from variable 5 (glucose concentration in the fermenter), which makes sense because it is directly related to variable 3. The rest of the corresponding contribution plots reveal variables that are affected sequentially as the fault continues. For instance, the second LV signals the fault later than SPE, hence there is enough time to see the effect of the fault on other variables such as variables 12 (temperature in the fermenter) and 13 (heat generated) while variable 3 is still having the maximum contribution (Figure 6.68(d)). T^2 chart signals out-of-control a little later than the second LV and at that point variables affected are variable 7 (biomass concentration in the fermenter), 13 (heat generated) and 14 (cooling water flow rate) (Figure 6.66(d)). An upward trend towards the out-of-control region can be seen in T^2 charts in Figure 6.66(c) when SPE chart detects the out-of-control situation. Variable contributions at 305th

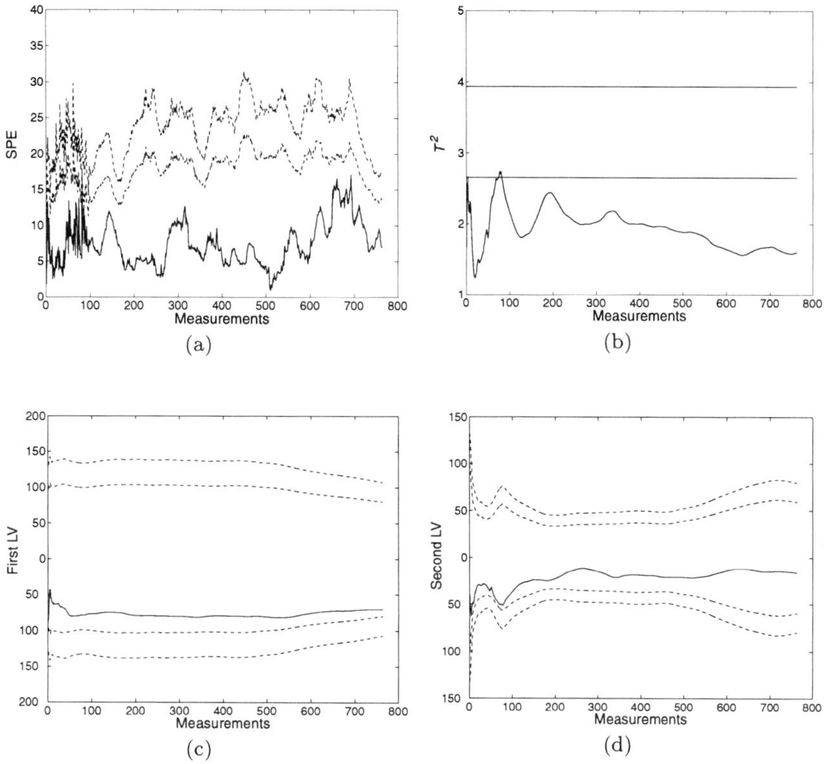

Figure 6.64. MPLS-based on-line monitoring results of a normal batch.

measurement are shown in Figure 6.67 to reveal the variables contributing to this deviation that is beginning to develop. As expected, variables 3 (glucose feed rate) and 5 (glucose concentration in the fermenter) are found responsible for the start of that upward trend towards the out-of-control region. End-of-batch product quality is also predicted (Figure 6.69). Significant variation is predicted from desired values of product quality variables (compare Figure 6.69 to Figure 6.65). The confidence intervals are plotted only until the SPE signals out-of-control status at 305th measurement because the model is only valid until that time. □

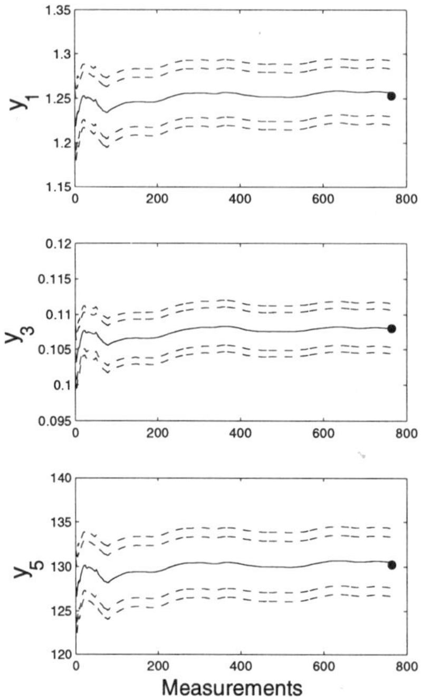

Figure 6.65. MPLS-based on-line predictions of end-of-batch product quality of an in-control NO batch. (•) represents the actual value of the end-of-batch product quality measurement.

6.5.2 Adaptive Hierarchical PCA

Hierarchical PCA provides a framework for dividing the data block $\underline{\mathbf{X}}$ into K two-dimensional blocks $(I \times J)$ and look at one time slice at a time [496] (see Figure 6.70). This gives separate score vectors $\mathbf{t}_{ak}$ for each individual time slice $\mathbf{X}_k$ where $a = 1, \cdots, A$, $k = 1, \cdots, K$. The initial step is to calculate a one-component PCA model for the first time slice, and to obtain the score and loading vector $(\mathbf{t}_{ak}, \mathbf{p}_{ak}, a = 1, k = 1)$ for the first block. The hierarchical part of the algorithm starts at $k = 2$ and continues for the rest of the batch $(k = K)$. The score and loading vectors are built iteratively, the score vector for the previous time slice model $\mathbf{t}_{a(k-1)}$ is used as the starting estimate for $\mathbf{t}_{ak}$. Then, $\mathbf{p}_{ak} = \mathbf{X}_{ak}^T \mathbf{t}_{ak}$ and the new score vector $\mathbf{r}_{ak}$ is calculated and normalized:

$$\mathbf{r}_{ak} = \mathbf{X}_{ak}\mathbf{p}_{ak}, \qquad \mathbf{r}_{ak} = d_k \mathbf{r}_{ak}/\|\mathbf{r}_{ak}\| \tag{6.125}$$

The weighting factor d_k balances the contributions of the new information

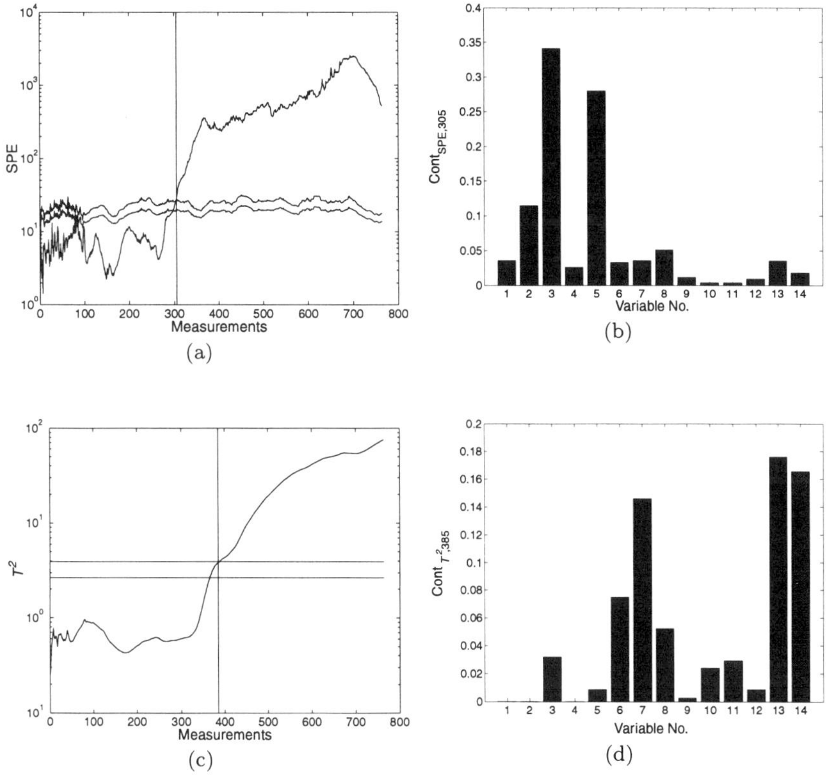

Figure 6.66. MPLS-based on-line monitoring results of a faulty batch.

($\mathbf{r}_{ak}$) at time k and the recent history ($\mathbf{t}_{a(k-1)}$), playing a role similar to that of the exponential weighting factor in an EWMA model. The consensus matrix $\mathbf{R}_{ak}$ is formed from $\mathbf{t}_{a(k-1)}$ and $\mathbf{r}_{ak}$ column vectors and the weight vector $\mathbf{w}_{ak}$ is computed as ($\mathbf{w}_{ak} = \mathbf{R}_{ak}^T \mathbf{t}_{ak}$) for calculating the new score vector $\mathbf{t}_{ak} = \mathbf{R}_{ak}\mathbf{w}_{ak}$. Then, $\mathbf{t}_{ak}$ is normalized and checked for convergence. If convergence is achieved the $\mathbf{X}_{ak}$ blocks are deflated as $\mathbf{X}_{(a+1)k} = \mathbf{X}_{ak} - \mathbf{t}_{ak}\mathbf{p}_{ak}^T$ to calculate the next dimension (a is increased by 1). The converged latent vectors are computed for a given a for all k, then a is incremented by 1 and the process is repeated until $a = A$. The model generated can be used to monitor future batches by storing $\mathbf{p}_{ak}$, $\mathbf{w}_{ak}$, and d_k for $a = 1, \cdots, A$, $k = 1, \cdots, K$.

As data are collected from the new batch and stored as row vectors $\mathbf{x}_k^T$, the values for $\mathbf{r}_{ak}$, $\mathbf{t}_{ak}$, and $\mathbf{x}_{(a+1)k}$ are computed at time k for $a = 1, \cdots, A$

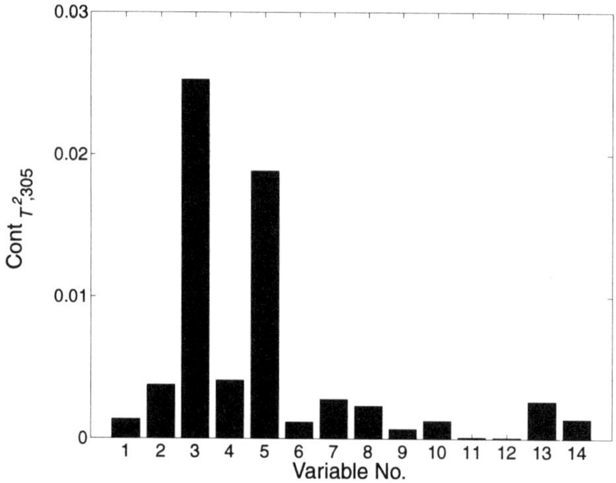

Figure 6.67. Variable contributions to T^2 when SPE chart signals out-of-control status at the 305th measurement.

by using

$$\mathbf{r}_{ak} = \mathbf{x}_{ak}^T \mathbf{p}_{ak}, \qquad \mathbf{t}_{ak} = [\mathbf{t}_{a(k-1)} \;\; d_k \mathbf{r}_{ak}] \mathbf{w}_{ak}, \tag{6.126}$$

$$\mathbf{x}_{(a+1)k}^T = \mathbf{x}_{ak}^T - \mathbf{t}_{ak} \mathbf{p}_{ak}^T \tag{6.127}$$

The prediction error is computed as

$$\mathbf{e}_k = \mathbf{x}_k - \sum_{a=1}^{A} \mathbf{t}_{ak} \mathbf{p}_{ak}^T \tag{6.128}$$

The score and error values at each k can be plotted for MSPM of the batch. Since no missing data estimation is required in AHPCA, the control limits are calculated directly using the residuals and scores from the model building stage.

Example. AHPCA-based SPM framework is illustrated using the same simulated data set of fed-batch penicillin production presented in Section 6.4.1. Two main steps of this framework can be expressed as *model development stage* using a historical reference batch database that defines normal operation and *process monitoring stage* by making use of the model developed for monitoring a new batch.

AHPCA model development stage: AHPCA model is developed from a data set of equalized/synchronized (Figures 6.42 and 6.43), unfolded and scaled 37 good batches (each containing 14 variables 764 measurements resulting

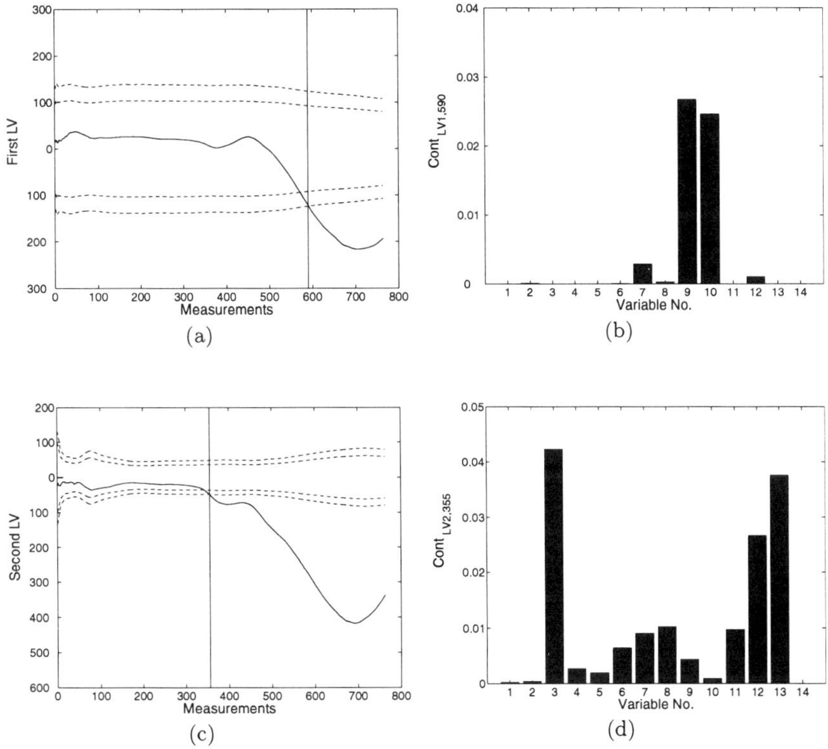

Figure 6.68. MPLS-based on-line scores and contribution plots of a faulty batch.

in a three-way array of size $\underline{\mathbf{X}}(37 \times 14 \times 764)$. A subset of 37 batches is chosen from the original 41 batches. After unfolding by preserving the batch direction, size of the resulting matrix $\mathbf{X}$ becomes (37×10696)). AHPCA-based empirical modeling is performed on the unfolded array $\mathbf{X}$ with three principal components and the weighting factor d is chosen as 0.35 (for all the sampling intervals). Explained variability on $\mathbf{X}$ block by AHPCA model is summarized in Figure 6.71(b) and Table 6.13. The explained variability even with 3 PCs is higher than that of 4 PC MPCA model presented in Section 6.4.3 (Figure 6.45(e)).

Process monitoring stage: The adaptive model developed is used to monitor new batches on-line. The batch fault scenario with 10% step decrease in agitator power input between the 140th and 180th measurements (Figure 6.44 and Table 6.8) is used. New batch data are processed with

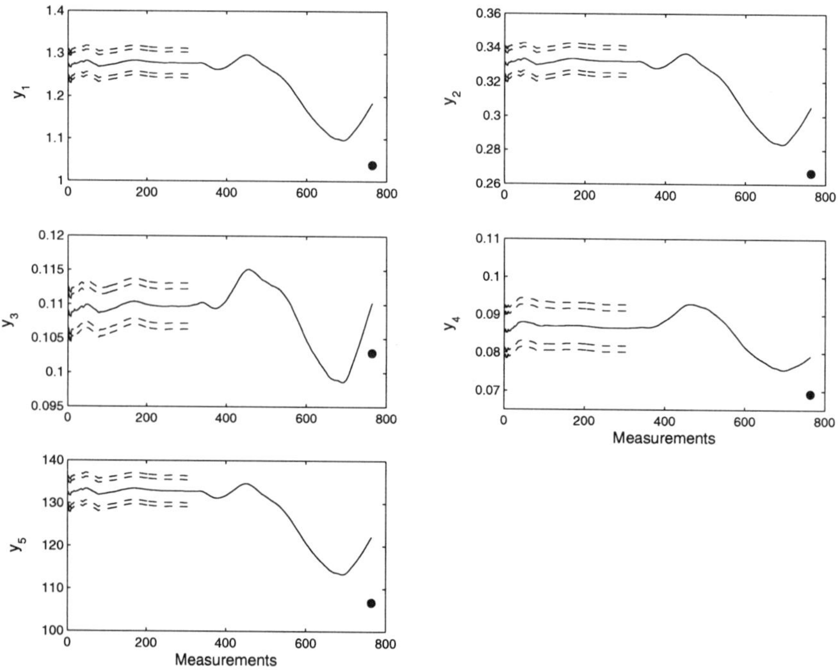

Figure 6.69. MPLS-based on-line predictions of end-of-batch product quality. (•) represents the actual value of the end-of-batch product quality measurement.

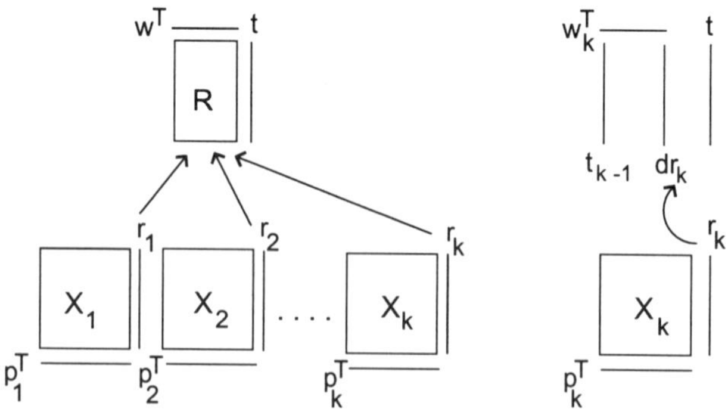

Figure 6.70. Adaptive hierarchical PCA scheme [496].

Table 6.13. Percent variance captured by AHPCA model

PC no.	X-block	
	This PC	Cumulative
1	28.25	28.25
2	20.05	48.30
3	10.29	58.59

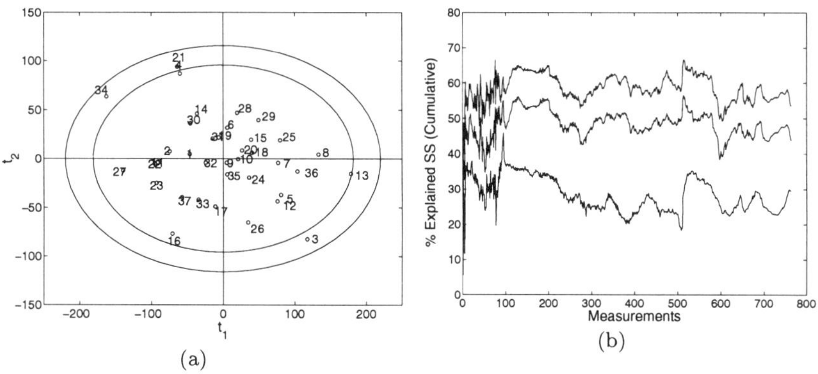

Figure 6.71. AHPCA model (with three PCs) statistics. (a) biplots of 37 reference runs, (b) cumulative explained variance with one, two and three PCs.

an AHPCA model as shown in Eqs. 6.126-6.128 after proper equalization/synchronization, unfolding and scaling, resulting in multivariate SPM charts (Figure 6.72). Both SPE and T^2 charts signal on-line when the process is out-of-control. The variable responsible for this deviation from NO is diagnosed to be variable 2 (agitator power input) by using on-line contribution plots to SPE and T^2 in Figures 6.72(b) and 6.72(d). Contributions are calculated for the interval of the fault occurrence (140th and 180th measurements). Variables 7, 13 and 14 (biomass concentration, heat generated and cooling water flow rate) are also contributing to deviation. This is due to the decrease in oxygen transfer rate during that short interval. □

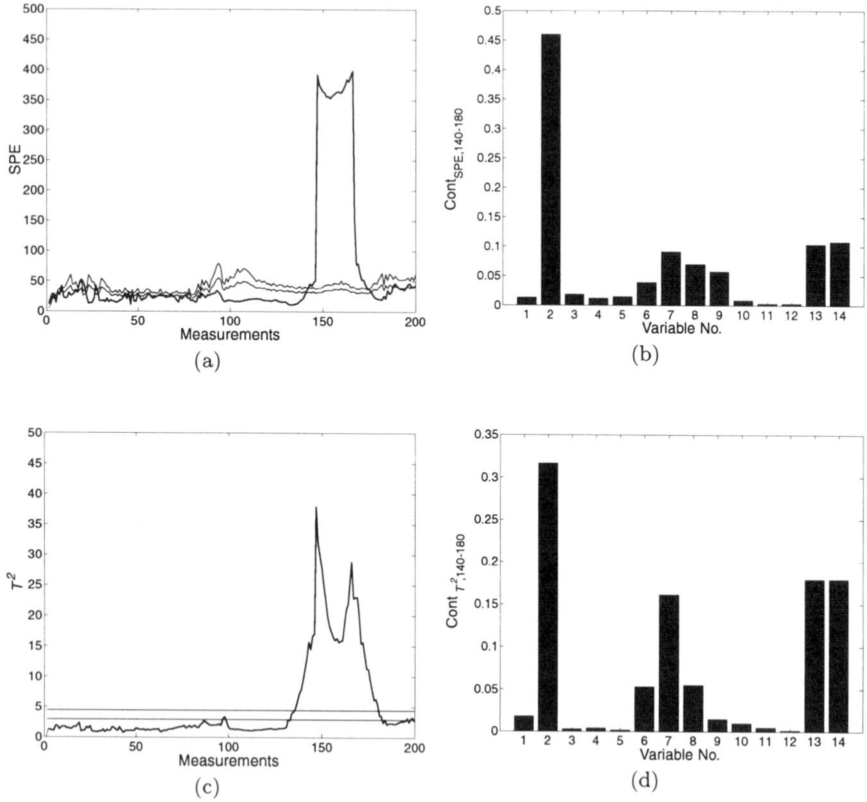

Figure 6.72. On-line monitoring of a faulty batch using AHPCA. The subscript "140-180" in figures (b) and (d) indicate that contributions are averaged between 140th and 180th measurements.

6.5.3 Online MSPM and Quality Prediction by Preserving Variable Direction

A different online MSPM framework can be established by unfolding the three-way data array by preserving variable direction [203, 232, 663]. In this MSPM framework, it is not necessary to estimate the future portions of variable trajectories. MPCA or MPLS models can be developed and used for online monitoring. A new methodology has been proposed based on developing an MPLS model between process variable matrix that is unfolded in the variable direction and local time stamp to use in the alignment of

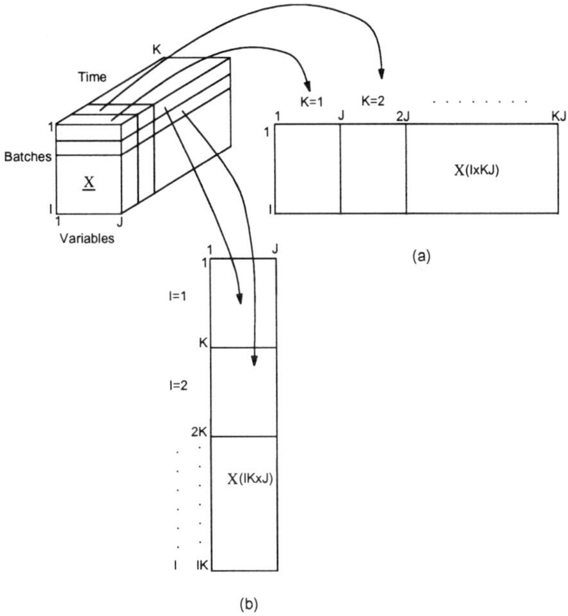

Figure 6.73. Three-way array formation and unfolding, (a) by preserving batch direction, (b) by preserving variable direction.

trajectories [663].

Process measurements array $\underline{\mathbf{X}}$ can be unfolded to $\mathbf{X}$ $(IK \times J)$ by preserving the variable direction [232, 552, 663]. In this case, $\underline{\mathbf{X}}$ can be thought of as a combination of slices of matrices of size $(K \times J)$ for each batch (Figure 6.73(a)). $\mathbf{X}$ is formed after rearrangement of these slices. This type of unfolding suggests a different multivariate modeling approach [232, 606, 663]. Batch evolution can be monitored by developing an MPLS model between $\mathbf{X}$ $(IK \times J)$ and a time stamp vector $\mathbf{z}$ $(IK \times 1)$ (Figure 6.75(b)). In this case, MPLS decomposes $\mathbf{X}$ and $\mathbf{z}$ into a combination of scores matrix $\mathbf{T}$ $(IK \times R)$, loadings matrix $\mathbf{P}$ $(J \times R)$ and vector $\mathbf{q}$ $(R \times 1)$ and weight matrix $\mathbf{W}$ $(J \times R)$ with different sizes compared to conventional MPLS decomposition discussed in Section 4.5.2

$$\begin{aligned} \mathbf{X} &= \mathbf{T}\mathbf{P}^T + \mathbf{E} \\ \mathbf{z} &= \mathbf{T}\mathbf{q} + \mathbf{f} \end{aligned} \tag{6.129}$$

where $\mathbf{E}$ and $\mathbf{f}$ are the residuals matrix and vector, respectively.

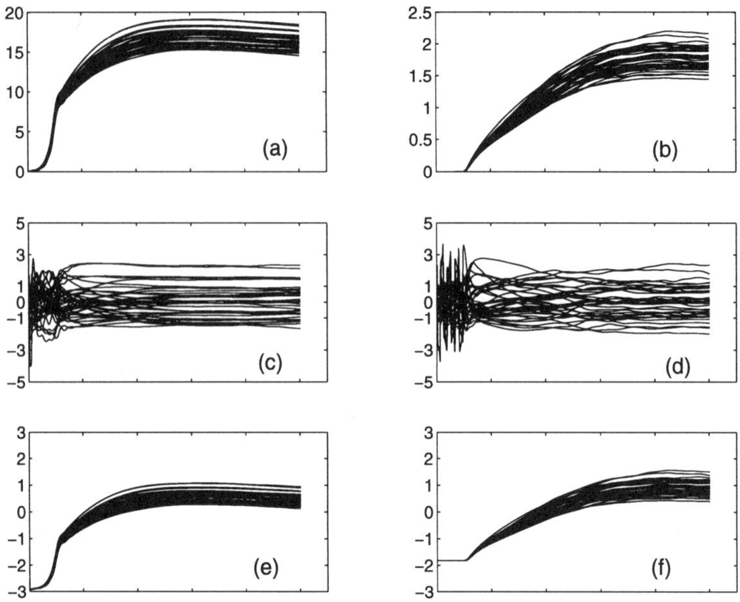

Figure 6.74. Comparison of different scalings applied to $\mathbf{X}$ matrix. Biomass (left figures) and penicillin concentration profiles (right figures). Raw profiles (a)-(b), mean-centered (by subtracting mean trajectories) and unit variance scaled profiles (c)-(d), mean-centered (by subtracting variable means) and unit variance scaled profiles (e)-(f).

During the progress of a new batch, a vector $\mathbf{x}_{\text{new}}$ of size $1 \times J$ becomes available at each time interval k. After applying the same scaling to new observations vector as reference sets, scores can be predicted for time instant k by using the MPLS model parameters

$$\hat{\mathbf{t}} = \mathbf{x}_{\text{new}} \mathbf{W} (\mathbf{P}^T \mathbf{W})^{-1}. \tag{6.130}$$

Since the size of the resulting matrix from the operation $\mathbf{W}(\mathbf{P}^T\mathbf{W})^{-1}$ is $J \times R$, online monitoring of the new batch can be performed without any future value estimation.

In the pre-processing step, $\mathbf{X}$ is mean-centered by subtracting variable means and usually scaled to unit variance. This pre-processing differs from the conventional approach (Figure 6.74(c)-(d)) in that the dynamic nonlinear behavior of trajectories in $\mathbf{X}$ is retained (Figure 6.74(e)-(f)). This technique can also be combined with conventional MPLS for predicting product quality after the completion of the batch run [606, 663].

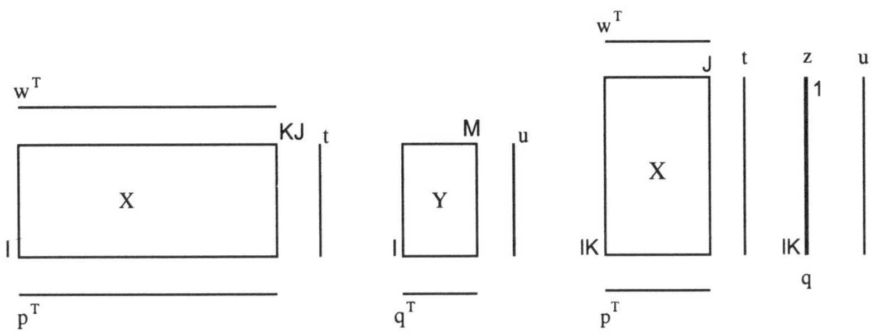

(a) MPLS model blocks for predicting product quality

(b) MPLS model blocks for predicting progress of the batch

Figure 6.75. MPLS modeling with using different unfolding approaches.

Development of Online Process Performance Monitoring Framework. This online monitoring framework in this work is based on unfolding a three-way array by preserving the variable direction. However, it is also incorporated with the conventional MPLS technique when online/offline quality prediction and end-of-batch monitoring are aimed. To differentiate the two MPLS techniques depending on different type of unfolding, the conventional technique that preserves batch direction is called MPLSB (Figure 6.75(a)) and the one that preserves variable direction is called MPLSV (Figure 6.75(b)). MPLSV relies on information at a specific time. The history of the batch up to that time is not considered in contrast to the MPLSB framework. In this respect, it is similar to a Shewhart chart while MPLSB is similar to a CUSUM chart.

A reference data set that contains good batches presenting normal operation is used in the model development. Equalization and alignment of trajectories are required if batches in this reference set are of different lengths using alignment techniques discussed in Section 6.3. Data alignment using an indicator variable (IV) can be performed in different ways. If there exists an indicator variable that other process variables can be measured against its percent completion, variable trajectories in the reference set are re-sampled by linear interpolation techniques with respect to this indicator variable. As an alternative method, (especially when such an indicator variable is not available), an MPLSV model can be developed between the process measurements matrix $\mathbf{X}$ and local time stamps vector $\mathbf{z}$ of the individual batches in the reference set (Figure 6.75(b), $\mathbf{y} = \mathbf{z}$).

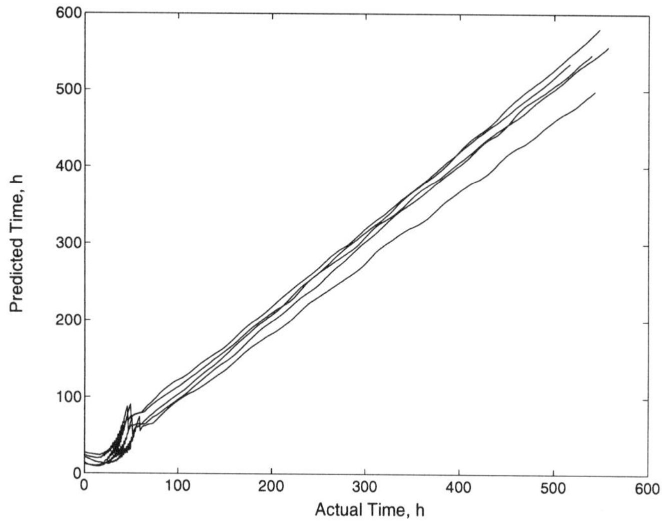

Figure 6.76. Predicted local batch time for the entire process duration. The peak corresponds to switching from batch to fed-batch operation.

This model provides information about the relationship between time and the evolution of process variable trajectories. The predicted time stamp vector $\mathbf{z}_{\text{pred}}$ can then be used as an indicator variable such that process variables are re-sampled on percent increments of this derived variable. It is assumed that variable trajectories contain sufficient information to fairly predict batch time in MPLSV modeling. This assumption implies that variable trajectories somewhat linearly increase or decrease in each time region. Local batch time prediction produces weak results when there are discontinuities or there exists instances that variables have simultaneous piecewise linear dynamics during the evolution of the batch. As illustrated in Figure 6.76 with fed-batch penicillin fermentation data, predicted time shows non-increasing or decreasing behavior in the region around the discontinuity which makes it inappropriate for data alignment. Similar results were also reported for industrial data [641].

A solution is proposed to this problem by partitioning the entire process into major operational phases [606]. Two different data alignment methods are used. For the general case when batches in the reference data set are of unequal length and no appropriate indicator variable is found, an MPLSV model is developed between $\mathbf{X}$ and local time stamps vector $\mathbf{z}$ for each process phase. Process variable trajectories are then re-sampled with respect to the percent completion of predicted local batch time vector $\mathbf{z}_{\text{pred}}$. A vector of $\mathbf{t}_{\mathrm{max}}$ containing predicted termination times of reference batches is used to calculate percent completion on $\mathbf{z}_{\mathrm{pred}}$. The second type of alignment does not use MPLSV modeling. Appropriate indicator variables are chosen for aligning batch variables in each phase of operation. The discontinuity occurs in the transition from batch to fed-batch operation in penicillin fermentation (Figure 1.4). Consequently, there are two operational phases and two indicator variables are used. In this case, process termination is determined according to a maturity indicator such as a preset percent conversion level or a certain total amount of a component fed. Both cases are common in industrial operations.

Once the reference data set of good batches is aligned to give an equal number of measurements in each batch and synchronized variable profiles, an MPLSV model is developed between the aligned process variables set and predicted percent completion of the batch run, $\mathbf{z}_{\mathrm{pred}}$. Model parameters from this step are used to construct MSPM charts as outlined earlier in Section 6.4.2.

Figure 6.77 shows aligned biomass concentration profiles of the reference batches in each phase of the batch run using indicator variables. As a result of the alignment procedure, temporal variation of process events is minimized so that similar events can be compared. A very useful byproduct of the alignment procedure is that the number of measurements in each batch on each variable is also equalized. $\mathbf{z}_{\mathrm{pred}}$ profiles in each phase of the reference batches are shown in Figure 6.78 along with their control limits. $\mathbf{z}_{\mathrm{pred}}$ of a new batch can be used as a maturity indicator. It can be inferred that if its value is smaller than the observed value, the process is progressing slower than the reference batches. Limits are used to detect an unusual deviation from the expected time course of the batch.

When used as is, MPLSV modeling produces nonlinear estimated scores. Control limits can be calculated as

$$\bar{\mathbf{t}} \pm 3\boldsymbol{\sigma} \tag{6.131}$$

where $\bar{\mathbf{t}}$ are average estimated scores and $\boldsymbol{\sigma}$ their standard deviations [663]. When a new batch is monitored with the model parameters of MPLSV, estimated scores of this new batch will also be nonlinear. After proceeding with mean-centering of these scores that reduces the nonlinearity, it is possible to construct tighter control limits by using Eq. 6.95. This modification allows faster fault detection as discussed in case studies. When an out-of-control status is detected with either type of score plots, variable contributions are checked for fault diagnosis.

Online Prediction of Product Quality. It is advantageous to use MPLSV type models for online monitoring because it is not necessary to

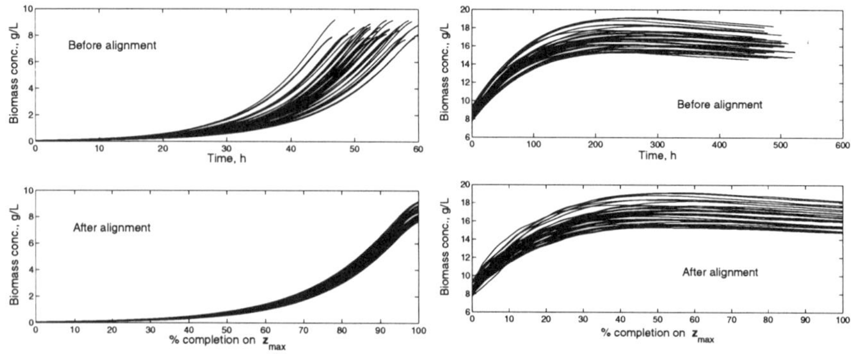

(a) Phase 1.

(b) Phase 2 (The 0 on the time axis corresponds to the end of Phase 1 time).

Figure 6.77. Results of the alignment procedure for biomass concentration profiles.

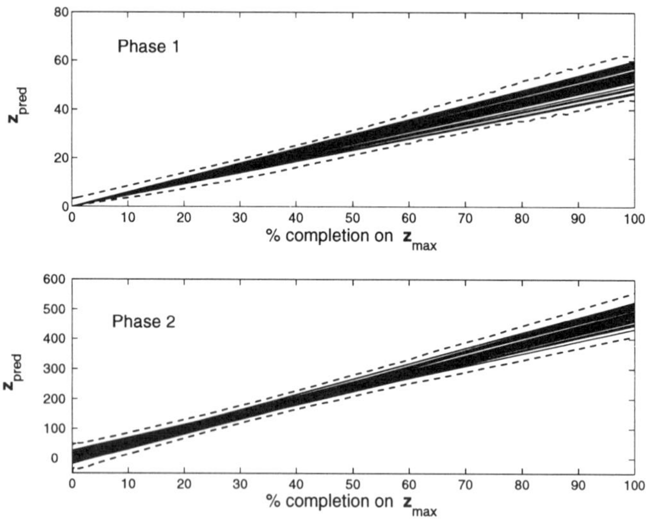

Figure 6.78. Predicted local batch times ($\mathbf{z}_{\text{pred}}$) in Phase 1 and 2 with control limits (dashed lines).

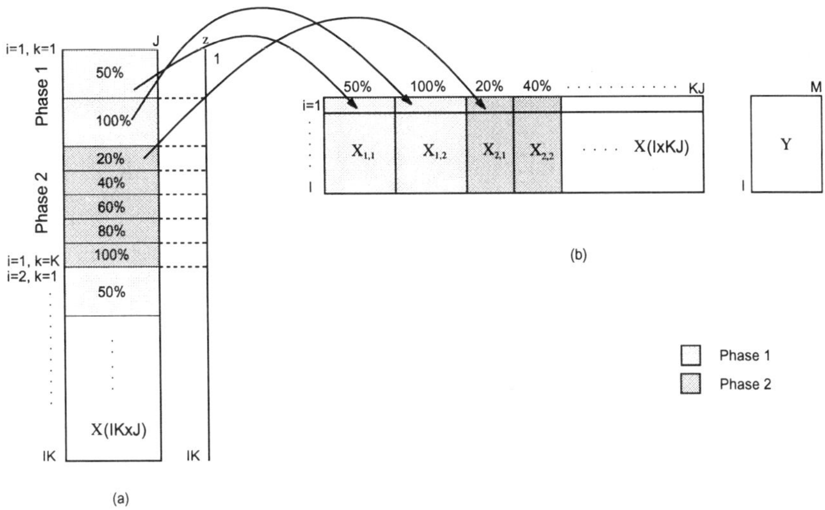

Figure 6.79. (a) Partitioning of process measurements space and (b) restructuring for online quality prediction framework.

estimate future portions of variable trajectories. However, this technique lacks online prediction of end-of-batch quality in real-time. A two-step integrated modeling approach is proposed to account for online quality prediction [606]. The first step is similar to MPLSV modeling discussed earlier. After reference batch data are aligned using IV technique, batch progress is determined according to percent increments on local batch time (or another IV) so that batches in the reference set are partitioned based on these increments that are chosen arbitrarily such as 10%, 20% of $\mathbf{z}_{\text{pred}}$ (Figure 6.79(a)). Each partition of $\mathbf{X}$ $(IK \times J)$ is rearranged and inserted into matrix $\mathbf{X}$ $(I \times KJ)$ as shown in Figure 6.79(b). This is similar to transition between MPLSV and MPLSB modeling. The difference is that whenever a partition is rearranged, i.e. some percent of the batch is completed, an MPLSB model is developed between this partial data and the final product quality matrix $\mathbf{Y}$. This gives an opportunity to predict end-of-batch quality on percent progress points reflected by partitions. The number of quality predictions will be equal to the number of partitions in this case.

Example. MPLS-based SPM and quality framework is illustrated using the simulated data set of fed-batch penicillin production presented in Section 6.4.1. Because of the modeling concerns about discontinuity, the data set is preprocessed for partitioning according to process phases. First, the batch/fed-batch switching point is found for each batch and data are divided into two sets as phase 1 and phase 2. Because the third variable (substrate feed) is zero in batch phase, only 13 variables are left in this first set. Data alignment is performed by using the IV technique. Since an indicator variable is not available for the entire batch run, separate indicator variables are selected for each phase. Variable 9, the culture volume decrease, is a good candidate to be chosen as an indicator variable for phase 1. A new variable called 'percent substrate fed' is calculated from variable 3 (substrate feed) and used as an indicator variable for phase 2 data set. This variable is added as the 15th variable to the data set of phase 2. It is assumed that fed-batch phase is completed when 25 L of substrate is added to the fermenter. Data are re-sampled by linear interpolation at each 1 percent completion of volume decrease for phase 1 and at each 0.2 percent of total substrate amount added for phase 2. Data alignment is achieved yielding in equal number of data points in each phase such that the data lengths are $K1 = 101$ and $K2 = 501$, respectively.

MPLS model development stage: Model development includes two stages. In the first stage, an MPLSV model is developed between process variables matrix (unfolded in variable direction) and an indicator variable. This model is used for online SPM purposes. The second stage involves developing predictive MPLSB models between available data partitions matrix (rearranged process variables matrix in batch direction) and end-of-batch quality matrix.

An MPLSV model is developed for phase 1 between autoscaled $\mathbf{X1}$ $(IK1 \times J1)$ and the IV vector $\mathbf{z1}$ $(IK1 \times 1)$ by using 5 latent variables. The number of latent variables should be chosen large enough to explain most of the information in $\mathbf{z1}$ block because the MPLSV model is used to predict batch evolution. Cross validation is used to determine the number of latent variables. $\mathbf{X1}$ $(IK1 \times J1)$ can be rearranged into matrix $\mathbf{X1}$ $(I \times K1J1)$ to develop an MPLSB model to obtain an estimate of end-of-batch quality at the end of phase 1. Since all $K1$ measurements of the first phase have been recorded by the beginning of the second phase, there would be no estimation of variable trajectories required and $I \times KJ$ partitioning can be used for modeling. Autoscaled $\mathbf{X1}$ $(I \times K1J1)$ and product quality matrix $\mathbf{Y}$ $(I \times M)$ are used as predictor and predicted blocks, respectively. Similarly, another MPLSV model is developed for phase 2 between autoscaled $\mathbf{X2}$ $(I \times K2J2)$ and IV vector $\mathbf{z2}$ $(IK2 \times 1)$.

In the second modeling stage, quality prediction models are developed. To develop the first MPLSB model, data are collected in 50% increment of phase 1 resulting in two data partitions $\mathbf{X}_{1,1}$ and $\mathbf{X}_{1,2}$ (Figure 6.79b). A similar approach is followed in phase 2 for every 20% increase in phase 2 evolution resulting in five data partitions ($\mathbf{X}_{2,n}$, $n = 1, \ldots, 5$). MPLSB

Table 6.14. Explained variance of MPLSB models for online quality prediction

Model no.	**X**-block	**Y**-block
1	61.57	68.85
2	61.27	71.27
3	58.85	89.21
4	60.62	95.07
5	63.10	97.31
6	63.35	98.39
7	63.39	98.89

type of modeling is performed between the rearranged **X** matrix which is augmented as a new data partition becomes available. As more data become available local models are developed (model no. 1...7 in Table 6.14) and explained variance in **Y** block increases with each local model as shown in Table 6.14.

Process variables for the new batch are sampled at percent increments of volume decrease for phase 1. After the completion of phase 1, the sampling rate is switched to percent completion of the amount of substrate added. New data vector $\mathbf{x}_{\text{new}}$ $(1 \times J)$ is monitored by using the following algorithm for each sampling point from $k = 1$ to $k = K1,\ K2$ for both phases

1. For $k = 1 \dots K$
2. New batch data: $\mathbf{x}_{\text{new}}$ $(1 \times J)$
3. Calculate new batch scores, SPE, T^2 and variable contributions to these statistics by using the information generated by MPLSV model
4. Compute $\mathbf{z}_{\text{pred}}$
5. Check MV control charts for abnormalities
6. End.

Process monitoring and quality prediction stage: A small drift of magnitude -0.018% h^{-1} was introduced into substrate feed rate from the start of fed-batch operation at 50 h until the end of the batch run as a test case. There are significant differences in fault detection times and out-of-control signal generation by different charts (Table 6.15). T^2 detected the fault fastest (Figure 6.80). Second fastest detection is obtained by the linear score control chart of latent variable 2 (Figures 6.81 and 6.82). The last four latent variables give out-of-control signals for both linear and nonlinear score matrices. Although SPE is in-control throughout the course

Table 6.15. Fault detection times

Type	% completed IV	Time, h
T^2	48.4	269
Linear Score LV 2	50	276
Linear Score LV 5	51.6	283
Nonlinear Score LV 2	51.6	283
Nonlinear Score LV 5	52	285
Linear Score LV 4	59.4	319
Nonlinear Score LV 4	60.6	324
Linear Score LV 3	70.2	368
Nonlinear Score LV 3	84.2	433
SPE	-	-

of the batch, the contribution plot for SPE signals an unusual situation for variable 3 (Figure 6.80c). Variable 3 and 11 are found to be the most affected variables because of the fault according to T^2 contribution plot. Deviation from average batch behavior plot is ineffective in indicating the most affected variable(s) in this case (Figure 6.83a).

Quality prediction ability of the integrated MSPM framework is also tested via two cases. A normal batch is investigated first. As expected, SPE plot produced no out-of-control signal and final product quality on all five variables (shown as a solid star) is successfully predicted (Figure 6.84). The prediction capability is somewhat poor in the beginning because of limited data, but it gets better as more data become available. In the second case, where a drift of magnitude -0.05% h^{-1} is introduced into substrate feed rate at the beginning of the fed-batch phase until the end of operation, SPE plot signaled out-of-control right after the sixth quality prediction point (80% completion of phase 2). Because MPLSB model is not valid beyond this point no further confidence limit is plotted (Figure 6.85). Although the predictions of MPLSB model might not be accurate for the seventh (and final) value, the framework generated fairly close predictions of the inferior quality. Predicting the values of end-of-batch quality during the progress of the batch provided a useful insight to anticipate the effects of excursions from normal operation on final quality. □

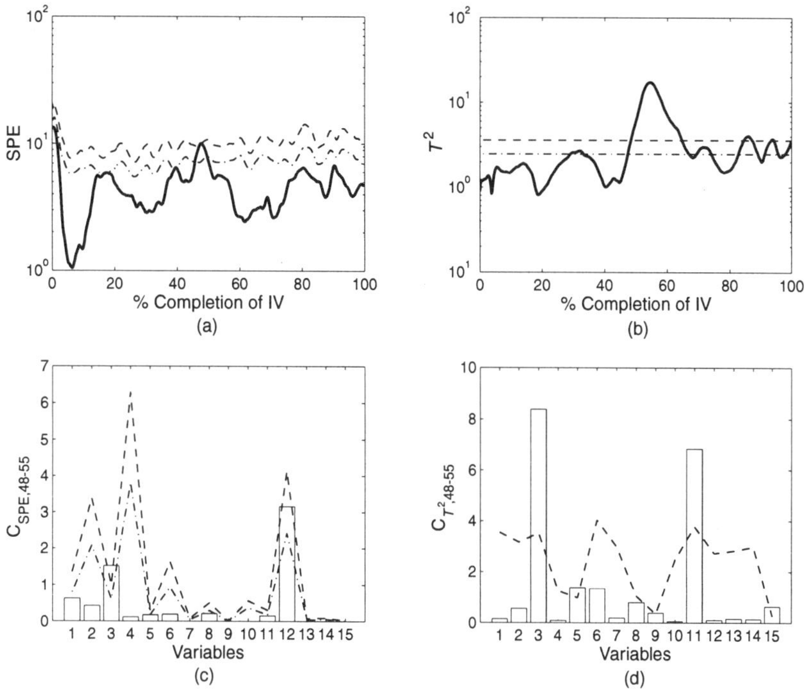

Figure 6.80. Control charts for SPE, T^2 for the entire process duration and contributions of variables to SPE and T^2 for a selected interval after out-of-control signal is detected in Phase 2 with 95% and 99% control limits (dashed-dotted and dashed lines).

6.5.4 Kalman filters for Estimation of Final Product Quality

Information on final product quality complements the information obtained from process variable trajectories. A model of the batch process and process variable measurements can be used to estimate final product properties before the completion of the batch. Consider the differential-algebraic nonlinear equation system that describes the batch process and its final state:

$$\frac{d\mathbf{x}}{dt} = \mathbf{f}_x(\mathbf{x}, \mathbf{u}, \mathbf{v}) \qquad \mathbf{y} = \mathbf{f}_y(\mathbf{x}, \mathbf{w}) \quad \mathbf{q} = \mathbf{f}_q(\mathbf{x}_{t_f}) \tag{6.132}$$

where $\mathbf{x}$ are the state variables, $\mathbf{u}$ the manipulated inputs, $\mathbf{v}$ and $\mathbf{w}$ the state and output disturbances, $\mathbf{y}$ the measured outputs, and $\mathbf{q}$ the final product quality at the end of the batch($t = t_f$). If a fundamental model

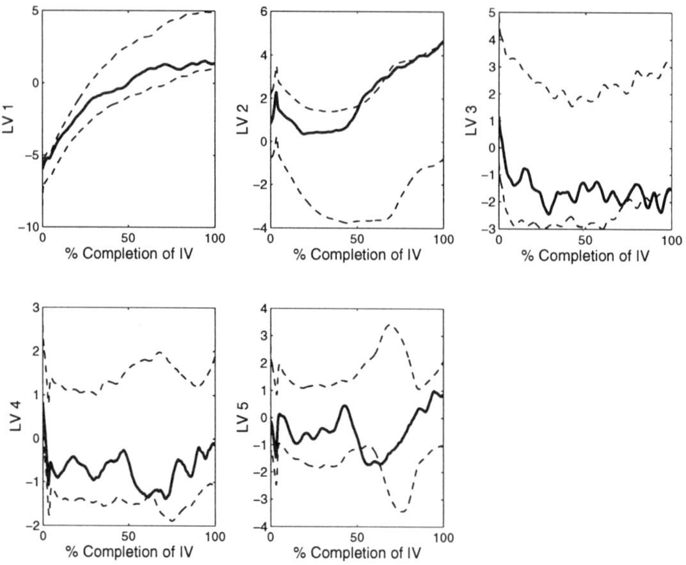

Figure 6.81. Nonlinear scores in Phase 2 with control limits (dashed lines).

of the process were available, the final product quality can be estimated by using an Extended Kalman filter. When a fundamental dynamic model is not available, an empirical model could be developed by using historical data records of successful batches. The problem may be cast as a regression problem where the measurements $\mathbf{y}$ upto the current time t_c, and inputs $\mathbf{u}$ upto the end of the batch are used at any time t_c to estimate $\mathbf{q}$. Note that the inputs at $t = t_c, \cdots, t_f$ have not been implemented yet and have to be assumed. A linear predictor for final product quality has been proposed by using a least squares estimator obtained through biased regression (by using PCA or PLS) and extended to recursive least squares prediction through a Kalman filter [531].

6.6 Monitoring of Successive Batch Runs

Batch process monitoring techniques presented in previous sections focus on detecting abnormalities in the current batch run by comparing it with performance criteria developed using a historical database. In some batch processes, disturbances may evolve over several batches, causing a gradual drift in product quality and eventually leading to significant quality deviation. MSPM methods that track changes in "between-batch" correlation

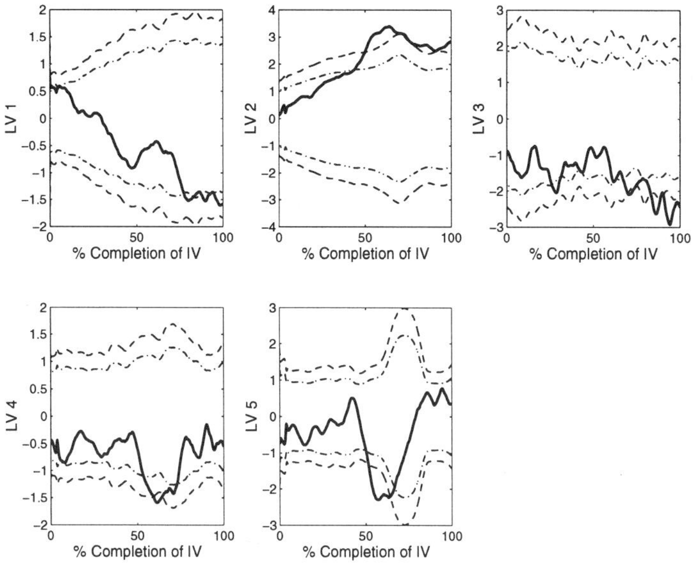

Figure 6.82. Linear scores in Phase 2 with 95% and 99% control limits (dashed-dotted and dashed lines).

structure, mean or drift are needed to track systematic variations in successive batch runs. Subspace identification (Section 4.3.2) where the time index is replaced by a batch index can be used to develop a framework for monitoring batch to batch (between-batch) variations [134].

Consider measurement data generated in a batch run arranged as I batches, J variables, and K sampling instants. Let $\mathbf{y}_{i,k}$ denote the vector of mean centered and scaled J process measurements of batch i at sampling time k. Collecting all process measurement vectors for $k = 1, \cdots, K$, the $J \times K$ data are unfolded to vectors of length JK for each k (called as lifting in subspace identification literature):

$$\mathcal{Y}_i = [\mathbf{y}_{i,1}^T,\ \mathbf{y}_{i,2}^T, \cdots,\ \mathbf{y}_{i,K}^T]^T \tag{6.133}$$

This is repeated for all batches $(1, \cdots, I)$ of the data set used for model building. The stochastic process model describing batch-to-batch variation is extracted from the data using subspace identification (Section 4.3.2):

$$\begin{aligned} \mathcal{X}_{i+1} &= \mathbf{F}\mathcal{X}_i + \mathbf{H}\mathcal{E}_i \\ \mathcal{Y}_i &= \mathbf{C}\mathcal{X}_i + \mathcal{E}_i \end{aligned} \tag{6.134}$$

Just as the state $\mathbf{x}_k$ in Section 4.3.2 was holding relevant process information from sampling times $k-1, \cdots, 1$ for predicting future process behavior

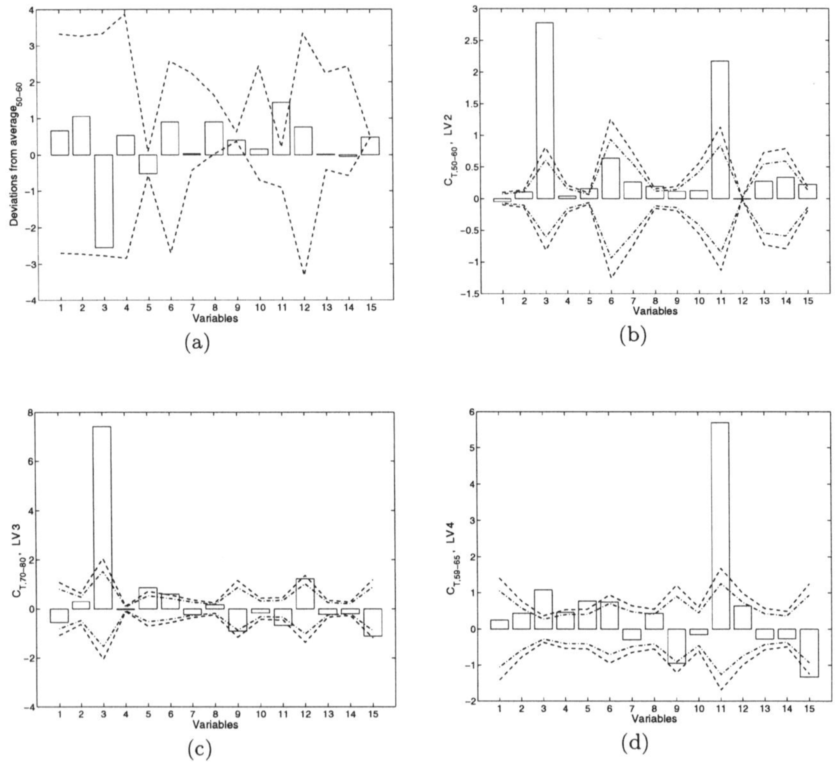

Figure 6.83. Phase 2 variable contributions to (a) deviations from average batch behavior at the interval of 50-60% completion of IV , (b) to linearized LV2 score at the interval of 50-60% completion of IV, (c) to linearized LV3 score at the interval of 70-80% completion of IV, (d) to linearized LV4 score at the interval of 59-65% completion of IV (dashed-dotted and dashed lines).

in time, the state $\mathcal{X}_i$ is holding valuable information about earlier batches $i-1, \cdots, 1$ for predicting future batches. Consequently, a monitoring algorithm can be developed using state variables $\mathcal{X}_i$ to detect undesirable behavior from batch-to-batch.

The simplest monitoring problem would be end-of-batch monitoring based on off-line final quality measurements. In this case, $\mathcal{Y}_i$ contains only the end-of-batch quality data. A more comprehensive monitoring problem would include both process variable data collected during the batch and end-of-batch quality variable data. In this case, Eq. (6.134) is augmented

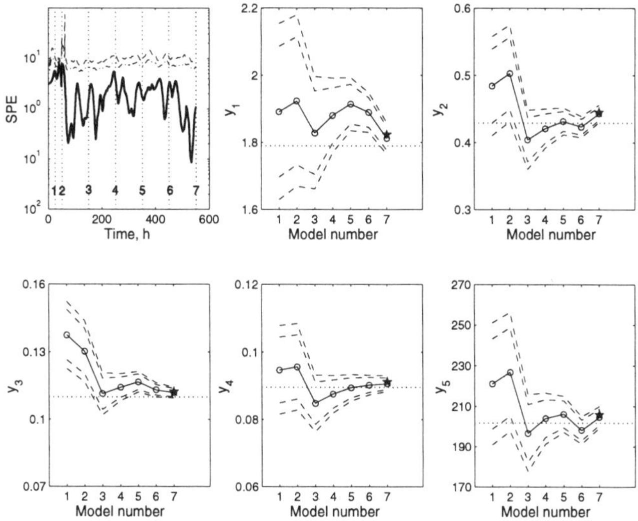

Figure 6.84. Online predictions for end-of-batch quality values for a normal batch. Dotted straight line indicates the average value of a quality variable based on reference batches.

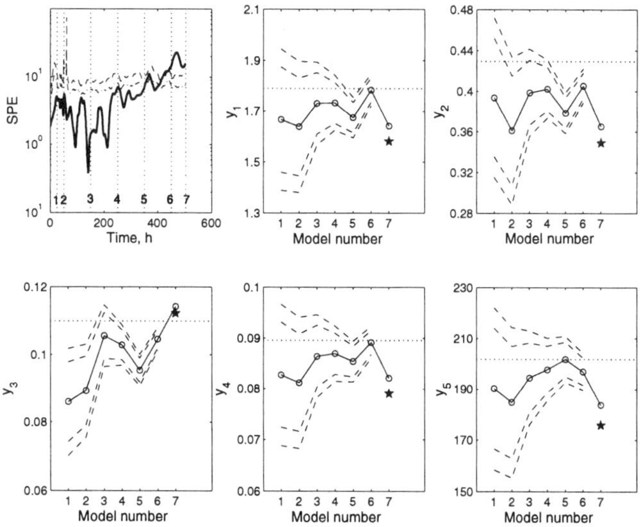

Figure 6.85. Online predictions for end-of-batch quality for the faulty batch.

to include the final product quality variables $\mathbf{q}_i$:

$$
\begin{aligned}
\mathcal{X}_{i+1} &= \mathbf{F}\mathcal{X}_i + \mathbf{H}\begin{bmatrix} \mathcal{E}_\mathcal{Y} \\ \mathcal{E}_q \end{bmatrix}_i \\
\begin{bmatrix} \mathcal{Y}_i \\ q_i \end{bmatrix} &= \begin{bmatrix} \mathbf{C}_\mathcal{Y} \\ \mathbf{C}_q \end{bmatrix}\mathcal{X}_i + \begin{bmatrix} \mathcal{E}_\mathcal{Y} \\ \mathcal{E}_q \end{bmatrix}_i
\end{aligned}
\tag{6.135}
$$

The size of the lifted output vector $\mathcal{Y}$ may be too large to develop subspace models quickly. This high dimensionality problem can be alleviated by applying PCA prior to subspace identification [134]. Since $\mathcal{Y}_k$ may have a high degree of colinearity, there is a potential to reduce the number of variables significantly. If the number of principal components is selected correctly, the residuals are mostly noise that tends to be batchwise uncorrelated, and the principal components will retain the important features of batch-to-batch behavior. Applying PCA to project $\mathcal{Y}$ of length JK to a lower dimensional space $\underline{\mathcal{Y}}$ of size a such that $a \ll JK$, the state-space model based on $\underline{\mathcal{Y}}$ is

$$
\begin{aligned}
\mathcal{X}_{i+1} &= \mathbf{F}\mathcal{X}_i + \mathbf{H}\underline{\mathcal{E}}_i \\
\underline{\mathcal{Y}}_i &= \mathbf{C}\mathcal{X}_i + \underline{\mathcal{E}}_i
\end{aligned}
\tag{6.136}
$$

where $\underline{\mathcal{Y}}$ is defined by

$$
\mathcal{Y} = \mathbf{\Theta}\underline{\mathcal{Y}} + \mathbf{E} \tag{6.137}
$$

with the columns of matrix $\mathbf{\Theta}$ being the principal directions (loadings) and $\mathbf{E}$ being the PCA residuals matrix.

Several monitoring charts can be developed to detect abnormal batch-to-batch behavior. T^2 and Q charts of principal components would be one alternative. But T^2 charts of states $\mathcal{X}_i$ and prediction errors $\underline{\mathcal{E}}_i$ offer better alternatives. The use of a small window CUSUM chart of prediction error T^2 has also been proposed [134]. It filters out high frequency variation in $\underline{\mathcal{E}}_i$ over i, and enhances the trends by accumulating the deviation over a number of batches. A window size of 5 batches provides a good compromise between capturing trends and delay in indicating big increases in the prediction error of one batch run.

7

Process Control

7.1 Introduction

It should be evident from the discussion and various illustrations in Chapter 2 that even in its simplest form, the representation of dynamic and steady-state behavior of a bioprocess is multivariate in nature. Even a simple unstructured kinetic representation for biomass formation would require knowledge / monitoring / prediction of a minimum of two variables, namely concentrations / amounts of biomass and at least one specie (substrate) which leads to production of cell mass. Recognizing that biomass formation is the sum total of a large number of intracellular chemical reactions, each of which is catalyzed by an enzyme, and that activity of each enzyme is very sensitive to intracellular pH and temperature, one can appreciate that this simplest black box representation would be applicable only if the intracellular pH and temperature, and therefore indirectly the culture (composite of abiotic and biotic phases) pH and temperature were kept invariant. Considering that the pH and temperature in the biotic portion of the culture and culture as a whole, if left uncontrolled, would vary with time because of the large number of intracellular chemical reactions, it is obvious that maintaining the culture pH and temperature at desired values would require addition of an acid or a base and addition/removal of thermal energy (heating/cooling) as appropriate. Thus, even in the simplest scenario where the focus in the *forefront* is on formation of biomass and consumption of a single substrate, one must consider in the *background* manipulation of rates of acid/base addition and heating/cooling to keep culture pH and temperature at the desired values.

Having realized that one must always deal with multivariate problems when dealing with biological reactors, the dimension of the system representation will depend on the nature of that kinetic representation employed (complexity of the kinetic model if one is available or complexity of the bioprocess under consideration if a kinetic model is not available), mode of operation of bioreactor [whether batch, fed-batch, or continuous, with/without recycle (after selective removal of a portion of the bioreactor contents using a separation technique)], and other external influences, such as addition/removal of thermal energy, acid/base addition for pH control, mechanical agitation to promote mixing, and circulation of a gas phase to provide oxygen (in an aerobic bioprocess) or nitrogen (to maintain oxygen-free atmosphere in an anaerobic bioprocess) and remove carbon dioxide.

The three popular modes of operation of mechanically agitated reactors for cell cultivation are batch, fed-batch and continuous operations. Irrespective of the mode of operation with respect to culture, the bioreactors are always operated in continuous mode with respect to gas phase (gas phase continuously entering and leaving the reactor). A batch culture operation is characterized by no addition to and withdrawal from the culture of biomass, fresh nutrient medium and culture broth (with the exception of gas phase). A fed-batch culture operation is characterized by predetermined or controlled addition of nutrient medium in an otherwise batch operation (no withdrawal of culture). This operation allows for temporal variation in the supply of nutrients, thereby allowing tighter control of various cellular processes such as cell growth, nutrient uptake and production of target metabolites. The feed conditions (volumetric flow rate and composition) can be varied in a predetermined fashion (open-loop control) or by using feedback control. In a continuous culture operation, nutrients essential for growth are continuously fed and a portion of the culture is continuously withdrawn. The culture volume is controlled using a level controller. A continuous culture is usually preceded by a batch or fed-batch culture. If the mass flow rates of the bioreactor feed and bioreactor effluent are identical and time-invariant, a time-invariant (steady state) operation can be realized after sufficient time has elapsed from the start of continuous culture operation. As in fed-batch culture, the feed rate to a continuous bioreactor can be varied in a temporal sense in a predetermined fashion or using feedback control. Since the culture conditions (in a global sense) can be kept time-invariant, continuous cultures are easier to monitor and control. Unlike the operation of continuous processes employed for production of chemicals, long-term operation of continuous cultures is subject to many operating difficulties, including risks of contamination and loss in productivity due to cell washout in case of unanticipated disturbances and substantial changes in characteristics of the biotic phase. For this reason, batch and fed-batch culture operations are more common than continuous culture operations.

While the batch and fed-batch operations of a bioreactor are inherently transient, transients are also encountered in continuous bioreactor operations before attaining a steady state or during transition from one steady state (established under one set of feed and culture conditions) to another steady state (established under a different set of feed and culture conditions).

The effectiveness of the operation of a biological reactor (cell cultivation) depends on the outcome of the operation and what went into such operation. Some indicators of the outcome are characteristics associated with metabolites which are desired and/or are generated in significant amounts, biomass (cells), and significant material resources employed (substrates). These characteristics typically are amounts of these species in the bioreactor in batch and fed-batch operations and mass flow rates of these in continuous operation, both of which are related to concentrations of these species in the culture. The inventory of *what went into the bioreactor operation* will typically include cost of material resources, operating costs for the bioreactor, and the costs associated with separation and recovery of the desired product from the culture. The cost of raw material resources is proportional to the amount of nutrient medium (substrates) supplied to the culture. The operating costs will take into consideration costs associated with mechanical agitation, pumping of different fluids (feeding of nutrient medium and addition of acid/base and antifoam solutions), supply of a suitable gas phase (aeration in the case of an aerobic bioprocess) to the bioreactor, and control of certain culture parameters such as pH and temperature. The costs associated with downstream processing to recover the desired product at desirable concentration will depend on the culture composition and hence on the outcome of the bioreactor operation. The cost of separation and recovery of the desired product is always related inversely to its concentration in the culture.

The effectiveness of bioreactor operation is assessed via an objective function or a performance index which takes into account the price of the target product and the costs associated with generating that product (most or at least the prominent entries in *what went into the bioreactor operation*). For cost-effective operation of a bioprocess, one is interested in maximizing the objective function. The outcome of the bioreactor operation being decisively dependent on the trajectories of variables affecting the kinetics of the bioprocess (information contained in $\mathbf{f}$ in Eq. 7.1), these trajectories strongly influence the magnitude of the objective function. The trajectories that lead to maximization of the objective function can be attained using feedback control with appropriate controllers. Guiding the trajectories of influential culture variables at or very near their optimal values is accomplished by appropriate manipulation of some of the input variables. (All of these may not be the physical inputs to the process.) Identification of the optimal trajectories of the manipulated inputs can be accomplished using the optimal control theory.

The controlled variables are the output variables that influence the outcome of the process, which is assessed in terms of an objective function or performance index. Those inputs to the process which have the strongest influence on the controlled outputs are usually chosen as manipulated variables. The remaining inputs are either used for process optimization or cannot be influenced or even measured (disturbance variables). In this work, we will assume that there are m manipulated variables, m_d disturbances and no additional inputs used for optimization. In a multivariate system such as a biological reactor, one normally encounters multiple input variables ($m_t = m + m_d$) and multiple output variables which decide the process outcome (p). To control the p outputs, it is essential to take into account how each of the m_t inputs influences each of the p outputs in an uncontrolled process. One would anticipate that input(s) which have the greatest influence on a particular output should be manipulated by appropriate controllers to control the particular output. Control can be implemented by using multivariable or multi-loop controllers. In multivariable control, information from all controlled variables and disturbances is used together to compute all manipulated variable moves. In multi-loop control, many single-input, single-output (SISO) controllers are developed by pairing the appropriate manipulated and controlled variables. The collective operation of these SISO controllers controls the multivariable process. Multivariable controllers, such as linear quadratic Gaussian controllers (LQCs) and model predictive control (MPC) systems, are more effective than multi-loop controllers and their use has increased in recent years.

The number of controllers involved in multi-loop control is min(m_t, p). The issues to be resolved in multi-loop control are (1) how to pair input and output variables and (2) how to design the individual single-loop controllers. The decision on input-output pairings is based on the nature of process interactions (effect of an an input on multiple outputs). Even with the best possible input-output pairings, functioning of individual controller loops may be influenced by other control loops due to process interaction. Ideally, one would like all control loops to function independently. This requires use of decouplers (additional elements inserted between single-loop controllers and the process), so that the output from a controller used to control a particular output influences not only the manipulated input that is paired with that output, but also other inputs in order to eliminate the effects of interaction. The idea behind the use of decouplers is to make the controllers function independently in entirety.

This chapter starts with determination of optimal trajectories during bioprocess operation. Given a process model, this can be done by solving an appropriate open-loop optimization problem to maximize a particular objective function or performance index. The general procedure for identification of optimal open-loop control policies for nonlinear bioprocess models is provided in Section 7.2, which is followed by a detailed case study involving identification of optimal feeding policies for fed-batch cultures with varying complexity of kinetics. The general procedure discussed in Section 7.2 is applicable to both dynamic and steady-state operations of batch / fed-batch / continuous bioprocesses. A related problem, that of further potential improvement in performance of steady-state continuous bioprocesses via periodic variations in one or more inputs, is considered in Section 7.3. The discussion on criteria for superiority of periodic forcing is followed by a case study. Closed-loop feedback control based on state-space models involving multiple single-input, single-output (SISO) controllers is considered in Section 7.4. Methods for selection of multi-loop controller configuration and minimizing / eliminating effects of bioprocess interaction are considered in this section, the ultimate goal being independent functioning of various control loops. Multivariable control is discussed next in Sections 7.5 and 7.6. Identification of optimal feedback control strategies based on optimal open-loop trajectories (Section 7.2) is the focus of Section 7.5. The optimal feedback controllers have the traditional proportional, integral, and derivative modes of action, with controller parameters being functions of time. Finally, the more powerful and increasingly popular model-based multivariable control, the Model Predictive Control (MPC), is the subject of Section 7.6. The discussion of the general recipe for MPC is followed by a specific illustration of one of the MPC schemes, namely the Dynamic Matrix Control (DMC) and an introduction to nonlinear MPC.

Several review papers and books provide overviews of batch process control and its implementation in chemical process industries that complement this chapter. An early comparative assessment of the effectiveness of simple control techniques and dynamic optimization in the 1980s favors simple control tools for controlling fed-batch fermentation processes [261]. Industrial practice in control and diagnosis of batch processes has been reported [429]. The progress and challenges in batch process control have been discussed in various review papers [51, 265, 510] and assessed in the context of scheduling and optimization of batch process operations [501].

7.2 Open-Loop (Optimal) Control

7.2.1 Nonlinear Models of Bioreactor Dynamics

A popular form of operation of bioreactors employing living cells involves the use of a well-mixed reactor. The uniformity of composition and temperature in the reactor allows its representation as a lumped parameter system. The reactor dynamics can be described succinctly as

$$\frac{d\mathbf{x}}{dt} = \mathbf{f}(\mathbf{x}, \mathbf{u}, \mathbf{d}), \qquad \mathbf{x}(0) = \mathbf{x_0}, \tag{7.1}$$

with $\mathbf{x}$ denoting the state variables which represent the status of the cell culture in the bioreactor, and $\mathbf{u}$ and $\mathbf{d}$ representing the input variables which indirectly influence the status of the cell culture. The input variables are further classified into manipulated inputs ($\mathbf{u}$) and disturbance variables ($\mathbf{d}$). Let n, m and m_d denote the number of state variables, manipulated inputs and disturbance variables. Not all state variables can be measured. Some of the state variables, which cannot be measured or can be measured less frequently, are estimated from measurements of other variables that are measured frequently by using estimators (Section 4.3). It must therefore be realized that only some of the state variables may be monitored or estimated. The set of variables which can be measured will be referred to as bioreactor outputs, $\mathbf{y}$, with the number of outputs being p. The relations among the state variables and the output (measured) variables can then be succinctly stated as

$$\mathbf{y} = \mathbf{h}(\mathbf{x}). \tag{7.2}$$

The functions $\mathbf{f}(\cdot)$ and $\mathbf{h}(\cdot)$ are in general nonlinear. But for mathematical convenience in developing the control equations, these functions are linearized. Linear state-space equations are discussed in Section 7.4.

7.2.2 Background on Optimal Control Theory

Whether the bioreactor is operating at a steady state or is exhibiting transients, one is interested in maximizing an appropriate objective function for cost-effectiveness of the operation. For the sake of generality, we consider here a transient bioreactor operation. During the time interval $(0,\ t_f)$, one may be interested in maximizing an objective function

$$J = G(\mathbf{x}(t_f)) + \int_0^{t_f} g(\mathbf{x}, \mathbf{u})dt. \tag{7.3}$$

The objective function in Eq. 7.3 is sufficiently general for a wide variety of practical problems. $G(\cdot)$ denotes the benefit generated at the end of the operation (t_f) and $g(\cdot)$ the benefit materialized during the operation. The optimization may be cast as a minimization problem by defining $G(\cdot)$ and $g(\cdot)$ as costs. The objective function may also be written in terms of output variables $\mathbf{y}$. Thus, for example, if

$$J = G_1\left(\mathbf{y}(t_f)\right) + \int_0^{t_f} g'\left(\mathbf{y}, \mathbf{u}\right) dt, \tag{7.4}$$

then in view of relation 7.2, the objective function in Eq. 7.4 can be restated as in Eq. 7.3. Besides the constraints imposed by the conservation equations (state equations) in Eqs. 7.1, the maximization of J may have to be achieved subject to the following integral constraints:

$$\frac{1}{t_f}\int_0^{t_f} \boldsymbol{\phi}(\mathbf{x},\mathbf{u})dt = \mathbf{0} \quad \text{and} \quad \frac{1}{t_f}\int_0^{t_f} \boldsymbol{\psi}(\mathbf{x},\mathbf{u})dt \leq \mathbf{0}. \tag{7.5}$$

where $\boldsymbol{\phi}(\cdot)$ and $\boldsymbol{\psi}(\cdot)$ are appropriate linear or nonlinear functions. Integral constraints expressed in terms of outputs $\mathbf{y}$, such as

$$\frac{1}{t_f}\int_0^{t_f} \boldsymbol{\phi}'(\mathbf{y},\mathbf{u})\,dt = \mathbf{0} \quad \text{and} \quad \frac{1}{t_f}\int_0^{t_f} \boldsymbol{\psi}'(\mathbf{y},\mathbf{u})\,dt \leq \mathbf{0} \tag{7.6}$$

can be readily expressed as in Eq. 7.5 in view of the relations in Eq. 7.2. As an example of the equality constraint in Eq. 7.5, consider a continuous culture operation. It is usually of interest to identify an optimal feed composition (for example, the substrate feed concentration, S_F) which will lead to maximization of an objective function, such as yield or productivity of the target metabolite. Economic considerations would dictate that this identification be done while keeping the throughput rate of the substrate (G_S) fixed. Different candidate continuous culture operations with variable S_F then would be subject to the integral constraint

$$\frac{1}{t_f}\int_0^{t_f} (FS_F - G_S)\,dt = 0 \quad \rightarrow \quad \frac{1}{t_f}\int_0^{t_f} FS_F dt = G_S. \tag{7.7}$$

The integral constraint in Eq. 7.7 is applicable for situations involving fixed F and S_F in an individual operation as well as time-varying F and S_F in an individual operation, as is the case of forced periodic operation of a continuous culture. As an example of the inequality constraint in Eq. 7.5, consider a batch, a fed-batch or a composite of batch and fed-batch culture operation. It may be desired to maximize an objective function such as the amount or yield of a target metabolite. This may have to be accomplished with a limited amount of substrate (M_S). The candidate operations would then be subject to the constraint

$$\int_0^{t_f} \left(FS_F - \frac{M_S}{t_f}\right) dt \leq 0 \quad \rightarrow \quad \int_0^{t_f} FS_F dt \leq M_S. \tag{7.8}$$

In a strictly batch operation, the nutrient medium containing substrate is added rapidly at the start of operation, i.e., at $t = 0$, the volumetric flow rate F being an impulse function in this operation.

Because of the considerable complexity that the constraints of the form in Eqs. 7.5 add to process optimization, we consider first situations involving constraints only on the manipulated inputs $\mathbf{u}$. Each manipulated input then is considered to be bounded from above and below as

$$\mathbf{u}_{\max} \geq \mathbf{u} \geq \mathbf{u}_{\min} \tag{7.9}$$

where $\mathbf{u}_{\min}$ and $\mathbf{u}_{\max}$ denote the lower and upper bounds, respectively. Maximization of the objective function is then accomplished via maximization of the Hamiltonian H with respect to $\mathbf{u}$. The Hamiltonian is defined as [84]

$$H = g(\mathbf{x}, \mathbf{u}) + \boldsymbol{\lambda}^T \mathbf{f}(\mathbf{x}, \mathbf{u}), \tag{7.10}$$

with $\boldsymbol{\lambda}$ being the vector of adjoint variables associated with Eqs. 7.1. The variation in $\boldsymbol{\lambda}$ with time is described by

$$\frac{d\boldsymbol{\lambda}^T}{dt} = -\frac{\partial H}{\partial \mathbf{x}}. \tag{7.11}$$

The influence of system equations Eq. 7.1 on H (and J) is transmitted by the adjoint variables. Eqs. 7.1 and 7.11 represent a set of ordinary differential equations. Their solution requires knowledge of each state and adjoint variable at some t (usually at $t = 0$ or t_f). Let $\mathbf{x}^*$, $\mathbf{u}^*$, t_f^*, and J^* denote the optimal values of $\mathbf{x}$, $\mathbf{u}$, t_f and J, respectively. Let the variations in $\mathbf{x}$, $\mathbf{u}$, t_f and J for an arbitrary operation from their respective values for the optimal operation be expressed as $\delta t_f = t_f - t_f^*$, $\delta J = J - J^*$, $\delta\mathbf{x} = \mathbf{x} - \mathbf{x}^*$, and $\delta\mathbf{u} = \mathbf{u} - \mathbf{u}^*$. Identification of optimal $\mathbf{u}$ then involves expressing the variation in J (δJ) entirely in terms of variation in $\mathbf{u}$ ($\delta\mathbf{u}$). In general, δJ depends on $\delta\mathbf{x}(0)$, $\delta\mathbf{x}(t_f)$ and δt_f as well. Trivializing the influences of these on δJ leads to the following conditions [84]:

$$\lambda_i(0) = 0 \text{ if } x_i(0) \text{ is not specified.} \tag{7.12}$$

$$\lambda_i(t_f) = \frac{\partial G}{\partial x_i} \text{ if } x_i(t_f) \text{ is not specified.} \tag{7.13}$$

$$H(t_f) = 0 \text{ if } t_f \text{ is not specified.} \tag{7.14}$$

The conditions in Eqs. 7.12 and 7.13 are applicable for once-through operations, i.e., process operations where $\mathbf{x}(0)$ and $\mathbf{x}(t_f)$ are independent (i.e., $\delta\mathbf{x}(0)$ and $\delta\mathbf{x}(t_f)$ are not identical). In cyclic operation of a bioreactor, the operation modes under consideration here (batch, fed-batch and continuous) and certain sequences of these are repeated, with t_f being the duration of a cycle. In this case, $\mathbf{x}(t)$ satisfy the periodic boundary conditions

$$\mathbf{x}(0) = \mathbf{x}(t_f) \Rightarrow \delta\mathbf{x}(0) = \delta\mathbf{x}(t_f). \tag{7.15}$$

The boundary condition for the adjoint variables are then obtained as

$$\lambda_i(t_f) = \frac{\partial G}{\partial x_i} + \lambda_i(0) \quad \text{if } x_i(t_f) \text{ is not specified.} \tag{7.16}$$

In view of the conditions above (Eqs. 7.12-7.14 and 7.16), δJ can be expressed as

$$\delta J = \int_0^{t_f} \left[\left(\frac{\partial H}{\partial \mathbf{u}} \right) \delta \mathbf{u}(t) \right] dt. \tag{7.17}$$

If some components of $\mathbf{u}^*(\mathrm{t})$ include segments (sections) where $u_i = (u_i)_{\min}$ or $(u_i)_{\max}$, then $\delta u_i(\mathrm{t})$ must be positive at $(u_i)_{\min}$ and $\delta u_i(\mathrm{t})$ must be negative at $(u_i)_{\max}$. Since δJ is expected to be non-positive, the following conditions must be satisfied on the optimal trajectory for u_i, $u_i^*(\mathrm{t})$:

$$u_i^*(t) = (u_i)_{\min} \ \text{if} \ \frac{\partial H}{\partial u_i} \leq 0, \tag{7.18}$$

$$u_i^*(t) = (u_i)_{\max} \ \text{if} \ \frac{\partial H}{\partial u_i} \geq 0, \tag{7.19}$$

$$(u_i)_{\min} < u_i^*(t) < (u_i)_{\max} \ \text{if} \ \frac{\partial H}{\partial u_i} = 0. \tag{7.20}$$

Further details on the derivation of Eqs. 7.10 through 7.20 are discussed in Bryson and Ho [84]. As long as H is a nonlinear function of u_i, Eq. 7.20 provides an explicit expression for $u_i^*(\mathrm{t})$.

7.2.3 Singular Control

If the Hamiltonian H varies linearly with u_i, i.e., if

$$H = h_0(\mathbf{x}, \mathbf{u}', \boldsymbol{\lambda}) + h_i(\mathbf{x}, \mathbf{u}', \boldsymbol{\lambda}) u_i(t), \tag{7.21}$$

with $\mathbf{u}'$ being the vector obtained from $\mathbf{u}$ by excluding u_i, then $u_i(t)$ cannot be obtained explicitly from the condition in Eq. 7.20 if h_i is trivial over a finite time interval ($t_1 \leq t \leq t_2$). The control over each such finite interval is referred to as *singular control* and the time interval is referred to as *singular control interval.* The singular control problems are especially difficult to handle due to difficulties associated with identification of *singular arc* (trajectory of $u_i^*(\mathrm{t})$), and estimation of when to transit from boundary control [$u_i^* = (u_i)_{\min}$ or $(u_i)_{\max}$] to singular control and vice versa. Triviality of h_i over a finite time interval implies triviality of first and higher derivatives of h_i with respect to time over the entire time interval. As will become evident in the illustrations presented later, this property is used to identify u_i^* in a singular control interval. Admissibility of singular control is related to the kinetics of the process being optimized, i.e., the elements of **f**. If the bioprocess kinetics is such that singular control is not admissible, then the optimal control policy (trajectory of a manipulated input u_i) would involve operation at the lower or upper bounds for u_i $[(u_i)_{\min}$ or $(u_i)_{\max}]$ or a composite of operations at the lower and upper bounds such that

$$u_i^*(t) = (u_i)_{\min} \text{ if } h_i < 0 \tag{7.22a}$$

$$u_i^*(t) = (u_i)_{\max} \text{ if } h_i > 0 \tag{7.22b}$$

The trajectory of $u_i^*(\text{t})$ then may involve one or more switches from the lower bound to upper bound and vice versa. The values of t at which such switches occur are called *switching times*.

7.2.4 Optimal Control

Next we consider situations where integral constraints in Eqs. 7.5 are applicable. Let there be a equality constraints and b inequality constraints. The original vector of state variables can be augmented by additional $(a+b)$ state variables satisfying the following relations

$$\frac{dx_j}{dt} = \phi_j(\mathbf{x},\ \mathbf{u}),\ x_j(0) = 0,\ j = (n+1), (n+2), \ldots, (n+a) \tag{7.23}$$

$$\frac{dx_j}{dt} = \psi_j(\mathbf{x},\ \mathbf{u}),\ x_j(0) = 0,\ j = (n+a+1), (n+a+2), \ldots, (n+a+b) \tag{7.24}$$

with $\mathbf{x} = [x_1\ x_2\ \ldots\ x_n]^T$ being the original vector of n state variables. The vector of state variables may have to be further augmented if the objective function cannot be directly expressed in the form displayed in Eq. 7.3. Consider for example batch and/or fed-batch operation of a bioprocess. For cost-effective operation, it may be of interest to maximize productivity of the target product P. The objective function in this case would be

$$J = [P(t_f)V(t_f) - P(0)V(0)]/t_f. \tag{7.25}$$

For a single-cycle operation, $P(0)$ and $V(0)$ will be known (specified). The objective function above can be expressed as in Eq. 7.3 by augmenting the vector of state variables by an additional variable satisfying the following

$$dx_j/dt = 1,\quad x_j(0) = 0,\ j = (n+a+b+1). \tag{7.26}$$

The objective function in Eq. 7.25 can now be expressed as in Eq. 7.3 with $g(\mathbf{x}, \mathbf{u})$ being trivial. The application of optimal control theory then follows as discussed before with the vector of state variables now consisting of n process variables and additional state variables satisfying relations such as Eqs. 7.23, 7.24 and 7.26. Since the Hamiltonian in Eq. 7.10 is independent of x_j $(j = n+1, n+2, \ldots, n+a+b)$, it follows from Eq. 7.11 that the corresponding adjoint variables, λ_j $(j = n+1, n+2, \ldots, n+a+b)$, are time-invariant. For a steady-state operation, all adjoint variables λ_j $(j = 1, 2, \ldots, n+a+b+1)$ are time-invariant and still provided by Eq. 7.11.

While integral constraints can be handled by augmenting the state variable space, if the process to be optimized is subject to algebraic constraints of the type

$$\boldsymbol{\phi}_1(\mathbf{x},\ \mathbf{u}) = \mathbf{0} \quad \text{and} \quad \boldsymbol{\psi}_1(\mathbf{x},\ \mathbf{u}) \leq \mathbf{0}, \tag{7.27}$$

then the Hamiltonian in Eq. 7.10 must be appended to account for these constraints. The case study which follows provides an illustration of this.

7.2.5 Case Study - Feeding Policy in Single-Cycle and Repeated Fed-Batch Operations

Fed-batch operation is used for production of a variety of biochemicals [452, 642]. Formation of the desired product(s) can be optimized by proper manipulation of the feed conditions (feed rate and feed composition). Here we consider a bioprocess represented by an unstructured model. Assuming that formation of biomass (X), utilization of the limiting substrate (S), and formation of the desired non-biomass product (P) represent the key rate processes and the biomass-specific rates of these can be expressed exclusively in terms of concentrations of cell mass, limiting substrate and the target non-biomass product, the dynamics of the bioreactor can be described by the following total and species material balances:

$$\frac{dV}{dt} = F, \quad V(0) = V_0 \tag{7.28}$$

$$\frac{dX}{dt} = \mu X - \frac{F}{V}X, \quad X(0) = X_0 \tag{7.29}$$

$$\frac{dS}{dt} = \frac{F}{V}(S_F - S) - \sigma X, \quad S(0) = S_0 \tag{7.30}$$

$$\frac{dP}{dt} = \epsilon X - \frac{F}{V}P, \quad P(0) = P_0 \tag{7.31}$$

The bioreactor volume (V), volumetric feed rate (F), and the substrate feed concentration (S_F) are constrained as

$$V(t) \leq V_m, \tag{7.32}$$

$$0 \leq F(t) \leq F_m, \tag{7.33}$$

and

$$0 \leq S_F(t) \leq S_{Fm}. \tag{7.34}$$

Constraint 7.32 is necessary to prevent flooding. It is desired to maximize the objective function in Eq. 7.3 with g being considered constant (a special case) with $\mathbf{x} = [V\ X\ S\ P\ t]^T$, the time variation of x_5 being described by Eq. 7.26 with $n = 4$ and $a = b = 0$. The Hamiltonian can be expressed as

$$H = h_0 + h_1' F + h_2' F S_F, \tag{7.35}$$

where

$$h_0 = (\lambda_2 \mu - \lambda_3 \sigma + \lambda_4 \epsilon) X + \lambda_5 + g + \eta_0 (V - V_m) \tag{7.36}$$

$$h_1' = \lambda_1 - [\lambda_2 X + \lambda_3 S + \lambda_4 P]/V, \quad h_2' = \lambda_3 / V. \tag{7.37}$$

η_0 is a Lagrangian multiplier to take care of the algebraic inequality constraint, Eq. 7.32 ($\eta_0(t) = 0$ when $V < V_m$ and $\eta_0(t) \geq 0$ when $V = V_m$) [392, 447, 454]. The variations in adjoint variables can be expressed as in Eq. 7.11, which assume the form

$$\frac{d\lambda_1}{dt} = -\frac{F}{V^2}[\lambda_2 X - \lambda_3 (S_F - S) + \lambda_4 P] - \eta_0 \tag{7.38}$$

$$\frac{d\lambda_2}{dt} = \frac{F}{V}\lambda_2 - (\mu + X\mu_X)\lambda_2 + (\sigma + X\sigma_X)\lambda_3 - (\epsilon + X\epsilon_X)\lambda_4 \tag{7.39}$$

$$\frac{d\lambda_3}{dt} = \frac{F}{V}\lambda_3 - (\lambda_2 \mu_S - \lambda_3 \sigma_S + \lambda_4 \epsilon_S) X \tag{7.40}$$

$$\frac{d\lambda_4}{dt} = \frac{F}{V}\lambda_4 - (\lambda_2 \mu_P - \lambda_3 \sigma_P + \lambda_4 \epsilon_P) X \tag{7.41}$$

$$\frac{d\lambda_5}{dt} = 0 \Rightarrow \lambda_5 = \text{constant}. \tag{7.42}$$

The subscripts X, S and P used in the above and elsewhere in the case studies in this chapter denote partial derivatives of a quantity (such as a specific rate or ratio of two specific rates) with respect to X, S and P, respectively. The Hamiltonian must be constant ($= H^*$) on the optimal path, H^* being zero when t_f is not specified (Eq. 7.14).

The trajectory of $\mathbf{x}(t)$ will, in general, comprise an interior arc ($V < V_m$) and a boundary arc ($V = V_m$). When the orders of the boundary control and singular control are both unity, there is no jump in the adjoint variables at each junction point of the boundary and interior arcs [392, 447, 454]. Once the bioreactor is full ($V = V_m$), its operation continues on the boundary arc ($F = 0$) until $t = t_f$ [392].

It is evident from Eqs. 7.35 and 7.37 that the Hamiltonian is linear in both F and S_F. Admissibility of singular control must therefore be examined. The conditions for admissibility of singular control can be obtained from Eq. 7.20 as

$$h_1 = h_1' + h_2' S_F = 0 \quad \text{in} \quad t_1 \leq t \leq t_2 \quad \text{for} \quad \partial H/\partial F = 0 \tag{7.43}$$

$$F\lambda_3 = 0 \quad \text{in} \quad t_3 \leq t \leq t_4 \quad \text{for} \quad \partial H/\partial S_F = 0. \tag{7.44}$$

For the sake of illustration, we consider here $\mathbf{u} = F$ (i.e., S_F is not manipulated and kept time-invariant) and designate the control policy in singular control as F_s. The values of manipulated inputs in the singular control interval(s) are dependent on both the state and adjoint variables. For some specific cases of bioprocess kinetics, it is possible to express the optimal control policy, $\mathbf{u}$, entirely in terms of state variables. These cases are considered here for the purpose of illustration. Such control policy can be easily implemented in a feedback mode.

Case 1. The three specific rates are related to one another by two linear relations as

$$\sigma = p\mu \quad \text{and} \quad \epsilon = c\mu \tag{7.45}$$

with p and c being constant. Substituting these relations and Eq. 7.28 into Eqs. 7.29-7.31, eliminating the specific cell growth rate between any two of these and integrating the resulting equations, it can be deduced that the bioreactor state moves along the intersection of the hyperplanes

$$(S_F - S - pX)\,V = c_1 = (S_F - S_0 - pX_0)\,V_0 \tag{7.46}$$

and

$$\left(S_F - S - \frac{p}{c}P\right)V = c_2 = \left(S_F - S_0 - \frac{p}{c}P_0\right)V_0 \tag{7.47}$$

Satisfaction of relations in Eq. 7.45 implies that the bioreactor dynamics can be completely described by two algebraic relations, Eqs. 7.46 and 7.47 and two differential equations among Eqs. 7.28-7.31. Utilizing triviality of h_1 and dh_1/dt, it can be deduced that the bioreactor trajectories must lie on the surface

$$g(X,S,P) = X\mu_X - (S_F - S)\mu_S + P\mu_P = 0 \tag{7.48}$$

during singular control intervals. The feed rate during the singular control interval is then obtained from triviality of d^2h_1/dt^2 as

$$\frac{F_s}{V} = \frac{(g_X - pg_S + cg_P)\,\mu X}{Xg_X - (S_F - S)g_S + Pg_P}. \tag{7.49}$$

If X_0, S_0 and P_0 for a typical cycle lie on the line

$$X = \frac{S_F - S}{p} = \frac{P}{c} \tag{7.50}$$

then X, S and P lie on the same line during that cycle (Eqs. 7.46 and 7.47). The expressions for the singular surface and control policy during singular control, Eqs. 7.48 and 7.49, then reduce to

$$\frac{d\mu}{dS} = \mu_S - \frac{1}{p}\mu_X - \frac{c}{p}\mu_P = 0 \tag{7.51}$$

and

$$\frac{F_s}{V} = \mu_i = \mu(S_i) \tag{7.52}$$

where S_i is the substrate concentration at which conditions in Eq. 7.50 and 7.51 are satisfied. In this special situation, S, X and P remain time-invariant at S_i, X_i and P_i, respectively (S_i, X_i and P_i satisfy relations in Eq. 7.50). After substitution of relation Eq. 7.52 into total mass balance, Eq. 7.28, and integration of the same, one can deduce that both F_s and V vary exponentially with time.

In a strictly batch operation ($F = 0$), it can be deduced from Eqs. 7.29-7.31 and 7.45 that the concentration trajectories will lie on the line in Eq. 7.50. Further, in another related operation, viz., a continuous culture at steady state, X, S and P also lie on the line in Eq. 7.50.

In a typical cycle of a fed-batch operation, increase in V implies that the bioreactor state moves closer to the line defined by Eq. 7.50 if not already on it at the beginning of that cycle (see Eqs. 7.46 and 7.47). The feed point ($S = S_F$, $X = P = 0$) also lies on the line defined by Eq. 7.50. In a cyclic operation, the bioreactor contents are partially or completely withdrawn at the termination of a cycle; this is followed by addition of fresh feed. The initial state for the reactive portion of the next cycle, (X_{in}, S_{in}, P_{in}), therefore moves closer to the line defined by Eq. 7.50 if not already on it. It follows then that in a cyclic fed-batch operation with reproducible cycles, the concentration trajectories lie on the line in Eq. 7.50.

Relations 7.45 imply that among the three rate processes under consideration, only one (for example, cell growth) is independent. Optimal singular control interval therefore involves maximization of the specific cell growth rate, as indicated by Eqs. 7.48 and 7.51.

In general, μ_X and μ_P are non-positive. Singular control is therefore feasible only when $\mu_S \leq 0$ (Eqs. 7.48 and 7.51). Since μ increases with increasing S at low values of the same, this then requires that μ exhibit non-monotonic behavior with respect to S. Singular control is therefore not admissible if μ is a monotonically increasing function of S, this being the case with some fermentations producing alcohols [47, 239, 325, 338, 544].

Case 2. The three specific rates are related to one another by a single linear relation as

$$A\mu + B\sigma + C\varepsilon = 0 \tag{7.53}$$

with A, B and C being constants, at least two of which are non-zero. In view of Eq. 7.53, it can be deduced that, in a typical cycle, the bioreactor state lies on the hyperplane

$$[AX + B(S_F - S) + CP]V = c_3 = [AX_0 + B(S_F - S_0) + CP_0]V_0 \tag{7.54}$$

We consider here the special case where g is trivial and G is independent of t_f in Eq. 7.3, and t_f is not specified (therefore $H = \lambda_5 = 0$). It then follows that h_0 (Eqs. 7.35 and 7.36) must be trivial in a singular control interval. Two types of linear relations among the three rate processes under consideration are commonly encountered in bioprocess kinetics. The first type arises when cell growth and synthesis of the target non-biomass product account almost entirely for substrate utilization, i.e., when

$$\sigma = a\mu + b\varepsilon \quad (a = -A/B \text{ and } b = -C/B, B \neq 0). \tag{7.55}$$

In the other type, synthesis of the target non-biomass product is associated with and proportional to cell growth, with cell growth and substrate utilization being linearly independent rate processes, i.e.,

$$\varepsilon = c\mu \quad (c = -A/C, B = 0). \tag{7.56}$$

When Eq. 7.55 is satisfied, the following necessary and sufficient condition for admissibility of singular control is obtained in view of the triviality of h_0 and dh_1/dt in a singular control interval.

$$X\left(\frac{\epsilon}{\mu}\right)_X - (S_F - S)\left(\frac{\epsilon}{\mu}\right)_S + P\left(\frac{\epsilon}{\mu}\right)_P = 0 \tag{7.57}$$

The above relation provides the description of the singular surface in the three dimensional concentration space (X, S, P). The feeding policy during singular control is obtained from the triviality of d^2h_1/dt^2 as

$$\frac{F_s}{V} = \frac{[\delta\mu - \phi\sigma + \psi\epsilon]X}{[\delta X - \phi(S_F - S) + \psi P]} = \eta(X, S, P) \tag{7.58a}$$

where

$$\delta = \alpha + X\alpha_X - (S_F - S)\beta_X + P\gamma_X \tag{7.58b}$$

$$\phi = \beta + X\alpha_S - (S_F - S)\beta_S + P\gamma_S \tag{7.58c}$$

$$\psi = \gamma + X\alpha_P - (S_F - S)\beta_P + P\gamma_P \tag{7.58d}$$

$$\alpha = \mu\epsilon_X - \epsilon\mu_X, \quad \beta = \mu\epsilon_S - \epsilon\mu_S, \quad \gamma = \mu\epsilon_P - \epsilon\mu_P \tag{7.58e}$$

When Eq. 7.56 is applicable, the following necessary and sufficient condition for admissibility of singular control, which also describes the singular surface in the $(X,\ S,\ P)$ plane, can be obtained in view of triviality of h_o and dh_1/dt.

$$X\left(\frac{\mu}{\sigma}\right)_X - (S_F - S)\left(\frac{\mu}{\sigma}\right)_S + P\left(\frac{\mu}{\sigma}\right)_P = 0. \tag{7.59}$$

The feeding policy during the singular control is obtained as in Eq. 7.58a (since d^2h_1/dt^2 is trivial), with α, β and γ being defined as

$$\alpha = \sigma\mu_X - \mu\sigma_X, \quad \beta = \sigma\mu_S - \mu\sigma_S, \quad \gamma = \sigma\mu_P - \mu\sigma_P. \tag{7.60}$$

If X_0, S_0 and P_0 in a typical cycle lie on the plane

$$AX + B(S_F - S) + CP = 0, \tag{7.61}$$

then X, S and P lie on the same during that cycle (see Eq. 7.54). The bioreactor trajectories can then be completely described in a two-dimensional phase-plane ($S - X$ if $B = 0$ or $X - P$ if $B \neq 0$) with the singular arc (X_i, S_i and P_i moving along the singular arc) being the intersection of the singular surface in Eq. 7.57 (if B is non-zero) or Eq. 7.59 (if $B = 0$) with the plane in Eq. 7.61.

The feed point ($S = S_F$, $X = P = 0$) lies on the plane in Eq. 7.61. Further, in a strictly batch operation ($F = 0$), it can be deduced from Eqs. 7.29-7.31 and 7.53 that the concentration trajectories will lie on the plane in Eq. 7.61. Moreover, for a continuous culture at steady state, X, S and P also lie on the plane in Eq. 7.61. In a typical cycle of a fed-batch operation, increase in V implies that the bioreactor state (in terms of concentrations) moves closer to the plane in Eq. 7.61 if not already on it at the start of that cycle (see Eq. 7.54). In a cyclic fed-batch operation, the bioreactor contents are partially or completely withdrawn at the end of each cycle, which is followed by rapid addition of fresh feed. The initial state of the reactive portion of the next cycle, $(X_0,\ S_0,\ P_0)$, therefore moves closer to the plane in Eq. 7.61 if not already on it. One can conclude then that in a repeated fed-batch operation with reproducible cycles, all concentration trajectories will lie on this plane. The trajectories in a batch operation in the two-dimensional phase-plane ($S - X$ if $B = 0$ or $X - P$ if $B \neq 0$) will in general be nonlinear. These can in some cases have inflection points. The locus of inflection points is described by one of the following surfaces [447]:

$$\mu\left(\frac{\epsilon}{\mu}\right)_X - \sigma\left(\frac{\epsilon}{\mu}\right)_S + \epsilon\left(\frac{\epsilon}{\mu}\right)_P = 0, \quad \text{if } \sigma = a\mu + b\epsilon \tag{7.62}$$

$$\mu\left(\frac{\mu}{\sigma}\right)_X - \sigma\left(\frac{\mu}{\sigma}\right)_S + \epsilon\left(\frac{\mu}{\sigma}\right)_P = 0, \quad \text{if } \epsilon = c\mu. \tag{7.63}$$

The intersections of surfaces in Eqs. 7.57 and 7.62 or those of surfaces in Eqs. 7.59 and 7.63, as appropriate, are of special significance in a fed-batch operation. At each such intersection,

$$\frac{\mu}{X} = \frac{\sigma}{(S_F - S)} = \frac{\epsilon}{P}. \tag{7.64}$$

These intersections are discrete points on the stoichiometric plane in Eq. 7.61. These points are referred to as limit points or singular inflection points. At a limit point, the feeding policy in Eqs. 7.58 and 7.60 reduces to

$$\frac{F_s}{V} = \mu_i = \mu(X_i, S_i, P_i) \tag{7.65}$$

with X_i, S_i and P_i satisfying Eq. 7.64 and being obtained from solutions of Eqs. 7.57 or 7.59, as appropriate, and Eq. 7.61. It follows from Eqs. 7.64 and 7.65 that X_i, S_i and P_i are equilibrium (time-invariant) solutions of Eqs. 7.29-7.31. Upon reaching the limit point, F_s varies exponentially with time until transition to boundary control ($F = 0$) occurs upon saturation of the bioreactor volume ($V = V_m$). The bioreactor operation at the limit point is a quasi-steady state operation since X, S, and P remain time-invariant while V increases with t.

In view of the nonlinear dependence of the three specific rates on X, S and P, multiplicity of limit points for a fixed feed composition cannot be ruled out. In a repeated fed-batch operation, the bioreactor trajectories during singular control interval of each cycle must terminate at a locally asymptotically stable (accessible) limit point. The necessary and sufficient conditions for local accessibility of a limit point can be obtained via linearized stability analysis of Eqs. 7.29-7.31. These conditions have been reported [447].

The first example of kinetics belonging to this case considered by Parulekar [447] pertains to binary quasi-linear relations among μ, σ and ε; $\mu = \mu(X,\ S,\ P)$, $\sigma = p\mu + q$, $\varepsilon = c\mu + e$; c, e, p and q are constants (at least e or q is non-zero). This kinetics is applicable for bioprocesses where the limiting substrate is utilized for cell growth and maintenance and/or product synthesis, the synthesis of the target product being partially growth associated and partially non-growth associated. Eqs. 7.57 ($e \neq 0$) and 7.59 ($e = 0$) reduce in this case to condition in Eq. 7.48. In general, μ_X and μ_P are non-positive. Singular control is therefore feasible only when $\mu_S \leq 0$ (Eqs. 7.48 and 7.51). Since μ increases with increasing S at low values of the same, this then requires that μ exhibit non-monotonic behavior with respect to S. Specific examples of this kinetics include (i) production of propionic acid by *Propionibacterium shermanii* [230] for which $\sigma = p\mu$, $e = c\mu + e$; and (ii) production of ammonium lactate by *Lactobacillus* species [575] for which $\sigma = b\varepsilon$, $\varepsilon = c\mu + e$, μ being function of S and P in both cases. For both bioprocesses, μ_S is positive and μ_P is negative for all S and P. Singular control is therefore inadmissible for either process.

The second example pertains to ethanol production from glucose by *Saccharomyces cerevisiae* [8, 9]. Two different forms of kinetics of cell growth, substrate utilization and product (ethanol) formation have been proposed for this bioprocess, these being

$$\mu = \mu_1(S)\mu_2(P),\ \epsilon = \epsilon_1(S)\epsilon_2(P),\ \sigma = p\mu,\ \mu_1(S) = \frac{\mu_m S}{(K_1 + S)},$$
$$\mu_2(P) = \frac{K_3}{(K_3 + P)},\ \epsilon_1(S) = \frac{\epsilon_m S}{(K_2 + S)},\ \epsilon_2(P) = \frac{K_4}{(K_4 + P)}, \quad (7.66)$$
$$\mu_2(P) = exp(-\alpha_1 P), \quad \epsilon_2(P) = exp(-\beta_1 P). \quad (7.67)$$

The third example pertains to ethanol production from cellulose hydrolysate by *S. cerevisiae*. The following kinetic expressions have been used for description of this bioprocess [190, 600].

$$\mu = \mu_m \left(1 - \frac{P}{P_m}\right) \frac{S}{K_S + S + S^2/K_I},$$
$$\epsilon = \epsilon_m \left(1 - \frac{P}{P'_m}\right) \frac{S}{K'_S + S + S^2/K'_I},\ \sigma = \frac{\epsilon}{Y_{P/S}}. \quad (7.68)$$

For these examples, optimization of single-cycle (once-through) and cyclic batch and fed-batch operations of these bioprocesses has been investigated in detail by Parulekar [447]. Here, we import numerical illustrations for the second example.

The values assigned for the kinetic parameters in Eqs. 7.66 and 7.67 were [8, 9]

$$K_1 = 0.22\ g/L,\ K_2 = 0.44\ g/L,\ \mu_m = 0.408\ h^{-1},\ p = 10,\ \varepsilon_m = 1.0\ h^{-1},$$
$$K_3 = 16\ g/L,\ K_4 = 71.5\ g/L,\ \alpha_1 = 0.028\ L/g,\ \beta_1 = 0.015\ L/g. \quad (7.69)$$

The performance index considered was the product (ethanol) yield at the end of the bioreactor operation, viz., $J = P_f/S_F$, $P_f = P(t_f)$. The performance of the optimal fed-batch operation was compared for two types of cyclic batch operation. Each cycle of a cyclic batch operation consists of filling the reactor rapidly ($F_m \to \infty$) with feed to increase the reactor volume from an initial volume V_0 to V_m ($0 \leq V_0 \leq V_m$), followed by a batch operation until the objective function J is maximized and then terminating the cycle by rapid withdrawal of the reactor contents to reduce the bioreactor volume from V_m to V_0. A batch operation is normally continued until the stoichiometrically limiting nutrient (limiting substrate here) is completely utilized and/or product synthesis is terminated, for this ensures that the bioreactor contents at the end of each cycle will have the maximum product concentration for a given feed composition. The batch reactor trajectories would terminate at (X_f, S_f, P_f) starting from (X_0, S_0, P_0) with the feed point being (0, S_F, 0). The relations among the three concentration variables at these points are

$$\frac{X_0}{X_f} = \frac{S_F - S_0}{S_F - S_f} = \frac{P_0}{P_f} = \frac{V_0}{V_m}. \tag{7.70}$$

The cyclic batch operations with $V_0 = 0$ are referred to as operations without recycle (from one batch to the next) while those with non-zero V_0 are termed as operations with recycle.

For the kinetics described in Eq. 7.66, a unique and locally asymptotically stable limit point on the singular arc is guaranteed for all S_F with the exception of very low S_F (Figure 7.1(a)). In a cyclic operation, the bioreactor trajectories in fed-batch mode terminate at a limit point and the trajectories approach the limit point in a single-cycle operation. The overall product-to-substrate yield for each of the three operations under consideration and the substrate and product concentrations at the limit point, S_i and P_i, respectively, all increased with increasing S_F (Figure 7.1). It is evident from Figure 7.1(b) that for the kinetics under consideration, the cyclic fed-batch operations are superior to cyclic batch operations with recycle, which in turn are superior to cyclic batch operations without recycle. The differential in the product yield between the optimal (fed-batch) operation and the two suboptimal (batch) operations increases as S_F is increased (Figure 7.1(b)). The maximum theoretical yield of ethanol based on glucose is 0.5111. For the kinetic parameters considered, Eq. 7.69, the overall ethanol yields for cyclic fed-batch operations exceed the maximum theoretical yield for S_F in excess of 76.6 g/L. The magnitudes of some of the kinetic parameters in this S_F range are therefore suspect.

Results for the kinetics described in Eqs. 7.66, 7.67, and 7.69 are presented in Figure 7.2. The singular control in this case has a richer variety. The number of limit points for the singular arc is (i) zero if $S_F \leq 3.1957\ g/L$ or if $S_F > 125.317\ g/L$, (ii) one if $3.1957\ g/L < S_F \leq 8.0894\ g/L$, and (ii) two if $8.0894\ g/L < S_F \leq 125.317\ g/L$. The critical S_F ($125.317\ g/L$)

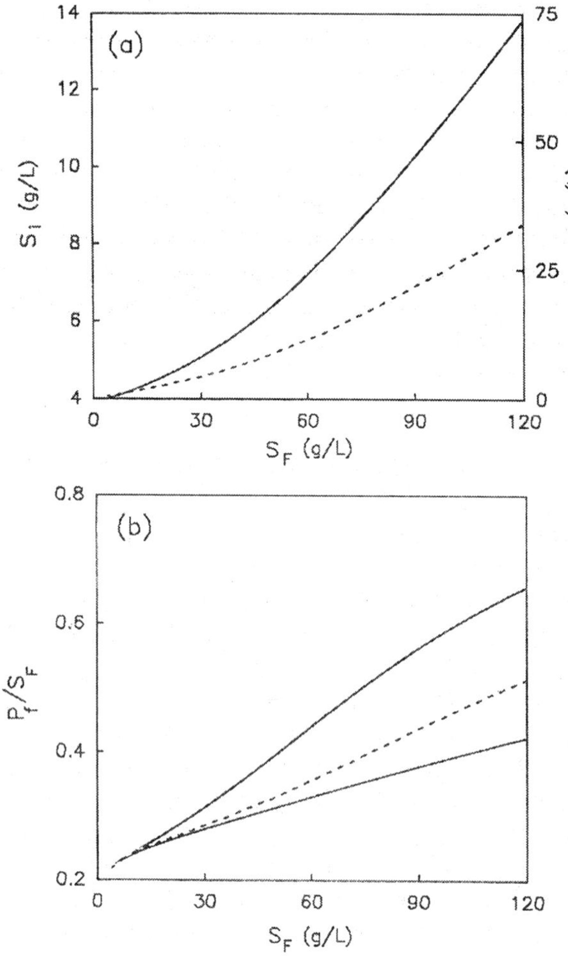

Figure 7.1. Profiles of (a) S_i (- - -) and P_i (—), and (b) overall product yields for repeated batch operation without recycle (lower solid curve) and repeated fed-batch operation (upper solid curve) for fermentation described by Eq. 7.66. The dashed curve in (b) represents the upper (open) bound on the profiles of the overall product yield for repeated batch operations with recycle [447].

also represents the bifurcation point for the limit-point curve (Figure 7.2). Limit points lying on the lower branch (portion CDE) of the limit-point

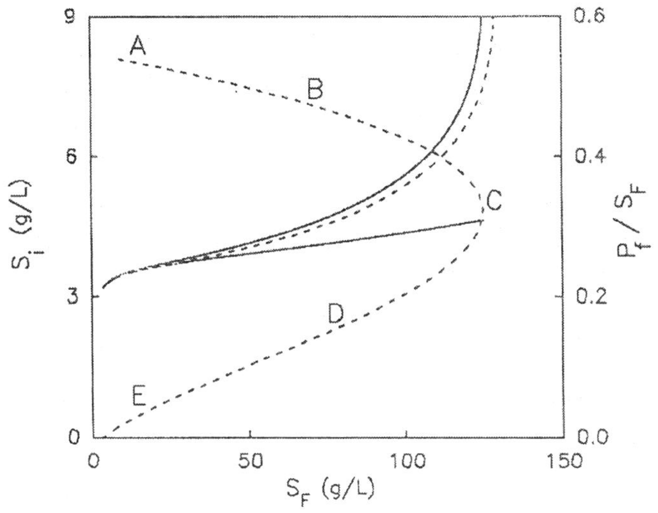

Figure 7.2. Profiles of S_i (dashed curve ABCDE) and overall product yields for repeated batch operation without recycle (lower solid curve) and repeated fed-batch operation (upper solid curve). Singular control in each cycle of a repeated fed-batch operation terminates at a locally stable limit point [S_i lying on the upper branch (portion ABC) of the limit point curve ABCDE]. The non-labeled dashed curve represents the upper (open) bound on the profiles of the overall product yield for repeated batch operations with recycle [447].

curve are unstable. Parulekar [447] has established that fed-batch operations terminating at an unstable limit point are not feasible. The profiles of overall product yield in Figure 7.2 illustrate the superiority of cyclic fed-batch operation with singular control terminating at the stable limit point over cyclic batch operations. These profiles also reveal the substantial improvement in yield that can be obtained with recycle in a cyclic batch operation.

The optimizations based on the highly lumped models such as the ones considered in Eqs. 7.66 and 7.67 may be sensitive to variations in the kinetic parameters in these, some of which have significant uncertainty. For the kinetic parameters considered in Eq. 7.69, the predicted maximum product yield exceeded the theoretical maximum yield beyond certain S_F. This indicates that these parameter values are not accurate enough to be used for fed-batch optimization. Sensitivity of the objective function (maximum product-to substrate yield at the termination of a cycle in a cyclic fed-batch operation, P_f/S_F) to variations in the kinetic parameters was therefore examined. The effects of ±25% variations in each of the kinetic parameters on the maximum product yield are illustrated in Table 7.1. The kinetic parameters, with the exception of the parameter being varied, were assigned the values listed in Eq. 7.69. In each case, a repeated fed-batch operation provided the maximum P_f for a given S_F. For both examples of kinetics, the maximum product yield increased with increases in K_1, ε_m and biomass-to-substrate yield ($Y_{X/S} = 1/p$) and decreases in K_2 and μ_m. The maximum product yield increased with increased cell growth inhibition by the desired product (signified by a decrease in K_3 or an increase in α_1) and with reduced inhibition/repression of product synthesis by the desired product (signified by an increase in K_4 or a decrease in β_m). The maximum product yield is very sensitive to μ_m, ε_m, p, K_3 (example 1), and α_m (example 2), moderately sensitive to K_4 (example 1) and β_m (example 2), and less sensitive to K_1 and K_2. The results in Table 7.1 clearly demonstrate the need for accurate estimation of the maximum specific cell growth and product formation rates, biomass-to-substrate yield, and product-inhibition coefficients and the necessity of frequent updating/retuning of kinetic parameters via on-line estimation when lumped kinetic models are employed for bioprocess optimization.

Case 3. The three specific rates are functions of S and P but have no linear relations among them.

The objective function in Eq. 7.3 is considered to be independent of X_f and t_f, both of which are not specified, and λ_5 is trivial as a result. Further, g in Eq. 7.3 is considered to be trivial. Termination of bioreactor operation in a particular cycle must occur in singular control or batch mode, the final reactor volume being V_m in either case. It has been shown that λ_2 is trivial during singular control. Triviality of h_0 and dh_1/dt during singular control then provides the following necessary and sufficient condition for admissibility of singular control and description of singular arc

$$(S_F - S)\left(\frac{\epsilon}{\sigma}\right)_S = P\left(\frac{\epsilon}{\sigma}\right)_P. \tag{7.71}$$

The feeding policy during singular control (obtained from $d^2h_1/dt^2 = 0$) is described in Eq. 7.58 with α, β and γ being defined as

$$\alpha = 0, \quad \beta = \sigma\epsilon_S - \epsilon\sigma_S, \quad \gamma = \epsilon_P\sigma - \epsilon\sigma_P. \tag{7.72}$$

The projections of the batch bioreactor trajectories on an $S - P$ plane will in general be nonlinear due to the nonlinear nature of σ and ε and may

Table 7.1. Sensitivity of the maximum product yield (P_f/S_F) to kinetic parameters in Eqs. 7.66, 7.67 and 7.69. Only one parameter was varied at a time. Base yields correspond to the appropriate parameter values provided in the text [447].

	Example 1		Example 2	
Variation (%)	+25	-25	+25	-25
S_F (g/L)	50	50	70	70
Base yield	0.39662	0.39662	0.30398	0.30398
Parameter	Maximum Product Yield		Maximum Product Yield	
K_1 (g/L)	0.40391	0.3901	0.30776	0.3006
K_2 (g/L)	0.38523	0.41175	0.298	0.3119
K_3 (g/L)	0.34085	0.52213	–	–
K_4 (g/L)	0.42313	0.3629	–	–
α_1 (L/g)	–	–	0.38941	0.26232
β_1 (L/g)	–	–	0.2789	0.33981
μ_m (1/h)	0.28681	0.6148	0.22863	0.46262
ϵ_m (1/h)	0.55642	0.26195	0.41764	0.21137
$Y_{X/S} = 1/p$	0.55642	0.26195	0.41764	0.21137

admit one or more inflection points ($d^2P/dS^2 = 0$ at these points). The locus of the inflection points is provided by

$$\epsilon \left(\frac{\epsilon}{\sigma}\right)_P - \sigma \left(\frac{\epsilon}{\sigma}\right)_S = 0 \tag{7.73}$$

The intersections of the singular arc and the locus of inflection points have special significance with respect to singular control. These intersections are discrete points ($S = S_i,\ P = P_i$) where

$$\frac{\sigma}{(S_F - S)} = \frac{\epsilon}{P} = \rho(S, P). \tag{7.74}$$

The feeding policy defined in Eqs. 7.58 and 7.72 reduces to

$$\frac{F_s}{V} = \rho(S_i, P_i)X. \tag{7.75}$$

at each such intersection.

It follows from Eqs. 7.74 and 7.75 that S_i and P_i are equilibrium (time-invariant) solutions of Eqs. 7.30 and 7.31, respectively. Singular control operation at such points is characterized by time invariance of S and P and a gradual approach of X to the equilibrium value X_i ($X_i = \mu(S_i,\ P_i)/\rho(S_i,\ P_i)$) unless already at it. Upon reaching such a quasi-steady state ($(X,\ S,\ P) = (X_i,\ S_i,\ P_i)$, V varies with t) in a cycle, the bioreactor will be operated in an exponential fed-batch mode until a transition to the boundary control ($F = 0$, batch operation) occurs upon saturation of bioreactor volume.

In view of the nonlinear dependence of the three specific rates on S and P, multiplicity of limit points for a fixed feed composition cannot be ruled out. In a repeated fed-batch operation, the bioreactor trajectories during singular control interval of each cycle must terminate at a locally asymptotically stable (accessible) limit point. The necessary and sufficient conditions for local accessibility of a limit point can be obtained via linearized stability analysis of Eqs. 7.29-7.31 [447].

7.3 Forced Periodic Operations

The performance of optimal steady state continuous chemical processes can be improved in some cases by forced periodic operation of these processes. Significant experimental and theoretical effort has been undertaken to identify operations that lead to improved productivity, yield and/or selectivity of chemical reactors by periodic variations in one or more reactor inputs [30, 34, 102, 103, 104, 115, 124, 136, 141, 237, 240, 256, 318, 319, 385, 442, 443, 444, 445, 509, 550, 561, 571, 572, 573, 632, 633, 634, 653]. Optimization of batch and fed-batch operations of bioprocesses has received much more attention compared to optimization of continuous operations. With the increasing significance of continuous bioreactor operations, it is important to know how these can be operated more effectively. The effect of cycling of feed conditions on behavior of continuous cultures has been examined experimentally and theoretically in few prior studies [4, 5, 74, 322, 448, 449, 450, 451, 455, 460, 468, 469, 525, 549, 568, 569, 570, 610, 636, 678, 688].

A matter of primary concern in the periodic control problem is whether and when forced periodic control is superior to steady state control. The three major approaches taken for analysis of forced periodic operations [202, 571] are: (1) the Hamilton-Jacobi approach based on the maximum principle [34, 202, 358, 431], (2) a frequency-domain approach using second-variations methods [66, 634], and (3) numerical approach based on, among other things, vibrational control [102, 103, 104, 442, 443, 444, 445, 509]. Sufficient conditions for optimality of periodic control have been proposed using the Hamilton-Jacobi approach, relaxed steady state analysis, and second-variations methods. For periodic operations employing high frequencies, the sufficient condition is either based on the maximum principle or relaxed steady state analysis [30, 34]. In very low frequency periodic operations, description of process dynamics is based on the quasi-steady state assumption. For the intermediate frequency range, the sufficient condition, based on second-variations methods, is provided by the π-criterion [66, 202]. While the relaxed steady state analysis allows for strong variations in control variables, both the π-criterion and the maximum principle are applicable only for weak variations in control variables.

A generalized π-criterion based on perturbations around arbitrary steady states which are locally, asymptotically stable has been proposed [573]. The development of the generalization, which is based on a line of reasoning similar to that outlined by Bryson and Ho [84] and the averaging result of Tikhonov et al. [589], allows application of the π-criterion to a broader range of problems. Continuous processes may not always operate at an optimal steady state and in some situations, optimal steady states may not be admissible [448, 449, 573]. The generalized π-criterion is useful in these cases to explore the possibility of improving the process performance via forced periodic operation.

7.3.1 Preliminaries on the π-Criterion

Consider the steady τ-periodic operation of a continuous process described by

$$\frac{d\mathbf{x}}{dt} = \mathbf{f}(\mathbf{x},\ \mathbf{u}),\ \mathbf{x}(0) = \mathbf{x}(\tau). \tag{7.76}$$

The optimal periodic control problem is to maximize a scalar objective function

$$J = \frac{1}{\tau}\int_0^{\tau} h(\mathbf{x},\ \mathbf{u})dt, \tag{7.77}$$

subject to the integral constraints in Eq. 7.5 with $t_f = \tau$. Let there be a steady-state solution $\mathbf{x}^*$ of Eq. 7.76 corresponding to $\mathbf{u} = \mathbf{u}^*$ at which the performance index in Eq. 7.77 is maximized, subject of course to satisfaction of constraints in Eq. 7.5. The forced periodic control is said to be proper if the objective function for a periodic operation exceeds that for the steady-state operation. A sufficient (but not necessary) condition for this is provided by the π-criterion. This criterion, originally developed for forced periodic operations around an optimal steady state [66, 202], has been generalized to be applicable to forced periodic operation around an arbitrary steady state [571, 572, 573]. Assuming that $\mathbf{f}$, h, $\boldsymbol{\phi}$ and $\boldsymbol{\Psi}$ are continuously differentiable in $\mathbf{x}$ and $\mathbf{u}$, the Hermitian matrix $\boldsymbol{\Pi}(\omega)$ is defined as

$$\boldsymbol{\Pi}(\omega) = \mathbf{G}^c(j\omega)\mathbf{P}\mathbf{G}(j\omega) + \mathbf{Q}^T\mathbf{G}(j\omega) + \mathbf{G}^c(j\omega)\mathbf{Q} + \mathbf{R} \tag{7.78}$$

where

$$\mathbf{G}(s) = (s\mathbf{I} - \mathbf{A})^{-1}\mathbf{B}, \quad \mathbf{A} = \mathbf{f_x}(\mathbf{x},\ \mathbf{u}), \quad \mathbf{P} = H_{\mathbf{xx}}(\mathbf{x},\ \mathbf{u},\ \boldsymbol{\lambda},\ \boldsymbol{\rho},\ \boldsymbol{\eta}),$$
$$\mathbf{B} = \mathbf{f_u}(\mathbf{x},\ \mathbf{u}),\ \mathbf{Q} = H_{\mathbf{xu}}(\mathbf{x},\ \mathbf{u},\ \boldsymbol{\lambda},\ \boldsymbol{\rho},\ \boldsymbol{\eta}),\ \mathbf{R} = H_{\mathbf{uu}}(\mathbf{x},\ \mathbf{u},\ \boldsymbol{\lambda},\ \boldsymbol{\rho},\ \boldsymbol{\eta}) \tag{7.79}$$

and the Hamiltonian H is defined as

$$H(\mathbf{x},\ \mathbf{u},\ \boldsymbol{\lambda},\ \boldsymbol{\rho},\ \boldsymbol{\eta}) = h + \boldsymbol{\lambda}^T\mathbf{f} + \boldsymbol{\rho}^T\boldsymbol{\phi}. \tag{7.80}$$

In Eqs. 7.78-7.80, $\mathbf{x}$ and $\mathbf{u}$ assume their respective values at a steady state, viz., $\mathbf{x}_0$ and $\mathbf{u}_0$, respectively. The superscript c in Eqs. 7.78, 7.83 and 7.84 denotes the complex conjugate. The adjoint variable vectors $\boldsymbol{\lambda}$, $\boldsymbol{\rho}$, and $\boldsymbol{\eta}$ satisfy the conditions

$$Z_{\mathbf{x}}(\mathbf{x},\ \mathbf{u},\ \boldsymbol{\lambda},\ \boldsymbol{\rho},\ \boldsymbol{\eta}) = \mathbf{0}^T, \quad Z = H + \boldsymbol{\eta}^T\boldsymbol{\Psi},$$
$$\boldsymbol{\eta} \leq \mathbf{0}, \quad \boldsymbol{\eta}^T\boldsymbol{\Psi}(\mathbf{x},\ \mathbf{u}) = 0, \tag{7.81}$$

at an arbitrary steady state. At the optimal steady state, the following additional condition must be satisfied

$$Z_{\mathbf{u}}(\mathbf{x},\ \mathbf{u},\ \boldsymbol{\lambda},\ \boldsymbol{\rho},\ \boldsymbol{\eta}) = \mathbf{0}^T, \quad \mathbf{x} = \mathbf{x}^*, \quad \mathbf{u} = \mathbf{u}^*. \tag{7.82}$$

For superior performance of forced periodic operation of continuous bioreactors vis-a-vis operation of the same at a steady-state, it is sufficient (but not necessary) to have positive-definite $\boldsymbol{\Pi}$ for some values of ω $(0 < \omega < \infty)$. For small variations in the control variables relative to their values at a steady state, the differential between the magnitudes of the objective function in a forced periodic operation and the corresponding steady-state operation can be expressed as [572]

$$\delta J = \frac{1}{2\tau}\int_0^{\tau}(\boldsymbol{\delta u}^c)^T\boldsymbol{\Pi}(\omega)\boldsymbol{\delta u}dt,\ \boldsymbol{\delta u} = \mathbf{u} - \mathbf{u}_0,\ \delta J = J - J_0,\ J_0 = h(\mathbf{x}_0,\ \mathbf{u}_0), \tag{7.83}$$

ω being the overall frequency of the periodically perturbed system.

Periodic variation in only two inputs is considered here since the case study to be discussed later pertains to two inputs. The results obtained here can be extended to higher dimensions, with tedious algebraic manipulations [449, 450]. Since the domain of $\mathbf{\Pi}$ is complex vectors, $\boldsymbol{\delta u}$ is assigned the following form:

$$\boldsymbol{\delta u} = \begin{bmatrix} r_1 e^{j\theta_1} \\ r_2 e^{j\theta_2} \end{bmatrix}, \quad \boldsymbol{\delta u}^c = \begin{bmatrix} r_1 e^{-j\theta_1} \\ r_2 e^{-j\theta_2} \end{bmatrix}, \quad \boldsymbol{\delta u} = \mathbf{u} - \mathbf{u}_0. \tag{7.84}$$

The complex exponential notation used here simplifies the analysis [549, 632, 634]. Let $\rho_{ij}(\omega)$ $(i,\ j = 1, 2)$ denote the individual elements of $\mathbf{\Pi}(\omega)$. Then Eq. 7.84 can be deduced to have the form $[\rho_{21}(\omega) = \rho_{12}^c(\omega)]$

$$\delta J = \frac{1}{2}\sum_{i=1}^{2} \rho_{ii}(\omega) r_i^2 + \frac{1}{\tau}\int_0^{\tau} Z_{21} r_1 r_2 dt,$$
$$Z_{21} = [Re(\rho_{21})\cos(\theta_2 - \theta_1) + Im(\rho_{21})\sin(\theta_2 - \theta_1)]\,. \tag{7.85}$$

In what follows, we examine the forms Eq. 7.85 reduces to when the number of inputs subject to periodic variation is 1 or 2.

Periodic Variation in Single Input

In bioreactor operations involving periodic variation in only one feed parameter (r_1 or r_2 is positive), Eq. 7.85 reduces to $\delta J = \rho_{ii}(\omega) r_i^2/2$ $(i = 1$ or $2)$. It follows then that for superiority of forced periodic operation vis-a-vis steady-state operation, $\rho_{ii}(\omega)$ $(i = 1$ or $2)$ must be positive for some ω.

Periodic Variations in Two Inputs

In bioreactor operations involving periodic variations in two feed parameters (i.e., $r_1 \neq 0$ or $r_2 \neq 0$), let the frequencies of variations in u_1 and u_2 be the same or different with $\theta_k = 2\pi\omega_k t + \phi_k$ $(k = 1,\ 2,\ \phi_1 = 0)$.

Unequal forcing frequencies.
When $\omega_1 \neq \omega_2$, let the maximum of ω_1 and ω_2 be an integral multiple of the minimum of ω_1 and ω_2, i.e., $\max(\omega_1,\ \omega_2)/\min(\omega_1,\ \omega_2) = n$ $(n > 1,\ n$ an integer). Then

$$\theta_1 = 2\pi n_1 \omega t, \quad \theta_2 = 2\pi n_2 \omega t + \phi, \quad n_1 = \omega_1/\omega \geq 1, \quad n_2 = \omega_2/\omega \geq 1,$$
$$\omega = \min(\omega_1,\ \omega_2), \quad \tau = \max(\tau_1,\ \tau_2), \quad \omega_j \tau_j = \omega\tau = 1, \quad j = 1,\ 2 \tag{7.86}$$

where n_1 and n_2 are integers.

In this case, Eq. 7.85 reduces to

$$\delta J = \frac{1}{2}\left[\rho_{11}(\omega_1)r_1^2 + \rho_{22}(\omega_2)r_2^2\right], r_j = 0 \text{ if } \rho_{jj}(\omega_j) \leq 0,\ j = 1, 2. \quad (7.87)$$

It is evident from Eq. 7.87 that the interaction between the control variables u_1 and u_2 vanishes when $\omega_1 \neq \omega_2$. Simultaneous periodic variations in u_1 and u_2 may provide improvement in process performance vis-a-vis periodic variations in u_1 or u_2 alone for those intervals of ω where both ρ_{11} and ρ_{22} are positive. For a particular η $(= r_2/r_1)$, the optimum frequency (ω_0) then is the frequency at which $(\rho_{11} + \rho_{22}\eta^2)$ is maximized.

Equal forcing frequencies.

When $\omega_1 = \omega_2 = \omega$, Eq. 7.85 assumes the form

$$\begin{aligned} \delta J &= \frac{1}{2}\left[\rho_{11}(\omega)r_1^2 + \rho_{22}(\omega)r_2^2 + 2fr_1r_2\right], \\ f(\phi,\ \omega) &= \left[Re(\rho_{21})\cos(\phi) + Im(\rho_{21})\sin(\phi)\right]. \end{aligned} \quad (7.88)$$

The third term on the right side of Eq. 7.88 represents the interaction between the control variables u_1 and u_2. A positive effect of interaction between the two control variables in forced periodic operation involving perturbations in both u_1 and u_2 requires that f be positive. Maximization of δJ for a particular steady state requires that f be maximized. Since

$$\begin{aligned} f(\phi,\ \omega) = |\rho_{21}|\cos(\phi - \chi), \quad &\frac{Im(\rho_{21})}{\sin(\chi)} = \frac{Re(\rho_{21})}{\cos(\chi)} = |\rho_{21}|, \\ |\rho_{21}| = \left[\{Re(\rho_{21})\}^2 + \{Im(\rho_{21})\}^2\right]^{1/2}, \quad &f_{max} = |\rho_{21}|, \end{aligned} \quad (7.89)$$

a maximum in f $(f_{\max})$ occurs when $\phi = \chi$. Positivity of f requires that ϕ lie in the interval $-\pi/2 < (\phi - \chi) < \pi/2$. Where possible, optimum ratios of the amplitudes of weak perturbations in u_1 and u_2 which lead to maximization of δJ are identified next.

Fixed r_2, variable r_1.

The necessary and sufficient condition for occurrence of a local maximum in δJ and its value are

$$\delta J = \frac{1}{2}r_2^2\left(\rho_{22} - \frac{f^2}{\rho_{11}}\right) \text{ at } \eta = \frac{r_1}{r_2} = -\frac{f}{\rho_{11}} \text{ if } \rho_{11} < 0. \quad (7.90)$$

The upper bound on δJ in Eq. 7.90 corresponds to $f = |\rho_{21}|$ (defined in Eq. 7.89). The necessary condition for a positive maximum in δJ is that $|\rho_{21}|^2 > \rho_{11}\rho_{22}$.

If weak perturbations in u_2 alone lead to improved process performance ($\rho_{22} > 0$), then it is evident from Eq. 7.90 that simultaneous variations in u_1 and u_2 lead to further improvement in the process performance. When $\rho_{22} > 0$, forced periodic operation involving variations in u_1 and u_2 is superior to steady-state operation only in the range $0 \leq \eta < \eta_2$ [$\eta = r_1/r_2$, Figure 7.3(a)], with η_2 being

$$\eta_2 = \frac{\left[f + \sqrt{f^2 - \rho_{11}\rho_{22}}\right]}{(-\rho_{11})}, \quad 0 < f \leq |\rho_{21}|. \tag{7.91a}$$

If $\rho_{22} = 0$, an improvement in performance of steady-state operation is feasible only in the range $0 < \eta < \eta_2$ [Figure 7.3(a)]. If $\rho_{22} < 0$, a positive δJ occurs in the range $\eta_1 < \eta < \eta_2$ [$\eta = r_j/r_i$, Figure 7.3(a)], with

$$\eta_1 = \frac{\left[f - \sqrt{f^2 - \rho_{11}\rho_{22}}\right]}{(-\rho_{11})}, \quad 0 < f \leq |\rho_{21}|, \tag{7.91b}$$

provided $f^2 > \rho_{11}\rho_{22}$.

Fixed r_1, variable r_2.

The necessary and sufficient condition for occurrence of a local maximum in δJ and its value are

$$\delta J = \frac{1}{2} r_1^2 \left(\rho_{11} - \frac{f^2}{\rho_{22}}\right) \text{ at } \zeta = \frac{r_2}{r_1} = -\frac{f}{\rho_{22}} \text{ if } \rho_{22} < 0. \tag{7.92}$$

The upper bound on δJ in Eq. 7.92 corresponds to $f = |\rho_{21}|$ (defined in Eq. 7.89). The necessary condition for a positive maximum in δJ is that $|\rho_{21}|^2 > \rho_{11}\rho_{22}$.

If weak perturbations in u_1 alone lead to improved process performance ($\rho_{11} > 0$), then it is evident from Eq. 7.92 that periodic variations in u_2 as well as u_1 lead to further improvement in the process performance. When $\rho_{11} > 0$, forced periodic operation involving variations in u_1 and u_2 is superior to steady-state operation only in the range $0 \leq \zeta < \zeta_2$ [$\zeta = r_1/r_2$, Figure 7.3(a)], ζ_2 being

$$\zeta_2 = \frac{\left[f + \sqrt{f^2 - \rho_{11}\rho_{22}}\right]}{(-\rho_{22})}, \quad 0 < f \leq |\rho_{21}|. \tag{7.93a}$$

If $\rho_{11} = 0$, a forced periodic operation involving variations in u_1 and u_2 is superior to steady-state operation ($\delta J > 0$ in Eq. 7.88) only in the range

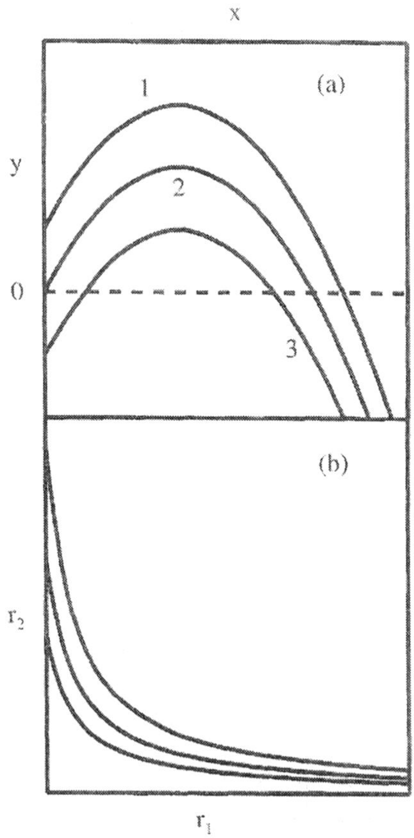

Figure 7.3. (a) Portraits of G_1 and η for $\rho_{22} < 0$ and portraits of G_2 and ζ for $\rho_{11} < 0$. $G_1 = \rho_{11} + 2f\eta + \rho_{22}\eta^2$, $G_2 = \rho_{22} + 2f\zeta + \rho_{11}\zeta^2$. [$(x, y) = (\eta,\ G_1)$ and $(x, y) = (\zeta,\ G_2)$.] The profiles 1, 2 and 3 correspond to $\rho_{11} > 0$, $\rho_{11} = 0$ and $\rho_{11} < 0$, respectively, when $(x, y) = (\eta,\ G_1)$ and to $\rho_{22} > 0$, $\rho_{22} = 0$ and $\rho_{22} < 0$, respectively, when $(x, y) = (\zeta,\ G_2)$. (b) Profiles of $\delta J = c$ (c an arbitrary constant) when $\min(\rho_{11},\ \rho_{22}) \geq 0$, $\max(\rho_{11},\ \rho_{22}) > 0$ and $f > 0$ [449].

$0 < \zeta < \zeta_2$ [Figure 7.3(a)]. If $\rho_{11} < 0$, a positive δJ occurs in the range $\zeta_1 < \zeta < \zeta_2$ [$\zeta = r_2/r_1$, Figure 7.3(a)], with

$$\zeta_1 = \frac{\left[f - \sqrt{f^2 - \rho_{11}\rho_{22}}\right]}{(-\rho_{22})}, \quad 0 < f \leq |\rho_{21}|, \tag{7.93b}$$

provided $f^2 > \rho_{11}\rho_{22}$.

When both ρ_{11} and ρ_{22} are non-negative, on the family of curves $\delta J = c$ (c is an arbitrary constant), an increase (a decrease) in r_2 is accompanied by a decrease (an increase) in r_1 [Figure 7.3(b)], since

$$\frac{dr_2}{dr_1} = -\frac{(\rho_{11}r_1 + fr_2)}{(fr_1 + \rho_{22}r_2)} < 0, \ f > 0 \ \text{ for } \ \delta J = c. \tag{7.94}$$

There therefore are infinite sets of r_1 and r_2 which lead to the same value of δJ for a particular J_s when $\min(\rho_{11}, \ \rho_{22}) \geq 0$, $\max(\rho_{11}, \ \rho_{22}) > 0$ and $f > 0$. An optimum amplitude ratio, r_1/r_2, is as a result not admissible.

The contribution of the off-diagonal elements of $\mathbf{\Pi}$, ρ_{jk} ($j \neq k, \ j, \ k =$ 1, 2 in the present case), to δJ is more significant than that of the diagonal elements, ρ_{jj} ($j = 1, \ 2$) [449, 450, 571, 572, 573]. In the case study that follows, the forcing frequencies of inputs subject to periodic variation are therefore considered to be equal.

7.3.2 Case Study - Forced Periodic Operations

A unified analysis of optimality of forced periodic operation of continuous cultures producing a wide range of products and subject to periodic variation in dilution rate and/or feed concentration of the limiting substrate has been reported by Parulekar [448, 449]. It was established that very low frequency periodic operations around the optimal steady state, where admissible, are non-optimal. Conditions for properness of periodic control and expressions for frequency ranges where periodic control is proper and optimum cycling frequency were obtained analytically. It was established that subjecting a bioprocess to simultaneous periodic variations in dilution rate and substrate feed concentration does always lead to improved performance, at least at high frequencies [449].

Problem Formulation

The dynamics of continuous bioprocesses of interest (continuous pure cultures) is considered to be described adequately by the conservation equations for cell mass (biomass), limiting substrate and the desired non-biomass product. The mass balances for cell- and product-free feed are provided in Eqs. 7.29-7.31 with F/V being referred to as the dilution rate (D) for continuous culture. In situations where the desired product is excreted to a large extent and is subject to degradation in the abiotic phase, ε can be expressed in terms of the cell mass-specific product synthesis rate (ε_0) and the volume-specific product degradation rate (R_d) as $\varepsilon = \varepsilon_0 - R_d/X$. It is of interest to maximize the performance index of the type [448, 449]

$$J = \frac{1}{\tau}\int_0^{\tau} D(P + \upsilon X - wS_F)dt \tag{7.95}$$

with w ($w \geq 0$) being the cost of limiting substrate relative to the price of the desired product. The term involving the coefficient v in Eq. 7.95 accounts for the difference between the price of those products (other than the desired product) whose formation is associated with cell growth and the cost of separation of cell mass from the desired product relative to the price of the desired product. The objective function in Eq. 7.95 is therefore appropriate for optimizing operation of continuous bioprocesses that generate growth-associated and non-growth associated products and incorporates costs associated with separation of the desired product from cell mass. In both steady-state and periodic operations of continuous cultures, the input variable space is defined by the inequality constraints

$$0 \leq D \leq D^* \quad \text{and} \quad 0 \leq S_F \leq S_F^* \tag{7.96}$$

with D^* being the dilution rate beyond which retention of cells is not possible in a steady state continuous culture and S_F^* the maximum permissible concentration of the limiting substrate in the bioreactor feed (usually decided by solubility limits of the substrate in the feed medium).

Since the performance index considered here (Eq. 7.95) is non-positive for steady-state operations at $D = 0$, $D = D^*$ or $S_F = 0$ ($P = X = 0$ for $D \geq D^*$ or $S_F = 0$), the optimal steady state solutions cannot lie on the boundaries $D = 0$, $D = D^*$ and $S_F = 0$ of the control variable space. The optimal steady state solutions may therefore lie strictly in the interior of the control variable space (defined by the inequality constraints in Eq. 7.96) or on the boundary $S_F = S_F^*$.

The expressions for the scalars and vectors involved in evaluation of $\mathbf{\Pi}(\omega)$ have the form

$$\mathbf{x} = \begin{bmatrix} X \\ S \\ P \end{bmatrix}, \quad \mathbf{F} = \begin{bmatrix} f_1 \\ f_2 \\ f_3 \end{bmatrix}, \quad \boldsymbol{\lambda} = \begin{bmatrix} \lambda_1 \\ \lambda_2 \\ \lambda_3 \end{bmatrix}, \quad h = D\left[P + vX - wS_F\right],$$

$$\mathbf{B} = \begin{bmatrix} -X & 0 \\ (S_F - S) & D \\ -P & 0 \end{bmatrix}, \quad \mathbf{R} = \begin{bmatrix} 0 & (\lambda_2 - w) \\ (\lambda_2 - w) & 0 \end{bmatrix}$$

$$\mathbf{u} = \begin{bmatrix} D \\ S_F \end{bmatrix}$$

$$\mathbf{A} = \begin{bmatrix} (X\mu_X + \mu - D) & X\mu_S & X\mu_P \\ -(X\sigma_X + \sigma) & -(D + X\sigma_S) & -X\sigma_P \\ (X\varepsilon_X + \varepsilon) & X\varepsilon_S & (X\varepsilon_P - D) \end{bmatrix}$$

$$\mathbf{P} = \{p_{ij},\ i,j = 1,2,3;\ p_{ij} = p_{ji},\ i \neq j\}, \quad H = h + \sum_{i=1}^{3} \lambda_i f_i, \tag{7.97}$$

with f_1, f_2, and f_3 being the right sides of Eqs. 7.29-7.31, respectively, and the elements of $\mathbf{P}$ being

$$\begin{aligned}
p_{11} &= (2\mu_X + \mu_{XX}X)\,\lambda_1 - (2\sigma_X + \sigma_{XX}X)\,\lambda_2 + (2\varepsilon_X + \varepsilon_{XX}X)\,\lambda_3, \\
p_{12} &= (\mu_S + \mu_{XS}X)\,\lambda_1 - (\sigma_S + \sigma_{XS}X)\,\lambda_2 + (\varepsilon_S + \varepsilon_{XS}X)\,\lambda_3, \\
p_{31} &= (\mu_P + \mu_{XP}X)\,\lambda_1 - (\sigma_P + \sigma_{XP}X)\,\lambda_2 + (\varepsilon_P + \varepsilon_{XP}X)\,\lambda_3, \\
p_{22} &= (\mu_{SS}\lambda_1 - \sigma_{SS}\lambda_2 + \varepsilon_{SS}\lambda_3)\,X, \\
p_{23} &= (\mu_{SP}\lambda_1 - \sigma_{SP}\lambda_2 + \varepsilon_{SP}\lambda_3)\,X, \\
p_{33} &= (\mu_{PP}\lambda_1 - \sigma_{PP}\lambda_2 + \varepsilon_{PP}\lambda_3)\,X. \qquad (7.98)
\end{aligned}$$

The adjoint variables at a steady state are obtained from solution of Eq. 7.81, which in this case assume the form

$$\begin{aligned}
& m_1\lambda_1 - m_2\lambda_2 + m_3\lambda_3 = -\upsilon D, \\
& m_4\lambda_1 - m_5\lambda_2 + m_6\lambda_3 = 0, \\
& m_7\lambda_1 - m_8\lambda_2 + m_9\lambda_3 = -D, \\
& m_1 = \mu - D + X\mu_X, \ m_2 = \sigma + X\sigma_X, \ m_3 = \varepsilon + X\varepsilon_X, \\
& m_4 = X\mu_S, \ m_5 = D + X\sigma_S, \ m_6 = X\varepsilon_S, \\
& m_7 = X\mu_P, \ m_8 = X\sigma_P, \ m_9 = X\varepsilon_P - D. \qquad (7.99)
\end{aligned}$$

For the problem formulation described by Eqs. 7.29-7.31, depending on the nature of relations among the three rate processes, various bioprocesses can be classified into three types as [448, 449]: (I) bioprocesses where σ and ε are each related linearly to μ, (II) bioprocesses where μ, σ and ε are related by a single linear relation, and (III) bioprocesses where μ, σ and ε are not related linearly. For type I bioprocesses, the relations in Eq. 7.45 are applicable. In steady-state operations and forced periodic operations with variation in D alone, the state variables X, S, and P satisfy the stoichiometric relations in Eq. 7.50 ($S_F = S_{F0}$). For type II bioprocesses, the three specific rates are related linearly as in Eq. 7.53. In steady-state operations and forced periodic operations with variation in D alone, the state variables X, S, and P satisfy the stoichiometric relation in Eq. 7.61 ($S_F = S_{F0}$). The analysis of forced periodic operation is simplified considerably if the bioreactor state (X, S, P) in steady-state and forced periodic operations satisfies Eq. 7.50 or Eq. 7.61, as appropriate [448, 449]. For details of the analysis of the forced periodic operations of the three bioprocess types, the reader should refer to [448, 449].

Forced periodic operations of continuous cultures may in some situations extend the regions of the operating parameter space where non-washout solutions are admissible [322]. The extension of the regions of admissibility of the meaningful states of continuous cultures via forced periodic operations subject to weak variations in inputs can be investigated by applying the π-criterion to the washout steady state ($X = P = 0,\ S = S_F$) when it is locally, asymptotically stable, the necessary and sufficient condition for which is that $\mu(X_F,\ S_F,\ P_F) < D$ [448]. It is established in [448] that weak periodic perturbations in D and S_F will allow for cell retention under conditions where such retention is not possible in steady-state continuous culture operation as long as the phase difference between the perturbations in the two inputs lies between 90° and 270°. For the performance index under consideration (Eq. 7.95), in view of the form of h in Eq. 7.97, R_{11} and R_{22} (diagonal elements of $\mathbf{R}$) are trivial and $\mathbf{P}$ and $\mathbf{Q}$ (Eq. 7.79) do not depend on w. For the three types of bioprocesses therefore, $\rho_{11}(\omega)$ and $\rho_{22}(\omega)$ are independent of w [448].We import here numerical results from specific examples in [449].

Example 1.
This example pertains to type I bioprocesses (Eq. 7.45) with μ being dependent exclusively on S. For a locally, asymptotically stable steady state ($\mu_S > 0,\ \mu_{SS} < 0$), improvement in bioprocess performance via periodic forcing in D or S_F alone is not possible since $\rho_{11} \leq 0$ and $\rho_{22} \leq 0$ for all ω ($0 \leq \omega < \infty$) [448]. Superiority of forced periodic operations subject to simultaneous variations in D and S_F over steady-state operation at a locally, asymptotically stable non-trivial steady state is guaranteed at low and high frequencies (Figure 7.4). This observation is valid in the entire portion(s) of the $S_F - D$ space where stable non-trivial steady states are admissible. The results in Figure 7.4 are for Monod kinetics [$\mu = \mu_m S/(K_S + S)$] with $\mu_m = 1.0\ h^{-1},\ K_S = 0.05\ gL^{-1}$, and $w = 0$.

Example 2.
This example pertains to type I bioprocesses with μ being dependent on S, X, and P (Eq. 7.50), μ_X and μ_P being negative. Case studies of these bioprocesses include fermentations producing alcohols [47, 239, 325, 326, 338, 395, 544]. For numerical illustration, μ is expressed as [326, 395]

$$\mu = \mu_1(S)\mu_2(P),\ \mu_1 = \frac{\mu_m S}{K_S + S + S^2/K_I},\ \mu_2 = \left(1 - \frac{P}{P_m}\right)^n \tag{7.100}$$

with

$$\begin{aligned} &\mu_m = 1.0\ h^{-1},\ K_S = 0.4\ g/L,\ K_I = 48.1\ g/L, \\ &P_m = 70\ g/L,\ n = 0.94,\ p = 1/0.048,\ \text{and}\ c = 1/0.097. \end{aligned} \tag{7.101}$$

For the parameters listed in Eq. 7.101, the maximum number of non-trivial steady states is two. When two non-trivial steady states are admissible, one of these is locally, asymptotically stable and the other is unstable. Where

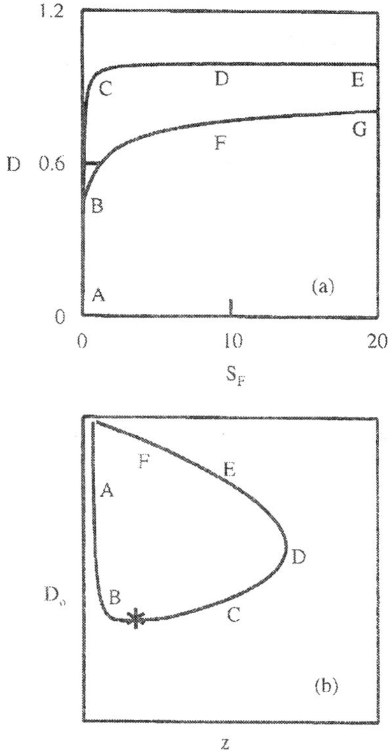

Figure 7.4. (a) Operating diagram for Example 1 with the cell growth following Monod kinetics ($\mu_m = 1.0\ h^{-1}$, $K_S = 0.05\ gL^{-1}$, and $K_I \to \infty$ in Table 3, D in h^{-1} and S_F in gL^{-1}). $\mu(S_F) = D$ on the curve ABCDE. Forced periodic operation with variations in D and S_F is superior to steady-state operation (i) at all frequencies in region I [(S_{F0}, D_0) lying below the curve ABFG], (ii) for $0 \leq \omega < \omega_1$ and $\omega_2 < \omega < \infty$ ($\omega_2 > \omega_1$) in region II [(S_{F0}, D_0) lying between the curves BFG and BCDE], and (iii) for all ω except $\omega = \omega_*$ for (S_{F0}, D_0) lying on the curve BFG ($f^2_{max} = \rho_{11}\rho_{22}$ at $\omega = \omega_*$). (b) For a particular S_{F0}, forced periodic operation involving variations in D and S_F is superior to steady-state operation for $(z,\ D_0)(z = \omega^2)$ lying outside the curve ABCDEF [$f^2_{max} = \rho_{11}\rho_{22}$ on the curve ABCDEF, $f^2_{max} < \rho_{11}\rho_{22}$ for $(z,\ D_0)$ enclosed by the curve ABCDEF, and $f^2_{max} > \rho_{11}\rho_{22}$ for $(z,\ D_0)$ lying outside the curve ABCDEF]. Regions I and II in (a) correspond to $D_0 < D_*$ and $D_0 > D_*$, respectively. D_* corresponds to asterisk [449].

$D_0 < \mu_1(S_{F0})$ [under the curve ABCEFG in Figure 7.5(a)], a unique locally, asymptotically stable non-trivial steady state is admissible, the global asymptotic stability of which is also assured since the washout state is unstable. Two non-trivial steady states are admissible in a portion of the $S_F - D$ space where $D_0 > \mu_1(S_{F0})$ [$(S_{F0},\ D_0)$ lying inside the envelope CEFHC in Figure 7.5(a)]. One of the non-trivial steady states and the washout state are locally stable in this portion. On the curve CEF [excluding points C and F, $\mu_1(S_{F0}) = D_0$], a unique non-trivial steady state, which is globally, asymptotically stable, is admissible.

Since positive J is of interest, it follows that $(v+c)$ must be positive (Eq. 7.97). In the entire region of the $S_F - D$ space where a stable non-trivial steady state is admissible [below the curve ABCHFG in Figure 7.5(a)], periodic control with variation in D alone is not proper. When $\mathbf{u} = S_F$, $\rho_{22}(\omega)$ is positive for some ω in region I [$(S_{F0},\ D_0)$ lying to the right of the curve FIJ and below the curve FG, Figure 7.5(a)] and negative for all ω in region II [$(S_{F0},\ D_0)$ lying below the curve ABCHFIJ] and for $(S_{F0},\ D_0)$ lying on the curve FIJ excluding point F. In region I, forced periodic operations subject to weak variations in S_F will yield superior performance vis-a-vis steady-state operation.

The effect of simultaneous periodic variations in D and S_F on the bioreactor performance was examined for $v = w = 0$ (Eq. 7.95). In these operations, δJ is positive at all frequencies in region I and for $\omega_1 < \omega < \infty$ ($\omega_1 \neq 0$, ω_1 depends on D_0 and S_{F0}) in region II. On the interface between the two regions [$(S_{F0},\ D_0)$ lying on the curve FIJ excluding point F], δJ is positive for $\omega > 0$.

Following conditions must be satisfied at the optimal steady-state ($D = D^*,\ S_F = S_F^*$)

$$\frac{\partial J}{\partial D} = 0, \quad \frac{\partial J}{\partial S_F} = 0, \quad \frac{\partial^2 J}{\partial D^2} < 0, \quad \frac{\partial^2 J}{\partial S_F^2} < 0. \tag{7.102}$$

For $v = w = 0$ and parameters in Eq. 7.101, the optimal steady-state lies in region II [Figure 7.5(a)]. The performance of the optimal steady-state operation cannot therefore be improved by weak periodic variations in D or S_F alone. Weak periodic variations in both D and S_F (around D^* and S_F^*, respectively) will lead to improved performance for $\omega > 4.165$ cycles h^{-1}.

An analytical expression for the optimum frequency (ω_0) for maximal improvement in performance of a steady-state operation via weak periodic variations in S_F alone (in region I) is provided in [448]. For $D_0 = 0.3045\ h^{-1}$, a comparison between the maximum improvement attainable (vis-a-vis steady-state operation) in forced periodic operations involving weak variations in S_F alone and in both D and S_F (δJ as in Eq.

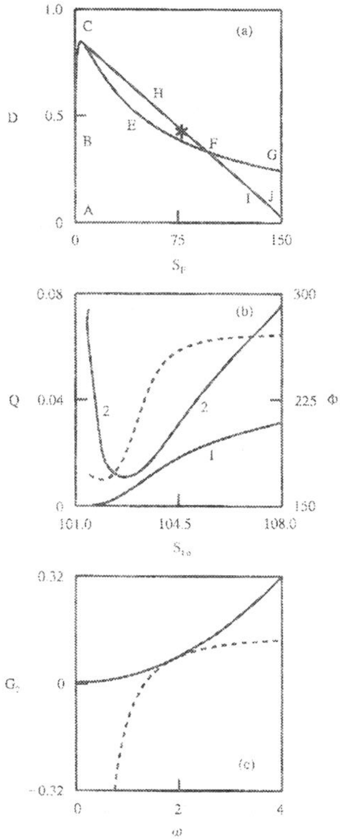

Figure 7.5. Results for Example 2. (a) Operating diagram for Example 2 (D in h^{-1} and S_F in gL^{-1}). $\mu_1(S_F) = D$ on the curve ABCEFG. Non-trivial steady-states are admissible for (S_F, D) lying below the curve ABCHFG. The number of non-trivial steady-states is (i) one for (S_F, D) lying below the curve ABCEFG and on the curve CEF (excluding points C and F) and (ii) two for (S_F, D) lying inside the envelope CEFHC. Periodic operations involving weak variations in S_F are superior to steady-state operation only in region I [(S_{F0}, D_0) lying to the right of the curve FIJ and below the curve FG]. For $v = w = 0$, periodic operations involving weak variations in S_F and D are superior to periodic operations involving weak variations in S_F and steady-state operation (i) at all frequencies in region I and for (S_{F0}, D_0) lying on the curve FIJ (excluding point F) and (ii) for $\omega_1 < \omega < \infty$ ($\omega_1 > 0$) in region II [(S_{F0}, D_0) lying below the curve ABCHFIJ]. The asterisk denotes the optimal steady-state for $v = w = 0$. (b) Portraits of Q [curve 1: $Q = \rho_{22}(\omega_0)$, curve 2: $Q = G_{2u}(\omega_0)$] and S_{F0} and $\phi_1(\omega_0)$ and S_{F0} (dashed curve) for $D_0 = 0.3045\ h^{-1}$. (c) Portraits of G_2 for $\zeta = 0.0417$ and ω (dashed curve) and G_{2u} and ω (solid curve) for $D_0 = 0.3045\ h^{-1}$ and $S_{F0} = 101.5\ gL^{-1}$. $\zeta = 0.0417$ is the optimum amplitude ratio ($G_2 = G_{2u}$) at $\omega = 1.904$ *cycles* h^{-1}. $G_2 = \rho_{22} + 2f\zeta + \rho_{11}\zeta^2$, $G_{2u} = \rho_{22} - f_{max}^2/\rho_{11}$ [449].

7.90 with $f = f_{max}$) is provided in Figure 7.5(b) for various S_{F0}'s in region I ($\omega = \omega_0$ in both types of operations). The benefit of simultaneous variation in D and S_F over variation in S_F alone is self-evident. The differential in the maximum improvement attainable in the two forced periodic operations is significantly sensitive to S_{F0}. For certain sets of operating parameters, (S_{F0}, D_0) therefore, periodic operations involving variation exclusively in S_F can be substantially inferior to those involving variations in both D and S_F. In the narrow range of S_{F0} considered in Figure 7.5(b), there is substantial variation in the optimum phase difference ($\phi = \chi$) that leads to maximum positive interaction between D and S_F ($f = f_{max}$). The optimum frequency ω_0 for forced periodic operation involving variation in S_F alone decreases with increasing S_{F0} (profile not shown).

For $D_0 = 0.3045\ h^{-1}$ and $S_{Fo} = 101.5\ gL^{-1}$, variations in G_2 ($G_2 = \rho_{22} - f^2/\rho_{11}$, $\rho_{11} < 0$, Eq. 7.90) for the optimal amplitude ratio (G_2 for $f = f_{max}$) and G_2 for a fixed amplitude ratio ($\zeta = 0.0417$) are presented in Figure 7.5(c). The amplitude ratio in the latter case is the optimal amplitude ratio only at $\omega = 1.904$ cycles h^{-1}. Periodic operations employing this amplitude ratio are suboptimal at other frequencies [Figure 7.5(c)]. The difference between the performance of periodic operation employing optimal amplitude ratio and that of the periodic operation employing a fixed amplitude ratio increases as the deviation of ω from the frequency for which the fixed amplitude ratio is the optimal one [$\omega = 1.904$ cycles h^{-1} in Figure 7.5(c)] increases.

Example 3.

The expressions for μ, σ and ε are provided in Eq. 7.68, with the parameter values being [190, 600]

$$\begin{aligned} &n = 1,\ \mu_m = 0.4\ h^{-1},\ \varepsilon_m = 1.4\ h^{-1},\ K_S = 0.476\ g/L, \\ &K_S' = 0.666\ g/L,\ P_m = 87\ g/L,\ P_m' = 114\ g/L,\ K_I = 203.49\ g/L, \\ &K_I' = 303.03\ g/L,\ Y_{P/S} = 0.47. \end{aligned} \tag{7.103}$$

A unique non-trivial steady state is admissible in that portion of the $S_F - D$ space where $\mu_1(S_{F0}) > D_0$ [(S_{F0}, D_0) lying below the curve ABCDEF in Figure 7.6]. The non-trivial steady state does not undergo any Hopf bifurcations and since the washout state is unstable when $\mu_1(S_{F0}) > D_0$, the non-trivial steady state is globally, asymptotically stable.

Application of π-criterion was considered for $v = 0$ (Eq. 7.95). Periodic control was found not to be proper when $\mathbf{u} = D$ in the entire region where $\mu_1(S_{F0}) > D_0$. When $\mathbf{u} = S_F$, $\rho_{22}(\omega)$ is positive for some ω in region I [(S_{F0}, D_0) lying to the right of the curve DGH and below the curve DEF] and negative for all ω in region II [(S_{F0}, D_0) lying below the curve ABCDGH] and for (S_{F0}, D_0) lying on the curve DGH (excluding point D)

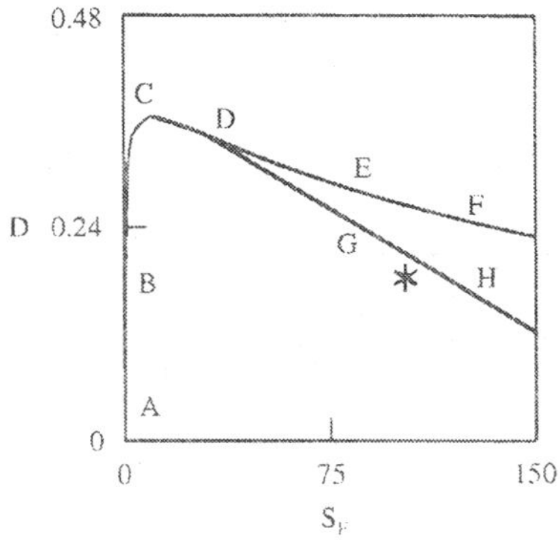

Figure 7.6. Operating diagram for Example 3 (D in h^{-1} and S_F in gL^{-1}). $\mu_1(S_F) = D$ on the curve ABCDEF. A unique non-trivial steady-state is admissible for (S_F, D) lying below the curve ABCDEF. For $v = 0$, periodic operations involving weak variations in S_F are superior to steady-state operation only in region I [(S_{F0}, D_0) lying to the right of the curve DGH and below the curve DEF]. For $v = w = 0$, periodic operations involving weak variations in S_F and D are superior to periodic operations involving weak variations in S_F and steady-state operation (i) at all frequencies in region I and for (S_{F0}, D_0) lying on the curve DGH (excluding point D) and (ii) for $\omega_1 < \omega < \infty$ $(\omega_1 > 0)$ in region II [(S_{F0}, D_0) lying below the curve ABCDGH]. The asterisk denotes the optimal steady-state for $v = w = 0$ [449].

(Figure 7.6). The performance of steady-state operation can be improved via periodic variations in S_F in region I. The effect of simultaneous periodic variations in D and S_F on the bioreactor performance was examined for $v = w = 0$ (Eq. 7.103). In such operations, δJ is positive (i) at all frequencies in region I and for (S_{F0}, D_0) lying on the curve DGH and (ii) for $\omega_1 < \omega < \infty$ ($\omega_1 \neq 0$, ω_1 depends on D_0 and S_{F0}) in region II. For $v = w = 0$ and the parameters in Eq. 7.103, the optimal steady-state (subject to Eq. 7.102) lies in region II (Figure 7.6). The performance of the optimal steady-state operation cannot therefore be improved by weak periodic variations in D or S_F alone. Weak periodic variations in both D and S_F around D^* and S_F^*, respectively, will lead to improved performance only for large ω ($\omega >$ 2921 cycles h^{-1}, $\tau < 1.23$ s). The rapid cycling required may need to be restricted to weak perturbations since large and very rapid perturbations in the extracellular environment may not be suitable for cellular metabolism.

Additional examples are discussed in [448, 449, 450]. Unstructured models, such as those considered in this section, predict a faster response to changes in operating parameters, such as D and S_F, than that observed experimentally [192]. This is presumably due to the inherent assumption in these models of no time lag between changes in the extracellular environment (abiotic phase) and adjustment of cellular metabolism. This assumption may be relaxed by considering that the specific rates (such as μ, σ, and ε) are functions not only of the current substrate concentration but also of previous substrate concentrations. Delay models for cell growth that account for this have been used previously [5, 440, 448]. For Example 1, periodic forcing in D or S_F does not lead to any improvement in bioprocess performance. However, accounting for the lag between changes in extracellular environment and alteration in cell growth rate revealed that forced periodic operations involving variations in D or S_F alone provide superior performance vis-a-vis steady-state operation [$\rho_{11}(\omega) > 0$ for $\omega_* < \omega < \infty$ ($\omega_* > 0$) and $\rho_{22}(\omega) > 0$ for $\omega' < \omega < \infty$ ($\omega' \neq 0$)] [448].

The generalized π-criterion, being based on weak variations around a steady state, provides a sufficient condition (and not a necessary one) for superiority of a forced periodic operation over a steady-state operation. Satisfaction of the criterion guarantees superiority of periodic operations involving both weak and strong variations in process inputs. It is anticipated that stronger input variations will lead to higher enhancement in performance under these conditions. Violation of the criterion does not necessarily rule out such superiority when strong variations in process inputs are considered. When strong variations in one or more of the bioreactor inputs, viz., D and S_F, are considered, the performance of the bioreactor subject to periodic forcing must be evaluated via solution of Eqs. 7.29-7.31 subject to the periodic boundary conditions in Eq. 7.76 and the objective function in Eq. 7.95 for particular forms of variations in the input(s). Application of the generalized π-criterion allows one to identify the regions in the multidimensional operating parameter space ($S_{F0} - D_0$ space for this case study) where periodic forcing may lead to improved process (bioreactor in the present case) performance. One may anticipate enlargement of these regions in bioprocess operations involving strong variations in the feed conditions (D and S_F).

As in the case of steady-state continuous bioprocesses, it is possible that in the inherently transient batch and fed-batch bioprocesses, periodic perturbations in one or more inputs around their optimal trajectories may lead to improvement in bioprocess performance. Identification of optimal trajectories for which such improvement occurs will however be a numerically challenging task, since unlike the continuous steady-state cultures where the variations in inputs are around a fixed point in the multidimensional space of inputs $\mathbf{u}$, the systematic variations in inputs are around trajectories in this space in the case of batch and fed-batch operations.

7.4 Feedback Control

7.4.1 State-Space Representation

The majority of techniques for design of controllers for multivariable systems apply to linear systems. The system represented by Eqs. 7.1 and 7.2 is a nonlinear multivariable system. The behavior of such a system in a close neighborhood of a reference state, $(\mathbf{x_r}, \mathbf{u_r}, \mathbf{d_r})$, can be represented by linearizing Eqs. 7.1 and 7.2 using the approach discussed in Section 4.7.2. Following Eqs. 4.90-4.95,

$$\frac{d\bar{\mathbf{x}}}{dt} = \mathbf{A}(t)\bar{\mathbf{x}}(t) + \mathbf{B}(t)\bar{\mathbf{u}}(t) + \mathbf{E}(t)\bar{\mathbf{d}}(t) \tag{7.104}$$

where $\bar{\mathbf{x}} = \mathbf{x} - \mathbf{x}_r$, $\bar{\mathbf{u}} = \mathbf{u} - \mathbf{u}_r$, $\bar{\mathbf{d}} = \mathbf{d} - \mathbf{d}_r$, and

$$\mathbf{A}(t) = \frac{\partial \mathbf{f}}{\partial \mathbf{x}} \qquad \mathbf{B}(t) = \frac{\partial \mathbf{f}}{\partial \mathbf{u}} \qquad \mathbf{E}(t) = \frac{\partial \mathbf{f}}{\partial \mathbf{d}}. \tag{7.105}$$

The output is

$$\bar{\mathbf{y}}(t) = \mathbf{C}(t)\bar{\mathbf{x}}(t), \qquad \mathbf{C}(t) = \frac{\partial \mathbf{h}}{\partial \mathbf{x}}, \qquad \bar{\mathbf{y}}(t) = \mathbf{y}(t) - \mathbf{y}_r. \tag{7.106}$$

$\mathbf{A}(t)$, $\mathbf{B}(t)$, $\mathbf{C}(t)$ and $\mathbf{E}(t)$ are the appropriately dimensioned system matrices with the respective multiplying vectors, the elements of which are partial derivatives evaluated at the reference state $(\mathbf{x_r},\ \mathbf{u_r},\ \mathbf{d_r})$. If the reference state happens to be a steady state (admissible only in a continuous bioreactor operation), then the system matrices are time-invariant. In that case, the state-space representation in Eqs. 7.104 and 7.106 can be transformed into transfer function representation by applying Laplace transform to Eqs. 7.104 and 7.106.

$$\bar{\mathbf{y}}(s) = \mathbf{G}(s)\bar{\mathbf{u}}(s) + \mathbf{G}_d(s)\bar{\mathbf{d}}(s), \tag{7.107}$$

with the transfer functions having the form

$$\mathbf{G}(s) = \mathbf{C}(s\mathbf{I} - \mathbf{A})^{-1}\mathbf{B} \qquad \mathbf{G}_d(s) = \mathbf{C}(s\mathbf{I} - \mathbf{A})^{-1}\mathbf{E}. \tag{7.108}$$

As mentioned previously, the process models for biological reactors are inherently nonlinear due to large number of chemical reactions occurring in a typical cell. Where kinetic descriptions are available, the values of model

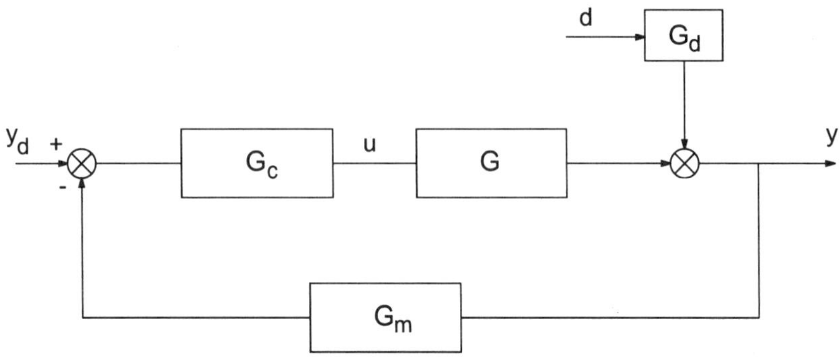

Figure 7.7. Multivariable block diagram.

parameters (kinetic and equilibrium coefficients) may be subject to substantial uncertainty due to complexity of the reaction scheme and difficulty to account for all reactions. Further, batch and fed-batch operations, which are inherently transient operations, are more common than continuous operations which permit steady-state operation. Taking cognizance of these, the transfer function representation is considered here since it can still provide valuable guidelines on control of transient multivariable processes.

7.4.2 Multi-Loop Feedback Control

In multivariable systems, all the output variables ($\mathbf{y}$) are measured and the information is sent to the controllers assigned to the task of regulating each output. The control decisions by the controllers are implemented as appropriate changes (manipulations) in certain process inputs ($\mathbf{u}$). The transfer functions for a multivariable feedback-controlled (closed-loop) process can be obtained from the concise representation of the block diagram for the process in Figure 7.7 as follows. By substituting the following input-output relations for the measuring devices and the controllers

$$\bar{\mathbf{y}}_{\mathbf{m}}(\mathrm{s}) = \mathbf{G}_{\mathbf{m}}(\mathrm{s})\bar{\mathbf{y}}(\mathrm{s}), \;\; \bar{\mathbf{y}}(\mathrm{s}) = \mathbf{G}_{\mathbf{c}}(\mathrm{s})\bar{\varepsilon}(\mathrm{s}), \;\; \bar{\varepsilon}(\mathrm{s}) = \bar{\mathbf{y}}_{\mathbf{d}}(\mathrm{s}) - \bar{\mathbf{y}}_{\mathbf{m}}(\mathrm{s}), \quad (7.109)$$

in Eqs. 7.107 and 7.108, one can obtain

$$(\mathbf{I} + \mathbf{G}\mathbf{G}_{\mathbf{c}}\mathbf{G}_{\mathbf{m}})\bar{\mathbf{y}}(\mathrm{s}) = \mathbf{G}\mathbf{G}_{\mathbf{c}}\bar{\mathbf{y}}_{\mathbf{d}} + \mathbf{G}_{\mathbf{d}}\bar{\mathbf{d}}. \quad (7.110)$$

In Eqs. 7.109 and 7.110, $\mathbf{G}_{\mathbf{c}}(s)$ and $\mathbf{G}_{\mathbf{m}}(s)$ represent the transfer function matrices for the controllers and measuring devices, respectively, and $\bar{\mathbf{y}}_{\mathbf{m}}(s)$ and $\bar{\mathbf{y}}_{\mathbf{d}}(s)$ the Laplace transforms of the vector of measured outputs and the vector of set-points for the measured outputs, respectively. In view of relation 7.110, one obtains the following relations among the outputs and inputs for the feedback-controlled process

$$\bar{\mathbf{y}}(s) = \mathbf{G}_1(s)\bar{\mathbf{y}}_d(s) + \mathbf{G}_2(s)\bar{\mathbf{d}}(s), \tag{7.111}$$
$$\mathbf{G}_1(s) = (\mathbf{I} + \mathbf{G}\mathbf{G}_c\mathbf{G}_m)^{-1}\mathbf{G}\mathbf{G}_c, \quad \mathbf{G}_2(s) = (\mathbf{I} + \mathbf{G}\mathbf{G}_c\mathbf{G}_m)^{-1}\mathbf{G}_d.$$

For an uncontrolled process with p outputs, m manipulated inputs and m_d disturbances, the dimensions of $\mathbf{G}$, $\mathbf{G_m}$, $\mathbf{G_c}$, and $\mathbf{G_d}$ are $(p \times m)$, $(p \times p)$, $(m \times p)$, and $(p \times m_d)$, respectively. It is evident from the dimension of $\mathbf{G_c}$ that a maximum of mp controllers will be needed to control p outputs by manipulating m inputs. Such controller configuration will be the most complex one for the given number of process inputs and outputs.

Following feedback control of SISO systems, the simplest controller configuration will involve control of one process output by manipulating only one process input. This one-to-one input-output pairing will require the least number of controllers, viz., $\min(m_t,\ p)$, with m_t $(= m + m_d)$ being the total number of process inputs (manipulated and non-manipulated).

Relative Gain Array

When using minimum number of single-loop controllers, an important consideration is the input-output pairing. The decision on the input-output pairing is based on how a particular output that is to be controlled is affected by each of the inputs that are being manipulated. In the vicinity of a steady state, by invoking the final value theorem $(s \to 0)$, one can relate the deviations in the outputs $(\bar{\mathbf{y}})$ to deviations in the manipulated inputs $(\bar{\mathbf{u}})$ as

$$\bar{\mathbf{y}}(t) = \mathbf{K}\bar{\mathbf{u}}(t), \quad \mathbf{K} = \mathbf{G}(0). \tag{7.112}$$

The elements of $\mathbf{K}$ are referred to as the steady-state gains. The $(i,\ j)$th element of the gain matrix $\mathbf{K}$ represents the ratio of change in the output y_i to change in the input u_j., i.e., $(\mathbf{K})_{ij} = \partial y_i/\partial u_j$.

The most widely used measure of interaction has been the relative gain array (RGA) introduced by Bristol [81]. For q manipulated inputs $[q = \min(m_t,\ p)]$, the array (denoted as $\mathbf{\Lambda}$) is a square matrix of dimension q. The $(i,\ j)$th element of RGA, λ_{ij}, is the ratio of gain between output y_i and input u_j $(\partial y_i/\partial u_j)$ when no control is implemented (the so-called open loop gain) and gain between output y_i and input u_j when all control loops except the y_i - u_j loop are functioning. Let $\mathbf{R}$ be the transpose of the inverse of the gain matrix $\mathbf{K}$ with elements r_{ij}. The elements of the relative gain array (λ_{ij}) are then related to the elements of $\mathbf{K}$ and $\mathbf{R}$ as per the relation [81, 438]

$$\lambda_{ij} = (\mathbf{K})_{ij} r_{ij}, \quad r_{ij} = (\mathbf{R})_{ij}, \quad \mathbf{R} = (\mathbf{K}^{-1})^T. \tag{7.113}$$

The relative gain array (RGA) has some interesting properties, which are listed below.

1. RGA is a symmetric matrix.

2. The elements of RGA in any row or any column add up to unity.

3. The elements of RGA are dimensionless.

4. The gain in the open loop pairing y_i with u_j when all other loops are closed (operating), K_{ij}^*, is related to the open-loop gain for this pair (K_{ij}) as

   $$K_{ij}^* = \frac{K_{ij}}{\lambda_{ij}}. \tag{7.114}$$

   The open-loop gain is thus altered by a factor of $1/\lambda_{ij}$ when all controllers except that for the y_i- u_j loop are active. This alteration is due to action from other control loops, which may be complementary or retaliatory. The sign of λ_{ij} then assumes special significance.

5. If $\mathbf{K}$ is a diagonal, an upper triangular, or a lower triangular matrix and if not, can be arranged into one via appropriate switches of rows or columns, then RGA is an identity matrix. The process under consideration then is *non-interactive*.

Recommendations for Input-Output Pairings

Based on the magnitudes of λ_{ij}, the recommendations for pairing and implications for interactions among control loops are discussed briefly.

1. $\lambda_{ij} = 1$. The input u_j can control y_i without interference from the other control loops. Pairing u_j with y_i is therefore recommended. This always is the case for non-interactive processes (property 5 of RGA).

2. $\lambda_{ij} = 0$. Since u_j has no direct influence on y_i, pairing u_j with y_i is absolutely not recommended.

3. $0 < \lambda_{ij} < 1$. In absolute values, the closed loop gain (all control loops except the $y_i - u_j$ loop closed) is larger than the open loop gain. The increase in gain is due to complementary effect from other active control loops. The complementary effect becomes increasingly pronounced as λ_{ij} is reduced. At the critical value of $\lambda_{ij} = 0.5$, the

direct effect of u_j on y_i is identical to the complementary effect of other control loops. As a result, the pairing $u_j - y_i$ is recommended when $0.5 < \lambda_{ij} < 1$ and should be avoided when $0 < \lambda_{ij} < 0.5$.

4. $\lambda_{ij} > 1$. Here, the open-loop gain between y_i and u_j exceeds the corresponding closed-loop gain. This is due to retaliatory effect of other control loops. The direct effect is still dominant. The retaliatory effect is enhanced as λ_{ij} is increased. The higher the λ_{ij}, the greater is the opposition u_j experiences from the other control loops in trying to control y_i. As a result, pair y_i with u_j as long as λ_{ij} is not very large and where possible, avoid pairing y_i with u_j if λ_{ij} is very large.

5. $\lambda_{ij} < 0$. When all loops except the $u_j - y_i$ loop are closed, a particular change in u_j will produce a change in y_i in opposite direction to that when all loops are open (uncontrolled process). The retaliatory effect of the other control loops is in opposition to the direct effect of u_j on y_i and is the dominant of the two effects. The $y_i - u_j$ pairing is potentially unstable and should be avoided.

6. In summary, one should pair input and output variables that have positive RGA elements that are closest to unity.

Further Comments on RGA

The relative gain array is based on the gain matrix for a process under consideration. The kinetics of processes of interest here (bioprocesses) being highly nonlinear, the elements of a steady-state gain matrix are based on the linearized version of the nonlinear process model. As a result, the elements of the process gain matrix as well as the elements of RGA will be functions of steady-state operating conditions for the process. The input-output pairings based on RGA analysis will therefore be dependent on the process operating conditions and may be altered as the operating conditions are changed. The concept of the relative gain array can be extended, with appropriate caution, to dynamic processes [173]. For process operation in the vicinity of a steady-state, the system matrices $\mathbf{A},\ \mathbf{B},\ \mathbf{C}$ and $\mathbf{E}$ are considered to be time-invariant (Eqs. 7.104 and 7.106) since the reference state, $(\mathbf{x_r},\ \mathbf{u_r},\ \mathbf{d_r})$, is time-invariant. The linearized version of the description of a nonlinear process (Eqs. 7.1 and 7.2) is a reasonable approximation in a small interval of t $(t - \Delta t < t' < t)$. The reference state, $(\mathbf{x_r},\ \mathbf{u_r},\ \mathbf{d_r})$, then must lie in this interval. The system matrices $\mathbf{A},\ \mathbf{B},\ \mathbf{C}$ and $\mathbf{E}$ (Eqs. 7.104 and 7.106) will then change from one time interval to another as the reference state, $(\mathbf{x_r},\ \mathbf{u_r},\ \mathbf{d_r})$, is altered. Obtaining information on dynamic process gains from these system matrices is not straightforward. Alternately, for each time interval, one can obtain an equivalent process gain matrix from the nonlinear process model (Eqs. 7.1 and 7.2), the individual gains, $(\mathbf{K})_{ij}$ (between y_i and u_j), being obtained as

$$(\mathbf{K}_{ij}) \approx [y_i(t) - y_i(t - \Delta t)]/[u_j(t) - u_j(t - \Delta t)]. \tag{7.115}$$

The choice of Δt is somewhat arbitrary. Witcher [657] has recommended Δt to be 20 to 100% of the dominant time constant in the process. The magnitude of Δt is reduced by the process time delay, if any, in effect of u_j on y_i, d_{ij} [173]. One can then proceed with obtaining RGA as described earlier (Eq. 7.114). This equivalent RGA has been referred to as the dynamic relative gain array. We will continue to refer to it as RGA. During the transient operation of a bioprocess from an initial state to a final state (this may be a steady state for continuous bioprocess operation) in a single operation (run or experiment), the elements of the process gain matrix and hence the elements of RGA may alter significantly. The input-output pairings therefore may not be the same throughout the operation and may have to be switched on one or more occasions.

It should be apparent from (Eq. 7.114) that even though the elements of RGA involve comparison of open-loop gain between an input u_j and an output y_i with the closed-loop gain for this pair (when all other control loops except the loop controlling y_i by manipulating u_j are closed), RGA can be estimated solely from the open-loop gains. Although the discussion related to estimating the interaction among inputs and outputs thus far has been based on availability of a mathematical description of the process, the so-called process model, one should not be under the impression that availability of a model is essential for estimation of RGA and decision on input-output pairing (controller configuration). When process models are not available or when available are reliable only in a narrow region of operating conditions, it is still possible to obtain the RGAs from experimental data. In an uncontrolled process, one can implement changes in an input (one input at a time) and observe the changes in various output variables. The elements of the process gain matrix, $\mathbf{K}$, can then be obtained, similar to Eq. 7.115 as

$$(\mathbf{K})_{ij} \approx \Delta y_i/\Delta u_j, \quad u_k \text{ fixed}, \; k \neq j. \tag{7.116}$$

Generation of RGA then would follow as per Eq. 7.113.

A system where $p > m_t$ is an underdefined system since there are not enough input variables to control all output variables. Based on economic considerations, one must decide which m_t of the p output variables are the most important. These will be paired with the m_t inputs and the remaining outputs ($p - m_t$ in number) will have to be left uncontrolled. Multiple independent sets (subsystems) of input-output pairing are candidates in this case, the exact number of sets being ${}^{p}C_{m_t}[= p!/\{m_t!(p - m_t)!\}]$. The relative gain arrays for all sets must be obtained. Comparison of the RGAs for these subsystems will reveal which subsystem has RGA closest to the ideal situation (elements corresponding to particular input-output pairing as close to unity as possible) and therefore will provide the best possible control.

A system where $m_t > p$ is an overdefined system since there are not enough output variables to be controlled with the available input variables which can be manipulated (m_t). The number of controllers in this situation is p and only p inputs can be manipulated. The remaining $(m_t - p)$ inputs would therefore not be manipulated and can be used for process optimization. If they cannot be regulated then they will be classified as disturbance. Multiple independent sets (subsystems) of input-output pairing are candidates in this case, the exact number of sets being ${}^{m_t}C_p[= m_t!/\{p!(m_t-p)!\}]$. The relative gain arrays for all sets must be obtained. Comparison of the RGAs for these subsystems will reveal which subsystem has RGA closest to the ideal situation (elements corresponding to particular input-output pairing as close to unity as possible) and therefore will provide the best possible control.

Decoupling Controllers for Interaction Compensation

The idea behind the use of minimum number of controllers and RGA-based selection of the input-output pairings is to have the controller loops be essentially independent. This occurs only when the RGA elements corresponding to all input-output pairings are either unity or very close to unity. When the RGA element corresponding to an input-output pairing (λ_{ij}) is substantially farther off from unity, there will be significant interaction from other control loops while controlling y_i. The interaction from the other control loops is due to process interactions. Consider as an example a process with two manipulated inputs u_1 and u_2 and two controlled outputs y_1 and y_2. Let the process transfer function matrix or the gain matrix be a full matrix. Let y_1 be paired with u_1 and y_2 with u_2. Controller for the $y_1 - u_1$ loop may change u_1 subject to information feedback on y_1. This change in u_1 would then lead to change not only in y_1 (the controlled variable for the $u_1 - y_1$ loop), but also in y_2. The alteration in y_2 due to action of controller 1 would then be fed back to controller 2. The action of controller 2 (manipulation of u_2) is thus influenced by action of controller 1. The change in u_2 will lead to change in not only y_2 (the controlled variable for the $u_2 - y_2$ loop), but also in y_1. The change in y_1 would then result in change in u_1. The action of controller 1 is thus influenced by action of controller 2. One can see that the two loops would be continually interacting.

The interaction among the minimum number of control loops can be minimized by use of appropriate decouplers. The decouplers, which are placed between the controllers and the process, try to compensate for the interaction in the process and therefore are also referred to as interaction compensators. The use of decouplers is intended to make the control loops independent. The controller-decoupler combination is also referred to as decoupling controller.

After RGA analysis, let the input-output pairings be such that u_j be paired with y_j, $j = 1, 2, \ldots, q$, $q = \min(m_t, p)$. If the RGA analysis suggests otherwise, then $\mathbf{u}$ or $\mathbf{y}$ and $\mathbf{G}$(s) or $\mathbf{K}$ may have to be reconfigured. As an example of this reconfiguration, consider a process with three manipulated inputs and three controlled outputs. If the pairings based on RGA are $u_1 - y_3$, $u_2 - y_1$ and $u_3 - y_2$, then (i) the output vector should be reconfigured as $(\mathbf{y})_{new} = [y_3\ y_1\ y_2]^T$ and the third, first and second rows of $\mathbf{G}$(s) or $\mathbf{K}$ should appear as the first, second and third rows, respectively, in the reconfigured $\mathbf{G}$(s) or $\mathbf{K}$; or (ii) the input vector should be reconfigured as $(\mathbf{u})_{new} = [u_2\ u_3\ u_1]^T$ and the second, third, and first columns of $\mathbf{G}$(s) or $\mathbf{K}$ should appear as the first, second and third columns in the reconfigured $\mathbf{G}$(s) or $\mathbf{K}$. Let the controller action (controller outputs) be denoted as $\mathbf{v}$. These serve as the inputs to decouplers, the outputs from the decouplers being the manipulated process inputs $\mathbf{u}$. The controlled outputs and the controller inputs can then be related as

$$\mathbf{y}(s) = \mathbf{G}(s)\mathbf{G}_I(s)\mathbf{G}_c(s)\boldsymbol{\varepsilon}(s) \tag{7.117}$$

in the s domain and by analogy as

$$\Delta\mathbf{y}(t) = \mathbf{K}(t)\mathbf{K}_I(t)\mathbf{K}_c(t)\Delta\boldsymbol{\varepsilon}(t) \tag{7.118}$$

in the time domain in terms of steady-state or dynamic gains. In the above, $\mathbf{G_I}$ and $\mathbf{K_I}$ represent the transfer function matrix and gain matrix, respectively, for the interaction compensators (decouplers). The number of controlled outputs being equal to the number of manipulated inputs in the simple multi-loop controller configuration under consideration, considering the situation where no decoupler is employed $[\mathbf{G_I}(s) = \mathbf{K_I}(t) = \mathbf{I}]$, it can be deduced from Eqs. 7.117 and 7.118 that $\mathbf{G_C}(s)$ and $\mathbf{K_C}(t)$ are diagonal matrices. It should then be evident from Eqs. 7.117 and 7.118 that for the control loops to perform independently of one another, $\mathbf{G}(s)\mathbf{G_I}(s)$ and $\mathbf{K}(t)\mathbf{K_I}(t)$ must be diagonal matrices.

Generalized Decoupler

The elements of $\mathbf{G}(s)\mathbf{G_I}(s)$ and $\mathbf{K}(t)\mathbf{K_I}(t)$ can be selected in multiple ways. The more general way assigns the non-trivial elements of these matrices to be the diagonal elements of $\mathbf{G}(s)$ or $\mathbf{K}(t)$, as applicable. The interaction compensator may then be obtained as

$$\mathbf{G_I}(s) = [\mathbf{G}(s)]^{-1}\text{diag } \mathbf{G}(s), \quad [\mathbf{G}(s)]^{-1} = \text{adj } \mathbf{G}(s)/|\mathbf{G}(s)| \tag{7.119}$$

and

$$\mathbf{K_I}(t) = [\mathbf{K}(t)]^{-1}\text{diag } \mathbf{K}(t), \quad [\mathbf{K}(t)]^{-1} = \text{adj } \mathbf{K}(t)/|\mathbf{K}(t)| \tag{7.120}$$

In Eqs. 7.119 and 7.120, adj $\mathbf{M}$ denotes the adjoint matrix of $\mathbf{M}$. Some words of caution are in order here. Perfect decoupling is possible only if the process model is perfect and reliable. However, even with imperfect process models, decoupling can be applied with considerable success. The dynamic decouplers being based on model inverses (Eqs. 7.119 and 7.120), these can be implemented only if the inverses are causal and stable. For further discussion of this and other related issues, the reader should refer to Ogunnaike and Ray [438].

Limitations of Decouplers - Ill-Conditioned Processes

It should be evident from Eqs. 7.118 and 7.120 that if the determinant of the process gain matrix is very small, the system will be extremely sensitive to any errors in the process model and decoupling will be difficult to achieve. Small changes in $\boldsymbol{\epsilon}$ or $\mathbf{v}$ will lead to large changes in $\mathbf{y}$. The process is said to be ill-conditioned when $|\mathbf{K}(t)|$ is very small. It is virtually impossible to achieve decoupling in an ill-conditioned process. There are situations where $|\mathbf{K}(t)|$ is not small, yet the process is poorly conditioned. Examples of these situations have been discussed in [438]. The most reliable indicator of the conditioning of a process is the condition number of the process gain matrix, $\kappa(\mathbf{K})$, which is provided by the ratio of the largest singular value of this matrix to the smallest singular value. The singular values of the real-valued $\mathbf{K}(t)$ are the the square root of the eigenvalues of the matrix $\mathbf{K}^T(t)\mathbf{K}(t)$. Since $\mathbf{K}^T(t)\mathbf{K}(t)$ is a symmetric matrix with real elements, its eigenvalues and therefore the singular values of $\mathbf{K}(t)$ are non-negative. When the input-output pairings are based on RGA, all singular values are positive since $\mathbf{K}(t)$ is not singular. As $|\mathbf{K}(t)|$ is reduced in absolute value, so is the smallest singular value of $\mathbf{K}(t)$ and the condition number is increased. A very small $|\mathbf{K}(t)|$ is indicative of the degeneracy of the process. Such degeneracy is only a special case of ill-conditioning. If the condition number of the process gain matrix is quite large, then the process is said to be poorly conditioned. Use of a decoupler in such situations would do more harm than good and should be avoided.

Decoupling Based on Singular Value Decomposition

The singular value decomposition (SVD) of the process gain matrix provides a much more general approach to decoupling. SVD allows for extension of matrix diagonalization to non-square process gain matrices, since $\mathbf{K}^T(t)\mathbf{K}(t)$ is a square matrix irrespective of whether or not $\mathbf{K}(t)$ is.

Singular Value Decomposition

Let $r(\mathbf{K})$ be the rank of $\mathbf{K}$ $[r = r(\mathbf{K}) < q,\ q = \min(m,p)]$, which is an $p \times m$ matrix. Then only r singular values of $\mathbf{K}$ (denoted as $\sigma_j,\ j = 1,2,\ldots,r$) are non-trivial and the remaining $(m - r)$ singular values are trivial. Let the non-trivial singular values be arranged as $\sigma_1 \geq \sigma_2 \geq \ldots \sigma_r$. For any matrix such as the process gain matrix, there exist orthogonal (i.e., unitary) matrices $\mathbf{W}$ and $\mathbf{V}$ such that

$$\mathbf{W^T K V} = \mathbf{\Sigma} \tag{7.121}$$

with $\mathbf{W}$, $\mathbf{V}$ and $\mathbf{\Sigma}$ being $p \times p$, $m \times m$ and $p \times m$ matrices related to $\mathbf{K}$ as follows. The p columns of $\mathbf{W}$, denoted as $\mathbf{w_i}$ $(i = 1,2,\ldots,p)$, are the orthonormal eigenvectors of $\mathbf{KK^T}$. Thus,

$$\mathbf{W} = [\mathbf{w_1}\ \mathbf{w_2}\ \cdots\ \mathbf{w_p}]. \tag{7.122}$$

Similarly, the m columns of $\mathbf{V}$, denoted as $\mathbf{v_i}$ $(i = 1,2,\ldots,m)$, are the orthonormal eigenvectors of $\mathbf{K^T K}$. Thus,

$$\mathbf{V} = [\mathbf{v_1}\ \mathbf{v_2}\ \cdots\ \mathbf{v_m}]. \tag{7.123}$$

Since $\mathbf{V}$ and $\mathbf{W}$ are composed of orthonormal vectors, these are orthogonal (or unitary) matrices, i.e., $\mathbf{V^T V} = \mathbf{VV^T} = \mathbf{I}$ (dimension of $\mathbf{I} = m$) and $\mathbf{W^T W} = \mathbf{WW^T} = \mathbf{I}$ (dimension of $\mathbf{I} = p$). It then follows that

$$\mathbf{V^T} = \mathbf{V^{-1}} \quad \text{and} \quad \mathbf{W^T} = \mathbf{W^{-1}}. \tag{7.124}$$

In view of the above, upon pre-multiplication by $\mathbf{W}$ and post-multiplication by $\mathbf{V^T}$, Eq. 7.121 can be restated as

$$\mathbf{K} = \mathbf{W}\Sigma\mathbf{V^T}. \tag{7.125}$$

The eigenvectors $\mathbf{v_i}$ of $\mathbf{K^T K}$ and $\mathbf{w_i}$ of $\mathbf{KK^T}$ are related to each other as per the following general pair of expressions.

$$\mathbf{Kv_i} = \sigma_i \mathbf{w_i} \quad \text{and} \quad \mathbf{K^T w_i} = \sigma_i \mathbf{v_i}. \tag{7.126}$$

With the singular values of $\mathbf{K}$ being arranged in descending order, the only non-trivial elements (Σ_{ij}) of the $p \times m$ matrix $\mathbf{\Sigma}$ appear for $i,j = 1,2,\ldots,r$, i.e.,

$$\Sigma_{ij} = \sigma_i,\ i = j = 1,2,\ldots,r; \quad \Sigma_{ij} = 0 \text{ otherwise.} \tag{7.127}$$

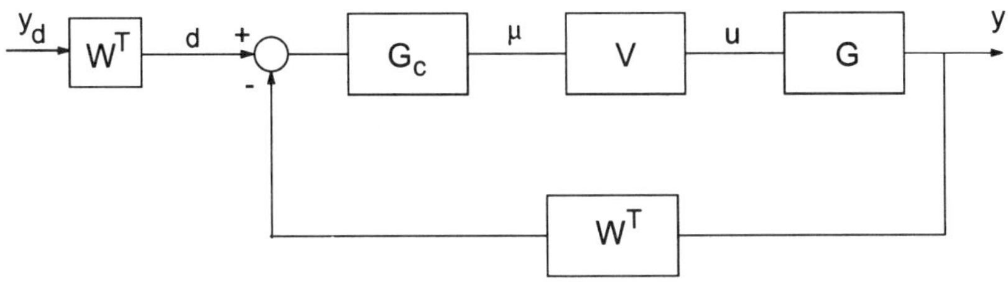

Figure 7.8. Block diagram of the multivariable controller based on SVD technique [438].

Decoupler Design

Substitution of the singular value decomposition of $\mathbf{K}$, Eq. 7.125, into the input-output relations for the process leads to the following

$$\Delta \mathbf{y}(t) = \mathbf{W \Sigma V^T} \Delta \mathbf{u}(t). \tag{7.128}$$

Pre-multiplication of the above by $\mathbf{W^T}$ and use of relation in Eq. 7.124 leads to the following restatement of Eq. 7.128

$$\Delta \boldsymbol{\eta}(t) = \boldsymbol{\Sigma} \Delta \boldsymbol{\mu}(t), \quad \Delta \boldsymbol{\mu}(t) = \mathbf{V^T} \Delta \mathbf{u}(t), \quad \Delta \boldsymbol{\eta}(t) = \mathbf{W^T} \Delta \mathbf{y}(t). \tag{7.129a}$$

Since the non-trivial elements of $\boldsymbol{\Sigma}$ lie on a diagonal (Eq. 7.127), the system is totally decoupled, with η_i being paired with μ_i ($i = 1, 2, \ldots, r$). The block diagram of the feedback-controlled multivariable process utilizing singular value decomposition is shown in Figure 7.8. The process outputs are mixed according to Eq. 7.129a to obtain $\boldsymbol{\eta}$, the information on which is then fed to the comparators to obtain controller inputs. The manipulated inputs $\mathbf{u}$ are obtained from the controller outputs, $\boldsymbol{\mu}$, via the mixing rule in Eq. 7.129a, i.e.,

$$\Delta \mathbf{u}(t) = \mathbf{V} \Delta \boldsymbol{\mu}(t) \tag{7.129b}$$

The application of the relative gain array method involves pairings among actual process inputs and outputs. For non-square process gain matrices (the number of inputs not being the same as the number of outputs), use of minimal controller configuration implies that either some inputs cannot be manipulated (overdefined system) or some outputs cannot be controlled (underdefined system). This problem does not arise when the controller configuration is based on SVD since all inputs and outputs are involved in the feedback control (Eq. 7.129).

7.5 Optimal Linear-Quadratic Feedback Control

A classical problem in optimal control theory, the linear-quadratic problem, is instrumental in identification of optimal feedback control strategies for both linear and nonlinear systems [84]. Since bioprocesses without any exception are nonlinear systems, we consider the nonlinear optimal control problem described by Eqs. 7.1 and 7.3-7.9. Let there be open-loop trajectories of $\mathbf{u}$(t) and $\mathbf{x}(t)$, $\mathbf{u}^*(t)$ and $\mathbf{x}^*(t)$, respectively, for a particular initial condition, $\mathbf{x}(0) = \mathbf{x}_0^*$, at which the necessary conditions for open-loop optimality, Eqs. 7.11, 7.13, 7.14 and 7.20, are satisfied, with the Hamiltonian H being defined in Eq. 7.10. It is assumed here that $\mathbf{x}(0)$ is specified and t_f and $\mathbf{x}(t_f)$ are unspecified. After a second-order expansion of the objective function J in Eq. 7.3 around the optimal open-loop trajectories of the state variables and the manipulated inputs, having adjoined the constraints in Eq. 7.1 and employed the necessary conditions for optimality listed above, the variation in δJ can be expressed as [498]

$$\delta J = \frac{1}{2}\int_0^{t_f} \left[(\delta\mathbf{u})^T\mathbf{R}\delta\mathbf{u} + 2(\delta\mathbf{x})^T\mathbf{Q}\delta\mathbf{u} + (\delta\mathbf{x})^T\mathbf{P}\delta\mathbf{x}\right] dt + \frac{1}{2}(\delta\mathbf{x}(t_f))^T\mathbf{S}_f\delta\mathbf{x}(t_f) \qquad (7.130)$$

with

$$\delta\mathbf{x}(t) = \mathbf{x}(t) - \mathbf{x}^*(t), \ \ \delta\mathbf{u}(t) = \mathbf{u}(t) - \mathbf{u}^*(t), \ \ \text{and} \ \ \delta J = J - J^*, \qquad (7.131)$$

and

$$\mathbf{P}(t) = \frac{\partial^2 H}{\partial \mathbf{x}^2}, \ \mathbf{Q}(t) = \frac{\partial^2 H}{\partial \mathbf{x}\partial \mathbf{u}}, \ \mathbf{R}(t) = \frac{\partial^2 H}{\partial \mathbf{u}^2}, \ \mathbf{S}_f = \left(\frac{\partial^2 G}{\partial \mathbf{x}^2}\right)_{t=t_f}. \qquad (7.132)$$

Notice that the definitions of $\mathbf{P}$, $\mathbf{Q}$ and $\mathbf{R}$ are the same as those in Eqs. 7.79. The matrices $\mathbf{P}(t)$, $\mathbf{Q}(t)$, $\mathbf{R}(t)$ and $\mathbf{S}_f$ are evaluated at the optimal trajectories of $\mathbf{x}$ and $\mathbf{u}$, viz., $\mathbf{x}(t) = \mathbf{x}^*(t)$ and $\mathbf{u}(t) = \mathbf{u}^*(t)$. The vector of state variables $\mathbf{x}$ considered here includes the n process variables which influence the process kinetics and additional up to $(a+b+1)$ state variables, the time-variance of which is described by Eqs. 7.23, 7.24 and 7.26. $\mathbf{P}$, $\mathbf{R}$ and $\mathbf{S}_f$ are symmetric matrices. One can then work with the following perturbation equations obtained from Eq. 7.1 via linearization around the open-loop optimal policy $[\mathbf{u}(t) = \mathbf{u}^*(t), \ \mathbf{x}(t) = \mathbf{x}^*(t)]$ for a fixed initial condition stated in Eq. 7.1, viz., $\mathbf{x}(0) = \mathbf{x}_0^*$.

$$\frac{d(\delta\mathbf{x}(t))}{dt} = \mathbf{A}(t)\delta\mathbf{x} + \mathbf{B}(t)\delta\mathbf{u}, \quad \delta\mathbf{x}(0) = \delta\mathbf{x}_0 \qquad (7.133)$$

with

$$\mathbf{A}(t) = \frac{\partial \mathbf{f}}{\partial \mathbf{x}}, \quad \mathbf{B}(t) = \frac{\partial \mathbf{f}}{\partial \mathbf{u}}. \tag{7.134}$$

The equation above represents the process behavior for initial conditions in a close neighborhood of $\mathbf{x}_0^*$. Definitions of system matrices $\mathbf{A}$ and $\mathbf{B}$ are the same as those in Eqs. 7.79 and 7.104. The variation in the objective function in Eq. 7.130 can be arranged in the following quadratic form

$$\delta J = \frac{1}{2}\{\delta\mathbf{x}(t_f)\}^T \mathbf{S}_f \{\delta\mathbf{x}(t_f)\} + \frac{1}{2}\int_0^{t_f} [\delta\mathbf{x}^T \delta\mathbf{u}^T] \begin{bmatrix} \mathbf{P}(t) & \mathbf{Q}(t) \\ \mathbf{Q}^T(t) & \mathbf{R}(t) \end{bmatrix} \begin{bmatrix} \delta\mathbf{x} \\ \delta\mathbf{u} \end{bmatrix} dt. \tag{7.135}$$

The objective of the optimal feedback control is then to minimize the degradation in the process performance ($\delta J < 0$) due to perturbations in $\mathbf{x}$ and $\mathbf{u}$. Maximization of δJ then requires solution of Eq. 7.133 and the associated adjoint variable equations. The boundary conditions for $\delta\mathbf{x}(t)$ are provided at $t = 0$, while those for the adjoint variables $\boldsymbol{\lambda}(t)$ are known at $t = t_f$. The solution to the resulting two-point boundary value problem can be conveniently expressed using the Riccati transformation wherein the adjoint variables and the corresponding state variables are related as [498, 560]

$$\boldsymbol{\lambda}(t) = \mathbf{S}(t)\delta\mathbf{x}(t). \tag{7.136}$$

For the objective functional in Eq. 7.135, the variation in the $n \times n$ matrix $\mathbf{S}(t)$ with t is described by the following Riccati equation

$$\frac{d\mathbf{S}}{dt} = -\mathbf{S}\mathbf{A} - \mathbf{A}^T\mathbf{S} + (\mathbf{S}\mathbf{B} + \mathbf{Q})\mathbf{R}^{-1}(\mathbf{Q}^T + \mathbf{B}^T\mathbf{S}) - \mathbf{P}, \quad \mathbf{S}(t_f) = \mathbf{S}_f. \tag{7.137}$$

The solution to Eq. 7.137 is then employed to relate the manipulated inputs to the state variables as per the following perturbation feedback control law [84, 498]

$$\mathbf{u}(t) = \bar{\mathbf{u}}(t) - \mathbf{K}(t)\left[\mathbf{x}(t) - \bar{\mathbf{x}}(t)\right], \tag{7.138}$$

with

$$\mathbf{K}(t) = \mathbf{R}^{-1}(\mathbf{Q}^T + \mathbf{B}^T\mathbf{S}). \tag{7.139}$$

For implementation of the feedback control policy outlined in Eqs. 7.137-7.139, knowledge of the optimal open-loop control policies is required. If the initial condition $\mathbf{x}(0)$ is altered, the entire nonlinear open-loop optimal control policy must be recalculated, since nonlinear optimal control problems, such as the ones encountered with bioprocesses, depend nonlinearly on the initial process conditions. A set of optimal open-loop control policies over a range of nominal initial conditions $\mathbf{x}_0^*$ must be calculated and stored prior to implementation of optimal feedback control. The corresponding trajectories of controller gains, $\mathbf{K}(t)$, based on solution of Riccati equation, Eq. 7.137, should be calculated and stored. The on-line feedback control can then be implemented by identifying the closest initial condition (among the stored values) to the actual initial condition and using the corresponding trajectory of proportional controller gain matrix, $\mathbf{K}(t)$, for feedback control. The procedure described here is useful for designing optimal proportional controllers with time-varying gains. Besides the proportional action, the other two controller actions, viz., the derivative and integral actions, can be built in with certain modifications of the problem considered earlier [135, 498]. For example, integral action can be added by inclusion of time derivative of $\mathbf{u}$ in the objective function J or by augmenting the state variables by p auxiliary state variables $\mathbf{z}(t)$ with

$$\frac{d\mathbf{z}}{dt} = \mathbf{Mx}, \quad \mathbf{z} = [z_1 \; z_2 \ldots \; z_p]^T. \tag{7.140}$$

In Eq. 7.140, $\mathbf{M}$ is an appropriate weight matrix and the p auxiliary variables correspond to those state variables for which integral action is desired. The state variable vector $\mathbf{x}(t)$ then would be comprised of the n process variables which influence the process kinetics, up to $(a + b + 1)$ auxiliary state variables which satisfy Eqs. 7.23, 7.24 and 7.26 and p auxiliary variables which satisfy Eq. 7.140. Derivative control action can similarly be incorporated through a different transformation [135].

7.6 Model Predictive Control

In the competitive global market, pharmaceutical and biotechnological companies are forced to increase the efficiency and productivity of well-established processes continuously. Changes amounting to a few percent in the final titers and product yields may lead to enormous benefits in large-scale cultivations. These goals can be achieved either by introducing more productive strains or by optimization of the cultivations. In the chemical and allied industries, the expectations for consistent attainment of high product quality, more efficient use of energy, and environmental impact of production activities have led to far stricter demands on control systems than can be met by traditional techniques alone. The response of these industries and academia to these challenges led to the development of a different control methodology, called Model Predictive Control (MPC). A built-in feature of MPC is the direct use of an explicit and separately identifiable process model. MPC is finding wide acceptance in applications in chemical and allied industries, because of its versatility, accommodation of nonlinear process models, and ability to handle variations in constraints in real time. MPC schemes use a process model for two key purposes, first to predict future process behavior explicitly, and the second to compute appropriate controller action required to drive the predicted outputs as close as possible to their respective desired values.

Industrial chemical, biotechnological and pharmaceutical processes are multivariable and nonlinear, and may exhibit difficult dynamic behavior due to time delays, inverse response, and open loop instability. Further, the process operations may be subject to constraints of all kinds to be satisfied by process inputs, process outputs, and certain state variables based on considerations for process economics and safety, environmental impact, and hardware (equipment) characteristics. An ideal controller required for optimal operation of these processes should be able to handle multivariable process interactions, time delays and other problematic dynamic behavior, input and output constraints, nonlinear process behavior, and influence of disturbance variables on the same, while optimizing the controller actions [438]. It should remain robust despite modeling errors and measurement noise and should be able to infer critical unmeasured information from whatever is available.

While no such ideal controller exists, the typical capabilities of MPCs come closest to the requirements for an ideal controller stated above. MPCs handle process interactions, time delays, inverse response and other difficult dynamics well. MPC utilizes a process model. A rigorous process model is not necessary since the MPC schemes can be based on non-parametric step- and impulse-response models. Being considered as an optimization, MPC is capable of meeting the control objectives by optimizing the control effort, while satisfying appropriate constraints. An attractive feature of MPC is compensation for the effect of measurable and unmeasurable disturbance variables. The compensation for measured disturbances is carried out in a feedforward mode, while that for unmeasured disturbances is carried out in a feedback fashion. A variety of references provide a good perspective of MPC [22, 174, 175, 176, 241, 376, 403, 420, 438, 477, 478]. MPC is best suited to processes with any of the following characteristics: (i) multiple input and output variables with significant interactions between single-input, single-output control loops, (ii) constraints in inputs and/or outputs, (iii) problematic dynamics such as long time delays, inverse response, and very large time constants. Although MPC is not inherently more or less robust than classical feedback control, it can be adjusted more easily for robustness.

Biotechnological processes have inherently slow dynamics. It therefore would take significant time for the full effect of each control action to be realized in the observable process outputs. It is therefore difficult to assess the full impact of the control actions taken in the past based only on the current output measurements. As a result, it is pertinent to consider how

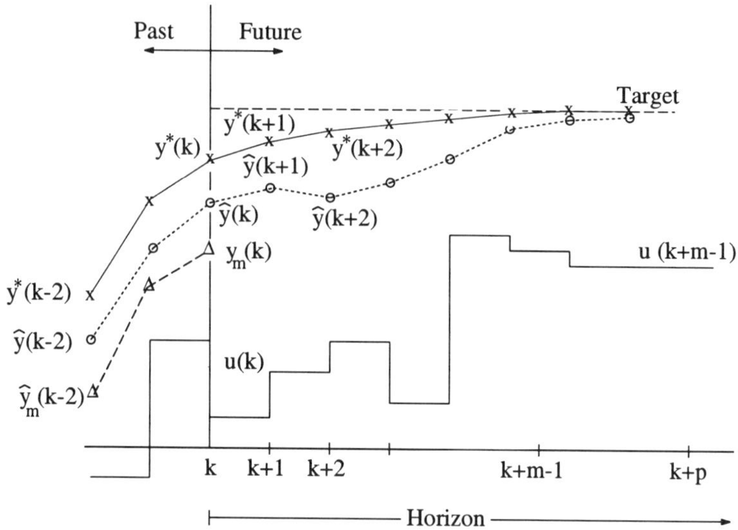

Figure 7.9. Example of elements in model predictive control: × − ×: reference trajectory, y^*; ○- - -○: predicted output, $\hat{y}$; △ −− △: measured output, y_m; ——: control action, u [438].
From BA Ogunnaike and WH Ray. Process Dynamics, Modeling, and Control. New York: Oxford University Press, Inc., 1994. Used by permission.

the process output will change in the future if no control action is taken (model-based prediction) and to target control action as a compensatory effect for what will need to be corrected after the full effects of the previously implemented control action have been completely realized. This is the motivation behind the MPC methodology.

The MPC design methodology consists of four elements: (i) specification of reference trajectories for the process outputs, (ii) model-based prediction of process outputs, (iii) model-based computation of control action, and (iv) update of error prediction for future control action. The variations in different MPC schemes are based primarily on how each element is implemented in the MPC scheme. The continuous-time process operation is comprised of successive time intervals. The four elements of MPC must be updated in each time interval. For this reason, it is convenient to work with discrete-time models for process and controllers. The discrete-time models are naturally well suited since most MPC schemes are implemented using digital computers. Techniques for transformation of continuous-time models into discrete-time models have been discussed earlier in Chapter 4.

The first element of MPC involves specification of the desired trajectories for the process outputs, $\mathbf{y}^*(k)$ (Figure 7.9). For an individual output, this can be a fixed set-point value or a trajectory. The second element involves prediction of trajectory of process outputs $\mathbf{y}$ in response to changes in the manipulated variables $\mathbf{u}$ in the absence of further control action. At the present time k ($t = kT$, T = sampling period), the behavior of the process is predicted over a horizon p. For discrete-time systems, this leads to prediction of $\hat{\mathbf{y}}(k+1)$, $\hat{\mathbf{y}}(k+2), \ldots, \hat{\mathbf{y}}(k+i)$ for i sample times into the future based on all actual past control actions $\mathbf{u}(k), \mathbf{u}(k-1), \ldots, \mathbf{u}(k-j)$ (Figure 7.9). In the third element of MPC, the same model as that used in the second element is employed to calculate control trajectories that lead to optimization of a specified objective function, which typically may include minimization of the predicted deviation of the process outputs from the target trajectories over the prediction horizon and minimization of the expense for control effort in driving the process outputs to their respective target trajectories. This is equivalent to constructing and utilizing a suitable model inverse to predict trajectories of the manipulated inputs. This optimization must of course be accomplished while satisfying pre-specified operating constraints. This element therefore involves prediction of the control sequence $\mathbf{u}(k), \mathbf{u}(k+1), \ldots, \mathbf{u}(k+m-1)$ required for achieving the desired output behavior p sampling times into the future [from $t = kT$ to $t = (k+p-1)T$] (Figure 7.9). Usually, the prediction horizon p is larger than the control horizon m. For computations, all control commands for times $(k+m)$ to $(k+p)$ are kept constant at their values at time $(k+m-1)$. This reduces the computational burden during real time optimization. The last element of MPC involves comparison of the output measurements $\mathbf{y}_m(k)$ to model-predicted values of the same, $\hat{\mathbf{y}}(k)$. The prediction error $\hat{\varepsilon}(k) = \mathbf{y}_m(k) - \hat{\mathbf{y}}(k)$ [not to be confused with the controller input, $\varepsilon(k) = \mathbf{y}_m(k) - \mathbf{y}^*(k)$] is then used to update future predictions $\hat{\mathbf{y}}(k+1), \ldots, \hat{\mathbf{y}}(k+i)$.

The conventional MPC schemes have relied on linear process models. Incorporation of nonlinear process models within the MPC framework is relatively recent [54]. Here we briefly review the three more commonly used forms of discrete models, the finite convolution models based on impulse- and step-response function, state-space models, and the transfer function models. For easier understanding, these models are presented below for SISO systems and can be readily extended to MIMO systems.

There are two entirely equivalent forms of *finite convolution models*, namely, the impulse-response model and the step-response model. The former can be expressed as

$$y(k) = \sum_{i=0}^{k} g(i)u(k-i), \tag{7.141}$$

with $g(i)$ being the impulse response functions of the process. The step-response model can be expressed as

$$y(k) = \sum_{i=0}^{k} \beta(i)\Delta u(k-i), \tag{7.142}$$

with $\beta(i)$ being the step-response functions for the process and $\Delta u(k) = u(k) - u(k-1)$. For all real, causal systems, both $g(0)$ and $\beta(0)$ are considered to be trivial, hence such systems will exhibit the mandatory one-step delay. For a process represented by the two model forms in Eqs. 7.141 and 7.142, the equivalency of the two models follows by equality of coefficients of $u(k-i)$, $i = 0, 1, \ldots, k$, leading to the following relations.

$$g(i) = \beta(i) - \beta(i-1), \qquad \beta(i) = \sum_{j=1}^{i} g(j). \tag{7.143}$$

Impulse or step response coefficients can be obtained directly from experimental data or from other parametric model forms, if available, such as those discussed in the following.

State-space models relate the variations in state variables and hence process outputs over time to the past history of state variables and to process inputs. The models may be in the form of differential equations (Eqs. 4.44, 4.45, 7.1 and 7.2) or difference equations (Eq. 4.51). The models in these forms are more advantageous for state estimation using Kalman filters and extended Kalman filters than the time series models. The reader should refer to Section 4.3 for discussion on continuous and discrete Kalman filters.

Time series models such as ARMA, provide relations among output values at the current and previous sampling times and input values at the current and prior sampling times

$$y(k) = \sum_{i=0}^{k} a(i)y(k-i) + \sum_{i=0}^{k} b(i)u(k-i-d), \tag{7.144}$$

The effect of time delay is included when $d \geq 1$. For real, causal processes, it follows that $a(0) = b(0) = 0$. The coefficients $a(i)$ and $b(i)$ and the time delay, d, in Eq. 7.144 must be identified by fitting the model to experimental process data. The linearized continuous-time versions of the nonlinear continuous-time state-space models, such as in Eqs. 7.104 and 7.106 obtained from linearization of Eqs. 7.1 and 7.2, can be transformed into time series models as in Eq. 7.144 with relative ease.

As the name suggests, in a *transfer function model*, the process outputs are related to the manipulated inputs by transfer functions. For a SISO process for example, the output y and the input u are related as

$$y(z) = \frac{z^{-d}B(z^{-1})}{A(z^{-1})}u(z). \tag{7.145}$$

$A(z^{-1})$ and $B(z^{-1})$ are appropriate polynomials in the z-transform variable, z^{-1}, and the term z^{-d} incorporating the effect of process time delay of d sampling times ($= dT$). The parameters of the transfer function model must be determined from experimental data for the process. As stated earlier, a rigorous process model is not necessary since non-parametric step- and impulse-response models can be readily employed for MPC in the absence of rigorous process models.

The "model inverse" required for prediction of manipulated inputs at future sampling times is carried out numerically as the solution of an appropriate optimization problem. Only the first computed change in the manipulated inputs is implemented. At time $k+1$, the computations for the four elements of MPC are repeated with the time horizon moved by one time interval (sampling time).

A variety of factors are responsible for the discrepancy observed between the actual output measurements and model-predicted values of the outputs. These include the effects of unmodeled and unmeasured disturbances, fundamental errors in model structure, and parameter uncertainties. Since it is difficult to independently assess these effects, it is a common strategy to attribute this discrepancy entirely to unmeasured disturbances, assume that this discrepancy will remain the same over the prediction horizon, until better information is available, and update model predictions by adding this discrepancy.

Two pioneering MPC schemes are the dynamic matrix control (DMC) and model algorithmic control (MAC) and derivatives of these, such as the quadratic dynamic matrix control (QDMC) [175] and IDCOM [383, 384, 503, 504, 505, 506]. In the following, we briefly discuss the basic formulation of DMC.

Dynamic Matrix Control (DMC). Let the current time instant be k. In the absence of further control action, let the process output take the following predicted values over the future horizon of p sampling times: $\hat{y}^o(k)$, $\hat{y}^o(k+1), \ldots, \hat{y}^o(k+p-1)$. Let the vector of the predicted p values be represented as

$$\hat{\mathbf{y}}^o(k) = [\hat{y}^o(k) \; \hat{y}^o(k+1) \; \ldots \; \hat{y}^o(k+p-1)]^T. \tag{7.146}$$

The argument in the vector above indicates the time origin of sequential predictions of the process outputs and the superscript ^{o} indicates that the

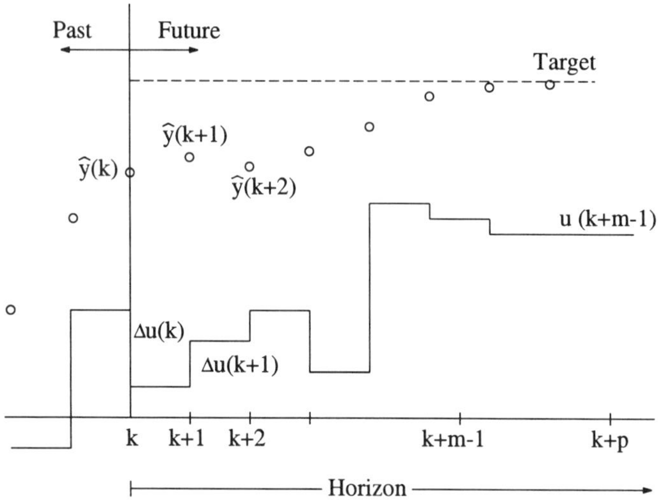

Figure 7.10. The elements of DMC: The "reference trajectory" is the set-point line [438].

From BA Ogunnaike and WH Ray. Process Dynamics, Modeling, and Control. New York: Oxford University Press, Inc., 1994. Used by permission.

output predictions are conditional on the absence of further control action. The case $k = 0$ corresponds to initial condition on the predicted process output. Sequences for $k > 0$ are obtained recursively using the predictions at the prior instant, $(k - 1)$. The vector notation used here is different from the one used earlier in this chapter (including this section). Earlier, $\mathbf{y}(t)$ or $\mathbf{y}(k)$ denoted the values of different process outputs at the same time instant, for the single-input, single-output system under consideration. Here, $\hat{\mathbf{y}}^o(k)$ denotes the predicted values of single output y at p sampling times including the current time k. An arbitrary sequence of m $(m < p)$ control actions, $\Delta u(k)$, $\Delta u(k+1), \ldots, \Delta u(k+m-1)$, will cause the process outputs to change from the initial conditions $\hat{\mathbf{y}}^o(k)$ to a new state (Figure 7.10)

$$\hat{\mathbf{y}}(k+1) = [\hat{y}(k+1)\ \hat{y}(k+2) \ldots\ \hat{y}(k+p)]^T . \tag{7.147}$$

Let the effect of unmeasured disturbances on the predicted output $\hat{y}(k+i)$ be represented as $w(k+i)$, $i = 1, 2, \ldots, p$. Then it follows from Eq. 7.142

that

$$\hat{y}(k+i) = \hat{y}^o(k+i-1) + \sum_{j=1}^{i} \beta(j)\Delta u(k+i-j) + w(k+i), \; i = 1, 2, \ldots, m, \tag{7.148}$$

$$\hat{y}(k+i) = \hat{y}^o(k+i-1) + \sum_{j=1}^{m} \beta(j+i-m)\Delta u(k+m-j) + w(k+i),$$

$i = m+1, \; m+2, \; \ldots \; , \; p.$

Eq. 7.148 may be rewritten succinctly as

$$\hat{\mathbf{y}}(k+1) = \hat{\mathbf{y}}^o(k) + \mathbf{w}(k+1) + \mathbf{X}\Delta\mathbf{u}(k) \tag{7.149}$$

with

$$\Delta\mathbf{u}(k) = [\Delta u(k) \; \Delta u(k+1) \ldots \; \Delta u(k+m-1)]^T, \tag{7.150a}$$

and

$$\mathbf{X} = \begin{bmatrix} \beta(1) & 0 & 0 & \cdots & 0 \\ \beta(2) & \beta(1) & 0 & \cdots & 0 \\ \beta(3) & \beta(2) & \beta(1) & \cdots & 0 \\ \cdots & \cdots & \cdots & \cdots & 0 \\ \beta(m) & \beta(m-1) & \beta(m-2) & \cdots & \beta(1) \\ \beta(m+1) & \beta(m) & \beta(m-1) & \cdots & \beta(2) \\ \vdots & \vdots & \vdots & \vdots & \vdots \\ \beta(p) & \beta(p-1) & \beta(p-2) & \cdots & \beta(p-m+1) \end{bmatrix} \tag{7.150b}$$

$\mathbf{X}$ is referred to as the Dynamic Matrix. The MPC methodology for the unconstrained problem is considered first.

The control problem expressed in Eq. 7.149 reduces to judiciously choosing and implementing the control sequence $\Delta\mathbf{u}(k)$ so that the predicted process output is driven to and remains at the desired trajectory

$$\mathbf{y}^*(k+1) = [y^*(k+1) \; y^*(k+2) \ldots \; y^*(k+p)]^T. \tag{7.151}$$

i.e., let $\hat{\mathbf{y}}(k+1) = \mathbf{y}^*(k+1)$. The difference between the desired output trajectory, $\mathbf{y}^*(k+1)$ and the current output prediction in the absence of further control action corrected for the effect of unmeasured disturbances on the process output, viz., $\hat{\mathbf{y}}^o(k) + \mathbf{w}(k+1)$, is the predicted error vector $\mathbf{e}(k+1)$, which is the input to feedback controllers. Eq. 7.149 therefore is restated as

$$\mathbf{e}(k+1) = \mathbf{X}\Delta\mathbf{u}(k). \tag{7.152}$$

The left hand side of Eq. 7.152 represents the predicted deviation of process output from the desired set-point trajectory in the absence of further control action and the right hand side the predicted change in the process output resulting from the control action, $\Delta \mathbf{u}(k)$.

The horizon over which control moves are computed is always smaller than the horizon chosen for output prediction (i.e., $m < p$). As a result, Eq. 7.152 represents an overdetermined system of equations. No exact solution exists for Eq. 7.152 as a result. A satisfactory "solution" to Eq. 7.152 then may be obtained by minimizing an appropriate metric that represents the difference between the left hand and right hand sides of Eq. 7.152. One such metric is described by the right hand side of Eq. 7.153.

$$\min_{\Delta \mathbf{u}(k)} J = [\mathbf{e}(k+1) - \mathbf{X}\Delta \mathbf{u}(k)]^T [\mathbf{e}(k+1) - \mathbf{X}\Delta \mathbf{u}(k)] + K[\Delta \mathbf{u}(k)]^T \Delta \mathbf{u}(k), \quad K \geq 0 \tag{7.153}$$

The second term on the right hand side of Eq. 7.153 reflects a penalty against excessive control action. The necessary condition for minimization of J with respect to $\Delta \mathbf{u}(k)$ is that the derivative vector $\partial J / \partial \Delta \mathbf{u}(k)$ be trivial. The application of this condition to Eq. 7.153 leads to the following feedback control law [438].

$$(\mathbf{X}^T \mathbf{X} + K\mathbf{I})\Delta \mathbf{u}(k) = \mathbf{X}^T \mathbf{e}(k+1) \;\Rightarrow\; \Delta \mathbf{u}(k) = (\mathbf{X}^T \mathbf{X} + K\mathbf{I})^{-1} \mathbf{X}^T \mathbf{e}(k+1). \tag{7.154}$$

The projected error vector requires the vector of future values of effects of unmeasured disturbances on the process output, values that are not available at the present time k. In the absence of any better information, $\mathbf{w}(k+1)$ is estimated as

$$\hat{w}(k+i) = y_m(k) - \hat{y}(k), \quad i = 1, 2, \ldots, p. \tag{7.155}$$

It is not advisable to implement the entire control sequence, $\Delta u(k), \Delta u(k+1), \ldots, \Delta u(k+m-1)$, as calculated from Eq. 7.154, in quick succession for the following reasons. It must be recognized that it is impossible to anticipate or predict precisely over the next m sampling intervals, the process-model mismatch and unmodeled disturbances which will cause the actual state of the process to differ from the model predictions used to compute this sequence of control actions. Additionally, there may be changes in the process set point at any time over the next m time intervals as better information becomes available on the status of the process. The pre-computed control sequence is inherently incapable of reflecting the changes which would occur after the computation. For these reasons, as mentioned earlier, the MPC strategy therefore is to implement only the first control action, $\Delta u(k)$, and repeatedly execute the following steps.

1. Update $\mathbf{y}^*(k+1)$ with any new desired set-point information.

2. Update $\hat{\mathbf{y}}^o(k)$ by adding the effect of implementation of $\Delta u(k)$ and assuming no further control move will be implemented.

3. Update $\mathbf{w}(k+1)$ as per Eq. 7.155.

4. Update the projected error vector $\mathbf{e}(k+1)[= \mathbf{y}^*(k+1) - \hat{\mathbf{y}}^o(k) - \mathbf{w}(k+1)]$.

5. Shift the origin of the prediction horizon from k to $(k+1)$. Obtain $\mathbf{y}^*(k+2)$, $\hat{\mathbf{y}}^o(k+1)$ and $\mathbf{w}(k+2)$ from $\mathbf{y}^*(k+1)$, $\hat{\mathbf{y}}^o(k)$, and $\mathbf{w}(k+1)$, respectively, by removing the first element in each of these updated vectors, advancing the other elements in order, and filling the last element by linear extrapolation.

6. Compute the new control action sequence using the updated and shifted vectors, implement $\Delta u(k+1)$ and repeat steps 1-6.

Application of DMC scheme to multiple-input, multiple-output (MIMO) processes is discussed next.

For multiple-input, multiple-output processes, step response models analogous to the one in Eq. 7.142 must be considered. For example, for a process with two inputs and three outputs, the impact of step changes in inputs u_1 and u_2 on output y_3 may be expressed by the step-response model

$$y_3(k) = \sum_{i=0}^{k} \beta_{31}(i)\Delta u_1(k-i) + \sum_{i=0}^{k} \beta_{32}(i)\Delta u_2(k-i), \tag{7.156}$$

with β_{31} and β_{32} being the parameters indicative of the sensitivity of y_3 to changes in u_1 and u_2, respectively. The procedure for application of the four elements of MPC for MIMO processes is the same as that for SISO processes, except that the matrices and vectors involved are much larger. For example, the relations between the predicted error vector and the control action in Eq. 7.152 are also applicable for MIMO processes with

$$\mathbf{e}(k+1) = [\mathbf{e}_1(k+1) \ \ldots \ \mathbf{e}_2(k+1)]^T, \quad \Delta\mathbf{u}(k) = [\Delta\mathbf{u}_1(k) \ \ldots \ \Delta\mathbf{u}_2(k)]^T,$$
$$\mathbf{e}_1(k+1) = \mathbf{y}_1^*(k+1) - \hat{\mathbf{y}}_1^o(k) - \mathbf{w}_1(k+1),$$
$$\mathbf{e}_2(k+1) = \mathbf{y}_2^*(k+1) - \hat{\mathbf{y}}_2^o(k) - \mathbf{w}_2(k+1),$$

$$\mathbf{X} = \begin{bmatrix} \mathbf{X}_{11} & \vdots & \mathbf{X}_{12} \\ \ldots & \vdots & \ldots \\ \mathbf{X}_{21} & \vdots & \mathbf{X}_{22} \end{bmatrix} \tag{7.157}$$

The dynamic matrices $\mathbf{X}_{\gamma\delta}$ in Eq. 7.157 (for example, $\gamma, \delta = 1,\ 2$) relate the dependence of the output y_γ on the input u_δ. For given m and p, these matrices are identical in structure to the dynamic matrix $\mathbf{X}$ for an SISO process (Eq. 7.150b), with the elements of $\mathbf{X}_{\gamma\delta}$ being obtained by replacing β by $\beta_{\gamma\delta}$ in Eq. 7.150b.

In multivariable processes, it must be realized that a change of certain magnitude (say a unit change) in one output may be more or less important than the change of the same magnitude in another output. One example is an output that is mole fraction of a specie, a second output that is temperature, and a third that is a flow rate. Since the three outputs will have different nominal values, a unit change in mole fraction spans the entire scale (mole fraction ranges between zero and unity), while a unit change in temperature may be a change of few percent and the same may be the case with flow rate. Appropriate scaling factors must therefore be incorporated when working with the error vector $\mathbf{e}(k+1)$ and the control action vector $\Delta\mathbf{u}(k)$ so that equally important changes in different outputs or inputs are treated equally. This is accomplished by pre-multiplying the projected error vector $\mathbf{e}(k+1)$ by a scaling matrix $\mathbf{W}$ and the control sequence vector $\Delta\mathbf{u}(k)$ by a scaling matrix $\mathbf{V}$ and employing the resulting scaled vectors in the definition of objective function J (an appropriate norm) to be minimized. The objective in Eq. 7.153 then may be modified as follows to reflect this.

$$\min_{\Delta\mathbf{u}(k)} J = [\mathbf{e}(k+1) - \mathbf{X}\Delta\mathbf{u}(k)]^T\mathbf{\Gamma}[\mathbf{e}(k+1) - \mathbf{X}\Delta\mathbf{u}(k)] + [\Delta\mathbf{u}(k)]^T\mathbf{\Lambda}\Delta\mathbf{u}(k)$$

$$\mathbf{\Gamma} = \mathbf{W}^T\mathbf{W}, \quad \mathbf{\Lambda} = \mathbf{V}^T\mathbf{V}. \tag{7.158}$$

The analytical, closed form solution for the classical least-squares problem in Eq. 7.158 is obtained by equating $\partial J/\partial\Delta\mathbf{u}(k)$ to zero. The control policy then is provided by

$$(\mathbf{X}^T\mathbf{\Gamma}\mathbf{X}+\mathbf{\Lambda})\Delta\mathbf{u}(k) = \mathbf{X}^T\mathbf{\Gamma}\mathbf{e}(k+1) \Rightarrow \Delta\mathbf{u}(k) = (\mathbf{X}^T\mathbf{\Gamma}\mathbf{X}+\mathbf{\Lambda})^{-1}\mathbf{X}^T\mathbf{\Gamma}\mathbf{e}(k+1). \tag{7.159}$$

The elements of the matrices $\mathbf{\Gamma}$ and $\mathbf{\Lambda}$ are best determined by choosing the elements of the scaling matrices $\mathbf{W}$ and $\mathbf{V}$. The scaling matrices $\mathbf{W}$ and $\mathbf{V}$ will be partitioned matrices with the elements of each submatrix being identical and equal to the scaling factor for a particular output or manipulated input, as appropriate.

When constraints are involved, the objective function ϕ used for computing the control action sequence $\Delta\mathbf{u}(k)$ must be augmented with the constraint equations. This prevents the development of closed-form controller equations such as Eq. 7.159. The resulting quadratic program must be solved as a real-time optimization problem to identify the recommended control action sequence [176, 477, 507, 681].

Batch and fed-batch bioprocesses are inherently transient operations. Although transients are also encountered in continuous bioprocesses, the focus usually is on operation at a desired steady-state. In a steady-state operation, the target trajectory reduces to a fixed set-point in the multidimensional space of output variables. For batch and fed-batch operations, which are more common compared to continuous operations, the target trajectories of the output variables will typically be time-variant, i.e., the set-point for a particular output will vary with time. One of the characteristics of industrial batch and fed-batch bioprocesses is that they are cyclic or repetitive. This characteristic allows the process operator and controllers the opportunity to make compensations based on errors from previous cycles. The target trajectories can therefore be updated from cycle to cycle as one gathers more knowledge of the bioprocess. Indeed, this idea, referred to by the generic name Iterative Learning Control (ILC) has been integrated into conventional model predictive control and applied to batch processes [327, 328, 682]. The integrated methodology is not only capable of eliminating persisting errors from previous cycles or runs, but also can respond to new disturbances as these occur during a particular cycle or run [327].

The DMC formulation uses a step response model which is limited to stable processes. To leverage the wealth of knowledge on state-space techniques, MPC algorithms based on state-space models were proposed by converting the step response models into state-space form [324, 339]. The state-space formulation for unstable linear processes has been addressed [405]. A related formulation called generalized predictive control is based on transfer function models with the aim of handling unstable processes [107]. Inequality constraints in an MPC lead to nonlinear feedback control laws. Considering the abundance of process nonlinearities in chemical and biochemical processes, the use of nonlinear process models in MPC formulation was a natural progress.

Nonlinear MPC

Many processes have significant nonlinearities that challenge successful implementation of *linear* MPC. This has motivated the development of *nonlinear* MPC (NMPC) which relies on the use of a nonlinear process model. NMPC has the potential of improving process operation, but it also provides challenging theoretical and practical problems mostly because of the nonlinear optimization problem that must be solved at each sampling instant in real time to compute the control moves.

Many nonlinear model representations were discussed in Section 4.3. Consider the general form expressed in Eqs. (4.44)-(4.45)

$$\frac{d\mathbf{x}}{dt} = \mathbf{f}(\mathbf{x}(t), \mathbf{u}(t)), \qquad \mathbf{y}(t) = \mathbf{h}(\mathbf{x}(t), \mathbf{u}(t)) \tag{7.160}$$

that can be written in discrete time as

$$\mathbf{x}(k+1) = \mathbf{f}(\mathbf{x}(k), \mathbf{u}(k)), \qquad \mathbf{y}(k) = \mathbf{h}(\mathbf{x}(k), \mathbf{u}(k)) \quad k = 0, 1, 2, \cdots \tag{7.161}$$

where $\mathbf{x}(k)$ is a condensed form of the terminology $\mathbf{x}(t_k)$ used in Section 4.3.2. The optimization problem in NMPC formulation can be expressed as finding the values of $\mathbf{u}$ to optimize the objective function J subject to constraints [234, 381]

$$\begin{aligned} \min_{\mathbf{u}(k|k), \mathbf{u}(k+1|k), \cdots, \mathbf{u}(k+m-1|k)} \quad J = & L_0 \left[\mathbf{y}(k+p|k)\right] \\ & + \sum_{j=0}^{p-1} L\left[\mathbf{y}(k+j|k), \mathbf{u}(k+j|k), \Delta\mathbf{u}(k+j|k)\right], \end{aligned} \tag{7.162}$$

where L_0 and L are nonlinear functions of their arguments, m is the control horizon, and p is the prediction horizon. $\mathbf{y}(k+j|k)$ denotes the value of the output $\mathbf{y}$ at time $k+j$ computed from information available at time k and $\Delta\mathbf{u}(k+j|k) = \mathbf{u}(k+j|k) - \mathbf{u}(k+j-1|k)$. The functions L_0 and L may represent a variety of objectives, including the minimization of the overall cost of process operation. For example, regulation to set points or tracking reference trajectories can be formulated as quadratic equations of the form

$$\begin{aligned} L_0 &= \left[\mathbf{y}(k+p|k) - \mathbf{y}_r(k)\right]^T \ \mathbf{Q} \ \left[\mathbf{y}(k+p|k) - \mathbf{y}_r(k)\right] \\ L &= \left[\mathbf{y}(k+j|k) - \mathbf{y}_r(k)\right]^T \ \mathbf{Q} \ \left[\mathbf{y}(k+j|k) - \mathbf{y}_r(k)\right] \\ & \quad + \left[\mathbf{u}(k+j|k) - \mathbf{u}_r(k)\right]^T \ \mathbf{R} \ \left[\mathbf{u}(k+j|k) - \mathbf{u}_r(k)\right] \\ & \quad + \Delta\mathbf{u}^T(k+j|k) \ \mathbf{S} \ \Delta\mathbf{u}(k+j|k) \end{aligned} \tag{7.163}$$

where $\mathbf{y}_r(k)$ and $\mathbf{u}_r(k)$ are the reference values for $\mathbf{y}$ and $\mathbf{u}$, and $\mathbf{Q}$, $\mathbf{R}$, and $\mathbf{S}$ are positive definite weighting matrices. These weighting matrices, m, p and the sampling time are the tuning parameters of the NMPC. The prediction of output values $\mathbf{y}(k+j|k)$ are based on state variables and the calculated input sequence. Hence, measurement or estimation of state variables is necessary. State estimation relies on the nonlinear model Eq. (7.161) and use of process information that reflects disturbance effects as discussed below. The solution of the NMPC problem yields the values for the input sequence $(\mathbf{u}(k|k), \mathbf{u}(k+1|k), \cdots, \mathbf{u}(k+m-1|k))$. Only the first input vector $\mathbf{u}(k|k)$ is implemented and the the real time optimization problem is solved again at the next sampling time.

Constraints Several constraints are imposed on inputs and outputs. Input (manipulated variable) constraints reflect actuator limitations such as saturation and rate-of-change restrictions such as rate of temperature increase:

$$\mathbf{u}_{\min} \leq \mathbf{u}(k+j|k) \leq \mathbf{u}_{\max} \qquad j = 0, m-1 \ , \tag{7.164}$$

$$\Delta\mathbf{u}_{\min} \leq \Delta\mathbf{u}(k+j|k) \leq \Delta\mathbf{u}_{\max} \qquad j = 0, m-1 \; . \tag{7.165}$$

where $\mathbf{u}_{\min}$ and $\mathbf{u}_{\max}$ denote the minimum and maximum values of the inputs. Output constraints are associated with operational limitations such as equipment, materials, product properties, and safety considerations:

$$\mathbf{y}_{\min} \leq \mathbf{y}(k+j|k) \leq \mathbf{y}_{\max} \qquad j = 0, p \; . \tag{7.166}$$

The nonlinear model in Eq. (7.161) is added as equality constraints:

$$\begin{aligned} \mathbf{x}(k+j+1|k) &= \mathbf{f}(\mathbf{x}(k+j|k), \mathbf{u}(k+j|k)) \qquad j = 0, p-1 \\ \mathbf{y}(k+j|k) &= \mathbf{h}(\mathbf{x}(k+j|k)) \qquad j = 1, p \end{aligned} \tag{7.167}$$

where $\mathbf{x}(k|k) = \mathbf{x}(\mathrm{k})$ if the state variables are measured.

State and Disturbance Estimation The objective of the NMPC system is to drive process outputs (and inputs) to their reference (target) values. If the reference values used in Eqs. (7.163) are not chosen properly, unmeasured disturbances and modeling errors would cause offset. The offset problem can be handled by designing a disturbance estimator that provides an implicit integral control action [234, 381]. A simple method for incorporating integral action is to modify the reference values $\mathbf{y}_r$ by shifting the set points with the disturbance estimates. The penalties on inputs are also eliminated in this method ($\mathbf{R} = 0$). The output references are computed as

$$\begin{aligned} \mathbf{y}_r(k) &= \mathbf{y}_{sp} - \hat{\mathbf{d}}(k) \\ \hat{\mathbf{d}}(k) &= \mathbf{y}(k) - \mathbf{y}(k|k) \end{aligned} \tag{7.168}$$

where $\mathbf{y}_{sp}$ are the set points of outputs, $\mathbf{y}(k)$ are the measured values of outputs, $\mathbf{y}(k|k)$ are output estimates obtained from the nonlinear model Eq. (7.161), and $\hat{\mathbf{d}}(k)$ are the estimated disturbances. This disturbance model assumes that plant-model mismatch is attributable to a step disturbance in the output that remains constant over the prediction horizon [234]. A method for incorporating integral action based on steady-state target optimization has been developed [381].

Simultaneous state and disturbance estimation can be performed by augmenting the state-space model:

$$\begin{aligned} \mathbf{x}(k+1) &= \mathbf{f}(\mathbf{x}(k), \mathbf{u}(k)) \\ \mathbf{d}(k+1) &= \mathbf{d}(k) \\ \mathbf{y}(k) &= \mathbf{h}(\mathbf{x}(k), \mathbf{u}(k)) + \mathbf{d}(k) \end{aligned} \tag{7.169}$$

where $\mathbf{d}(k)$ is a constant output disturbance. The augmented process model can be used for designing a nonlinear observer. A general theory for nonlinear observer design is not available, and input-output models are preferred over state-space models when full state feedback is not available. A list of NMPC applications with simulations and experimental studies is given in [234] along with a discussion of computational issues and future research directions.

Heuristic tuning guidelines are discussed in [381] and summarized in [234]. For stable systems, the sampling interval should be selected to provide a compromise between on-line computation load and closed-loop performance. There is an inverse relationship between sampling interval and allowable modeling error. Smaller control horizons (m) yield more sluggish output responses and more conservative input moves. Large values of m increase the computation burden. Large prediction horizons (p) cause more aggressive control and heavier computation burden. The weighting matrices ($\mathbf{Q}$, $\mathbf{R}$, $\mathbf{S}$) are dependent on the scaling of the problem. Usually they are diagonal matrices with positive elements. The parameter values can be tuned via simulation studies.

Computational constraints and stability of the controlled system are critical issues in NMPC. The need to solve the nonlinear programming problem in real time necessitate efficient and reliable nonlinear programming techniques and MPC formulations that have improved computational speed. Successive linearization of model equations, sequential model solution and optimization, simultaneous model solution and optimization are some of the approaches proposed in recent years [234, 381].

MPC of Batch Bioprocesses

Batch and fed-batch bioprocesses typically exhibit large variations in the operating conditions during a cycle or a run. During different phases of batch and fed-batch operations, culture parameters such as pH, temperature, and substrate availability may change and these changes would substantially alter parameters in the bioprocess model. The performance of MPC depends critically on the predictive abilities of the process model employed for prediction. For the reasons mentioned above, building a nonlinear model for batch and fed-batch cultures is a cumbersome task [156]. Empirical models are therefore appealing in model predictive control of batch and fed-batch bioprocesses. In fact, most of the practical applications of MPC have involved use of empirical models. Another way to circumvent the nonlinear first principles model development for application of MPC is to use artificial neural networks (ANNs). Models based on ANNs rely on data from previous runs or cultivations and are capable of extracting the relevant parameters and relationships from them [397]. Model predictive control with ANNs has been used for on-line optimization of riboflavin production in a fed-batch bioprocess [299]. A variant of this approach is the use of empirical reference trajectories and predictive models developed using the multivariate statistical methods discussed in Chapter 6. With landmark and trajectory alignment using dynamic time warping (DTW) and curve registration, the reference trajectories and predictive models developed show good promise for MPC.

8

Fault Diagnosis

Detection and diagnosis of faults in batch process operations is of great significance for productivity and product quality improvements. Several disciplines including statistics, systems science, signal processing, and computer science have contributed to the development of fault detection and diagnosis (FDD) techniques. FDD systems implement the following tasks [189]:

Fault detection: Indication of abnormal system behavior. This can be achieved by process monitoring techniques discussed in earlier chapters or by a number of other paradigms.

Fault isolation: Determination of the specific cause or location of the fault.

Fault identification: Determination of the magnitude of the fault.

The term "diagnosis" is used to refer to the combined isolation and identification tasks, but it can also be used as a synonym for isolation.

Faults are deviations from normal (expected) behavior in a process or its instruments. Faults may be grouped as sensor, actuator or process faults. *Sensor faults* are discrepancies between measured and actual values of process variables. *Actuator faults* are discrepancies between the control command received by an actuator and the actuator output. *Process faults* include all other faults. They may be *additive* such as leaks or *multiplicative* such as deterioration of process equipment like fouling of heat exchange surfaces. In general, additive faults are unknown inputs which are normally zero, and multiplicative faults are abrupt or gradual changes that affect the parameters of the process.

FDD methods can be classified as model-free and model-based methods. *Model-free FDD methods* do not utilize a mathematical model of the plant. They are based on limits on variables, physical redundancy or empirical process knowledge (mental models). *Model-based FDD methods* use a mathematical model of the process developed using by first principles or data-based empirical techniques. They either use the residuals between measured and estimated values of process variables or recursive estimates of model parameters to implement FDD. MSPM methods discussed in Chapter 6 for determining out-of-control status (fault detection) and contribution plots, or other statistical tools such as discriminant analysis are also model-based techniques.

The performance of fault detection and diagnosis methods is characterized by several benchmarks:

Sensitivity: Ability to detect and diagnose faults of a specific size. The magnitude of fault size to detect depends on process needs.

Discrimination power (isolation performance): Ability to discriminate the correct fault(s) when several faults occur simultaneously, masking each other.

Robustness: Ability to detect and diagnose a fault in the presence of noise, disturbances, and modeling errors.

Missed fault detections and false alarms: The number of faults that have not been detected and the number of alarms issued when there were no faults.

Detection and diagnosis speed: Time to detect and diagnose faults after their occurrence.

The first four benchmarks are related to Type I and Type II errors discussed in Section 6.1.

To check the correctness of measurements additional information is necessary. For example, the correctness of some temperature measurement reported by a sensor can be checked by using readings from a second temperature sensor (that measures the same temperature) or other relevant process information such as readings of other variables and energy balances. This information *redundancy* is a critical element of FDD. If duplicate sensors are used to measure the same variable and their readings are compared to detect presence of faults, there is *physical redundancy.* If a process model is used to estimate process variables and the difference between measured and estimated values forms the basis of diagnosis, there is *analytical* or *functional redundancy.* Since physical redundancy necessitates duplication of measurement systems, it is usually more expensive. Furthermore, it is usually focused on FDD of a single variable. Physical redundancy is considered when instantaneous FDD is needed for critical process equipment. Most modern FDD techniques focus on multivariable systems and use analytical redundancy that can leverage the correlation between various process variables.

One approach for FDD that appeals to plant personnel is to first identify process variables that have significant influence on an out-of-control signal issued by process monitoring tools, and then to reason based on their process knowledge about the possible source causes that affect these variables. The influence of process variables can be determined by *contribution plots* discussed in Section 8.1. The second stage of this indirect FDD approach can be automated by using *knowledge-based systems.* Many FDD techniques are based on direct pattern recognition and discrimination that diagnoses the fault directly from process data and models. Their foundations are built on signal processing, machine learning and statistics theory. In some techniques, trends in process variables are compared directly to a library of patterns that represent normal and faulty process behavior. The closest match is used to identify the status of the process. Statistical discrimination and classification analysis, and Fisher's discriminant function are some of the techniques drawn from statistical theory. They are discussed in Section 8.2. Other model-based FDD techniques are based on signal processing and systems science theory such as Kalman filters, residuals analysis, parity relations, hidden Markov models, and parameter estimation. They are introduced in Section 8.3. Artificial neural networks provide FDD techniques relying on fundamentals in statistics and computer science classification and machine learning, respectively. Knowledge-based systems (KBS) provide another group of FDD techniques that have roots in artificial intelligence. KBSs and their use in integrating and supervising various model-based and model-free FDD techniques are discussed in Section 8.4.

Faults can be classified as *abrupt* (sudden) faults and *incipient* (slowly developing) faults. Abrupt faults may lead to catastrophic consequences. They need to be detected quickly to prevent compromise of safety, productivity or quality. Incipient faults are usually associated with maintenance problems (heat exchange surfaces getting covered with deposits) or deviation trends in critical process activities from normal behavior (trends in cell growth in penicillin production). Incipient faults are typically small and consequently more difficult to detect. Multivariate techniques are more useful in their detection (See Chapter 6) since these techniques make use of information from all process measurements and can notice burgeoning trends in many variables and integrate that information to reach a decision. Quick detection may not be as critical for maintenance related problems, but deviations in critical process activities are usually time critical. The time behavior of faults can be grouped into a few generic types: *jump* (also called step or bias change), *intermittent*, and *drift* (Figure 8.1). Jumps in sensor readings are often caused by bias changes or breakdown. Wrong manual recordings of data entries or loose wire connections that lose con-

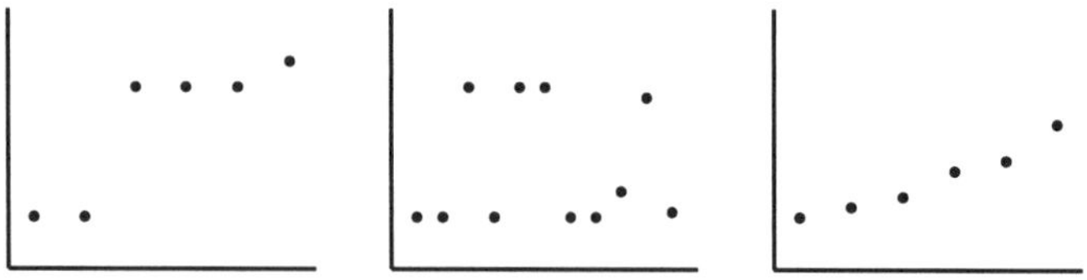

Figure 8.1. Typical fault functions [189].

tact would result in intermittent erroneous measurements. A measurement instrument that is warming up or an actuator that is wearing out would yield drift faults. *Disturbances* have the same types of time behavior. These faults and disturbances are usually slow and generate low frequency signals. In addition to faults, sensors, actuators, and process equipment are subjected to noise. *Noise* is usually assumed to be a random, zero mean, high frequency signal.

8.1 Contribution Plots

Multivariate quality control techniques use data from measurements of process variables, taking into account the correlation between process variables, to detect special causes affecting the process. Multivariate control charts such as SPE and T^2 charts indicate when the process goes out of control, but they do not provide information on the source causes of abnormal process operation. The engineers and plant operators need to determine the actual problem once an out-of-control situation is indicated. Miller et al. [388, 389] have introduced variable contributions and contribution plots concept to address this need. The diagnosis activity can be done by determining which process variables have contributed to inflate D-statistic (or T^2), squared prediction error Q-statistic (or SPE) and scores and use the knowledge of plant personnel to relate these process variables to various equipment failures and disturbances.

Contributions of process variables to the Q-statistic. Contribution to Q-statistic can either be calculated for the whole batch or for a time period during that batch. The Q-statistic for a new batch is calculated as

$$Q_i = (\mathbf{x}_{\text{new}} - \hat{\mathbf{x}}_{\text{new}})(\mathbf{x}_{\text{new}} - \hat{\mathbf{x}}_{\text{new}})^T = \sum_{jk=1}^{JK} (e_{\text{new},jk})^2 \tag{8.1}$$

where $\hat{\mathbf{x}}_{\text{new}}$ is the vector $(1 \times JK)$ of predicted values of the (centered and scaled) data for the new batch and $e_{\text{new},jk}$ is the residuals vector. An inflated Q-statistic suggests that the new observation does not follow the same covariance structure as that of the reference set that defines NO. This usually happens when there is a sensor failure or a shift in the process. If the Q-statistic for a batch represents an out-of-control situation, the process variables responsible for inflating the Q-statistic are diagnosed by computing the variable contributions to Q-statistic as

$$C_{JK}^{Q} = \sum_{c=(k-1)J}^{JK} [\mathbf{e}_{\text{new}}(c+1:c+J)]^2 \tag{8.2}$$

resulting in a $(1 \times J)$ vector of contributions from J variables over the entire batch. When deviations from NO are small and last for short periods of operation, this measure will not indicate the responsible variable(s) explicitly due to the masking effect from the contributions of other variables. To overcome this problem, the contribution C_{jk}^{Q} of process variable j at time period k to the Q-statistic is calculated as

$$C_{jk}^{Q} = (e_{\text{new},jk})^2 = (x_{\text{new},jk} - \hat{x}_{\text{new},jk})^2 \tag{8.3}$$

where $x_{\text{new},jk}$ is the jkth element of $x_{\text{new}}(1 \times JK)$, $\hat{x}_{\text{new},jk}$ is its prediction by the model, and $e_{\text{new},jk}$ is the vector of residuals.

Recently, control limits for variable contributions to Q-residuals were suggested by Westerhuis et al. [639] to compare the residuals of the new batch to the residuals of the NO data. If a particular variable has high residuals in the NO set, it can also be expected to have high residuals in the new batch. The control limits are calculated similar to those of the Q-statistic as discussed in Section 6.4.2 (Eqs. 6.104-6.111). The residuals matrix $\mathbf{E}$ of the reference set that is used to calculate contribution limits is obtained by "monitoring" each reference batch with one of the on-line SPM techniques discussed in Sections 6.5.1 and 6.5.2.

Contributions of process variables to the D-statistic. Two different approaches for calculating variable contributions to D-statistic have been proposed. The first approach introduced by Miller et al. [389] and by MacGregor et al. [355] calculates the contribution of each process variable to a separate score. The first step in this approach is to determine $\mathbf{t}$ score that is above its own confidence limits. Constructing confidence limits on individual scores is discussed and formulated in Section 6.4.2 (Eq. 6.95). The next step is to calculate the contribution of each element of the new batch run $x_{new,jk}$ on the rth score [389, 639]

$$C_{jk,r}^{t_r} = x_{\text{new},jk} p_{jk,r} \tag{8.4}$$

The sum of the contributions in Eq. 8.4 is equal to the $t_{\text{new},r}$ score of the new batch.

The second approach was proposed by Nomikos [432]. This approach calculates contributions of each process variable to the D-statistic instead contributions of separate scores.

$$C_{jk}^{D} = \sum_{r=1}^{R} \mathbf{S}_{rr}^{-1} t_{\text{new},r} x_{\text{new},jk} p_{r,jk} \tag{8.5}$$

In Eq. 8.5, the contribution of each element in $x_{\text{new},jk}$ to the D-statistic is summed over all r components. This formulation is valid for the case of orthogonal scores because $\mathbf{S}^{-1}$, which is the inverse of covariance matrix of reference set scores $\mathbf{T}$, then becomes diagonal and its diagonal elements are used. The loadings $\mathbf{P}$ of the MPCA model are also assumed to be orthogonal so that $\mathbf{P}^T\mathbf{P} = \mathbf{I}$. Westerhuis et al. [639] have extended Nomikos' [432] formulation to cases where scores and loadings are non-orthogonal. According to this generalization, D-statistic is calculated as follows:

$$\begin{aligned} D_{\text{new}} &= \mathbf{t}_{\text{new}}^T \mathbf{S}^{-1} \mathbf{t}_{\text{new}} = \mathbf{t}_{\text{new}}^T \mathbf{S}^{-1} \left[\mathbf{x}_{\text{new}}^T \mathbf{P} \left(\mathbf{P}^T \mathbf{P} \right)^{-1} \right]^T \\ &= \mathbf{t}_{\text{new}}^T \mathbf{S}^{-1} \sum_{jk=1}^{JK} \left[x_{\text{new},jk} \mathbf{p}_{jk}^T \left(\mathbf{P}^T \mathbf{P} \right)^{-1} \right]^T \\ &= \sum_{jk=1}^{JK} \mathbf{t}_{\text{new}}^T \mathbf{S}^{-1} \left[x_{\text{new},jk} \mathbf{p}_{jk}^T \left(\mathbf{P}^T \mathbf{P} \right)^{-1} \right]^T \\ &= \sum_{jk=1}^{JK} C_{jk}^{D}. \end{aligned} \tag{8.6}$$

Hence, the contribution of new observation vector $x_{\text{new},jk}$ of the new batch to the D-statistic is calculated as

$$C_{jk}^{D} = \mathbf{t}_{\text{new}}^T \mathbf{S}^{-1} \left[x_{\text{new},jk} \mathbf{p}_{jk}^T \left(\mathbf{P}^T \mathbf{P} \right)^{-1} \right]^T \tag{8.7}$$

where $\mathbf{t}_{\text{new}}^T = \mathbf{x}_{\text{new}}^T \mathbf{P} \left(\mathbf{P}^T \mathbf{P} \right)^{-1}$.

The control limits for variable contributions to D-statistic are also given [639]. These are computed by means of a jackknife procedure in which each of the NO batches is left out once, and variable contributions are calculated for each batch that is left out. The next step is to calculate the mean and variance of these contributions from I batches for each jth variable at kth time period. Westerhuis et al. [639] proposed to use an upper control limit (UCL) for contributions that is calculated as the mean of the variable contributions at each time interval plus three times the corresponding standard deviation. It is noted that UCL obtained by this calculation is not considered to have a statistical significance, but it is useful for detecting contributions that are higher than those of NO batches in the reference set. A lower control limit (LCL) can also be developed in the same manner. If it is preferred to sum contributions over all time instances or over all process variables, then the control limits are obtained by summing the means of the corresponding jackknifed contributions from the reference set. The standard deviation of these summed means can be calculated as [639]

$$\sigma_k = \sqrt{\sum_{j=1}^{J} \sigma_{jk}^2}, \qquad \sigma_j = \sqrt{\sum_{k=1}^{K} \sigma_{jk}^2} \tag{8.8}$$

where σ_k and σ_j are the standard deviations of the summed mean contributions over all process variables and all time instances, respectively. If the sum of the contributions over all variables at each time instance is used, one can zoom in the region(s) where summed contributions exceed the control limits that are calculated by using σ_k in Eq. 8.8.

It is always a good practice to check individual process variable plots for those variables diagnosed as responsible for flagging an out-of-control situation. When the number of variables is large, analyzing contribution plots and corresponding variable plots to reason about the faulty condition may become tedious and challenging. All these analyzes can be automated and linked with real-time diagnosis [436, 607] by means of knowledge-based systems.

Example. Consider a reference data set of 42 NO batches from fed-batch penicillin fermentation process (see Section 6.4.1). An on-line SPM framework is developed with that data set ($\mathbf{X}(42 \times 14 \times 764)$). The *model development stage* and the MPCA model developed are the same as in Section 6.4.3, except that the construction of control limits is performed by passing each batch data in the reference set through the estimation-based on-line SPM procedure. Estimation method 2 (the future values of disturbances being assumed to remain constant at their current values over the remaining batch period) discussed in Section 6.5.1 is chosen for on-line SPM. A new batch scenario with a small downward drift on glucose feed rate (variable 3) between 180th and 300th measurements (Figure 8.3(d)) is produced for illustration of contribution plots. Both SPE (Figure 8.2(a)) and T^2 (Figure 8.2(c)) charts have detected the out-of-control situation between 250th and 310th measurements and 270th and 290th measurements, respectively. Variable contributions are summed for the intervals of out-of-control for SPE and T^2 in Figures 8.2(b) and 8.2(d). Since these summations represent

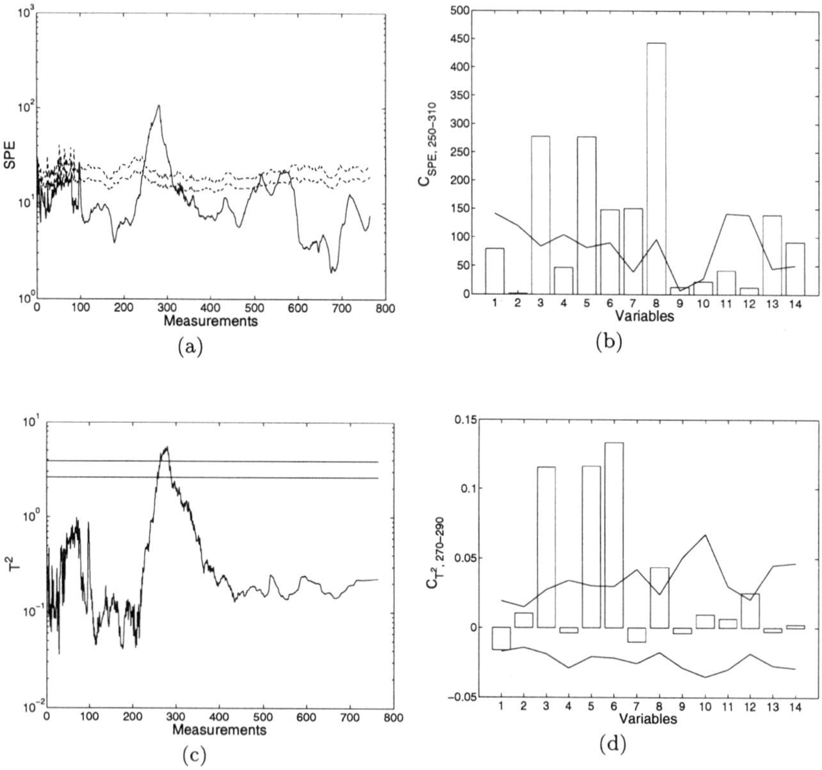

Figure 8.2. On-line monitoring results with contribution limits for a faulty batch.

faulty situation after the fault has developed long enough to affect related variables, most of the variable contributions in Figures 8.2(b) and 8.2(d) violate the control limits. The real fault is the drift in glucose feed rate (variable 3), which is highly correlated with glucose concentration (variable 5), dissolved oxygen concentration (variable 6), biomass concentration (variable 7), penicillin concentration (variable 8), culture volume (variable 9), heat generated (variable 13), and cooling water flow rate (variable 14). Note that penicillin concentration (variable 8) in Figure 8.2(b) and dissolved oxygen concentration (variable 6) in Figure 8.2(d) have the highest contributions to SPE and T^2 during the out-of-control period. Variable contributions to T^2 over all of the variables at each time instant are also presented in Figure 8.3(a) as another indicator for detecting out-of-control

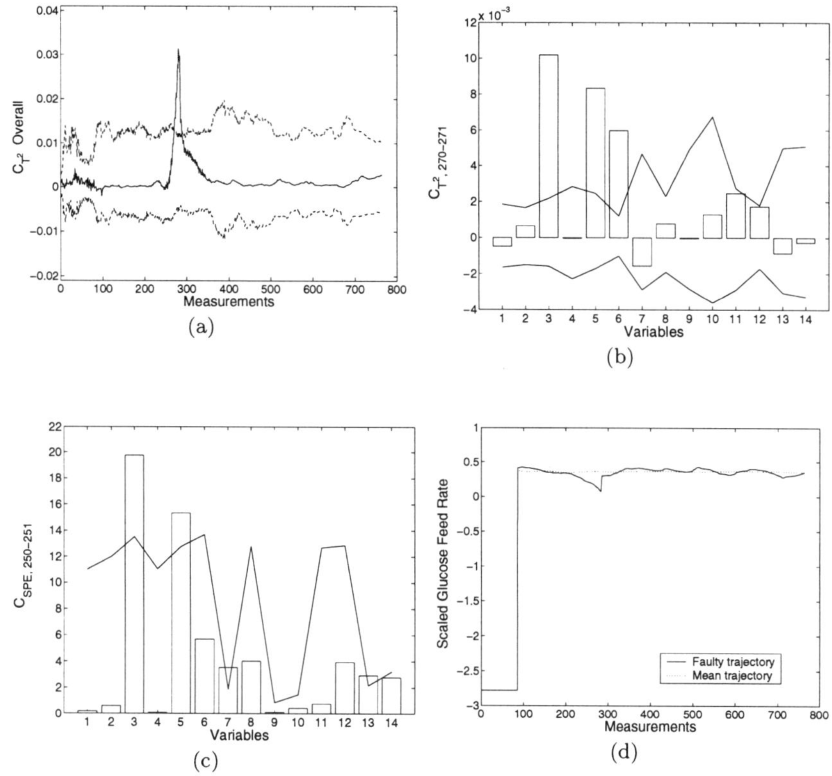

Figure 8.3. On-line monitoring results with contribution limits for a faulty batch.

situation. To reveal the root cause of the fault, variable contributions to SPE and T^2 are plotted along with the control limits right after the out-of-control situation is detected. Variable contributions to T^2 are summed for 270th and 271st measurements in Figure 8.3(b) and are summed for 250th and 251st measurements in the case of SPE in Figure 8.3(c). Both charts indicate that the glucose feed rate (variable 3) has the highest contribution, and therefore is the root cause to the deviation. Second highest contribution is that of the glucose concentration (variable 5) as expected. As a good practice, univariate chart of the variable that has the highest contribution is plotted. Figure 8.3(d) represents the glucose feed rate profile of the faulty batch superimposed on the reference glucose feed rate profile of NO.

All of the aforementioned tasks can be integrated into a real-time knowledge-based system for automated supervision and ease of interpretation. The details of implementation will be discussed and presented in Section 8.4.1.

8.2 Statistical Techniques for Fault Diagnosis

8.2.1 Statistical Discrimination and Classification

Statistical discrimination and classification are multivariate techniques that *separate* distinct sets of objects (or events), and *allocate* new objects (or events) into previously defined groups of objects, respectively [262]. Discrimination focuses on discrimination criteria (called discriminants) for converting salient features of objects from several known populations to quantitative information separating these populations as much as possible. Classification sorts new objects or events into previously labelled classes by using rules derived to optimally assign new objects to the labelled classes. A good classification procedure should yield few misclassifications. The probability of occurrence of an event may be greater if it belongs to a population that has a greater likelihood of occurrence. A good classification rule should take these "prior probabilities of occurrence" into consideration. A good classification procedure should also account for the costs associated with misclassification, classification of an event to a different class. Consider two hypothetical sensor faults, one necessitating process shutdown because without measuring and controlling that variable the process may produce hazardous products, and the other causing higher use of utilities. Their misclassification would yield different levels of hazards and damages, hence their costs of misclassification are different.

Consider a data set with g distinct events such as normal process operation and operation under $g-1$ different faults. The operation type (class) is determined on the basis of p measured variables $\mathbf{x} = [x_1 \; x_2 \; \cdots \; x_p]^T$ that are random variables. Denote the classes by π_i, $i = 1, \cdots, g$, their prior probability by p_i $i = 1, \cdots, g$ and their probability density functions by $f_i(\mathbf{x})$. While it is not necessary to assume that $f_i(\mathbf{x})$ be the multivariate normal density, in most derivations and in this discussion it will be assumed that it is, with population and sample means $\boldsymbol{\mu}_i$ and $\bar{\mathbf{x}}_i$, respectively and population and sample variances $\boldsymbol{\Sigma}_i$ and $\mathbf{S}_i$, respectively. Denote the *cost of misclassification* as $c(k|i)$, the cost of allocating an object to π_k (for $k = 1, \cdots, g$) when in fact it belongs to π_i (for $i = 1, \cdots, g$). If R_k is the set of $\mathbf{x}$'s classified as π_k, the probability of classifying an event as π_k when in reality it belongs to π_i is

$$P(k|i) = P(\text{classifying event as } \pi_k|\pi_i) = \int_{R_k} f_i(\mathbf{x})d\mathbf{x} \quad i,k = 1,\cdots,g \tag{8.9}$$

with $P(i|i) = 1 - \sum_{k=1,k\neq i}^{g} P(k|i)$. The conditional *expected cost of misclassification* (ECM) of an event in π_1 to any other class is

$$ECM(\pi_1) = \sum_{k=2}^{g} P(k|1)c(k|1) \tag{8.10}$$

This conditional expected cost of misclassifying an event belonging to π_1 occurs with prior probability p_1 (the probability of π_1). The conditional *overall* expected cost of misclassification is computed by multiplying each ECM with its prior probability and summing over all classes

$$\begin{aligned} ECM &= p_1 ECM(\pi_1) + \cdots + p_g ECM(\pi_g) \\ &= p_1 \sum_{k=2}^{g} P(k|1)c(k|1) + p_2 \sum_{k=1,\ k\neq 2}^{g} P(k|2)c(k|2) \\ &\quad + \cdots + p_g \sum_{k=1}^{g-1} P(k|g)c(k|g) \\ &= \sum_{i=1}^{g} p_i \left(\sum_{k=1,\ k\neq i}^{g} P(k|i)c(k|i) \right) \end{aligned} \tag{8.11}$$

Determination of the optimal classification procedure becomes selection of mutually exclusive and exhaustive classification regions $R_1, R_2, \cdots, R_g$ such that the ECM in Eq. (8.11) is minimized [262]. The classification regions that minimize Eq. (8.11) are defined by allocating $\mathbf{y}$ to that population π_k, $k = 1, \cdots, g$ for which

$$\sum_{i=1,\ i\neq k}^{g} p_i f_i(\mathbf{x})c(k|i) \tag{8.12}$$

is smallest [18, 262]. If all misclassification costs are equal, the event described by data $\mathbf{x}$ will be assigned to that population π_k for which $\sum_{i=1,\ i\neq k}^{g} p_i f_i(\mathbf{x})$ is smallest. This means that the omitted term $p_k f_k(\mathbf{x})$ is largest. Consequently, the minimum ECM rule for equal misclassification costs becomes [262]:

Allocate $\mathbf{x}$ to π_k if $p_k f_k(\mathbf{x}) > p_i f_i(\mathbf{x})$ for all $i \neq k$.

given prior probabilities, density functions, and misclassification costs (when they are not equal). This classification rule is identical to the one that maximizes the "posterior" probability $P(\pi_k|\mathbf{x})$ ($P(\mathbf{x})$ comes from π_k given that $\mathbf{x}$ was observed) where

$$P(\pi_k|\mathbf{x}) = \frac{p_k f_k(\mathbf{x})}{\sum_{i=1}^{g} p_i f_i(\mathbf{x})} = \frac{(\text{prior}) \times (\text{likelihood})}{\sum[(\text{prior}) \times (\text{likelihood})]} \quad k = 1, \cdots, g \tag{8.13}$$

If the populations follow Normal distributions with mean vectors $\boldsymbol{\mu}_i$, covariance matrices $\boldsymbol{\Sigma}_i$, and *generalized variance* $|\boldsymbol{\Sigma}_i|$ (determinant of the covariance), $f_i(\mathbf{x})$ is defined as

$$f_i(\mathbf{x}) = \frac{1}{(2\pi)^{p/2}|\Sigma_i|^{1/2}} \exp\left[-\frac{1}{2}(\mathbf{x} - \boldsymbol{\mu}_k)^T \boldsymbol{\Sigma}_k^{-1}(\mathbf{x} - \boldsymbol{\mu}_k)\right] \quad i = 1, \cdots, g \tag{8.14}$$

and all misclassification costs are equal, then $\mathbf{x}$ is allocated to π_k if

$$\begin{aligned} \ln p_k f_k(\mathbf{x}) &= \ln p_k - \frac{p}{2}\ln(2\pi) - \frac{1}{2}\ln|\boldsymbol{\Sigma}_k| - \frac{1}{2}(\mathbf{x} - \boldsymbol{\mu}_k)^T \boldsymbol{\Sigma}_k^{-1}(\mathbf{x} - \boldsymbol{\mu}_k) \\ &= \max_i \ \ln \ p_i F_i(\mathbf{x}) \ . \end{aligned} \tag{8.15}$$

The constant $p/2 \ \ln(2\pi)$ is the same for all populations and can be ignored in discriminant analysis. The *quadratic discrimination score* for the ith population $d_i^Q(\mathbf{x})$ is defined as [262]

$$d_i^Q(\mathbf{x}) = \ln p_i - \frac{1}{2}\ln|\boldsymbol{\Sigma}_i| - \frac{1}{2}(\mathbf{x} - \boldsymbol{\mu}_i)^T \boldsymbol{\Sigma}_i^{-1}(\mathbf{x} - \boldsymbol{\mu}_i) \quad i = 1, \cdots, g \ . \tag{8.16}$$

The generalized variance $|\boldsymbol{\Sigma}_i|$, the prior probability p_i and the Mahalanobis distance contribute to the quadratic score $d_i^Q(\mathbf{x})$. Using the discriminant scores, the minimum total probability of misclassification rule for Normal populations and unequal covariance matrices becomes [262]:

Allocate $\mathbf{x}$ to π_k if $d_k^Q(\mathbf{x})$ is the *largest* of all $d_i^Q(\mathbf{x})$, $i = 1, \cdots, g$.

In practice, population mean and covariances ($\boldsymbol{\mu}_i$ and $\boldsymbol{\Sigma}_i$) are unknown. Computations are based on historical data sets of classified observations, and sample mean ($\bar{\mathbf{x}}_i$) and covariance matrices ($\mathbf{S}_i$) are used in Eq. (8.16).

A simplification is possible if the population covariance matrices $\boldsymbol{\Sigma}_i$ are equal for all i. Then, $\boldsymbol{\Sigma}_i = \boldsymbol{\Sigma}$ and Eq. (8.16) reduces to

$$d_i^Q(\mathbf{x}) = \ln p_i - \frac{1}{2}\ln|\boldsymbol{\Sigma}| - \frac{1}{2}(\mathbf{x}^T \boldsymbol{\Sigma}^{-1}\mathbf{x}) + \boldsymbol{\mu}_i^T \boldsymbol{\Sigma}^{-1}\mathbf{x} - \frac{1}{2}\boldsymbol{\mu}_i^T \boldsymbol{\Sigma}^{-1}\boldsymbol{\mu}_i \tag{8.17}$$

Since the second and third terms are independent of i, they are the same for all $d_i^Q(\mathbf{x})$ and can be ignored in classification. Since the remaining terms consist of a constant for each i ($\ln p_i - 1/2\boldsymbol{\mu}_i^T \boldsymbol{\Sigma} \boldsymbol{\mu}_i$) and a linear combination of the components of $\mathbf{x}$, a *linear discriminant score* is defined as

$$d_i(\mathbf{x}) = \boldsymbol{\mu}_i^T \boldsymbol{\Sigma}^{-1} \mathbf{x} - \frac{1}{2} \boldsymbol{\mu}_i^T \boldsymbol{\Sigma}^{-1} \boldsymbol{\mu}_i + \ln p_i \tag{8.18}$$

An estimate of $d_i(\mathbf{x})$ can be computed based on the *pooled* estimate of $\boldsymbol{\Sigma}$ [262]:

$$\hat{d}_i(\mathbf{x}) = \bar{\mathbf{x}}_i^T \mathbf{S}_{pl}^{-1} \bar{\mathbf{x}} - \frac{1}{2} \bar{\mathbf{x}}_i^T \mathbf{S}_{pl}^{-1} \bar{\mathbf{x}}_i + \ln p_i \quad i = 1, \cdots, g \tag{8.19}$$

where

$$\mathbf{S}_{pl} = \frac{1}{n_1 + n_1 + \cdots + n_g - g} [(n_1 - 1)\mathbf{S}_1 + \cdots + (n_g - 1)\mathbf{S}_g] \tag{8.20}$$

and n_g denotes the data length (number of observations) in class g. The minimum total probability of misclassification rule for Normal populations with equal covariance matrices becomes [262]:

Allocate $\mathbf{x}$ to π_k if $\hat{d}_k(\mathbf{x})$ is the *largest* of all $\hat{d}_i(\mathbf{x})$, $i = 1, \cdots, g$.

FDD by Integrating PCA and Discriminant Analysis

An integrated statistical method was developed [488] for automated detection of abnormal process operation and discrimination between several source causes by utilizing PCA and discriminant analysis techniques for multivariable continuous processes. The method was developed for monitoring continuous processes deviating from their steady state operation. The lack of significant autocorrelation, stationarity, and ergodicity should be established before utilizing this method. The method does not rely on visual inspection of plots; consequently, it is suitable for processes described by large sets of variables. It can be extended to batch processes by making appropriate modifications, but such extensions have not been reported. The method was illustrated by monitoring the Tennessee Eastman industrial challenge problem [137].

Detection and diagnosis of *multiple simultaneous faults* is an important concern. In a real process, combinations of faults may occur. An intervention policy to improve process operation may need to take into account each of the contributing faults. Diagnosis should be able to identify major contributors and correctly indicate which, if any, secondary faults are occurring [487]. Most FDD techniques rely on the assumption of a single fault. Raich and Cinar proposed several statistical measures to assess the overlap between models describing process behavior caused by single faults. The similarity between models indicates the potential for confusion and masking of the effects (symptoms) of multiple faults. Quantitative measures to compare multivariable models permit decisions about their usefulness and discrimination capability. They also provide *a priori* information about faults that are likely to be masked by other faults.

PCA is used to develop a model describing variation under normal operation (NO). This PC model is used to detect outliers from NO, as excessive variation from normal target or unusual patterns of variation. Operation under various known upsets can also be modeled if sufficient historical data are available. These fault models are then used to isolate source causes of faulty operation based on similarity to previous upset behavior. Using PCs for several sets of data under different operating conditions (NO and with various upsets), statistics can be computed to describe distances of the current operating point to regions representing other areas of operation. Both scores distances and model residuals are used to measure such distance-based statistics.

Fault Diagnosis

PC models for specific faults can be developed using historical data sets collected when that fault was active. When current measurements exhibit out-of-control behavior, a likely cause for this behavior can be assigned by pattern matching by using scores, residuals or their combination.

Score Discriminant. Assuming that PC models retain sufficient variation to discriminate between possible causes in scores that have independent normal distributions, the maximum likelihood that data $\mathbf{x}$ are from fault model i is indicated by the minimum distance. This minimum can be determined by the maximum of d_i expressed for example by quadratic discrimination (Eq. 8.16)

$$d_i(\mathbf{t}) = \ln p_i - \frac{1}{2}\ln|\mathbf{\Sigma}_i| - \frac{1}{2}\mathbf{t}^T\mathbf{\Sigma}_i{}^{-1}\mathbf{t} \tag{8.21}$$

where $\mathbf{t} = \mathbf{x}\mathbf{P}_i$ is the location of original observation $\mathbf{x}$ in PC space for fault model i, $\mathbf{\Sigma_i}$ is the covariance along PCs for fault model i, and p_i is the adjustment for overall occurrence likelihood of fault i [262]. Figure 8.4 illustrates the fault isolation process. Score discriminants are calculated using PC models for the various known faults (Figure 8.4c); this semilog plot shows the negative of the discriminant. The most likely fault is chosen over time by selecting the fault corresponding to the maximum discriminant (curve with the lowest magnitude). Figure 8.4d reports the fault selected at each sampling time. Fault 3, which is the correct fault, has been reported consistently after the first 10 sampling times.

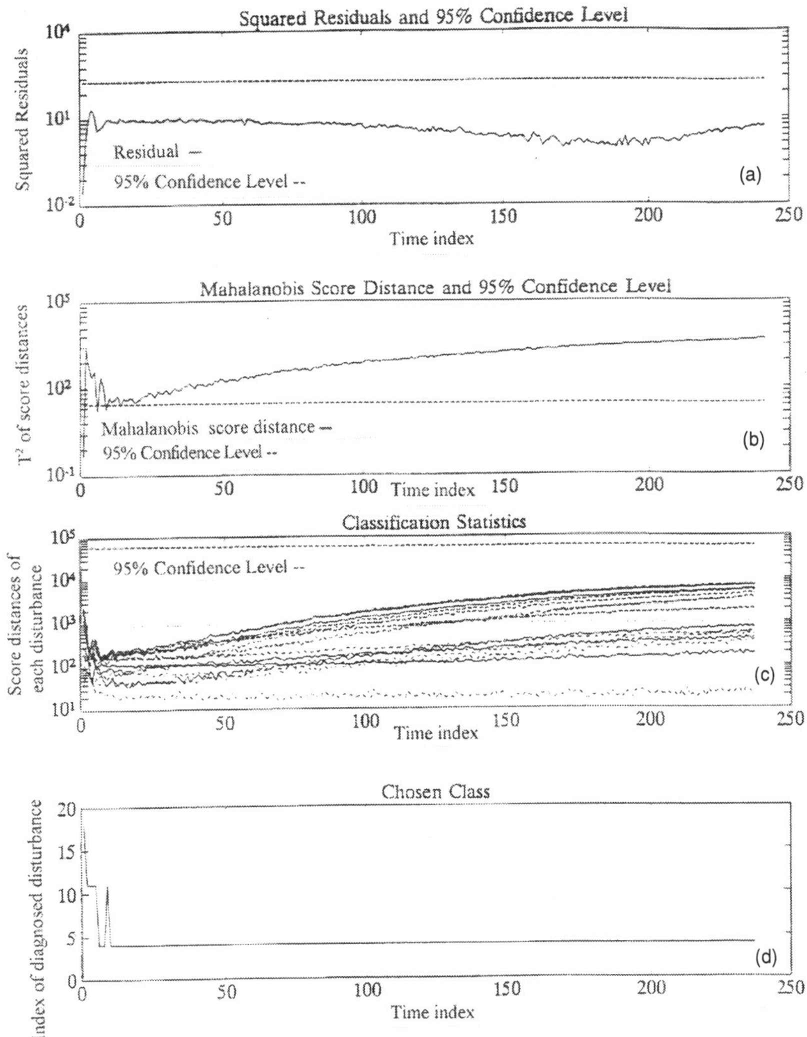

Figure 8.4. Detection and diagnosis of process upsets (a) Detection of outliers based on residuals, (b) detection based on T^2 test of scores, (c) diagnosis statistics considering each possible disturbance, (d) index of chosen disturbance for each observation [488].

Residual Discriminant. Assuming that observations will not be well described by PC models for other faults but will be within the residual threshold of their own class, it is most likely that $\mathbf{x}$ is from the fault model i with minimum

$$r_i/r_{i,\alpha} \qquad \text{where} \qquad r_i = \mathbf{t}_i^T(\mathbf{I} - \mathbf{P}\mathbf{P}^\mathbf{T})\mathbf{t}_i \tag{8.22}$$

r_i is the residual computed using the PCA model for fault i and $r_{i,\alpha}$ is the residual threshold at level 100α based on the PCA model for fault i.

Combined Discriminant. Combining the information available in scores and residuals usually improves the diagnosis accuracy [408]. Comparing the combined information to the confidence limits of each fault model, $\mathbf{x}$ is most likely to be from the fault model i with minimum

$$c_i\left(\frac{r_i}{r_{i,\alpha}}\right) + (1 - c_i)\left(\frac{s_i}{s_{i,\alpha}}\right) \tag{8.23}$$

where s_i and r_i are the score distance and residual based on the PC model, respectively, for fault i, $s_{i,\alpha}$ and $r_{i,\alpha}$ are the score distance and residual thresholds using the PC model, respectively, for fault i, and c_i is a weight between 0 and 1. To weigh scores and residuals according to the amount of variation in data explained by each, c_i is set equal to the fraction of total variance explained by scores. The combined discriminant value thus calculated gives an indication of the degree of certainty for the diagnosis; statistics less than 1 indicate a good fit to the chosen model. If no model results in a statistic less than 1, none of the models provide an adequate match to the observation.

The FDD system design includes development of PC models for NO and faulty operation, and computation of threshold limits using historical data sets collected during normal plant operation and operation under specific faults. The implementation of the FDD system at each sampling time starts with monitoring. The model describing NO is used with new data to decide if the current operation is in-control. If there is no significant evidence that the process is out-of-control, further analysis is not necessary and the procedure is concluded for that measurement time. If score or residual tests exceed their statistical limits, there is significant evidence that the process is out-of-control. Then, the PC models for all faults are used to carry out the score and residuals tests, and discriminant analysis is performed by using PC models for various faults to diagnose the source cause of abnormal behavior.

Discrimination and Diagnosis of Multiple Disturbances

In fault diagnosis, where process behavior due to different faults is described by different models, it is useful to have a quantitative measure of similarity or overlap between models, and to predict the likelihood of successful diagnosis. In comparing multivariate models, much work has been reported for testing significant differences between means when covariance is constant. Testing for differences in covariance is more difficult yet crucial; diagnosis can be successfully done, whether or not means are different, as long as there is a difference in covariance [171]. Testing for eigenvalue models of covariance adds new complications, since the statistical characteristics are not well known, even for common distributions. Simplifying assumptions for special cases can be made, with significant loss of generality [378].

Angles Between Different Coordinate Systems and Similarity Index. Raich and Cinar proposed a method based on the angles between principal coordinate directions of current data and regions corresponding to operation with different faults [489]. The method uses angles between different coordinate systems and a similarity index defined by using the angle information [308].

The similarity index has a range from 0 to 1, increasing as models become more similar. It provides a quantitative measure of difference in covariance directions between models and a description of overall geometric similarity in spread. The similarity index can be used to evaluate discrimination models by selecting a threshold value to indicate where mistakes in classification of data from the two models involved may occur. It can also be used to compare models built from different operating runs of the same process for monitoring systematic changes in process variation during normal operation. Another possible application is in batch processes, where use of the similarity index could provide a way to check if PC model orientation around a moving mean varies in time.

Overlap of Means. The other important statistical test in comparing multivariate models is for differences in means. This corresponds to comparison of origin of coordinates rather than the coordinate directions. Many statistical tests have been developed for testing means, but most of them can become numerically unstable when significant correlation exists between variables. In order to work around the instability, overlap between eigenvalue-based models can be evaluated. Target factor analysis can assign a likelihood on whether a candidate vector is a contributor to the model of a multivariate data set. A statistic is defined to test if a specific vector is significantly inside the confidence region containing the modeled data [361]. For overlap of means, the test can determine whether the mean from one model, μ_1, significantly overlaps the region of data from another (second) model [488]. Mean overlap analysis can be used to test if an existing PC model fits a new set of observations or if two PC models are analogous.

Comparison of models for individual faults and their combinations can provide information for extending the diagnosis methods to multiple simultaneous faults and masking of contributing faults. If there is no overlap between regions spanned by two different faults, two alternative schemes might handle multiple faults modeled by PCA. In one method, the combination fault is idealized as being located between the regions of the underlying component faults; allocations of membership to the different independent faults contributing to the combination may provide diagnosis of underlying faults. The second method is based on a more general extension of the discrimination scheme by introducing new models for each multiple-fault combination of interest. The measures of similarity in model center and direction of spread can be useful to determine the independence of the models used in diagnosis.

Masking of Multiple Faults. When the region spanned by the model for one (*outer*) fault contains the model for another (*inner*) fault, their combination will not be perfectly diagnosed. Idealizing the two fault regions as concentric spheres, the *inner* model region is enveloped by the *outer* model. As a result, only the *outer* fault will be diagnosed and the *inner* fault will be masked. Overlap of regions is likely to exist for most processes under closed-loop control, the multiple fault scenario is further complicated for such processes.

Random variation faults such as excessive sensor noise move a process less drastically off-target than step or ramp faults. Consequently, similarity measures should indicate that the random variation faults have more overlap with other models, particularly with each other. Ramp or step faults tend to be the *outer* models, this is consistent with moving the process off its control target or NO. As *outer* model, ramp or step fault masks secondary random variation faults.

Similarity measures serve as indicators of the success in diagnosing combinations of faults. They can identify combinations of faults that may be masked or falsely diagnosed, and provide information about the success rates of different diagnosis schemes incorporating single and combinations of faults. Using these guidelines, multiple faults occurring in a process can be analyzed *a priori* with respect to their components, and accommodated within the diagnosis framework described earlier.

8.2.2 FDD with Fisher's Discriminant Analysis

A problem that emerges when statistical techniques are used in multivariate classification and clustering is what Bellman calls the curse of dimensionality [139]. Principal components analysis (PCA) is discussed as a linear dimensionality reduction technique in Section 4.1. PCA is optimal in terms of capturing the variability among the data. Another technique called Fisher's discriminant analysis (FDA) is optimal in terms of maximizing the separation among the set of classes [153]. Suppose that there is a set of $n(= n_1 + n_2 + \cdots + n_g)$ p-dimensional samples $\mathbf{x}_1, \cdots, \mathbf{x}_n$ belonging to classes π_i, $i = 1, \cdots, g$. Fisher suggested to transform the multivariate observations $\mathbf{x}$ to another coordinate system that enhances the separation of the samples belonging to each class π_i. In this section, the FDA concept is illustrated first for separating data belonging to two classes π_1 and π_2. Then, FDA is generalized to process data with many classes. Finally, classification and diagnosis with FDA is discussed.

FDA for data belonging to two classes

Fisher suggested transformation of multivariate observations $\mathbf{x}$ to univariate observations z such that the z's derived from populations π_1 and π_2 are separated as much as possible. If the multivariate observations have more than two variables, additional z variables $(z_2, z_3, \cdots)$ may be necessary for enhancing the separation. The total scatter of data points $(\mathbf{S}_T)$ consists of two types of scatter, *within-class scatter* $\mathbf{S}_W$ and *between-class scatter* $\mathbf{S}_B$. The objective of the transformation proposed by Fisher is to maximize $\mathbf{S}_B$ while minimizing $\mathbf{S}_W$. Fisher's approach does not require that the populations have Normal distributions, but it implicitly assumes that the population covariance matrices are equal, because a pooled estimate of the common covariance matrix $(\mathbf{S}_{pl})$ is used (Eq. 8.20).

The transformation is based on a weighted sum of observations $\mathbf{x}$. In the case of two classes, the linear combination of the samples $(\mathbf{x})$ takes values $z_{11}, \cdots, z_{1p1}$ for the observations from the first population π_1 and the values $z_{21}, \cdots, z_{2p2}$ for the observations from the second population π_2. Denote the weight vector that transforms $\mathbf{x}$ to z by $\mathbf{w}$. FDA is illustrated for the case of two normal populations with a common covariance matrix in Figure 8.5. First consider separation using either x_1 or x_2 axis. The diagrams by the abscissa and ordinate indicate that several observations belonging to one class (π_1) are mixed with observations belonging to the other class (π_2). The linear discriminant function $z = \mathbf{w}^T\mathbf{x}$ defines the line in the upper portion of Figure 8.5 that observations are projected on to maximize the ratio of between-class scatter and within-class scatter [262, 139]. One may visualize changing the slope of the line to see how the number of observations of a specific class that move in the region of the other class changes.

The separation of the two sets of z's can be assessed in terms of the difference between $\bar{z}_1$ and $\bar{z}_2$ expressed in standard deviation units:

$$\text{separation} = \frac{|\bar{z}_1 - \bar{z}_2|}{s_z} \tag{8.24}$$

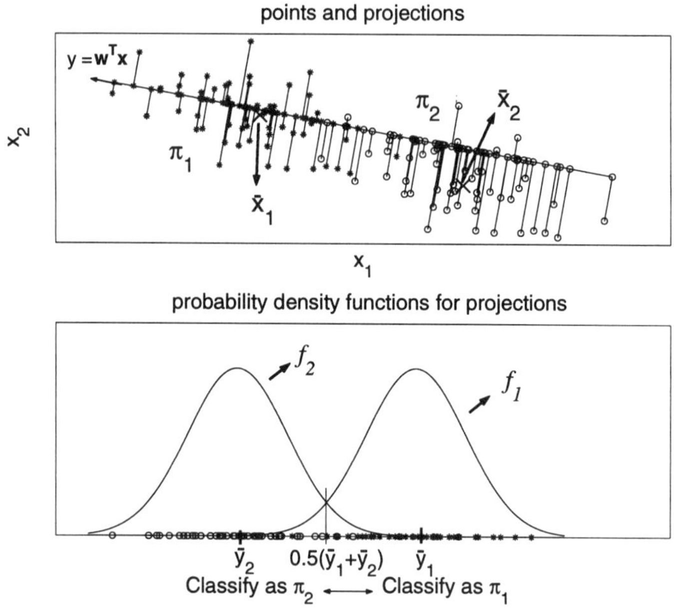

Figure 8.5. Fisher's discriminant technique for two populations ($g = 2$), $\pi_1(*)$ and $\pi_2(\circ)$, with equal covariances.

where s_z^2 is the pooled estimate of the variance

$$s_z^2 = \frac{1}{n_1 + n_2 - 2}\left[\sum_{j=1}^{n_1}(z_{1j} - \bar{z}_1)^2 + \sum_{j=1}^{n_2}(z_{2j} - \bar{z}_2)^2\right] \tag{8.25}$$

The linear combination that maximizes the separation is [262]

$$\hat{z} = \mathbf{w}^T\mathbf{x} = (\bar{\mathbf{x}}_1 - \bar{\mathbf{x}}_2)^T\,\mathbf{S}_{pl}^{-1}\mathbf{x} \tag{8.26}$$

which maximizes the ratio

$$\frac{(\bar{z}_1 - \bar{z}_2)^2}{s_z^2} = \frac{(\mathbf{w}^T\bar{\mathbf{x}}_1 - \mathbf{w}^T\bar{\mathbf{x}}_2)^2}{\mathbf{w}^T\mathbf{S}_{pl}\mathbf{w}} = \frac{(\mathbf{w}^T\mathbf{d})^2}{\mathbf{w}^T\mathbf{S}_{pl}\mathbf{w}} \tag{8.27}$$

over all possible coefficient vectors $\mathbf{w}$ where $\mathbf{d} = (\bar{\mathbf{x}}_1 - \bar{\mathbf{x}}_2)$. The maximum of the ratio in Eq. 8.27 is $T^2 = (\bar{\mathbf{x}}_1 - \bar{\mathbf{x}}_2)^T\mathbf{S}_{pl}^{-1}(\bar{\mathbf{x}}_1 - \bar{\mathbf{x}}_2)$ [262]. For two populations with equal covariances, FDA corresponds to the particular case of the minimum ECM rule discussed in Section 8.2.1. The first terms in Eqs. 8.18 and 8.19 are the linear function obtained by FDA that maximizes the univariate between-class scatter relative to the within-class scatter (Eq. 8.26) [262].

The allocation rule of a new observation $\mathbf{x}_0$ to classes π_1 or π_2 based on FDA is [262]

Allocate $\mathbf{x}_0$ to π_1 if

$$(\bar{\mathbf{x}}_1 - \bar{\mathbf{x}}_2)^T \mathbf{S}_{pl}^{-1} \mathbf{x}_0 \geq \frac{1}{2} (\bar{\mathbf{x}}_1 - \bar{\mathbf{x}}_2)^T \mathbf{S}_{pl}^{-1} (\bar{\mathbf{x}}_1 + \bar{\mathbf{x}}_2) \tag{8.28}$$

Allocate $\mathbf{x}_0$ to π_2 otherwise.

Separation of Many Classes $(g > 2)$

The generalization of the *within-class scatter matrix* $\mathbf{S}_W$ for g classes is

$$\mathbf{S}_W = \sum_{i=1}^{g} (n_i - 1) \mathbf{S}_i \tag{8.29}$$

where

$$\mathbf{S}_i = \frac{1}{n_i - 1} \sum_{j=1}^{n_i} (\mathbf{x}_{ij} - \bar{\mathbf{x}}_i)(\mathbf{x}_{ij} - \bar{\mathbf{x}}_i)^T \tag{8.30}$$

represents the covariance matrix for class i and the mean vector for class i is [139]

$$\bar{\mathbf{x}}_i = \frac{1}{n_i} \sum_{j=1}^{n_i} \mathbf{x}_{ij} \tag{8.31}$$

where n_i denotes the number of observations in class i. $\mathbf{S}_W / (n_1 + n_2 + \cdots + n_g - g) = \mathbf{S}_{pl}$ is an estimate of $\mathbf{\Sigma}$. The $\mathbf{w}$ that maximizes $\mathbf{w}^T \mathbf{S}_B \mathbf{w} / \mathbf{w}^T \mathbf{S}_{pl} \mathbf{w}$ also maximizes $\mathbf{w}^T \mathbf{S}_B \mathbf{w} / \mathbf{w}^T \mathbf{S}_W \mathbf{w}$.

Define the *between-class scatter matrix* $\mathbf{S}_B$ and the *total scatter matrix* $\mathbf{S}_T$ as [139, 246]

$$\mathbf{S}_B = \sum_{i=1}^{g} n_i (\bar{\mathbf{x}}_i - \bar{\mathbf{x}})(\bar{\mathbf{x}}_i - \bar{\mathbf{x}})^T \tag{8.32}$$

$$\mathbf{S}_T = \sum_{i=1}^{g} \sum_{j=1}^{n_i} (\mathbf{x}_{ij} - \bar{\mathbf{x}})(\mathbf{x}_{ij} - \bar{\mathbf{x}})^T \tag{8.33}$$

where $\bar{\mathbf{x}}$ is the *total mean vector*

$$\bar{\mathbf{x}} = \frac{1}{n} \sum_{i=1}^{g} n_i \bar{\mathbf{x}}_i = \frac{1}{n} \sum_{i=1}^{g} \sum_{j=1}^{n_i} \mathbf{x}_{ij} \tag{8.34}$$

and $n = \sum_{i=1}^{g} n_i$ denotes the total number of observations in all classes. Eq. 8.33 can be rewritten by adding $-\bar{\mathbf{x}}_i + \bar{\mathbf{x}}_i$ to each term and rearranging the sums so that the total scatter is the sum of the within-class scatter and the between-class scatter as [139]

$$\begin{aligned}
\mathbf{S}_T &= \sum_{i=1}^{g}\sum_{j=1}^{n_i}(\mathbf{x}_{ij} - \bar{\mathbf{x}}_i + \bar{\mathbf{x}}_i - \bar{\mathbf{x}})(\mathbf{x}_{ij} - \bar{\mathbf{x}}_i + \bar{\mathbf{x}}_i - \bar{\mathbf{x}})^T \\
&= \sum_{i=1}^{g}\sum_{j=1}^{n_i}(\mathbf{x}_{ij} - \bar{\mathbf{x}}_i)(\mathbf{x}_{ij} - \bar{\mathbf{x}}_i)^T + \sum_{i=1}^{g} n_i(\bar{\mathbf{x}}_i - \bar{\mathbf{x}})(\bar{\mathbf{x}}_i - \bar{\mathbf{x}})^T \qquad (8.35) \\
&= \mathbf{S}_W + \mathbf{S}_B.
\end{aligned}$$

The first FDA vector $\mathbf{w}_1$ maximizes the scatter between classes ($\mathbf{S}_B$) while minimizing the scatter within classes ($\mathbf{S}_W$) is obtained as

$$\max_{\mathbf{W} \neq 0} \frac{\mathbf{W}^T\mathbf{S}_B\mathbf{W}}{\mathbf{W}^T\mathbf{S}_W\mathbf{W}} \qquad (8.36)$$

under the assumption of $\mathbf{S}_W$ being invertible [139, 99]. The second FDA vector is calculated to maximize the scatter between classes while minimizing the scatter within classes among all axes perpendicular to the first FDA vector ($\mathbf{w}_1$). Additional FDA vectors are determined if necessary by using the same maximization objective and orthogonality constraint. These FDA vectors $\mathbf{w}_a$ form the columns of an optimal $\mathbf{W}$ that are the generalized eigenvectors corresponding to the largest eigenvalues in

$$\mathbf{S}_B\mathbf{w}_a = \lambda_a\mathbf{S}_W\mathbf{w}_a \qquad (8.37)$$

where the magnitude ordered eigenvalues λ_a indicate the degree of overall separability among the classes by linearly transforming the data onto $\mathbf{w}_a$ [139, 99]. The eigenvalues in Eq. 8.37 can be computed as the roots of the characteristic polynomial $\texttt{det}(\mathbf{S}_B - \lambda_a\mathbf{S}_W) = 0$ and then solving $(\mathbf{S}_B - \lambda_a\mathbf{S}_W)\mathbf{w}_a = 0$ directly for the eigenvectors $\mathbf{w}_a$ [139].

Classification with FDA

Consider a data set from a fermentation process with g distinct events such as normal process operation and operations under $g-1$ different faults. Each operation type (class) is determined on the basis of p measurements $\mathbf{x}_1, \cdots, \mathbf{x}_n$ that belong to one of the classes π_i, $i = 1, \cdots, g$ with prior probabilities of p_i, $i = 1, \cdots, g$ and probability density functions of $f_i(\mathbf{x})$. The objective of the fault diagnosis is to assign the new on-line out-of-control observations ($\mathbf{x}_0$) to the most likely fault class.

FDA is used to diagnose faults by modifying the *quadratic discrimination score* for the ith population defined in Eq. 8.16 in the FDA framework such that

$$d_i^Q(\mathbf{x}_0) = \ln p_i - \frac{1}{2}(\mathbf{x}_0 - \bar{\mathbf{x}}_i)^T \mathbf{W}_a \left(\mathbf{W}_a^T \mathbf{S}_i \mathbf{W}_a\right)^{-1} \mathbf{W}_a^T (\mathbf{x}_0 - \bar{\mathbf{x}}_i) - \frac{1}{2} \ln \left[\det \left(\mathbf{W}_a^T \mathbf{S}_i \mathbf{W}_a\right)\right] \quad (8.38)$$

where $\mathbf{W}_a$ contains the first a FDA vectors [99]. The allocation rule is:

Allocate $\mathbf{x}_0$ to π_k if $d_k^Q(\mathbf{x}_0)$ is the *largest* of all $d_i^Q(\mathbf{x}_0)$, $i = 1, \cdots, g$.

The classification rule in conjunction with Bayes' rule is used [262, 99] so that the *posterior* probability (Eq. 8.13) assuming $\sum_{i=1}^{g} P(\pi_k|\mathbf{x}) = 1$ that the class membership of the observation $\mathbf{x}_0$ is i. This assumption may lead to a situation where the observation will be classified wrongly to one of the fault cases which were used to develop the FDA discriminant when an unknown fault occurs. Chiang et al. [99] proposed several screening procedures to detect unknown faults. One of them involves FDA related T^2 statistic before applying Eq. 8.38 as

$$T_{i,a}^2 = (\mathbf{x} - \bar{\mathbf{x}}_i)^T \mathbf{W}_a \left(\mathbf{W}_a^T \mathbf{S}_i \mathbf{W}_a\right)^{-1} \mathbf{W}_a^T (\mathbf{x} - \bar{\mathbf{x}}_i) \quad (8.39)$$

so that it can be used to determine if the observation is associated with fault class i. The threshold for $T_{i,a}^2$ is defined as

$$T_{\alpha,a}^2 = \frac{a(n-1)(n+1)}{n(n-a)} F_\alpha(a, n-a) \quad (8.40)$$

where $F_\alpha(a, n-a)$ denotes the F-distribution with a and $n-a$ degrees of freedom [262]. Chiang et al. [99] introduce another class of data that are collected under NO to allow the class information in the known fault data to improve the ability to detect faults. The first step then becomes the detection of out-of-control situation. A threshold for NO class is developed based on Eq. 8.40 for detection; if $T_{i,a}^2 \geq T_{\alpha,a}^2$, there is an out-of-control situation. One proceeds with calculation at thresholds for each class i using Eq. 8.40. If $T_{i,a}^2 \geq T_{\alpha,a}^2$ for all $i = 1, \ldots, g$, then the observation $\mathbf{x}_0$ does not belong to any fault class i, and it is most likely associated with an unknown fault. If $T_{i,a}^2 \leq T_{\alpha,a}^2$ for some fault class i, then $\mathbf{x}_0$ belongs to a known fault class. Once this is determined, Fisher's discriminant score in Eq. 8.38 can be used to assign it to a fault class π_i with the highest $d_i^Q(\mathbf{x}_0)$ of all $d_i^Q(\mathbf{x}_0)$, $i = 1, \cdots, g$.

FDA and PCA can also be combined to avoid assigning an unknown fault to one of the known fault classes [246, 99, 530]. PCA is widely used for fault detection as discussed in Chapter 6. Chiang et al. [99] proposed two algorithms incorporating FDA and PCA. In the first algorithm (PCA/FDA), PCA is used to detect unknown faults and FDA to diagnose faults (by assigning them to fault classes). The NO class and classes with fault conditions are used to develop the PCA model. When a new observation $\mathbf{x}_0$ becomes available, T_a^2 value is calculated based on PCA as

$$T_a^2 = \mathbf{x}_0^T \mathbf{P}_a \lambda_a^{-1} \mathbf{P}_a^T \mathbf{x}_0 \tag{8.41}$$

where λ_a is $(a \times a)$ diagonal matrix containing eigenvalues and $\mathbf{P}$ are the loading vectors. A set of threshold values based on NO and the known fault classes using Eq. 8.40 is calculated. If $T_a^2 \leq T_{\alpha,a}^2$, it is concluded that this is a known class (either NO or faulty) and FDA assignment rule is used to diagnose the fault class (or NO class if it is in-control).

The second combined algorithm (FDA/PCA) deploys FDA initially to determine the most probable fault class i. Then it uses PCA T^2 statistic to find out if the observation $\mathbf{x}_0$ is truly associated with fault class i.

8.2.3 FDD with Neural Networks

Artificial neural network (ANN) applications in bioprocess and fermentation operations deal mostly with estimation, control and optimization. An outline and literature review on ANNs are presented in Section 4.6. ANNs have also been used for classification and FDD, which may be formulated as a classification problem. FDD in chemical processes with ANN was initially proposed by Hoskins and Himmelblau [242] and extended by various researchers [302, 618, 631]. ANN structures have been proposed to detect multiple simultaneous faults [620, 630].

The usual way to apply ANNs to FDD is to classify process operation states using data representing various states of operation of the process (normal or faulty). As discussed in Section 4.6, ANNs are well-suited to solve complicated classification problems especially in the case of highly nonlinear processes such as fermentations. In the most general case, a set of state or input variables are used as a measurement space (input space) and mapped onto a fault space (output space) where variables reflecting malfunctions reside (Figure 8.6). Data are scaled for faster convergence before training the network. Backpropagation is used in most applications as a training algorithm. Once the network is trained with data $(X_1, \ldots, X_n)$ for particular fault conditions $(F_1, \ldots, F_{g-1})$ as well as normal operating conditions, new observations can be used to classify process operation as faulty or normal using the trained ANN. Variants to this traditional approach have also been suggested. In one case, a two-stage ANN structure is used where an ANN for discriminating among the possible causes of faults

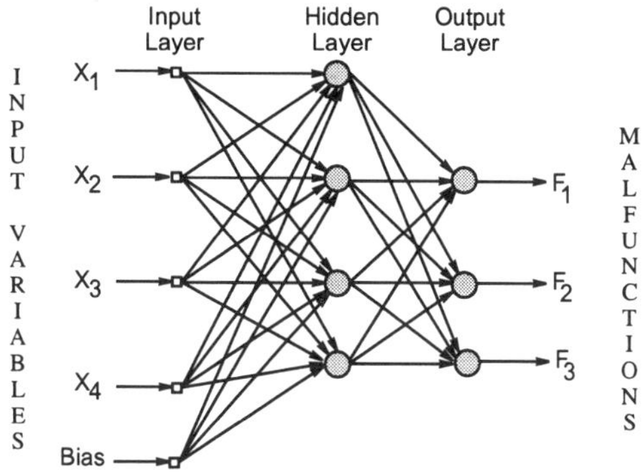

Figure 8.6. A three-layered feedforward ANN structure with four input, four hidden and three output units designed for FDD purposes.

(first stage) is followed by another ANN (second stage) that uses the outputs of the previous one to determine the level of deterioration (severity of the deviations) [631]. Such a network will have a number of outputs that is equal to the number of causes × the number of levels of deterioration, this considerably increases the computational requirement. A cascaded hierarchically layered network is also suggested for simultaneously detecting multiple faults [630]. Recently, an alternative two-stage framework was suggested for use of ANNs in FDD [360]. In this two-stage network, a primary network is trained to determine basic process trends (increasing, decreasing and steady) including the level of change. The secondary network receives the outputs from the primary network and assigns them to particular faults that it is trained for. It is reported that when network receives data for an unknown fault, it assigns the fault to either normal operation or untrained faults class [360].

Most ANN based FDD architectures assume that input-output pairs are available on-line. But in fermentation processes, very important state variables such as biomass and substrate concentrations are measured off-line in the laboratory while measurements on variables such as dissolved oxygen and carbon dioxide concentrations are available on-line. To develop a reliable ANN-based FDD scheme, values of infrequently measured (or off-line available) variables must be provided as well. This can be done by including some state observers or estimators such as Extended Kalman Filters (EKF) (Section 6.5.4) into the FDD framework. Such cascaded ANN-based fault diagnosis system (Figure 8.7) particularly designed for fermentation processes (glutamic acid fermentation in particular) is proposed by Liu [345]. A typical ANN architecture is used in the classifier that is a multi-layer feed-

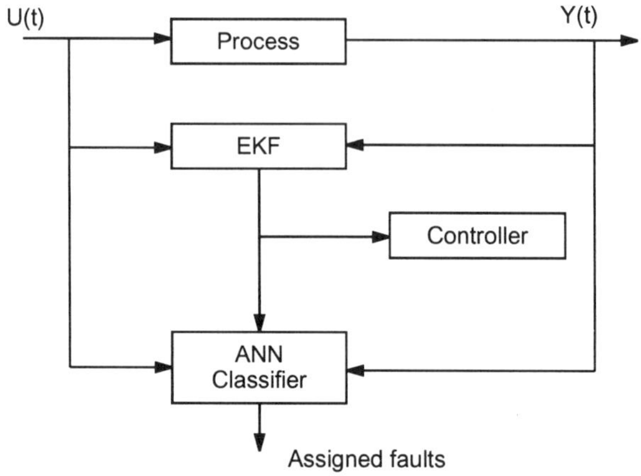

Figure 8.7. The structure of the cascade EKF-ANN-based FDD [345].

forward network although infrequently measured variables are estimated by the EKF. It is reported that once the classifier was trained with on-line measurements and estimates of off-line measurements, it achieved 89% fault diagnosis accuracy and it could be implemented in real-time.

8.2.4 Statistical Techniques for Sensor Fault Detection

Misleading process information can be generated if there is a *bias* change, *drift* or high levels of *noise* in measurements reported by some of the sensors. Erroneous information often causes decisions and actions that are unnecessary, resulting in the deterioration of product quality, safety and profitability. Identifying failures such as a broken thermocouple is relatively easy since the signal received from the sensor has a fixed and unique value. Incipient sensor failures that cause drift, bias change or additional noise are more difficult to identify and may remain unnoticed for extended periods of time. Auditing sensor behavior can warn plant personnel about incipient sensor faults and initiate timely repair and maintenance. Many approaches have been proposed for sensor fault detection and diagnosis using statistical methods, model-based fault diagnosis techniques (Section 8.3) such as parity space [189, 292], state estimators [456] and parameter change detection [686], artificial intelligence applications such as neural networks [631] and knowledge-based systems. In this section, sensor auditing methods based on functional redundancy generated by PLS or *canonical variate state space* (CVSS) models (Section 4.3.2) are presented. Integration of these statistical methods with knowledge-based systems to discriminate between sensor faults and process disturbances is discussed in [584]. Another method for FDD of sensors and discrimination of faults from process upsets relies on changes in correlation between data from various sensors [138]. Sensor FDD by wavelet decompositions of data followed by calculation of signal features and nonparametric statistical test has also been reported [353].

Sensor Auditing Using PLS and CVSS Models

The use of the mean and variance of residuals between measured and estimated sensor readings based on PCA and PLS models was proposed in late 1980s [656]. The authors cautioned the users about the corruption of estimates when erroneous sensor data were used when multiple sensors were faulty. Negiz and Cinar [412] developed a sequential PLS model development and sensor FDD method to reduce the effect of multiple faulty sensors and to discriminate between sensor bias, drift and noise. These methods use a data sequence to compare the mean and variance of data batches. They can be implemented to run repeatedly at frequent intervals and warn plant personnel about incipient sensor faults to initiate timely repair and maintenance. Both methods are based on interpreting the magnitudes of the mean and variance of the residuals between a data batch and their prediction from a process model. The PLS-based method is useful for process data with milder autocorrelation, while the CVSS-based version is more appropriate for processes with significant dynamic changes and autocorrelation in data. Industrial continuous processes have a large number of process variables and are usually operated for extended periods at fixed operating points under closed-loop control, yielding process measurements which are autocorrelated, cross correlated, and colinear. A CVSS model would be appropriate for modeling data generated from such processes. Once an accurate statistical description of the in-control variability of a continuous process is available, the next step is the design and implementation of the sensor monitoring SPM procedure.

Multipass PLS-Based Sensor FDD Method. The multipass PLS algorithm was developed for detecting simultaneous multiple sensor abnormalities. This is achieved by eliminating successively the corrupted measurements from both the calibration and test data sets and identifying a different smaller PLS submodel.

Assume that there are p sensors to be monitored and the *calibration* data set is of length N. The mean and the variance of the residuals for each variable is computed through the $N \times p$ residuals block matrix $\mathbf{R}$. Once the PLS (calibration) model is identified for the in-control data set, the statistics for the residuals are computed for setting the null hypothesis. Then, a test data block of size $N_t \times p$ is formed from new process measurements. The residual statistics for the test sample are then generated by using the PLS *calibration* model. The statistical test compares the residuals statistics of the test sample with the statistics of the *calibration* set for detecting any significant departures.

Denote by $\mathbf{R}_{\bullet i}$ the ith $N \times 1$ residual vector column from the $N \times p$ residual block matrix $\mathbf{R}$. The statistic for testing the null hypothesis of the equality of means from two normal populations with equal and unknown variances is

$$\frac{\bar{\mathbf{R}}_{\bullet i_{\mathrm{test}}} - \bar{\mathbf{R}}_{\bullet i_{\mathrm{model}}}}{\hat{\sigma}_{p_i}\sqrt{1/N + 1/N_t}} \sim t_{N+N_t-2} \tag{8.42}$$

where $\bar{\mathbf{R}}_{\bullet i_{\mathrm{test}}}$ and $\bar{\mathbf{R}}_{\bullet i_{\mathrm{model}}}$ denote the maximum likelihood estimates of the residual means for the variable i in the test sample and the *calibration* set, $\hat{\sigma}_{p_i}$ is the pooled standard deviation of the two residual populations for the i-th variable, N and N_t denote the sizes of the *calibration* and testing populations, and t_{N+N_t-2} is the t-distribution with $N + N_t - 2$ degrees of freedom [140].

The statistic for testing the null hypothesis of the equality of variances from two normal populations with unknown means is [140]

$$\frac{\hat{\sigma}^2_{i_{\mathrm{test}}}}{\hat{\sigma}^2_{i_{\mathrm{model}}}} \sim F_{N_t-1,N-1} \tag{8.43}$$

where $F_{N_t-1,N-1}$ is the F distribution with respective degrees of freedom. The level of the test for all the testing statistics is chosen to be 5% and two sided. This part of the procedure is similar to that given by [656].

The algorithm takes action when either the mean or variance of the residuals are out of the statistical limits (based on t and F probability distributions) for a particular variable. Since the corrupted variable affects the predictions of the remaining ones, false alarms might be generated unless the corrupted variable is taken out from both the *calibration* and *test* data blocks. The information loss due to taking the variable out of both the calibration and the test sample sets is not significant since the testing procedures are based on the *iid* assumption of the residuals and not on the minimum prediction error criterion by the model. The algorithm discards the variable with the highest corruption level by looking at the ratios of its residual variance and its residual mean to their statistical limits which are based on Eqs. 8.42–8.43.

Excluding variables and computing a new PLS model for the remaining variables is the key step of the sensor auditing and fault detection algorithm. The likelihood for all of the process sensors to become simultaneously faulty is extremely small. After several successive steps, if the mean and variance of the remaining residuals still indicate significant variation, then it is more likely that a disturbance is active on the system causing the in-control variability to change.

Multipass CVSS-Based Sensor FDD Method. A *multipass* CVSS technique similar to the *multipass PLS* algorithm is developed for detecting multiple sensor failures. This is achieved by eliminating successively the corrupted measurements from both the calibration and test data sets and identifying a different CVSS submodel. The algorithm discards the variable with the highest corruption level by looking at the ratios of its residual variance and its residual mean to their statistical limits which are based on Eqs. 8.42–8.43. Excluding variables and computing a new CV realization for the remaining variables, the algorithm proceeds in a manner similar to the PLS-based version. The application of the method for FDD of the sensors of a high-temperature short-time pasteurization process is reported in [417].

Real-time Sensor FDD by Statistical Methods. A sensor FDD that checks sensor status at each sampling instant can be developed by using T^2 and squared prediction error (SPE) charts. Once these charts indicate an out-of-control status, discrimination between sensor faults and disturbances should be made and the faulty sensor should be isolated. One approach used for discrimination of sensor faults and disturbances is the redundant sensor voting system [577] that utilizes backward elimination for sensor identification [578]. The backward elimination is similar to the multipass PLS approach, but remodeling is implemented at each time instant the SPE limit is violated. In this approach, once the SPE limit is violated at a specific sampling time, every sensor is sequentially removed from the model matrix and the control limit is recalculated. If the ratio $\mathrm{SPE}/\mathrm{SPE}_{\mathrm{limit}} < 1$, the search terminates and the sensors eliminated up to that point are declared faulty. Otherwise, the search continues by eliminating another sensor from the model. This approach has significant computational burden. In addition, sensor faults that do not inflate the SPE statistic cannot be detected. Incorporation of T^2 charts and use of correlation coefficient criterion were proposed to improve this method [138].

8.3 Model-based Fault Diagnosis Techniques

A mathematical model of the process is used in model-based fault diagnosis to describe the expected behavior of the process. In most model-based FDD techniques, measured values of process variables are compared to their estimated values. The estimations are based on a process model describing the expected (nominal) operation, past measurements of process variables, and noise/disturbance information. The difference between measured and estimated values are *residuals* that are subjected to statistical tests to detect significant magnitudes of residuals that indicate presence of faults. Various

FDD methods based on residuals are discussed in Section 8.3.1. Another group of model-based FDD techniques use parameter estimation. They are presented in Section 8.3.2. In this approach, model parameters are estimated for nominal operating conditions. They are estimated repeatedly as new measurement information is collected. Significant deviations in model parameter values are used for FDD. Hidden Markov models provide another FDD framework. Markov processes, hidden Markov models and their use in FDD are discussed in Section 8.3.3.

Model-based FDD has its origins in various engineering areas. Material and energy balance calculations were used for gross error detection and data reconciliation in chemical process operations [235, 359, 519]. FDD applications in aerospace systems were reported [652] leading to *parity relations* concepts [101, 187]. *Kalman filters* were used in aerospace and nuclear power industries for FDD [168, 650]. *Diagnostic observers* were also proposed for similar applications [105, 161, 163, 458]. FDD by *parameter estimation* has been used in manufacturing industries [50, 251]. Excellent review papers [44, 162, 457] and books [45, 189, 456, 518] report recent developments in the FFD theory and applications in many areas. The presentation of various model-based FFD techniques in this text is based on these resources and the research of Cinar and co-workers [292, 414].

Input-output relations for systems subject to faults. Consider a process that receives *measured inputs* $\mathbf{u}_M$ subject to sensor faults $\delta\mathbf{u}_M$, *controlled inputs* $\mathbf{u}_C$ subject to actuator faults $\delta\mathbf{u}_C$, *process faults* $\delta\mathbf{u}_P$ that are interpreted as additional inputs, and *measured outputs* $\mathbf{y}$ subject to sensor faults $\delta\mathbf{y}$ (Figure 8.8). *Additive faults* acting on the process include:

- input actuator faults $\delta\mathbf{u}_C(t)$
- input sensor faults $\delta\mathbf{u}_M(t)$
- output sensor faults $\delta\mathbf{y}(t)$
- plant faults $\delta\mathbf{u}_P(t)$.

In addition, there are *additive disturbances* acting on the process (these are usually unmeasured input disturbances) ($\mathbf{d}$) and various noises acting on measurements and the process:

- input actuator noise $\mathbf{v}_C(t)$
- input sensor noise $\mathbf{v}_M(t)$
- output sensor noise $\mathbf{v}_y(t)$
- plant noise $\mathbf{v}_P(t)$.

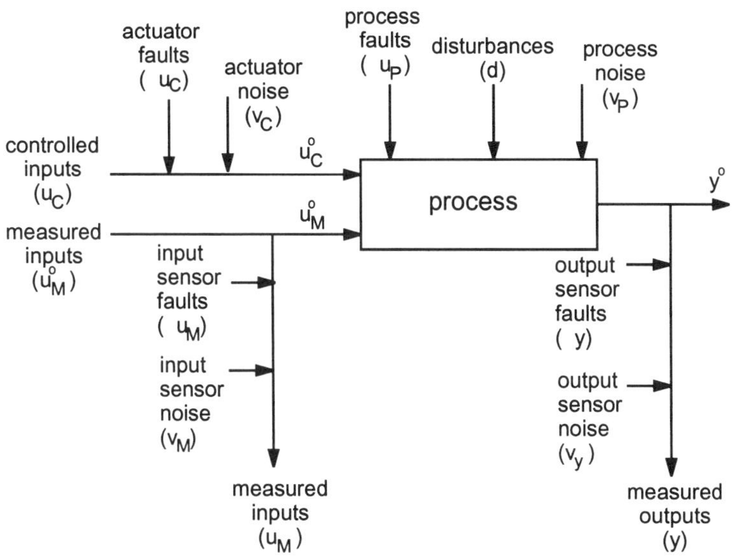

Figure 8.8. Additive faults and disturbances [189].

Consider a multiple-input, multiple-output system described by

$$\mathbf{y}(t) = \mathbf{G}(q)\mathbf{u}(t) \tag{8.44}$$

where $\mathbf{G}(q)$ denotes the multivariable input-output transfer function matrix (TFM) and q is the shift operator defined in Section 4.3.1. Denoting the actual inputs and outputs of the process with superscript ^{o}, they can be expressed as:

$$\begin{aligned} \mathbf{u}_C^{\text{o}}(t) &= \mathbf{u}_C(t) + \delta\mathbf{u}_C(t) + \mathbf{v}_C(t) \\ \mathbf{u}_M^{\text{o}}(t) &= \mathbf{u}_M(t) - \delta\mathbf{u}_M(t) - \mathbf{v}_M(t) \\ \mathbf{y}^{\text{o}}(t) &= \mathbf{y}(t) - \delta\mathbf{y}(t) - \mathbf{v}_y(t) \end{aligned} \tag{8.45}$$

The relation in Eq. (8.44) is between the nominal inputs and outputs. Expanding this relationship to show explicitly the faults, noises and process disturbances:

$$\mathbf{y}(t) = \mathbf{G}(q)\mathbf{u}(t) + \mathbf{S}_F(q)\mathbf{f}(t) + \mathbf{S}_D(q)\mathbf{d}(t) + \mathbf{S}_N(q)\boldsymbol{\nu}(t) \tag{8.46}$$

where $\mathbf{S}_F(q)$ is the combined fault FTM, $\mathbf{S}_N(q)$ is the combined noise TFM,

$\mathbf{S}_D(q)$ is the process fault TFM, and

$$\begin{aligned}
\mathbf{f}(t) &= \left[\, \delta\mathbf{u}_C^T(t) \quad -\delta\mathbf{u}_M^T(t) \;\; \delta\mathbf{u}_P^T(t) \quad -\delta\mathbf{y}(t)^T \, \right]^T \\
\boldsymbol{\nu}(t) &= \left[\, \mathbf{v}_C^T(t) \quad -\mathbf{v}_M^T(t) \;\; \mathbf{v}_P^T(t) \quad -\mathbf{v}_y(t)^T \, \right]^T \\
\mathbf{S}_F &= \left[\, \mathbf{G}_C(q) \;\; \mathbf{G}_M(q) \;\; \mathbf{S}_{PF}(q) \;\; \mathbf{I} \, \right] \\
\mathbf{S}_N &= \left[\, \mathbf{G}_C(q) \;\; \mathbf{G}_M(q) \;\; \mathbf{S}_{PN}(q) \;\; \mathbf{I} \, \right]
\end{aligned} \tag{8.47}$$

with $\mathbf{G}_C(q)$ denoting the actuator (controlled input) fault TFM, $\mathbf{G}_M(q)$ the input sensor fault TFM, $\mathbf{S}_{PF}(q)$ the plant fault TFM and $\mathbf{S}_{PN}(q)$ the plant noise TFM.

This framework with additive faults has to be augmented to consider multiplicative faults. Multiplicative faults may reflect a parametric fault resulting from the change in process operation (hence the model is not accurate anymore) or a modeling error such as inaccuracy in model structure or parameters resulting from approximating a nonlinear process with a linear model or a high ordered process with a low order model. The input-output model representation Eq. (8.46) is further expanded to incorporate multiplicative faults

$$\begin{aligned}
\mathbf{y}(t) &= \mathbf{G}(q)\mathbf{u}(t) + \mathbf{S}_F(q)\mathbf{f}(t) + \mathbf{S}_D(q)\mathbf{d}(t) + \mathbf{S}_N(q)\boldsymbol{\nu}(t) \\
&\qquad + \mathbf{N}_P(t)\boldsymbol{\phi}_P + \mathbf{N}_M(t)\boldsymbol{\phi}_M
\end{aligned} \tag{8.48}$$

where $\mathbf{N}_P(t)$ denotes the matrix of time-varying coefficients of multiplicative parametric faults, $\mathbf{N}_M(t)$ the coefficient matrix of multiplicative modeling faults, $\boldsymbol{\phi}_P$ the parametric faults and $\boldsymbol{\phi}_M$ the modeling errors. An important difference between additive and multiplicative faults is that the TFMs of additive faults are constant while the TFMs of multiplicative faults are time dependent. Whereas additive fault vectors are time dependent, multiplicative fault vectors are constant [189]. The remainder of this section will focus mainly on additive faults. Multiplicative faults are equally important, but the model-based techniques for addressing them are active research issues that can not be adequately treated in the framework of this text.

State-space relations for systems subject to faults. The relationship between $\mathbf{y}$ and $\mathbf{u}$ can also be written in state-space form. The state-space form equivalent to input-output relations in Eq. (8.46) is given in Eq. (8.85). Most early methods based on observers and Kalman filters use a simplified state-space representation that ignores noise as a separate factor, resulting in

$$\begin{aligned}
\mathbf{x}_{k+1} &= \mathbf{F}\mathbf{x}_k + \mathbf{G}\mathbf{u}_k + \mathbf{E}_F\mathbf{f}_k + \mathbf{E}_D\mathbf{d}_k \\
\mathbf{y}_k &= \mathbf{C}\mathbf{x}_k + \mathbf{F}_F\mathbf{f}_k + \mathbf{F}_D\mathbf{d}_k
\end{aligned} \tag{8.49}$$

where component and actuator faults are modeled by $\mathbf{E}_F\mathbf{f}_k$ and sensor faults are modeled by $\mathbf{F}_F\mathbf{f}_k$. The unknown inputs affecting the actuators and process dynamics are introduced by $\mathbf{E}_D\mathbf{d}_k$ and unknown inputs to sensors are introduced by $\mathbf{F}_D\mathbf{d}_k$. This modified representation is used in illustrating the use of Kalman filters and observers in subsequent sections.

8.3.1 Residuals-Based FDD Methods

Residuals, the difference between measured and model predicted values of process variables, carry valuable information. Unfortunately, this information is blended with measurement noise and prediction errors due to modeling accuracy. Robust FDD techniques are needed to interpret the residuals in spite of noise in data and modeling errors. Three different approaches for computation of residuals and their interpretation are discussed in this section: Parity equations, diagnostic observers and Kalman filters, and robust observers for unknown inputs. Often, statistical tests are conducted to assess the significance level of the magnitudes of residuals to detect or diagnose a fault. Two popular tests, χ^2 tests and *likelihood ratio* tests are introduced below.

χ^2 tests of residuals for fault detection. Fault:!detection!χ^2 tests of residuals for Statistical testing of residuals for fault detection can be cast as testing for the zero mean hypothesis. The null hypothesis ($\mathcal{H}_0$) is residual mean being zero (or having a nonsignificant magnitude), which indicates lack of evidence for a fault. The alternative hypothesis ($\mathcal{H}_1$) is large nonzero values of the residual mean, indicating the presence of a fault.

$$\begin{array}{lcll} \mathcal{H}_0 & : & \boldsymbol{\mu}_r = 0 & \textit{no fault} \\ \mathcal{H}_1 & : & \boldsymbol{\mu}_r \neq 0 & \textit{fault} \end{array} \tag{8.50}$$

where $\boldsymbol{\mu}_r$ is the mean of the residual vector. Because of limited data, the test is conducted using the sample mean of residuals $\bar{\mathbf{r}}$ instead of $\boldsymbol{\mu}_r$. The test may be conducted on a *single residual* at a given time ($r(t)$), a single residual over a time window l ($\vec{r}(t) = [r(t), \cdots, r(t-l)]^T$), or an average residual over the window l ($\tilde{r}(t,l) = [1/(l{+}1)]\sum_{j=0}^{l} r(t{-}j)$). The same tests can be conducted on a *vector of residuals* where $\mathbf{r}(t) = [r_1(t), \cdots, r_n(t)]^T$, $\vec{\mathbf{r}}(t) = [\mathbf{r}^T(t), \cdots, \mathbf{r}^T(t-l)]^T$, and $\tilde{\mathbf{r}}(t,l) = [1/(l+1)]\sum_{j=0}^{l} r(t-j)$. The tests are designed for a specified false alarm rate α, and Normal distribution and zero mean of residuals is assumed. χ^2 tests are used for fault detection. They can be developed for scalar or vector residuals. Detailed discussion of scalar and vector residuals tests is given in [189]. The tests for vector residuals are summarized below.

`Single observation of vector residual.` The joint density function for a single observation of vector residuals $\mathbf{r}(t)$ of length n is

$$f(\mathbf{r}) = \frac{1}{(2\pi)^{n/2}|\mathbf{\Sigma}_\mathbf{r}|^{1/2}} \, exp \left[-\frac{1}{2}\mathbf{r}^T \mathbf{\Sigma}_\mathbf{r}{}^{-1}\mathbf{r} \right] \tag{8.51}$$

where $\mathbf{\Sigma}_\mathbf{r}$ is the population covariance matrix of $\mathbf{r}$. The corresponding sample covariance matrix is $\mathbf{S}_\mathbf{r}$. The statistic

$$\rho_n(t) = \mathbf{r}^T(t)\mathbf{S}_\mathbf{r}^{-1}\mathbf{r}(t) \tag{8.52}$$

follows the χ^2 distribution with n degrees of freedom and can be used to detect faults with the hypotheses

$$\begin{aligned} \mathcal{H}_0 &: \quad \rho_n(t) < \chi^2_{n,\alpha} \qquad \text{no fault} \\ \mathcal{H}_1 &: \quad \rho_n(t) \geq \chi^2_{n,\alpha} \qquad \text{fault} \end{aligned} \tag{8.53}$$

`Vector residual sequence.` The joint density function of the vector sequence is

$$f(\vec{\mathbf{r}}) = \frac{1}{(2\pi)^{n(l+1)/2}|\mathbf{\Sigma}_{\vec{\mathbf{r}}}|^{1/2}} \, exp \left[-\frac{1}{2}\vec{\mathbf{r}}^T \mathbf{\Sigma}_{\vec{\mathbf{r}}}^{-1}\vec{\mathbf{r}} \right] \tag{8.54}$$

with the covariance matrix $\mathbf{\Sigma}_{\vec{\mathbf{r}}} = E[\vec{\mathbf{r}}^T\vec{\mathbf{r}}]$. The test statistic

$$\rho_{n(l+1)}(t) = \vec{\mathbf{r}}^T(t)\mathbf{S}_{\vec{\mathbf{r}}}^{-1}\vec{\mathbf{r}}(t) \tag{8.55}$$

can be tested against the threshold $\chi^2_{n(l+1),\alpha}$.

`Window average of vector residual.` The density function for the window average is

$$f(\tilde{\mathbf{r}}) = \frac{1}{(2\pi)^{n/2}|\mathbf{\Sigma}_{\tilde{\mathbf{r}}}|^{1/2}} \, exp \left[-\frac{1}{2}\tilde{\mathbf{r}}^T \mathbf{\Sigma}_{\tilde{\mathbf{r}}}^{-1}\tilde{\mathbf{r}} \right] \tag{8.56}$$

where

$$\mathbf{\Sigma}_{\tilde{\mathbf{r}}} = \frac{1}{l+1}\mathbf{\Sigma}_{\mathbf{rr}}(0) + \frac{1}{(l+1)^2}\sum_{j=1}^{l}(l+1-j)\left[\mathbf{\Sigma}_{\mathbf{rr}}(j)\mathbf{\Sigma}_{\mathbf{rr}}^T(j)\right] \tag{8.57}$$

with covariance matrix $\mathbf{\Sigma}_{\mathbf{rr}}(j) = E[\mathbf{r}_j(t)\mathbf{r}_j^T(t-j)]$ [189]. The corresponding fault detection test statistic is

$$\rho_n(t) = \tilde{\mathbf{r}}^T(t)\mathbf{S}_{\tilde{\mathbf{r}}}^{-1}\tilde{\mathbf{r}}(t) \tag{8.58}$$

with threshold $\chi^2_{n,\alpha}$.

Likelihood Ratio Tests for Change Detection. Consider a simple change detection problem, detecting the change in the mean of an i.i.d. random variable y_k from μ_0 to μ_1 and estimating the time of change (switching or jump time) q. If y_k is a sequence of n observations and ϵ_k is a white noise sequence with variance σ^2

$$y_k = \mu_k + \epsilon_k \tag{8.59}$$

where

$$\mu_k = \begin{cases} \mu_0 & \text{if} \quad k \leq q-1 \\ \mu_1 & \text{if} \quad k \geq q \end{cases}$$

The detection problem can be phrased as a hypothesis testing problem [44].

$$\begin{array}{lcll} \mathcal{H}_0 & : & q > k & \text{no change} \\ \mathcal{H}_1 & : & q \leq k & \text{change} \end{array} \tag{8.61}$$

This is an easy case since the new value of the mean (μ_1) is known and only the change time is investigated. The likelihood ratio between these two hypotheses is

$$\prod_{i=q}^{k} \frac{p_1(y_i)}{p_0(y_i)} \tag{8.62}$$

where $p_l(\cdot)$ is the Gaussian probability density function of y_i with mean μ_l, $l = 0, 1$

$$p_l(y_i) = \frac{1}{(2\pi)^{1/2}\sigma} \, exp \left[-\frac{(y_i - \mu_l)}{2\sigma^2} \right] . \tag{8.63}$$

The log-likelihood ratio is derived by taking the logarithm of the likelihood function Eq. (8.62) and noting that the constant term cancels out. After some algebraic manipulation, the *log-likelihood ratio* can be expressed as [44]

$$\begin{aligned} \Lambda_k(r) &= \frac{\mu_1 - \mu_0}{\sigma^2} \sum_{i=q}^{k} \left(y_i - \frac{\mu_0 + \mu_1}{2} \right) \\ &= \frac{1}{\sigma^2} S_q^k(\mu_0, \delta) \end{aligned} \tag{8.64}$$

where

$$S_m^n(\mu^*, \delta) = \delta \sum_{i=m}^{n} (y_i - \mu^* - \frac{\delta}{2}) \tag{8.65}$$

and $\delta = \mu_1 - \mu_0$ is *change magnitude* which is known in this case. The jump time q is not known. Consequently, q in likelihood ratio (Eq. 8.62) and log-likelihood ratio (Eq. 8.64) should be replaced by its maximum likelihood estimate $\hat{q}_k$ under hypothesis $\mathcal{H}_1$:

$$\begin{aligned} \hat{q}_k &= \arg \max_{1 \leq q \leq k} \left[\prod_{i=0}^{q-1} p_0(y_i) \prod_{i=q}^{k} p_1(y_i) \right] \\ &= \arg \max_{1 \leq q \leq k} S_q^k(\mu_0, \delta) \; . \end{aligned} \tag{8.66}$$

The resulting change detector with a threshold τ is

$$g_k = \Lambda_k(\hat{q}_k) = \max_q S_q^k(\mu_0, \delta) \tag{8.67}$$

where

$$\begin{aligned} \mathcal{H}_0 &: \quad g_k < \tau \quad \text{no change at time } k \\ \mathcal{H}_1 &: \quad g_k \geq \tau \quad \text{change at time } k \end{aligned} \tag{8.68}$$

Hence, the detector detects a jump of magnitude δ in the mean at the first time where

$$g_k = S_1^k(\mu_0, \delta) \;-\; \min_{1 \leq i \leq k} \; S_1^i(\mu_0, \delta) > \tau \tag{8.69}$$

which is called the Page-Hinkley Stopping Rule or the cumulative sum algorithm that may be computed recursively [44].

If the jump magnitude is unknown (μ_0 is known but not μ_1), one approach is to select a minimum jump magnitude in the mean that is desired to be detected and run two tests (increase and decrease in the mean). A second approach is to replace the unknown magnitude δ by its maximum likelihood estimate (MLE) and then run the likelihood ratio test. In this case,

$$g_{\delta,k} = \max_{1 \leq q \leq k} \max_{\delta} S_1^k(\mu_0, \delta) \tag{8.70}$$

and using Eq. (8.65)

$$\hat{\delta}_k = \arg \max_{\delta} S_q^k(\mu_0, \delta) = \frac{1}{k - q + 1} \sum_{i=q}^{k} (y_i - \mu_0) \tag{8.71}$$

which reduces the double maximization in Eq. (8.70) to a single maximization [44]. The likelihood ratio test becomes

$$\begin{aligned} \mathcal{H}_0 &: \quad g_{\delta,k} < \tau \quad \text{no change of } \delta_k \text{ at time } k \\ \mathcal{H}_1 &: \quad g_{\delta,k} \geq \tau \quad \text{change of } \delta_k \text{ at time } k \end{aligned} \tag{8.72}$$

These maximum likelihood tests for composite hypothesis testing problems (such as finding MLE of δ and the jump time q) are called *generalized likelihood ratio* (GLR) tests [45]. Likelihood ratio tests for more complex changes or models can be found in [45].

Maximum Likelihood and GLR Tests for Fault Diagnosis. When faulty operation is detected, the fault needs to be diagnosed. Diagnosis can be implemented by several approaches. Estimation of maximum likelihoods (ML) and the GLR test are popular techniques for model-based fault diagnosis.

Rewrite the joint density function in Eq. (8.51) in a generic form

$$f(\mathbf{z}(t), \boldsymbol{\mu}_{\mathbf{z}}(t)) = K\ exp\ \left[-\frac{1}{2}[\mathbf{z}(t) - \boldsymbol{\mu}_{\mathbf{z}}(t)]^T \boldsymbol{\Sigma}_{\mathbf{z}}^{-1} [\mathbf{z}(t) - \boldsymbol{\mu}_{\mathbf{z}}(t)]\right] \tag{8.73}$$

where K denotes the constant term preceding the exponential part (it remains the same for all hypotheses tested), $\mathbf{z}$ stands for the type of observation used ($\mathbf{r}, \vec{\mathbf{r}}, \text{or } \tilde{\mathbf{r}}$), and $\boldsymbol{\mu}_{\mathbf{z}}(t)$ the corresponding mean. The simplified log-likelihood function $L[\cdot]$ is defined as

$$log\ L[(\mathbf{z}(t), \boldsymbol{\mu}_{\mathbf{z}}(t))] = -\frac{1}{2}[\mathbf{z}(t) - \boldsymbol{\mu}_{\mathbf{z}}(t)]^T \boldsymbol{\Sigma}_{\mathbf{z}}^{-1} [\mathbf{z}(t) - \boldsymbol{\mu}_{\mathbf{z}}(t)] \tag{8.74}$$

where $log\ K$ is omitted because it will cancel out when the likelihood ratio is defined.

The ML test consists of the following procedure:

1. Compute the maximum likelihood estimates of the residual mean from observations under various hypotheses $\mathcal{H}_j$:

$$\hat{\boldsymbol{\mu}}_{\mathbf{z}j}(t) = arg\ \max_{\boldsymbol{\mu}_{\mathbf{z}}(t)}\ log\ L[(\mathbf{z}(t), \hat{\boldsymbol{\mu}}_{\mathbf{z}}(t)) \mid \mathcal{H}_j] \qquad j = 1, \cdots, f \tag{8.75}$$

   where $\mathcal{H}_j$ are the hypotheses about various possible faults that impose constraints on the estimates of the mean and f is the number of faults. The hypotheses are a function of the properties of the residuals generators such as directional residuals or structured residuals discussed below.

2. Compute the conditional likelihood functions using the observations and conditional estimates

$$log\ L_j(t) = log\ L[(\mathbf{z}(t), \hat{\boldsymbol{\mu}}_{\mathbf{z}j}(t))] \qquad j = 1, \cdots, f \tag{8.76}$$

The most likely fault is the fault that yields the highest log-likelihood value. Extensions of ML approach with additional checks to account for the uncertainty in the decision because of signal noise are discussed in [189].

FDD by Parity Equations

The basic idea is to check the parity (consistency) of the mathematical equations describing the process by using process measurement information. For illustration, consider a simple system with redundant information described by [162]

$$\mathbf{y} = \mathbf{C}\mathbf{x} + \delta\mathbf{y} \tag{8.77}$$

where $\mathbf{y}$ is the measurement vector of length q, $\mathbf{C}$ the $q \times n$ measurement matrix of rank n, $\mathbf{x}$ the unknown true measurements, and $\delta\mathbf{y}$ the error vector. If $\delta y_i > \tau_i$ where τ_i is the error threshold for variable i, there is a faulty operation indicated by the ith measured variable. Define a parity vector of dimension $q - n$ such that

$$\mathbf{p} = \mathbf{W}\mathbf{y} \tag{8.78}$$

The projection matrix $\mathbf{W}$ is of dimension $(q-n) \times q$ is determined such that the parity vector is only a function of $\delta\mathbf{y}$. To achieve this, $\mathbf{W}$ is determined such that

$$\mathbf{W}\mathbf{C} = 0 \qquad \mathbf{W}^T\mathbf{W} = \mathbf{I}_q - \mathbf{C}(\mathbf{C}^T\mathbf{C})^{-1}\mathbf{C}^T \qquad \mathbf{W}\mathbf{V}^T = \mathbf{I}_{q-n} \tag{8.79}$$

These conditions assure that the rows of $\mathbf{W}$ are orthogonal and $\mathbf{W}$ is the null space of $\mathbf{C}$. Consequently,

$$\mathbf{p} = \mathbf{W}\ \delta\mathbf{y} \tag{8.80}$$

Hence, parity equations are independent of $\mathbf{x}$ and contain only the errors $\delta\mathbf{y}$ caused by faults. Furthermore, the columns of $\mathbf{W}$ define q distinct fault directions, each associated with only one of the measurements. If there is significant increase in the ith direction of $\mathbf{p}$, it indicates faulty measurement y_i.

The residual vector $\mathbf{r} = \mathbf{y} - \mathbf{C}\hat{\mathbf{x}}$ is related to $\mathbf{p}$ as $\mathbf{r} = \mathbf{W}^T\mathbf{p}$ where $\hat{\mathbf{x}} = (\mathbf{C}^T\mathbf{C})^{-1}\mathbf{C}^T\mathbf{y}$ is the least squares estimate of $\mathbf{x}$. The FDD problem can be stated as a two-step procedure: (1) Find $\hat{\mathbf{x}}$ and compute $\mathbf{r}$; (2) Detect and diagnose the faulty measurements by parity checks. This concept is extended during the last three decades to handle more complex cases involving faults, disturbances, and noise. A short discussion of the formulation of residual generator and parity equations is given below.

Residual Generators. A residual generator is a linear discrete dynamic algorithm acting on observable variables [189]

$$\mathbf{r}(t) = \mathbf{V}(q)\mathbf{u}(t) + \mathbf{W}(q)\mathbf{y}(t) \tag{8.81}$$

where $\mathbf{r}(t)$ is the vector of residuals, and $\mathbf{V}(q)$ and $\mathbf{W}(q)$ are TFMs. Noting that $\mathbf{r}(t)$ must be zero when all inputs $\mathbf{u(t)}$ and $\mathbf{y(t)}$ are zero, and substituting $\mathbf{y}(t) = \mathbf{G}(q)\mathbf{u}(t)$ into Eq. (8.81) yields $(\mathbf{V}(q) + \mathbf{W}(q)\mathbf{G}(q))\mathbf{u}(t) = \mathbf{0}$. Hence, Eq. (8.81), the *computational form* of the residual generator can be written as

$$\mathbf{r}(t) = \mathbf{W}(q)[\mathbf{y}]t) - \mathbf{G}(q)\mathbf{u}(t)] \tag{8.82}$$

The term in brackets in Eq. (8.82) can be substituted using Eq. (8.48) to yield the *internal form* of the residual generator

$$\begin{aligned} \mathbf{r}(t) &= \mathbf{W}(q)\left[\mathbf{S}_F(q)\mathbf{f}(t) + \mathbf{S}_D(q)\mathbf{d}(t) + \mathbf{S}_N(q)\boldsymbol{\nu}(t)\right. \\ &\qquad \left. + \mathbf{N}_P(t)\boldsymbol{\phi}_P + \mathbf{N}_M(t)\boldsymbol{\phi}_M\right] \end{aligned} \tag{8.83}$$

Ideally, residuals $\mathbf{r}(t)$ should only be affected by faults. If specific unique residuals patterns for each fault could be generated, fault detection and isolation would reduce to checking the violation of limits of residuals and recognizing the patterns. However, disturbances, noise and modeling errors (nuisance inputs) contribute to residuals as well and interfere with FDD. The residual generator should be designed such that the effects of these nuisance inputs on the residuals are as small as possible, leading to robust residual generators. The differences in the properties of these three nuisance inputs determine the approach used in marginalizing them. Additive disturbances and modeling errors have similar temporal behavior to additive faults. Explicit decoupling of residuals from disturbances and modeling errors is necessary to improve the detection and diagnosis capability of the residuals.

Noises usually have much higher frequencies than fault signals and zero mean values. Therefore, filtering the residuals signals with low-pass filters reduces the effects of noise without affecting the fault signals significantly. In addition, testing the residuals against some threshold value as opposed to testing them for nonzero values reduces false alarms caused by noise. There is a tradeoff between the number of false alarms and the number of missed alarms which is affected by the level of thresholds selected (Type I and Type II errors).

The residual generator should be designed to improve fault isolation. The residual set should have different patterns for particular faults. Residual sets designed with the isolation objective are called enhanced residuals. There are two enhancement approaches, structured and directional. In *structured residuals*, each residual responds to a different set of faults and is insensitive to others. Threshold tests are applied to each element of the residual vector and the test results are converted to a fault code vector $\mathbf{s}(t)$ of binary digits. Defining a residual threshold vector ($\boldsymbol{\tau}$), $s_i(t) = 1$ if $|r_i(t)| \geq \tau_i$; $s_i(t) = 0$ otherwise. The pattern of the fault code vector (a binary string) is matched against the library of fault signatures for diagnosis. *Directional residuals* generate fault-specific vector directions β and the scalar transfer function $\gamma(q)$ in that direction indicates the dynamics of the fault [189]

$$\mathbf{r}(t|\mathbf{f}_j) = \beta_j \gamma_j(q) \mathbf{f}_j(t) \tag{8.84}$$

where β_j is the direction of the jth fault. Fault diagnosis is based on associating $\mathbf{r}(t|\mathbf{f})$ with the closest fault direction in the fault library.

The implementation of the residual generator may be done either in input-output form (Eq. (8.46)) or in the equivalent state-space form. (Note the conventional use of $\mathbf{G}$ in state-space representation which is different than its as a TFM $\mathbf{G}(q)$ and the difference between the conventional use of $\mathbf{F}$ in state-space representation and $\mathbf{F}_F$, $\mathbf{F}_D$, and $\mathbf{F}_N$.)

$$\begin{aligned} \mathbf{x}_{k+1} &= \mathbf{F}\mathbf{x}_k + \mathbf{G}\mathbf{u}_k + \mathbf{E}_F\mathbf{f}_k + \mathbf{E}_D\mathbf{d}_k + \mathbf{E}_N\boldsymbol{\nu}_k \\ \mathbf{y}_k &= \mathbf{C}\mathbf{x}_k + \mathbf{D}\mathbf{u}_k + \mathbf{F}_F\mathbf{f}_k + \mathbf{F}_D\mathbf{d}_k + \mathbf{F}_N\boldsymbol{\nu}_k \end{aligned} \tag{8.85}$$

The residual responses are specified such that detection and diagnosis are enhanced. For additive faults and disturbances (noise and multiplicative faults are neglected) define the specifications as

$$\mathbf{r}(t) = \mathbf{Z}_F(q)\mathbf{f}(t) + \mathbf{Z}_D(q)\mathbf{d}(t) \tag{8.86}$$

Comparing the internal residual expression Eq. (8.83) (ignoring the noise term) and the specification in Eq. (8.86), one can deduce that

$$\mathbf{W}(q)\ [\mathbf{S}_F(q) \quad \mathbf{S}_D(q)] = [\mathbf{Z}_F(q) \quad \mathbf{Z}_D(q)]\ . \tag{8.87}$$

The residual generator is obtained by solving Eq. (8.87) for $\mathbf{W}(q)$. Detailed examples in [189] illustrate the technique and its extensions with multiplicative faults and disturbances. Other extensions include integration of parity relation design and residual evaluation with GLR test and whitening filters for FDD of dynamic stochastic processes [464]. An implementation of this approach to continuous pasteurization systems and comparison of parity space approach with a statistical approach that combines T^2 and SPE tests with contribution plots illustrates the strengths and limitations of both techniques [292].

FDD with Kalman Filters and Diagnostic Observers

The basic concept is to generate residuals for FDD by comparing measurements and their estimates computed using Kalman filters or observers. The estimation errors of observers or innovations of Kalman filters are used as residuals. Consider an observer for the deterministic process without faults and disturbances, the Kalman filter being used for the stochastic case that includes noise

$$\begin{aligned} \mathbf{x}_{k+1} &= \mathbf{F}\mathbf{x}_k + \mathbf{G}\mathbf{u}_k \\ \mathbf{y}_k &= \mathbf{C}\mathbf{x}_k \end{aligned} \tag{8.88}$$

The observer with a gain matrix $\mathbf{K}_{ob}$ has a structure similar to Kalman filters discussed in Section 4.3.2, viz.,

$$\begin{aligned} \hat{\mathbf{x}}_{k+1} &= \mathbf{F}\hat{\mathbf{x}}_k + \mathbf{G}\mathbf{u}_k + \mathbf{K}_{ob}(\mathbf{y} - \mathbf{C}\hat{\mathbf{x}}_k) \\ \hat{\mathbf{y}}_k &= \mathbf{C}\hat{\mathbf{x}}_k \ . \end{aligned} \tag{8.89}$$

The relations for the state estimation error $\boldsymbol{\epsilon} = \mathbf{x} - \hat{\mathbf{x}}$ and the output estimation error $\mathbf{e} = \mathbf{y} - \hat{\mathbf{y}}$ for the system with faults and disturbances become

$$\begin{aligned} \epsilon_{k+1} &= (\mathbf{F} - \mathbf{K}_{ob}\mathbf{C})\epsilon_k + \mathbf{G}\mathbf{u}_k + \mathbf{E}_F\mathbf{f}_k + \mathbf{E}_D\mathbf{d}_k - \mathbf{K}_{ob}\mathbf{F}_F\mathbf{f}_k - \mathbf{K}_{ob}\mathbf{F}_D\mathbf{d}_k \\ \mathbf{e}_k &= \mathbf{C}\epsilon_k + \mathbf{F}_F\mathbf{f}_k + \mathbf{F}_D\mathbf{d}_k \end{aligned} \tag{8.90}$$

when Eq. (8.88) is augmented with faults and disturbances by adding $\mathbf{E}_F\mathbf{f}_k + \mathbf{E}_D\mathbf{d}_k$ to the state equation and $\mathbf{F}_F\mathbf{f}_k + \mathbf{F}_D\mathbf{d}_k$ to the output equation. Note that the term $\mathbf{G}\mathbf{u}_k$ drops out because of the subtraction in deriving ϵ, and the estimation errors are independent of deterministic inputs $\mathbf{u}_k$. The output estimation error $\mathbf{e}$ can be used as the residual $\mathbf{r}$ for FDD. In the absence of faults ($\mathbf{f} = \mathbf{0}$), $\mathbf{r}$ is influenced only by unknown inputs $\mathbf{d}$ and noise that is not included in the process model. Faults can be detected by setting up threshold values for $\mathbf{r}$ (greater than zero to avoid false alarms due to noise and small disturbances) and developing some FDD logic.

Various observer and Kalman filter configurations have been considered to detect and diagnose multiple faults. One configuration is based on *multiple hypotheses testing* where a bank of estimators are designed such that each estimator is designed for a different fault hypothesis. For example, $\mathcal{H}_0$ would be no faults, $\mathcal{H}_1$, bias in sensor 1, $\mathcal{H}_2$ zero output in sensor 1, etc. The hypotheses are tested in terms of likelihood functions. The *dedicated observer* configuration has multiple estimators where each estimator is driven by a different single sensor output to estimate as many components of the output vector $\mathbf{y}$ as possible. When a certain sensor fails, the output estimate given by the corresponding estimator will be erroneous. FDD of multiple simultaneous faults is carried out by checking values of structured sets of estimation errors [106]. The *generalized observer* approach uses a bank of estimators where each estimator is dedicated to a certain sensor. Each estimator receives process information from all other sensors except the sensor whose reading is being estimated. The residuals are checked using threshold logic to diagnose a faulty sensor. Reduced-order or nonlinear estimators can also be used to develop FDD systems with Kalman filters and diagnostic observers. The equivalence between parity relation based and diagnostic observer based FDD has been shown [162, 188].

FDD Using Robust Observers for Unknown Inputs

Deterministic observers and filters were used in the previous section to estimate state variables and outputs. The effect of disturbances and noise were accounted for by using nonzero threshold limits for residuals. Robust observers can be designed by including disturbances [163] and both disturbances and noise [427]. To illustrate the methodology and design challenges, robust residuals generation using unknown deterministic input (disturbance) observers [163] are discussed. Consider the process model

$$\begin{aligned} \mathbf{x}_{k+1} &= \mathbf{F}\mathbf{x}_k + \mathbf{G}\mathbf{u}_k + \mathbf{E}_F\mathbf{f}_k + \mathbf{E}_D\mathbf{d}_k \\ \mathbf{y}_k &= \mathbf{C}\mathbf{x}_k + \mathbf{F}_F\mathbf{f}_k + \mathbf{F}_D\mathbf{d}_k \end{aligned} \tag{8.91}$$

Define a linear transformation

$$\mathbf{z}_k = \mathbf{T}\mathbf{x}_k \tag{8.92}$$

for the fault free system and the robust unknown input observer

$$\mathbf{z}_{k+1} = \mathbf{R}\mathbf{z}_k + \mathbf{S}\mathbf{y}_k + \mathbf{J}\mathbf{u}_k \tag{8.93}$$

with the residual

$$\mathbf{r}_k = \mathbf{L}_1\mathbf{z}_k + \mathbf{L}_2\mathbf{y}_k \tag{8.94}$$

such that if $\mathbf{f}_k = \mathbf{0}$ then $\lim_{k\to\infty}\mathbf{r}_k = 0$ for all $\mathbf{u}$ and $\mathbf{d}$, and for all initial conditions $\mathbf{x}_0$ and $\mathbf{z}_0$. If $\mathbf{f}_k \neq \mathbf{0}$, then $\mathbf{r}_k \neq \mathbf{0}$. The estimation error equation for the observer is

$$\begin{aligned} \mathbf{e}_{k+1} &= \mathbf{z}_{k+1} - \mathbf{T}\mathbf{x}_{k+1} \qquad (8.95) \\ &= \mathbf{R}\mathbf{z}_k + \mathbf{S}\mathbf{y}_k + \mathbf{J}\mathbf{u}_k - \mathbf{T}\mathbf{F}\mathbf{x}_k - \mathbf{T}\mathbf{G}\mathbf{u}_k - \mathbf{T}\mathbf{E}_F\mathbf{f}_k - \mathbf{T}\mathbf{E}_D\mathbf{d}_k \end{aligned}$$

Substituting for $\mathbf{x}_k$ and $\mathbf{y}_k$, and imposing that the error should be independent of state variables, control inputs and disturbances, the following equations are established:

$$\begin{aligned} \mathbf{TF} - \mathbf{RT} &= \mathbf{SC} \\ \mathbf{J} &= \mathbf{TG} \\ \mathbf{TE}_D &= \mathbf{0} \\ \mathbf{SF}_D &= \mathbf{0} \\ \mathbf{TE}_F &\neq \mathbf{0} \\ \mathbf{SF}_F &\neq \mathbf{0} \end{aligned} \tag{8.96}$$

where the last two equations ensure that the residual is nonzero if there is a fault ($\mathbf{f}_k \neq \mathbf{0}$). The equations for $\mathbf{y}_k$ and $\mathbf{r}_k$ and Eq. (8.95) lead to

$$\begin{aligned} \mathbf{L}_1\mathbf{T} + \mathbf{L}_2\mathbf{C} &= \mathbf{0} \\ \mathbf{L}_2\mathbf{F}_D &= \mathbf{0} \\ \mathbf{L}_2\mathbf{F}_F &\neq \mathbf{0} \end{aligned} \tag{8.97}$$

The solution to Eqs. (8.96-8.97) is obtained by supplying the matrices $\mathbf{R}$, $\mathbf{S}$, $\mathbf{J}$, $\mathbf{L}_1$, and $\mathbf{L}_2$. The solution details are given in [456]. However, Eq. (8.94) (nonzero and zero values for $\mathbf{r}_k$) and Eq. (8.92) may not exactly be satisfied in many practical situations, and optimal approximations may be needed [163]. The procedure for finding optimal approximate solutions have been proposed [163].

8.3.2 FDD Based on Model Parameter Estimation

The parameters of a model describing the dynamic behavior of a process change when the operation of the process varies significantly. If a process model is developed for normal process operation, the model parameters can be re-estimated when new data are collected and compared with nominal values of model parameters. Significant deviations in the values of model parameters indicate the presence of faults or disturbances or modification of the operating point of the process. If the last two possibilities are eliminated, then changes in parameter values indicate faults.

The model of the process may be constructed from first principles. Then the parameters that depend on process operation should be determined and those parameters should be estimated when new data are collected [251]. Another alternative is to develop a time series type of model (Section 4.3.1). Then, changes in model parameters over time are monitored for FDD.

A procedure suggested for implementing this approach in a deterministic framework is outlined [251]. Consider a process model described by linear input/output differential equations with constant coefficients

$$a_n y^{(n)}(t) + \cdots + a_1 \dot{y}(t) + y(t) = b_m u^{(m)}(t) + \cdots + b_0 u(t) \tag{8.98}$$

where $y^{(n)}$ indicates the nth derivative of y. The model parameters are collected in a vector θ

$$\theta = [1 \; a_1 \; \cdots \; a_n \; b_0 \; \cdots \; b_m]^T \; . \tag{8.99}$$

Determine relationships between model parameters θ_i and physical parameters ϕ_j

$$\theta = \mathbf{f}(\phi) \; . \tag{8.100}$$

Identify model parameters θ from process data $(\mathbf{u}, \mathbf{y})$. Then, determine the physical parameter values using the inverse relationship $\phi = \mathbf{f}^{-1}(\theta)$ and compute changes in ϕ, $\Delta\phi$. Use threshold logic or other tools to determine the magnitude of changes in $\Delta\phi$ and presence of faults.

A more general framework can be established for modeling the changes in the eigenstructure of a data-based model in state space form ($\mathbf{F}$ matrix of discrete-time equation such as Eq. (8.49)) or time series form (AR or ARMA model). The version discussed below will provide detection of change in univariate systems. Extension to multivariable processes has been developed [45]. Additional steps are necessary for diagnosis if multiple faults are possible. For the case of additive changes, the cumulative sum to be computed becomes

$$S_m^n(p_{\theta_0}, p_{\theta_1}) = \sum_{k=m}^{n} \log \frac{p_{\theta_1}(y_k|\mathcal{Y}_{k-1})}{p_{\theta_0}(y_k|\mathcal{Y}_{k-1})} \tag{8.101}$$

where p_{θ_1} reflects the change of magnitude δ at time r and the stacked output values are

$$\mathcal{Y}_{k-1} = [y_{k-1}\ y_{k-2}\ \cdots\ y_1]^T\ . \tag{8.102}$$

The GLR is

$$\Lambda_k(p_{\theta_0}, p_{\theta_1}) = \max_{1\le r\le k} \max_{\theta_1} S_r^k(p_{\theta_0}, p_{\theta_1}) \tag{8.103}$$

and the GLR test becomes

$$\begin{aligned} \mathcal{H}_0 &: \quad \Lambda_k(p_{\theta_0}, p_{\theta_1}) < \tau \quad \text{no change at time } k \\ \mathcal{H}_1 &: \quad \Lambda_k(p_{\theta_0}, p_{\theta_1}) \ge \tau \quad \text{change at time } k \end{aligned} \tag{8.104}$$

Significant savings in computation time can be generated by using a two-model approach [50]. For illustration, consider a two-model approach for on-line detection of change in scalar AR models

$$y_k = \sum_{i=1}^{n} a_i^{(p)} y_{k-i} + \epsilon_k^{(p)} \tag{8.105}$$

where ϵ_k is Gaussian white noise with variance $\sigma_{(p)}^2$ and for $i = 1, \cdots, p$

$$a_i^{(p)} = \begin{cases} a_i^0 & \text{for} \quad n \le r-1 \\ a_i^1 & \text{for} \quad n \ge r \end{cases} \tag{8.106}$$

$$\sigma_{(p)}^2 = \begin{cases} \sigma_0^2 & \text{for} \quad n \le r-1 \\ \sigma_1^2 & \text{for} \quad n \ge r\ . \end{cases}$$

Define the parameter vectors

$$\theta^p = [a_1^p\ \cdots\ a_n^p\ \sigma_p^2] \qquad p = 0, 1\ . \tag{8.107}$$

The on-line change detection can be formulated as a GLR test using Eqs. (8.101-8.104). If the AR model M_0 (with parameter vector θ^0) for the no change hypothesis is not known, identify it with a recursive growing memory filter. For each possible change time r, identify the after change AR model M_1 using data for the time window [r,k] and compute the log-likelihood ratio S_r^k. Maximize S_r^k over r. Simplifications for saving computation time and other distance measures between models M_0 and M_1 are discussed in [50].

Parameter Change Detection (PCD) method for SPM of Strongly Autocorrelated Processes. The model parameter estimation paradigm is a powerful change detection method for strongly autocorrelated processes. The SPM framework based on time series model *forecasts* introduced by Alwan and Roberts [16] is one of the most widely used approaches to handle the SPC of processes with autocorrelated data [402, 220]. In this framework, a time series model that describes the autocorrelated process behavior is determined from either some preliminary information or a data set collected when the process was in a state of statistical control. The *residuals* are generated from the difference of actual measurements and one-step-ahead predictions computed by using this model. The estimation of the one-step-ahead minimum variance forecasts can also be formulated by using a state-space form of the time series model, where the states are the one-step-ahead forecasts. The optimal estimation is given by a Kalman filter. In this approach, change detection in the autocorrelated signal is converted to a change detection in residuals that have suitable statistical properties such as *iid* which permit the use of standard SPC charts. Generalized likelihood ratio was also used to develop process monitoring schemes based on residuals [45, 677]. Monitoring of forecast residuals have not proven to be very useful SPC tools especially for *highly* positively correlated time series models. The ability to make correct decisions gets worse particularly when the *AR* part of the model has roots close to the unit circle [220]. This is frequently encountered in process variables that are under feedback control. The behavior of the controlled variable is dominated by the closed-loop dynamics that include the feedback controller. Usually the controller has an integrator by design in order to compensate for steady-state offset, and the integral action yields roots of magnitude one.

An alternative SPM framework can be developed by monitoring the variations in model parameters that are updated at each new measurement instant. Sastri [534] used such an approach together with the concept of discounted recursive least-squares also known as recursive weighted least-squares (RWLS). An extension of this approach was called *parameter change detection* (PCD) method for monitoring autocorrelated processes

[414]. The new features of the PCD method include use of recursive variable weighted least squares (RVWLS) with adaptive forgetting, and an implicit parametrization scheme that estimates the process level at each sampling instant. RVWLS parameter updating with adaptive forgetting provides better tracking of abrupt changes in the process parameters than the usual RWLS updating, and it reduces the number of false detections of change as well. The detection capabilities of PCD method are superior to methods based on forecast residuals for highly positively correlated processes. As autocorrelation increases, the improvement of PCD over residuals based SPM methods becomes more significant. The PCD method possesses several attractive features for on-line, real time operation: its computations are efficient, its implementation is easy, and the resulting charts are clearly interpretable. The implicit parametrization feature of PCD provides a statistic for the process level (mean) which is used to detect and distinguish between changes in level and *eigenstructure* of a time series. Model eigenstructure is determined by the roots (or eigenvalues) of a model. It is related to the order and parameter values of AR or $ARMA$ models, and has a direct effect on the level (bias) of the variable described by the model and its variance (spread). Based on the values assigned by PCD to various indicators one can determine if an eigenstructure change has occurred and if so whether this involves a level change, a spread change, or both. The outcome of implicit parametrization confirms the existence or lack of a level change, and provides the magnitude of the level change [414, 411].

8.3.3 FDD with Hidden Markov Models

Hidden Markov models provide a modeling framework when the state of a system can be inferred from some measured variables (observations) without direct knowledge of the state variables. It is a double stochastic process in the sense that both the observations and the states are stochastic [279, 484]. The discussion focuses first on *discrete-time Markov processes.* Then, HMMs, their parameters, and the fundamental problems in developing an HMM are summarized. Finally, some applications are presented.

Discrete-Time Markov Processes

Consider a process that can be in any one of N distinct states $S = \{s_1, s_2, \cdots, s_N\}$ at any time. The process state changes at regularly spaced times indexed by t, and the actual state at time t is denoted by q_t. The evolving sequence of states are $Q = \{q_1, q_2, \cdots, q_t\}$ and q_t belongs to one of the states in S. A full probabilistic description of the process may necessitate specification of the current and some of the preceding states. A special case that is similar to state-space models requires only the immediate previous state. Called discrete-time, first order Markov chain, this special case is represented as $P[q_t = s_j | q_{t-1} = s_i]$. If the state transition probabilities a_{ij} from state i to state j are not time dependent

$$a_{ij} = P[q_t = s_j | q_{t-1} = s_i] \qquad 1 \leq i, j \leq N \tag{8.108}$$

with the constraints

$$a_{ij} \geq 0 \quad \text{for all } i, j \qquad \sum_{j=1}^{N} a_{ij} = 1 \quad \text{for all } i. \tag{8.109}$$

This process is an observable Markov model because the process outputs are the set of states and each state corresponds to a deterministically observable event. The outputs in any given state are not random. A simple Markov chain with three states is presented in Figure 8.9.

Hidden Markov Models

Consider the case where the stochastic process is observed only through a set of stochastic processes that produce the sequence of observations. The states are not directly observable, they are inferred from the observations. An example would be a process consisting of a few containers that are filled with marbles of multiple colors. Each container has marbles of all colors, but the fraction of marbles of a certain color in each container varies. At each observation time, a marble rolls out through a channel that connects all containers, but the observer does not see the container that dispenses the marble and does not know the rule that selects the container that dispenses the marble. The dispensing of marbles generates a finite observation sequence of colors which can be modeled as the observable output of an HMM. A simple HMM of this process would have each state corresponding to one of the containers and for which a marble color probability is defined for each state. The choice of containers is determined by the state transition matrix $\mathbf{A} = [a_{ij}]$ of the HMM.

The elements of the HMM as depicted in Figure 8.10, include [279, 484]:

N *The number of states* in the model. The model is called *ergodic* if any state can be reached from any other state. Generally all states are interconnected. Denote the state at time t as q_t.

M *The number of distinct observation symbols per state*, the alphabet size. The individual symbols are denoted as $\mathbf{V} = \{\mathbf{v}_1, \cdots, \mathbf{v}_M\}$

$\mathbf{A} = \{a_{ij}\}$ *The state probability distribution* where

$$a_{ij} = P[q_t = s_j | q_{t-1} = s_i] \qquad 1 \leq i, j \leq N \tag{8.110}$$

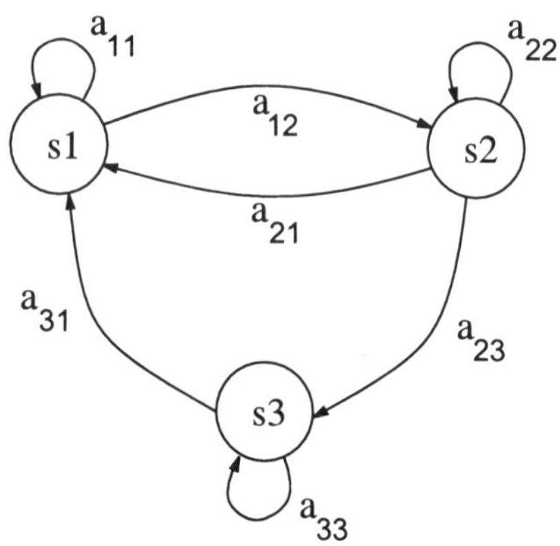

Figure 8.9. A Markov chain with three states (labelled $s1$, $s2$ and $s3$) and selected transitions (a_{13} and a_{32} set to 0).

$\mathbf{B} = \{b_j(k)\}$ *The observation symbol probability distribution*, where

$$b_j(k) = P[\mathbf{o}_t = \mathbf{v}_k | q_t = s_j] \qquad 1 \leq k \leq M \tag{8.111}$$

defines the symbol distribution in state s_j, for $1 \leq j \leq N$.

$\mathbf{C} = \{c_i\}$ *The initial state distribution* $\{c_i\}$ also called the initial state occupancy probability

$$c_i = P[q_1 = s_i] \qquad 1 \leq i \leq N \tag{8.112}$$

A complete specification of an HMM requires specification of two model parameters (N and M), observation symbols, and three sets of probability measures (**A**, **B**, and **C**). The set of probabilities is written compactly as $\lambda = (\mathbf{A}, \mathbf{B}, \mathbf{C})$. This parameter set defines a probability measure for **O**, $P(\mathbf{O}|\lambda)$.

Given the values of N, M, **A**, **B** and λ, the HMM can generate an observation sequence $\mathbf{O} = (\mathbf{o}_1 \ \cdots \ \mathbf{o}_T)$ where each observation $\mathbf{o}_t$ is one of the symbols in **V**, and T is the total number of observations in the sequence.

The three basic problems for HMMs to develop a useful model for FDD are:

1. Given the observation sequence $\mathbf{O} = (\mathbf{o}_1 \ \cdots \ \mathbf{o}_T)$ and a model $\lambda = (\mathbf{A}, \mathbf{B}, \mathbf{C})$, how can the probability of the observation sequence $P(\mathbf{O}|\lambda)$ be computed efficiently? The solution provides a measure of similarity (goodness of fit) between the observation sequence and a sequence

that would be generated with the given model. Hence, if these sequences are similar, one may accept that the model used is describing the process that generated these observations.

2. Given the observation sequence $\mathbf{O} = (\mathbf{o}_1 \; \cdots \; \mathbf{o}_T)$ and a model $\lambda = (\mathbf{A}, \mathbf{B}, \mathbf{C})$, how can the most likely state sequence $\mathbf{q} = (q_1 \; q_2 \; \cdots \; q_T)$ that "explains" best the given observation sequence $\mathbf{O}$ be determined? This is an attempt to find the hidden part of the model, the "correct" state sequence. In general, there is no analytic solution and usually a solution based on some optimality criteria is obtained.

3. How are the model parameters $\lambda = (\mathbf{A}, \mathbf{B}, \mathbf{C})$ adjusted to maximize $P(\mathbf{O}|\lambda)$? The parameter re-estimation is carried out using a set of observation sequences called training data in an iterative manner until the model parameters maximize $P(\mathbf{O}|\lambda)$.

The HMM is formulated in two stages: *training* stage that solves Problem 3, and *testing* stage that addresses Problems 1 and 2. Problem 1 is solved using either the forward or the backward computation procedure, Problem 2 is solved using the Viterbi algorithm, and Problem 3 is solved using the Expectation-Maximization (EM) (also called Baum-Welch) method [279, 484]. Details of the algorithms, implementation issues, and illustrations are given in both references.

Pattern recognition systems combined with finite-state HMMs have been used for fault detection in dynamic systems [557]. The HMM parameters are derived from gross failure statistics. Wavelet-domain HMMs have also been proposed for feature extraction and trend analysis [671]. A wavelet-based smoothing algorithm filters high-frequency noise. A trajectory shape analysis technique called triangular episodes converts the smoothed data into semi-qualitative mode and membership functions transforms the information to a symbolic representation. The symbolic data is classified with a set of sequence matching HMMs for trend analysis. This approach is extended to detection and classification of abnormal process situations using multidimensional hidden Markov *trees* [37, 579]. The case studies discussed in these publications illustrate the application of the method to various continuous processes.

8.4 Model-free Fault Diagnosis Techniques

The traditional model-free FDD method relies on *physical redundancy* created by multiple sensors that measure the same variable. It is used in the measurement of critical process variables. Significant difference between

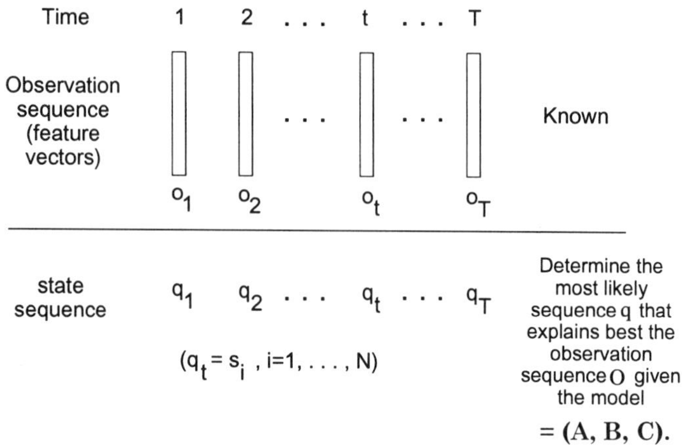

Figure 8.10. HMM structure.

different sensor readings indicate a sensor fault. When three or more sensors are used, a voting mechanism may be established to diagnose the faulty sensor(s). Another hardware-based FDD is self-diagnosing or smart sensors that can check the correctness of their own operation. These approaches are usually more expensive than FDD based on analytical redundancy, but the cost may be justifiable for mission critical measurements and equipment.

A simple model-free FDD is based on *limit checking*. Each measurement is compared to its upper and lower (preset) limits, and exceeding the limits indicates a fault. The limit checking approach can be made more elaborate by defining warning and alarm limits, and by monitoring time trends (run rules in univariate SPC in Section 6.1.1). An important disadvantage of the limit checking approach is the need to interpret the alarms generated. A single disturbance that travels through the process can generate many alarms. Extensive process knowledge and process operation experience is necessary to determine the source cause of the alarms. Knowledge-based systems (KBS) can automate alarm interpretation.

Logic reasoning using ladder diagrams and hard-wired systems have been useful for FDD in the latter part of the 20th century. Recently, fuzzy logic and FDD based on fuzzy logic has gained popularity. Fuzzy logic systems are discussed in Section 8.4.1 as an integral part of KBS.

Software based logic reasoning, especially real-time KBSs have become more abundant with the increase of computation power and reduction of computer costs. Object oriented real-time KBSs and their use in logic reasoning are discussed in Section 8.4.1. KBSs can also provide a supervisory operation to integrate various monitoring and diagnosis activities. Real-time supervisory KBSs that integrate statistical process monitoring, generation and interpretation of contribution plots, and FDD are presented in Section 8.4.2.

8.4.1 Real-time Knowledge-Based Systems (RTKBS)

Chemical process industries (CPI) require a high level of supervision in real-time. Supervision tasks may include scheduling processing stages, supervising data acquisition, distributed control systems, and alarm management. This means low level process operations such as adjustment of PID control settings and high level qualitative decisions such as implementing different operational policies and fault handling are to be dealt with together. All these activities are realized with accumulated expertise over the years. Experience of process operators and engineers is an invaluable asset and should be incorporated in an automated supervisory system. Real-time knowledge-based systems (RTKBS) provide such an environment where a high level automated process supervision can be achieved.

KBSs have been one of the rapidly growing applications of *Artificial Intelligence* (AI) in the scientific and engineering arena during the last two decades. A KBS is a computer program that emulates the decision-making ability of a human expert. The terms KBS and *Expert Systems* are often used synonymously. In this book we will use the term KBS.

Figure 8.11 illustrates the general framework and the common components of a KBS [191]. The knowledge-base contains facts, rules and heuristics that are used by the *inference engine* to draw conclusions. The user interfaces with the KBS by using an interface to input information or learn the conclusion reached by the KBS. Many algorithms have been proposed for inferencing by AI researchers [502, 428]. In the context of KBSs for

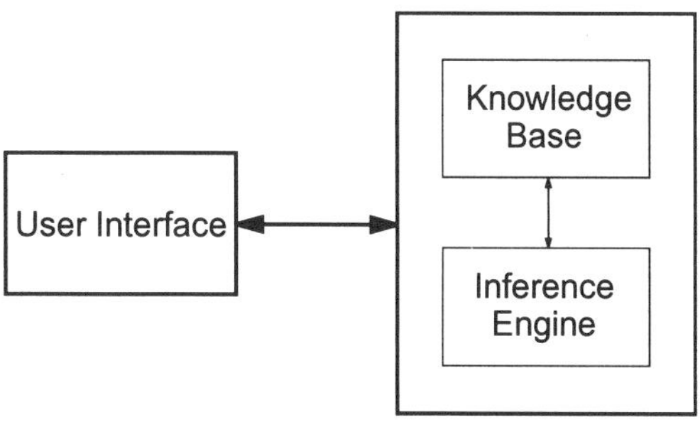

Figure 8.11. Basic structure of a KBS.

supervising process operations, *forward-chaining* is used to update all measurement information and derived variables when new process information is made available to the KBS. Then, *backward-chaining* is used to infer the status of process operation (normal or faulty), diagnose the source cause of abnormal operation, and formulate the proper intervention.

The process of building a KBS, developing and implementing a problem-solving strategy is called *knowledge engineering*. The knowledge of an expert or a group of experts is transferred into KBS by knowledge engineer. The customary way of performing this task is to repeat a cycle of interviewing the expert(s), constructing a prototype, testing and re-interviewing that is a very time consuming and laborious task. This task is called knowledge acquisition and elicitation [221]. Recently, more effective systematic techniques have been proposed for knowledge acquisition and elicitation [193] and techniques for automatic knowledge acquisition have been suggested [46, 513]. In the early days of the technology, knowledge engineers had to develop the entire system from scratch by using one of the available AI programming languages such as LISP (LISt Processing language), Prolog (*Prog*ramming in *Log*ic), Smalltalk, OP5 (its current version is OP83) and NASA's CLIPS software. Today's KBS development software such as Gensym's G2, make the development easier. One of the major differences between conventional programming languages such as FORTRAN and C and AI programming languages is that the former rely on the numbers and algorithms while the latter are designed over symbols, lists and searches. KBSs use inferences to achieve a reasonable solution that is the best that can be expected based on data, facts and rules currently available, in contrast to a numerical optimization approach based on an objective function, process model, constraints to equations, and numerical optimization algorithm.

Several types of knowledge are used in a KBS. Most of the early KBSs for CPI are developed using *shallow knowledge* which is based on empirical and heuristic knowledge [303, 314, 492, 617]. Heuristics are rules of thumb or empirical knowledge gained by experience which may provide a quick solution to a specific problem by relating the symptoms with causes without using a system model. *Deep knowledge* is based on the basic structure, function and behavior of a process such as underlying physiological phenomena about microbial activities in fermentations (a process model in a mathematical form). Shallow and compiled knowledge provide the basis for various kinds of knowledge. Rules formed from information derived from deep knowledge are called *compiled knowledge* and rules derived from shallow and compiled knowledge are called *rule-based knowledge*. Several KBSs have been proposed for FDD based on these knowledge abstractions in CPI [467, 619].

Rule-based knowledge representation became dominant in early KBSs developed in many fields. The initial number of rules to be retained in the KBS depends on the complexity of the process and the amount of knowledge acquired from experts. One of the advantages of a KBS is that rule library can be expanded by adding new rules as they become available. As the number of rules increases, the use of information becomes challenging. Some commercial KBSs had tens of thousands of rules for knowledge representation and inferencing, putting on an enormous burden on the execution of the software. Recent KBS shells such as G2 of Gensym Inc., and Nexpert Object of Neuron Data have adopted a hybrid structure based on object based systems. The object framework is used to develop classes, objects, and instances to represent knowledge, and rules are used for inferencing. This results in significant reduction in the number of rules and increase in computation speed. Rules are conditionally true and can be cast into IF-THEN statements such as

IF *the substrate consumption rate is lower than that expected and the fermentation is in the fed-batch operation mode*
THEN *the flow of substrate feed rate is high.*

Object-based knowledge representation is another technique in which field of knowledge representation, the object is the central notion. The design of G2 knowledge representation relies on this technique. Knowledge is here expressed by means of two kinds of objects: (1) *classes* (which describe families of individuals), and (2) *instances* (which describe the individuals). Classes are organized in hierarchies by a specialization relation upon which an inheritance mechanism is settled. This mechanism allows a more specific sub-class to inherit from all the properties of it super-class it does not redefine. Inference mechanisms are also proposed in order to complete knowledge; default value, classification, and procedural attachment. Classification is a central mechanism which determines for an instance the set of sub-classes of its current class to which it also could be linked. Procedural attachment consists in specifying a peace of code to be executed in order to obtain the value of a property in a class, if needed [183, 184].

A number of KBS are proposed in early nineties for knowledge-based (KB) process control and control systems design in CPI [43, 56, 274, 601]. KB control technologies are also proposed for bioprocesses and fermentation industries [1, 27, 236]. A review of knowledge-based control systems for fermentations is given by Konstantinov and Yoshida [288] where they summarize the functions of a supervisory KBS for fermentation control as

1. *Input data validation.* KB is structured such that contradictory measurements with respect to previous fermentation can be identified.

2. *Identification of the state of the cell culture.* On-line detection and evaluation of physiological phase of the cell population is one of most challenging tasks to be achieved. Phase specific control activities can then be performed.

3. *Detection and diagnostics of instrumentation faults.* Instrument measurements should be closely monitored and failures detected/diagnosed by the KBS.

4. *Supervision of conventional control.* Phase detection type of high level decisions are used to change low level control parameters.

5. *Communication with user.* KBS should be able to inform the user (operator) about the process, explain its activities and give advice.

6. *Plantwide supervision and scheduling.* KBS should be extended to perform supervision and scheduling activities for upstream and downstream processes.

On-line estimation of infrequently measured variables and prediction of product quality variables can also be added to the list above. Achieving all of these tasks in a KBS environment can be realized with the combined use of different techniques. For example, ANNs (Section 4.6), are integrated with fermentation KBS for estimating state variables such as biomass concentration [28, 490].

A variety of RTKBS applications can be found for bioprocesses including novel interface design [580], extended Kalman filter integration for on-line state estimation [399], use of qualitative physics for behavior monitoring [574] and simple rule-based intensive designs [205]. Successful development and implementation of RTKBS using Gensym's G2 for supervision of industrial fermentation plants are reported [13, 14]. This application which contained approximately 300 rules and integrated with large plant databases is credited for increasing the plant yield by 4%, reducing process variability by more than 10% and saving more than 10 production fermentation batches from total loss over a period of two years [14].

Fuzzy set theory (widely known as fuzzy logic (FL)) has also received attention during the last decade in control applications and integrated with RTKBSs [466, 543]. A brief introduction of FL is presented next in conjunction with its use in fermentation technologies and RTKBS integration. Techniques and applications for using integrated ANN-FL controllers have also been reported [280, 293, 342].

Fuzzy logic and its integration with KBS for supervision of fermentation processes

FL is inspired by the way human thinking deals with inexact uncertain information and can be interpreted as the generalization of classical set theory. In classical set theory, a set is comprised of a finite or infinite number of elements belonging to some specified set called *universe of discourse.* An element x of the universe of discourse (U) may belong to a set A which is included in U so that

$$A = \{(x, \mu_A(x))\,|x \in U\} \tag{8.113}$$

where the membership function (or characteristic function) is defined as (Figure 8.12(a))

$$\mu_A(x) = \begin{cases} 0, & x \in A \\ 1, & x \notin A. \end{cases} \tag{8.114}$$

This is a *crisp* or Boolean description. FL is based on Fuzzy Set Theory which was introduced by Zadeh in 1965 [680]. A *fuzzy set* is a generalization of a classic set so that it allows the degree of membership for each element in a range over, say closed unit interval [0, 1]. A fuzzy set $\tilde{A}$ (also called as *support set*) in the universe of discourse U can be defined as a set of ordered pairs,

$$\tilde{A} = \{(x, \mu_{\tilde{A}}(x))\,|x \in U\} \tag{8.115}$$

where $\mu_{\tilde{A}}(x)$ is called the *membership function* of set $\tilde{A}$ and it maps each element of the universe of discourse to its range space that is the unit interval in most cases [342]. A variety of membership functions illustrated in Figures 8.12(b) and 8.12(c) can be used.

Consider the classical Boolean description of the level of temperature in a fermenter: the temperature is *high* or *low* based on a reference point. In contrast, FL defines vague qualifiers such as *quite high, very high, rather high, rather low, very low, quite low* on temperature. Figure 8.12(d) illustrates how the linguistic (fuzzy) variable 'temperature' is mapped for a few of its values onto the universe of discourse (temperature scale in this example) for a range of [0, 100 ^oC] through linguistic descriptors and their assigned values. For the fuzzy value $Very_{Low}$ for instance, the mapping is described in terms of a set of positive integers in the range [0, 100 ^oC]. The support set ($\tilde{A}$) expresses the degree to which the temperature is considered $Very_{Low}$ over the range of all possible temperatures in discrete values specified in degrees Centigrade using $\mu_{\tilde{A}}(x)$ such that

$$\mu_{\tilde{A}}(0) = \mu_{\tilde{A}}(5) = 1, \quad \mu_{\tilde{A}}(10) = \mu_{\tilde{A}}(15) = 0.8 \quad \mu_{\tilde{A}}(20) = \mu_{\tilde{A}}(25) = 0.6,$$
$$\mu_{\tilde{A}}(30) = \mu_{\tilde{A}}(35) = \ldots, \mu_{\tilde{A}}(100) = 0. \tag{8.116}$$

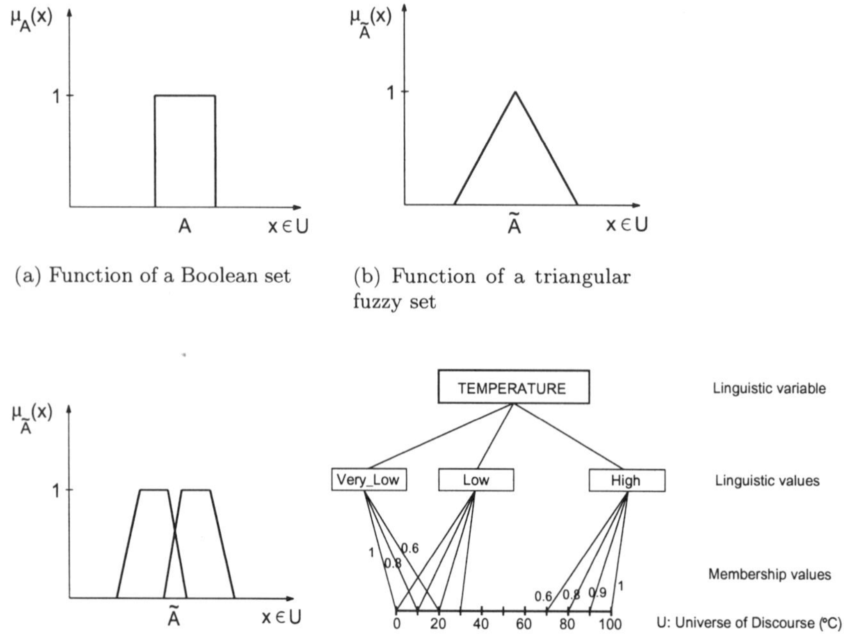

(a) Function of a Boolean set

(b) Function of a triangular fuzzy set

(c) Function of trapezoidal fuzzy sets

(d) The linguistic variable temperature and some of its values

Figure 8.12. Examples of membership functions and mapping of a linguistic variable (d) [280].

This relation is usually represented in the following more compact form

$$\tilde{A} = \mu_{\tilde{A}}(x_1)/x_1 + \mu_{\tilde{A}}(x_2)/x_2 + \ldots + \mu_{\tilde{A}}(x_n)/x_n = \sum_{i=1}^{n} \mu_{\tilde{A}}(x_i)/x_i \quad (8.117)$$

where '+' denotes the union of elements (/ does not indicate division) and $\mu_{\tilde{A}}(x_i)$ is the grade of membership of x_i for n membership values. For the temperature example, Eq. 8.116 becomes

$$\tilde{A} = 1/0 + 1/5 + 0.8/10 + 0.8/15 + 0.6/20 + 0.6/25 + \ldots + 0/100. \quad (8.118)$$

Theory and applications of FL, and integration of KBS and FL for control of fermentation processes are discussed in the literature [285]-[422].

Control of Fermentation Processes using an Integrated KBS-FL System. Konstantinov and Yoshida proposed a methodology (Figure 8.13) for detection of the physiological state of fermentation and control of bio-processes based on expert identification of the physiological state of a cell population [285, 286, 287] using FL, RTKBS and temporal reasoning [465]. The physiological state (PS) vector ($\mathbf{x}$) is defined quantitatively by a set of on-line measured variables ($\mathbf{u}$, $\mathbf{y}$) such as ammonia flow rate and substrate feed rate that are used to calculate variables such as specific oxygen to substrate consumption rate, forming the physiological state-space of the culture. Based on the practical experience on the process, a finite number of physiological situations (PSN) are defined where the physiological characteristics of the cell population and its reactions to different control actions are well known. The description of PSNs is mostly in qualitative terms such as "situation of optimal productivity." Hence, PSNs can be interpreted as fuzzy variables. When the process passes from one state to the next, it often exhibits variation in structure behavior and therefore proper alteration in control strategies is required for each state. Adaptive weighting of the membership functions is also introduced to give more importance to certain physiological states. The synthesis of physiological recognition algorithm consists of the development of a decision procedure in which qualitatively defined PSNs are related quantitatively to PS vector $\mathbf{x}$. State recognition is performed by means of expert decision rules using fuzzy sets defined over PSNs. An example of a decision rule relating the current PS vector $\mathbf{x}$ and PSNs is:

IF $\mathbf{x}_1$ *is high and* $\mathbf{x}_2$ *is low*
THEN *the current* $\mathbf{x}$ *belongs to* PSN_1 *with the possibility* $M_1 = 1$.

The fuzzy values used in the rules are described by fuzzy sets in the general form given in Eq. 8.117 leading to the following system of nonlinear decision functions

$$\mathbf{w}\boldsymbol{\mu}(\mathbf{x}) = \mathbf{M} \tag{8.119}$$

where $\mathbf{w}$ denotes the matrix of weights, $\boldsymbol{\mu}(\mathbf{x})$ matrix of fuzzy membership functions and $\mathbf{M}$ vector of possibilities for the recognition of the current PS as an element of PSN_i. M_i is a real number in the range [0, 1] and equal to $\sum_{j=1}^{m} w_j \mu_{ij}(x_j) = M_i$ [285]. Once the physiological state is determined, RTKBS uses another rule-base to decide on switching to appropriate control algorithm:

IF *PSN is* PSN_i
THEN *activate control algorithm* $\mathbf{a}_i$ *where the control action is defined as* $\mathbf{u}_i = \mathbf{a}_i(\mathbf{y},\ \mathbf{x})$.

They have also proposed variants of this methodology by developing temporal shape libraries for real-time detection of physiological phenomena in a KBS framework [289].

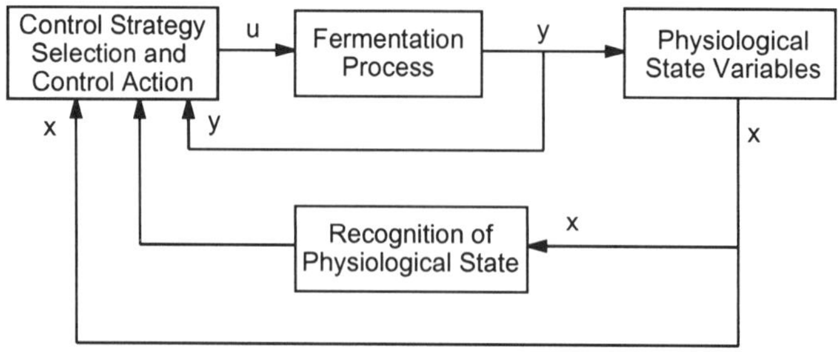

Figure 8.13. The structure of physiological state control system [285].

8.4.2 Real-time Supervisory KBS for Process Monitoring and FDD

KBS applications discussed in Section 8.4.1 for FDD and supervision of fermentation processes lacked multivariate statistical inference. MV statistical techniques are found to be very suitable for on-line SPM and FDD of fermentations processes as discussed in detail in Chapter 6 and Sections 8.1 and 8.2. There is a growing interest in the use of MV techniques in fermentation process modeling, monitoring and FDD [199, 248, 333, 608]. The synergistic integration KBS and MSPM tools offers advantage. Integrating MSPM and RTKBS enables the automated interpretation of MV charts during the abnormal situations and relate this information with process knowledge. The basic structure of the overall integrated framework based on Gensym's G2 KBS development environment is given in Figure 8.14 [184].

Research on developing integrated KBS-SPM for process supervision and FDD progressed during the last decade. Norvilas et al. proposed an intelligent SPM framework by interfacing KBS and MV techniques [337, 436] and demonstrated its performance with simulation studies. Integrated use of MSPM techniques and RTKBS for real-time on-line monitoring and FDD of fermentation processes is proposed by Undey et al. [607] and Glassey et al. [193]. Applications of the integrated RTKBS and MSPM techniques are also reported by industrial researchers [11]. Most of the recent applications are developed using Gensym's G2 software. G2 offers a graphical, object-oriented environment for creating intelligent applications that monitor, diagnose, and control dynamic events in on-line and simulated environments. It features a structured natural language for creating rules, models, and procedures. G2 includes concurrent execution of rules and procedures and the ability to reason about behavior over time. Communication between

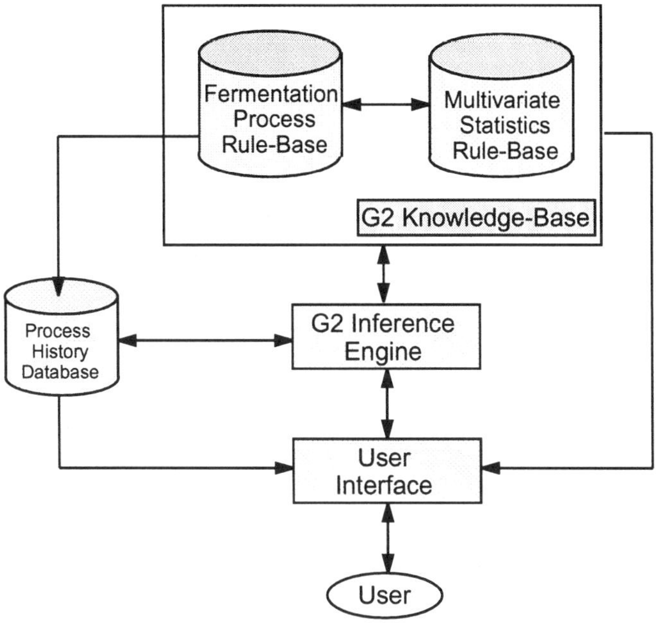

Figure 8.14. The structure of integrated KBS [607].

G2 and external programs such as executable C routines and real-time data systems including relational databases and distributed control systems is realized by using G2 Bridge Products provided by Gensym [185].

Recently, Undey et al. [609] have extended the integrated RTKBS framework by including process landmark detection and time alignment for a more refined SPM and FDD in their work for fed-batch fermentation processes. G2 knowledge-base is comprised of two kinds of rule-bases: (1) *Fermentation process rule-base*, where fermentation specific rules are stored such as physiological phase related heuristics, (2) *Multivariate statistics rule-base*, where interpretation about the MV charts are stored. The first rule-base is process-specific, hence the level of knowledge depends on the knowledge acquired from process experts, while the second rule-base is process-independent so that it can be used for different types of batch processes. Examples of statistical and process related rules are, respectively

IF *the T^2 chart is out-of-control*
THEN *start checking T^2 contribution plots and identify faulty variables whose contributions exceed contribution limits.*

IF *the glucose feed rate is diagnosed as out-of-control and fermentation*

is in the fed-batch operation mode
THEN *Check the condition of the glucose feed pump.*

These two rule-bases are also connected to an external database where process related data such as historical data sets and reference batches as well as statistical limits and parameters are stored. There is a continuous flow of information between the G2 KBS and external databases. MV statistical algorithms are first developed in Mathworks' MATLAB software environment because it allows faster prototyping. A number of software bridges are created by using G2 Standard Interface (GSI) to provide communication between the KBS and external statistical modules. Since GSI bridge development requires C code, the Matlab functions developed are compiled into C functions. Performing statistical calculations outside the G2 KBS environment allows faster execution and better computational performance.

Detection of abnormal process operation can be performed on-line in real-time by implementing one of the on-line SPM techniques discussed in Section 6.5. In this application, AHPCA technique (Section 6.5.2) is preferred due to its superior computational performance and elimination of the need to estimate future values of variable trajectories. Fault diagnosis is performed by means of contribution plots and inferencing. Statistical limits for variable contributions to T^2 and SPE are calculated as discussed in Section 8.1. When an out-of-control signal is observed on either T^2 or SPE charts, the corresponding contribution plots are investigated automatically and variable(s) exceeding control limits are diagnosed as major contributor(s) to the abnormal situation. The RTKBS helps operators on this interpretation. Once the MSPM rule-base detects and diagnoses the abnormal situation, the process specific rule-base is activated to further investigate the problem by emulating the reasoning of the human expert.

The penicillin fermentation simulator developed based on the unstructured mathematical model discussed in Section 2.7.1 is integrated with the RTKBS as a test-bed. It is run as an external C executable to provide fermentation data in real-time through another GSI bridge. G2 also allows a nice representation of the process flow chart. Each processing unit can be interpreted as objects that can inherit a general class information. Changes in the process operation can also be animated by means of changing color schemes of objects such as active pumps or showing attributes on each object such as temperature value in the fermentor next to fermenter object. A typical fed-batch fermentation for penicillin production flow chart is developed in a G2 workspace as a part of integrated G2 RTKBS (Figure 8.15). Figure 8.16 shows a case where a small downward drift is introduced in glucose flow rate. First, the RTKBS uses its statistical inference rule-base to detect the out-of-control situation and reports it along with the time of its occurrence. The RTKBS continues to use the statistical rule-base to find out the responsible variable(s) by analyzing contribution plots. Based on the contribution limit violations, a conclusion is reached on the responsible variable(s). At this point process expertise is required. Hence, the RTKBS turns to process specific rule-base to further investigate the situation and generate some advice to isolate the problem. In the example shown in Figure 8.16, the process rule-base is used by the RTKBS to infer that the problem is with the glucose feed. The RTKBS also checks certain variables that are highly correlated with glucose feed such as glucose and biomass concentrations in the fermenter to verify its conclusion. Since these variables are also affected (determined by analyzing their contribution values), the certainty about a potential glucose feed failure is high and these findings are reported to the operator by displaying them on the monitor. The messages include the time of detection of the deviation, the input variable(s) responsible, and the process variable(s) affected by the

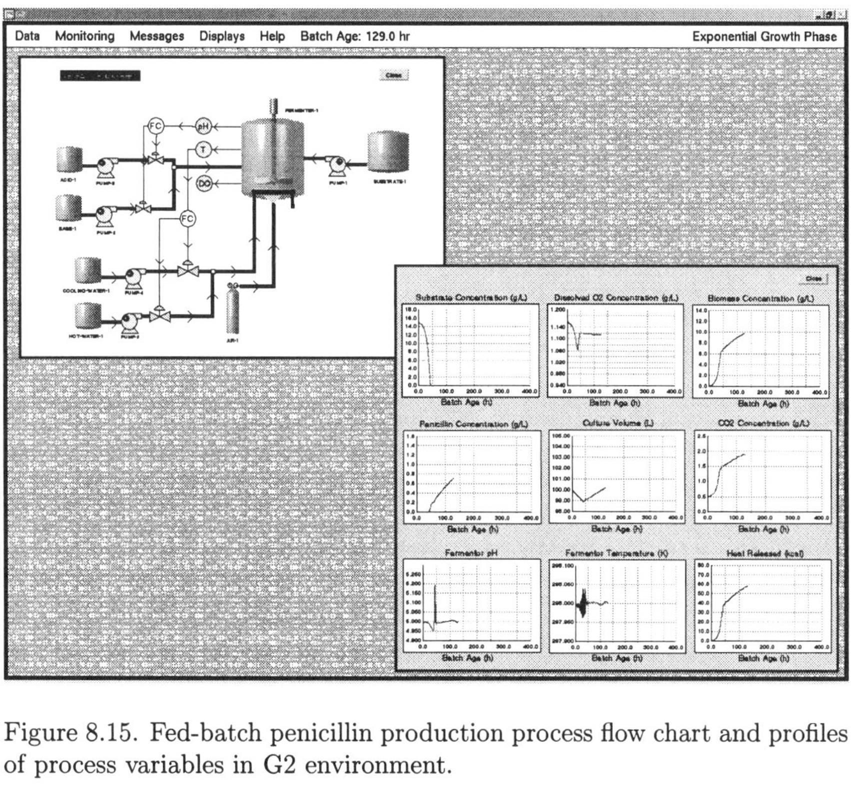

Figure 8.15. Fed-batch penicillin production process flow chart and profiles of process variables in G2 environment.

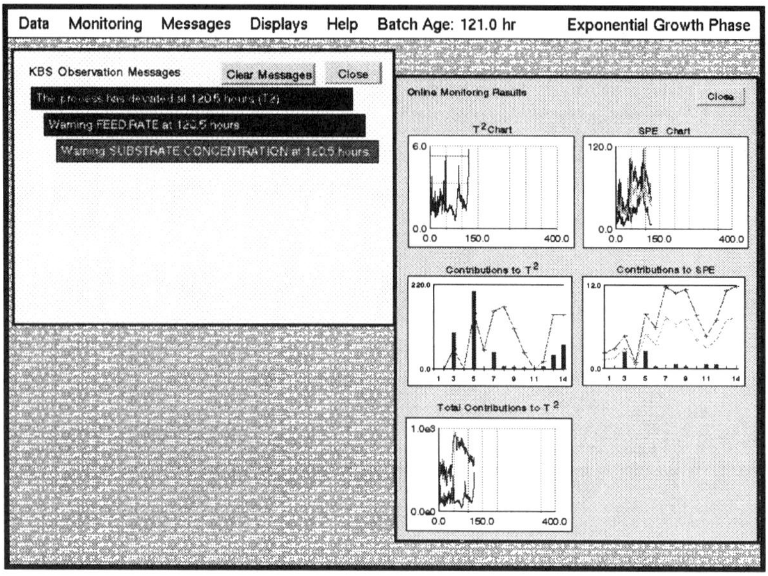

Figure 8.16. MV charts (and their interpretation by RTKBS) detecting out-of-control situation in real-time.

disturbance. As the batch proceeds, additional process variables may be affected. Figure 8.17 shows multivariate charts along with their RTKBS interpretation at the end of the batch. In addition to variable(s) that are diagnosed in real-time at the time of their deviation from NO (Figure 8.16), there are variables such as dissolved oxygen and penicillin concentrations diagnosed as having contributed to the deviation. This is due to after-effect of the root cause of the deviation that is a temporary drift in the glucose feed. If the detection is delayed, the diagnosis effort must sort out all this additional information, by considering the time that they were listed and other characteristics. This underlines the importance of early detection for easier diagnosis. All of the results about the performance of the batch run such as whether it has gone out-of-control and the time of out-of-control occurrence, variables responsible to deviation from NO and a list of productivity and yield related measures are reported conveniently for the review of process operator (Figure 8.18). These results can also be stored for future reference and integrated with other software that monitors plant performance, supply chain management, and profitability.

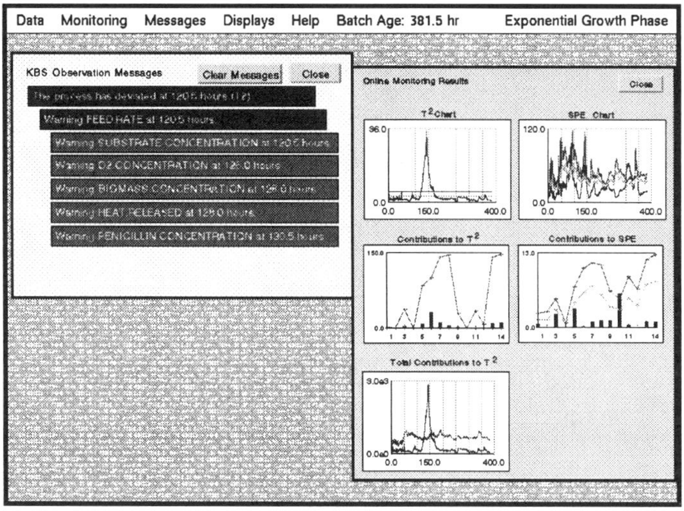

Figure 8.17. MV charts and RTKBS findings at the end-of-batch.

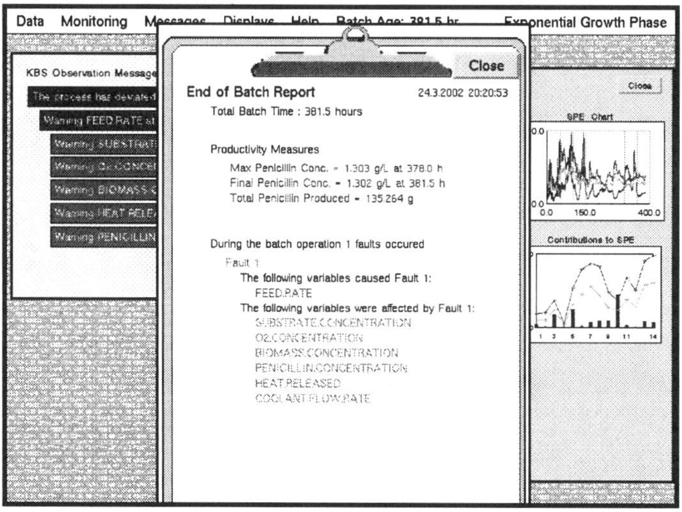

Figure 8.18. RTKBS report including interpretation of MV charts at the end-of-batch.

9

Related Developments

Increased use of batch processes in various industries has invigorated research and development in batch process design and operation. Recent developments in batch process modeling, monitoring, diagnosis, and control have been presented in earlier chapters of the book. Related developments in other aspects of batch processes are presented in this chapter, focusing on three areas:

- Process development and modeling using metabolic engineering and pathway analysis
- Dynamic optimization of batch process operations
- Integration of monitoring, control, and diagnosis activities using supervisory systems

The earlier chapters of this book presented many powerful techniques that have been developed for batch operations or extended from other fields to improve modeling, monitoring, diagnosis, and control of multivariable fermentation processes. This chapter introduces additional research directions and techniques that will have an impact on bioprocess operations, and in particular the operation of multivariable batch fermentation processes.

Advances in *process development* have been influenced by better understanding of fundamentals of fermentation processes and developments in metabolic engineering, focusing on metabolic pathway analysis and modification. The role of metabolic engineering in process improvement is discussed in Section 9.1.

Progress in *process modeling* can be discussed based on advances in various key areas. One influential factor is the interest in building layers of models, starting with the model of a cell. An ambitious plan in biomedical applications is to integrate the models of cells to build models of organs and integrate the models of organs to build models of the body for conducting computational experiments. The availability of such models for screening potential drug candidates in pharmaceutical industry will have significant impact on drug development time and cost. Advances in fundamental areas such as biology, biochemistry, mathematics, computer science and bioengineering have contributed to progress in the development of these multi-layer first principles models. Methods for developing first principles models of batch fermentation processes were introduced in Chapter 2. Information on metabolic pathways and integration of systems science methods and metabolic pathway analysis can provide the tools to add detailed knowledge to first principles models. Metabolic flux and control analysis are introduced in Section 9.2 to underline the use of sensitivity analysis in model development. The other alternative for describing a batch fermentation process is an empirical model discussed in Chapter 4. The existence of many nonlinearities in living systems has motivated researchers for developing nonlinear empirical models. Advances in statistics, computer science, mathematics, and systems science enabled development and application of nonlinear model development techniques in many fields. Section 4.7 presented extensions of linear model development techniques and Chapter 5 introduces many useful techniques for modeling and analyzing the dynamic behavior of nonlinear systems.

Progress in *dynamic optimization* of batch fermentation process operations is influenced by advances in modeling, optimization, statistical methods, and control theory. Model predictive control (MPC) presented in Section 7.6 relies on similar techniques and focuses on tracking a reference trajectory while rejecting the effects of disturbances. The performance of MPC systems is strongly related to availability of good process and disturbance models, and powerful optimization techniques. Dynamic optimization methods offer a variety of alternatives to select optimal values of process inputs and switching times to maximize productivity and yield. The alternatives in dynamic optimization of batch processes, current practice and emerging technologies are discussed in Section 9.3.

Software environments for efficient real-time operations and powerful computer hardware enable horizontal and vertical integration of various tasks in guiding batch process operations. Horizontal integration focuses on the coordination of monitoring, diagnosis, and control tasks. Vertical integration focuses on the coordination of process operations related tasks with higher level management tasks. Section 9.4 presents supervisory knowledge-based systems (KBS) to implement horizontal integration and introduces vertical integration paths with supply chain management and plantwide optimization.

9.1 Role of Metabolic Engineering in Process Improvement

For several decades, various industrial strains have been successfully developed by traditional mutagenesis and selection to improve the yield and productivity of native products synthesized by these strains. The development of molecular biological techniques for DNA recombination introduced a new dimension to metabolic pathway modification. Genetic engineering allowed precise modification of specific enzymatic reactions in metabolic pathways, leading to the construction of well-defined genetic background [564]. Soon after the feasibility of DNA recombination was established, the potential of directed pathway modification became apparent and various terms were coined to express the potential applications of this technology, such as, molecular breeding [273], *in vitro* evolution [590], microbial or metabolic pathway engineering [357, 593], cellular engineering [386], and metabolic engineering [31, 567]. The advent of recombinant DNA technology has enabled metabolic pathway modification by means of targeted genetic modifications.

Metabolic engineering can be defined as directed modification of cellular metabolism and properties through the introduction, deletion, inhibition and/or modification of metabolic pathways by using recombinant DNA and other molecular biology techniques [31, 331, 564]. The analysis aspect of metabolic engineering focuses on the identification of important parameters that define a physiological state, use of this information to elucidate the control architecture of a metabolic network, and propose targets for modification to achieve an appropriate objective [565]. The synthesis aspect of metabolic engineering examines the complete biochemical reaction network, focusing on pathway synthesis, thermodynamic feasibility, pathway flux, and flux control [565]. This multidisciplinary field embraces principles from chemical engineering, computational sciences, biochemistry, and molecular biology. Potential advantages of using genetically engineered organisms instead of natural isolates can be [546]:

- The pathway can be turned on in situations where it would normally be suppressed (e.g. degradation of a hazardous compound to a concentration lower than necessary to induce the pathway in the natural isolate),

- High levels of an enzyme in desired pathways can be obtained by the aid of strong promoters,

- A single promoter can be used to control the pathways moved from lower eucaryotes to bacteria, keeping in mind that each protein is controlled by a separate promoter in lower eucaryotes,

- Several pathways can be combined in a single recombinant organism by recruiting enzymes from more than one organism,
- A pathway can be moved from a slowly growing organism into a more easily cultured organism,
- The genetically engineered cell can be proprietary property.

Product biosynthesis ranging from primary to secondary metabolite production pathways of microorganisms are of highest interest in metabolic engineering. Biopharmaceutical production via plant and mammalian cell cultures are also of immediate interest due to potential uses in pharmaceutical industry.

Applications of Metabolic Engineering

Several reviews on metabolic engineering cover general ([86, 87, 149, 566] and specific organisms such as yeast [218], plants [114, 127], and *Escherichia coli* [52]. Bacteria and yeast have numerous applications in metabolic engineering because they are well-studied microorganisms and genetic tools for these are well-developed. Mathematical representations for their growth and substrate utilization, and product synthesis in these organisms are available in literature. Furthermore, their generation times are relatively small, allowing quick experimentation and development. Many practical applications of metabolic engineering are cited in various papers and books [87, 331, 564, 565]:

- Improvement of yield and productivity of native products synthesized by microorganisms. Examples include ethanol production by *Escherichia coli* [250], succinic acid production by *E. coli* [576], acetone and butanol production by *Clostridium acetobutylicum* [387], production of L-lysine, L-phenylalanine, and L-tyrosine by *Corynebacterium* sp. [143, 249, 611], L-proline production by *Serratia marcescens* [371].
- Expansion of the range of substrates for cell growth and product formation. Examples include ethanol production from xylose (and possibly from hemicellulose hydrolysates) by *Saccharomyces cerevisiae* [583], ethanol production from lactose (and possibly from whey) by *S. cerevisiae* [310], and ethanol production from starch [60, 238].
- Synthesis of products that are new to the host cell. Examples are various modified and novel polyketide antibiotics by *Saccharopolyspora erythraea* and *Streptomyces* sp. [247], production of 1,3 propanediol by *E.coli* [593], and polyhydroxyalkanoate production by a small oilseed plant, *Arabidopsis thaliana* [474].

- Design of improved or new metabolic pathways for degradation of various chemicals, especially xenobiotics. Examples are degradation of mixtures of benzene, toluene and xylene (BTX) by *Pseudomonas putida* [320] and degradation of polychlorinated biphenyls (PCBs) by *Pseudomonas* sp. [515].

- Modification of cell properties that facilitate fermentation and/or product recovery. Examples include better growth of *E. coli* and other microorganisms under microaerobic conditions [278], uptake of glucose without consuming phosphoenolpyruvate in *E.coli*, and ammonia transport without ATP consumption in *Methylophilus methylotrophus* [654].

These examples are a small subset of many success stories of metabolic engineering that have been reported. They illustrate the various types of approaches that can be undertaken experimentally:

- Extending an existing pathway to obtain a new product

- Amplifying a flux-controlling step

- Diverting flux at branch points ("nodes") to a desired product by circumventing a (feedback) control mechanism, amplifying the step initiating the desired branch (or the converse), removing reaction products, or manipulating levels of signal metabolites.

Examples of Industrially Important Products

The interest in metabolic engineering is stimulated by potential commercial applications in that improved methods are sought for developing strains which can increase production of useful metabolites. Recent endeavors have focused on the theme of using biologically derived processes as alternatives to chemical processes. Such manufacturing processes pursue goals related to "sustainable development" and "green chemistry" as well as positioning companies to exploit advances in the biotechnology field. Examples of these new processes include the microbial production of indigo (developed by Genencor) and propylene glycol (developed by DuPont) and other improvements in more traditional areas of antibiotic and amino acid production. The extension of metabolic engineering to produce desired compounds in plant tissues and to provide better understanding of genetically determined human metabolic disorders broadens the interest in this field beyond the fermentation industry and bodes well for increasing impact of this approach in the future [676].

A number of industrially important aromatic compounds, including the aromatic amino acids and other metabolites, can be produced in microorganisms through metabolic engineering of the aromatic pathway. Plastics and other synthetic polymers, whose desirable properties include chemical and biological inertness, have become essential for a multitude of applications in common consumer products. On the other hand, increasing concern about environmental pollution by these non-biodegradable polymers has created interest in the development of completely biodegradable polymers [387]. Polyhydroxyalkanoates (PHAs) are an important class of biodegradable polymers that can be produced by a number of microorganisms. However, the high production cost and some poor material properties are preventing the use of PHAs in a wide range of applications. Continued pressure to provide aromatic compounds with very low production costs will create new challenges to develop competitive biotechnological processes [331].

Traditional strain improvement methods as well as metabolic engineering strategies have been used for enhancing the production of antibiotics and production of novel antibiotics. A wide variety of microorganisms synthesize antibiotics while only clinically useful antibiotics are produced by the eubacteria, Actinomycetes, in particular *Streptomyces*, and the filamentous fungi. Metabolic engineering techniques are applied for strain improvement to increase the final amount of antibiotics produced in fermentation processes. A typical strain improvement program involves generation of genotype variants in the population, either by means of physically or chemically induced mutations or by recombination among strains [331]. A detailed case study in penicillin producing strain improvement is discussed in Nielsen [424].

Yeasts have been associated in a number of ways with mankind for the production of alcoholic beverages, baker's yeast, and recently for the production of ethanol, pharmaceutical proteins and enzymes. Other metabolites, including pyruvate, xylitol, carotenoids, and inositol, can be produced by metabolically engineered yeasts. Metabolic engineering strategies have been applied to modify the cellular properties of yeast to improve fermentation and product recovery processes as a result of extended range of substrate utilization. Renewable substrates for extension of substrate range include starch, the most abundant and readily extractable plant biomass, and cellulose, hemicellulose and pectin fractions in lignocellulosic materials, and whey lactose [331]. Traditional approaches include using mixed cultures or multistage operations such as physical and enzymatic pretreatment of substrates prior to fermentation. With the development of recombinant DNA technology, the introduction of heterologous genes into a host yeast facilitates one step conversion of substrates into useful end products (e.g., recombinant *Saccharomyces cerevisiae* containing α-amylase and glucoamylase genes that allow the yeast to grow on starch and convert it into ethanol [60]).

Another area of application of metabolic engineering is in the production of secondary metabolites that can be used as pharmaceuticals, including anticancer drugs vinblastine and more recently, taxol by plant cells [331]. Some major objectives are:

- Improving nutritional value of crops (e.g., essential amino acid supply for storage proteins, modifying lignin amount or type to enhance forage digestibility)
- Creating new industrial crops (e.g., modified fatty acid composition of seed triglycerides, pharmaceuticals, polyhydroxybutyrate synthesis, bioremediation)
- Altering photosynthate partitioning to increase economic yield
- Enhancing resistance to biotic and abiotic stresses
- Reduction of undesired (toxic or unpalatable) metabolites
- Using them as research tool to test basic ideas about metabolic regulation.

The Future of Metabolic Engineering

No single discipline can bring about the successful development and applications of metabolic engineering [331]. Metabolic engineering offers one of the best ways for meaningfully engaging chemical engineers in biological research for it allows the direct application of the core subjects of kinetics, transport, and thermodynamics to the reactions of metabolic networks [565]. With the advent of genomics and proteomics, enormous amounts of information on the genetic and protein makeup of various microorganisms are becoming available. As a consequence, bioinformatics will play an increasingly significant role in the evolution of metabolic engineering. Also, directed evolution of enzymes will become a powerful tool for the generation of enzymes or even metabolic pathways suitable for given tasks. These improvements in metabolic engineering will lead to reshaping the biotechnology endeavor, giving rise to more precise, more focused and more effective bioprocessing and intervention at the cellular and organismal levels. Furthermore, it will also bring more control at all levels (gene expression and protein translation, protein, metabolite, pathway, and flux levels).

9.2 Contributions of MFA and MCA to Modeling

Metabolic flux analysis (MFA) and metabolic control analysis (MCA) are mathematical tools that have become widely applicable in metabolic engineering. Both tools are interrelated and widely used in metabolic engineering research [331, 426, 565]. They are useful in developing models of metabolic activity in a biochemical system. They would be instrumental in developing detailed first principles models of fermentation processes.

The *flux* is a fundamental determinant of cell physiology and a critical parameter of a metabolic pathway [565]. The pathway is the sequence of feasible and observable biochemical reaction steps linking the input and output metabolites. Consider the linear metabolic pathway in Figure 9.1(a), where A is the input metabolite, B is the output metabolite, ν_i denotes the reaction rate of the ith reaction step and E_i the corresponding enzyme. The flux J of this linear pathway is equal to the rates of the individual reactions at steady state [565]:

$$\nu_1 = \nu_2 = \cdots = \nu_i = \cdots = \nu_L \tag{9.1}$$

For a branched pathway splitting at *intermediate* I (Figure 9.1(b)) to produce two output metabolites B and C, two additional fluxes are defined for the branching pathways. The flux of each branch (J_2 and J_3, in Figure

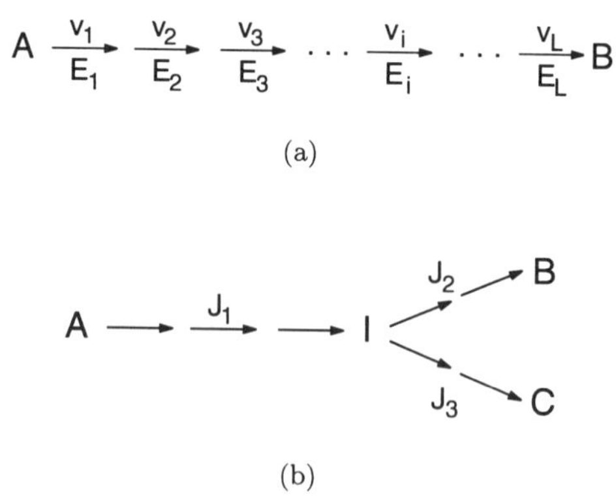

Figure 9.1. Linear metabolic pathway [565].

9.1(b)) is equal to individual reaction rates at the corresponding branches. At steady state, $J_1 = J_2 + J_3$. During a transient, the individual reaction rates are not equal and the pathway flux varies with time. Consequently, MFA can not be used to develop transient first principle models. But any dynamic model proposed to describe the transients in the metabolic pathway has to be consistent with the steady state model based on MFA. This provides a reliable reference for a dynamic model when it is reduced to a steady state description.

MFA is used for studying the properties and capabilities of metabolic networks in microorganisms. It allows stoichiometric studies of biochemical reaction networks and may be used for the determination of *stationary* metabolic flux distributions, if measurements of uptake and/or excretion rates of a cell culture in steady state are known. The result is a flux map that shows the distribution of anabolic and catabolic fluxes over the metabolic network. Based on such a flux map or a comparison of different flux maps, possible targets for genetic modifications might be identified, the result of an already performed genetic manipulation can be judged or conclusions about the cellular energy metabolism can be drawn. The MFA is also used to optimize the product yield by redirecting fluxes using genetic manipulations [282, 426, 565].

Metabolic control analysis (MCA) applies to steady-state or pseudo-steady-state conditions and relies on the assumption that a stable steady state is uniquely defined by the activities of enzymes catalyzing individual reactions in a metabolic pathway. Enzyme activities are considered to be *system parameters* along with concentrations of substrate for the first reaction and product of the last reaction in the metabolic pathway, while the flux through the pathway or intermediate metabolite concentrations are considered to be *system variables* [565]. MCA is a sensitivity analysis framework for the quantitative description of metabolism and physiology that allows the analysis and study of the responses of metabolic systems to changes in their parameters [151, 223, 229, 267, 272]. MCA relies on linear perturbations for the nonlinear problem of enzymatic kinetics of metabolic networks. Hence, MCA predictions are local and any extrapolations should be made with caution. Yet, MCA has been useful in providing measures of metabolic flux control by individual reactions, elucidating the concept of rate-controlling step in enzymatic reaction networks, describing the effects of enzymatic activity on intracellular metabolite concentrations, and coupling local enzymatic kinetics with the metabolic behavior of the system [565].

Consider a two-step pathway where the substrate S is converted to the product P via an intermediate X and enzymes activities E_1 and E_2

$$S \overset{E_1}{\longleftrightarrow} X \overset{E_2}{\longleftrightarrow} P \tag{9.2}$$

The flux of conversion of S to P at steady state is denoted by J. The steady-state is uniquely defined by the parameters of the system, the levels of enzyme activities E_1 and E_2, substrate concentration S and product concentration P [565]. Given the values of these parameters, intermediate metabolite concentration c_X and pathway flux J can be determined. If any parameter value is altered, a new steady state is reached and c_X and J are changed.

One objective of MCA is to relate the variables of a metabolic system to its parameters and then determine the sensitivity of a system variable to system parameters [565]. These sensitivities summarize the extent of systemic flux control exercised by the activity of an enzyme in the pathway. One can also solve for the concentrations of intracellular metabolites and determine their sensitivities to enzyme activities or other system parameters. The sensitivities are represented by *control coefficients* that indicate how a parameter affects the behavior of the system at steady state. The *flux control coefficients* (FCC) are the relative change in steady-state flux resulting from an infinitesimal change in the activity of an enzyme of the pathway divided by the relative change of the enzymatic activity [565]:

$$C^J = \frac{E}{J}\frac{dJ}{dE} = \frac{d\ln J}{d\ln E} \tag{9.3}$$

Because enzymatic activity is an independent system parameter, its change affects the flux both directly and indirectly through changes caused in other system variables, as indicated by the total derivative symbol in Eq. 9.3. FCCs are dimensionless and for linear pathways they have values from 0 to 1. For branched pathways, FCCs can be generalized to describe the effect of each of the L enzyme activities on each of the L fluxes through various reactions [565]:

$$C_i^{J_k} = \frac{E_i}{J_k}\frac{dJ_k}{dE_i} = \frac{d\ln J_k}{d\ln E_i} \qquad i,k = 1,\cdots,L \tag{9.4}$$

where J_k is the steady-state flux through the kth reaction in the pathway and E_i is the activity of the ith enzyme. A similar definition is developed based on the rate of the ith reaction (ν_i) [565]:

$$C_i^{J_k} = \frac{\nu_i}{J_k}\frac{dJ_k}{d\nu_i} = \frac{d\ln J_k}{d\ln \nu_i} \qquad i,k = 1,\cdots,L \tag{9.5}$$

The FCCs for branched pathways may have any positive or negative value. The normalization in the definition of FCCs leads to their sum being equal to unity, the *flux-control summation theorem*:

$$\sum_{i=1}^{L} C_i^{J_k} = 1 \qquad k = 1, \cdots, L \; . \tag{9.6}$$

The relative magnitudes of FCCs would depend on the structure of the system and length of the pathway. FCCs should only be compared with each other in the same pathway but not with FCCs of other pathways.

Sensitivities can also be defined for the effect of system parameters on intracellular metabolite concentrations. The concentration control coefficients (CCC) specify the relative change in the level of the jth intermediate X_j when the activity of the ith enzyme is changed:

$$C_i^{X_j} = \frac{E_i}{c_j}\frac{dc_j}{dE_i} = \frac{d\ln c_j}{d\ln E_i} \qquad i = 1, \cdots, L \; ; \; j = 1, \cdots, K \tag{9.7}$$

where c_j denotes the concentration of X_j. Because the level of any intermediate X_j remains unchanged when all enzyme activities are changed by the same factor, the sum of all CCCs for each of the K metabolites is equal to zero [565]:

$$\sum_{i=1}^{L} C_i^{X_j} = 0 \qquad j = 1, \cdots, K \; . \tag{9.8}$$

Eq. 9.8 implies that for each metabolite at least one enzyme exerts negative control. For example, in the two-step pathway of Eq. 9.2 the CCC C_2^X will normally be negative because c_X will decrease when the activity of E_2 is increased [565].

The control coefficients are systemic properties of the overall metabolic system. Local properties of individual enzymes in the metabolic network can be described by *elasticity coefficients* such as the sensitivities of reaction rates with respect to metabolite concentrations. The elasticity of the ith reaction rate with respect to the concentration of metabolite X_j is the ratio of the relative change in the reaction rate caused by an infinitesimal change in the metabolite concentration, assuming that none of the other system variables changed from their steady state values:

$$\epsilon_{X_J}^{i} = \frac{c_j}{\nu_i}\frac{\partial \nu_i}{\partial X_j} = \frac{\partial \ln \nu_i}{\partial \ln c_j} \qquad i = 1, \cdots, L \; ; j = 1, \cdots, K \tag{9.9}$$

Elasticity coefficients may also be defined for other compounds that influence a reaction rate that may not be pathway intermediates.

The relationship between FCCs and elasticity coefficients is expressed by the *flux-control connectivity theorem* that indicates how local enzyme kinetics affect flux control [267, 565]:

$$\sum_{i=1}^{L} C_i^{J_k} \epsilon_{X_j}^i = 0 \qquad i = 1, \cdots, L \; ; j = 1, \cdots, K \tag{9.10}$$

For the two-step pathway of Eq. 9.2, the connectivity theorem gives

$$C_1^J \epsilon_X^1 + C_2^J \epsilon_X^2 = 0 \tag{9.11}$$

or

$$\frac{C_1^J}{C_2^J} = -\frac{\epsilon_X^2}{\epsilon_X^1} \tag{9.12}$$

indicating that large elasticities are associated with small FCCs. For example, reactions operating close to thermodynamic equilibrium are normally very sensitive to variations in metabolite concentrations; their elasticities are large indicating that flux control for such reactions would be small [565]. Connectivity theorems have also been developed for CCCs.

9.3 Dynamic Optimization of Batch Process Operations

Discussion on optimal operation of fermentation processes in Chapter 7 focused on the search for open-loop optimal trajectories (Section 7.2) and regulation of process operation to track reference trajectories while rejecting disturbances by using optimal feedback control (Section 7.5) and model predictive control (MPC) (Section 7.6). These techniques rely on the availability of reliable dynamic models for the process and disturbances. Industrial practice involves following recipes developed in the laboratory that are modified to accommodate changes in equipment and scale. The 'educated trials' approach based on experience and heuristics is used often for recipe adjustment. The recipes and reference profiles are often non-optimal since the search is usually limited to the vicinity of a known 'acceptable' recipe and the reference profiles are somewhat conservative to assure feasible process operation in spite of process disturbances.

From an industrial perspective, there is a need to improve the performance of batch processes in spite of incomplete and/or inaccurate process models, few online measurements and estimates of process variables, large uncertainties (model inaccuracies, process disturbances, variations in raw material properties), and important operational and safety constraints. In a series of papers, Bonvin and co-workers assessed the industrial perspective in batch process operation and the batch process optimization problem

[71, 73, 348], and proposed an 'Invariant-Based Optimization' approach that docs not require an accurate process model [73].

Industrial Perspective and Practice The operational objectives of batch processes are high productivity, reproducible product quality, process and product safety, and short time to market. These objectives could be posed as an optimization problem, but the implementation of optimization through mathematical modeling and optimization techniques is not widespread. The popular approach is to develop recipes in the laboratory for safe implementation in production environment, then empower plant personnel to adjust the process based on heuristics and experience for incremental improvements from batch to batch [622]. Various organizational and technical reasons are cited for this practice [73].

The organizational reasons hindering the adoption of a rigorous dynamic optimization approach include process registration and validation, low levels of interaction between design of individual steps in multi-step processes, and separation between design and control tasks [73]. Process registration and validation with regulatory agencies such as the U.S. Food and Drug Administration is mandatory in production of active compounds in pharmaceutical and food processing. Because this is a time-consuming and costly task, it is performed simultaneously with the research and development work of a new process. Consequently, the main operational parameters are fixed within conservative limits at an early stage of process development. Changes in process operation may require revalidation and registration; a costly venture. The second reason is related to use of different design teams for different steps of the process. While each step is optimized by introducing the appropriate conservatism to account for uncertainty, the process as a whole may become too conservative. The third reason stems from the practice of treating design and control as separate tasks; a legacy from the times when automatic process control was considered mostly as the installation of hardware for process control and tuning individual controllers. This prevents the use of systems science, control theory, and optimization tools to develop a better design that is easier to optimize and control.

A number of technical reasons has influenced the administrative decisions to favor this conservative optimization practice. Lack of reliable first principles models, absence of on-line quality measurements, uncertainty due to variations in feedstock properties, assumptions during process scale-up, and modeling errors, and constraints caused by equipment limitations, operational limits on variables and end-points are some of these reasons [73]. These reasons also hint that improvements in model development and measurements that are coupled with powerful optimization techniques may generate significant improvements in batch process operation, productivity, product quality, and safety.

Dynamic Optimization Problem Batch process optimization is a *dynamic optimization* problem that involves dynamic and static constraints. Various types of optimization problems are formulated depending on the assumptions used for uncertainty and availability of measurement information, and the method used for updating the optimal values of inputs [73]. If it is assumed that there is no uncertainty, the problem is reduced to nominal optimization and the computational load is lighter. But the solution computed may not be feasible when implemented in a real application that invariably has some uncertainty. Uncertainty necessitates the adoption of conservative operation (control) strategies. The uncertainty is taken into account in robust optimization by considering the range of possible values for uncertain parameters and the optimization is performed by considering the worst-case scenarios (selecting the best solution for the worst conditions assures that the solution would be feasible under better scenarios) or by using expected values. The availability of process measurements reduces uncertainty and consequently less conservative process operation strategies can be adopted. Various types of dynamic optimization problems and their major disadvantage are classified in [73] (Figure 9.2). If quality mea-

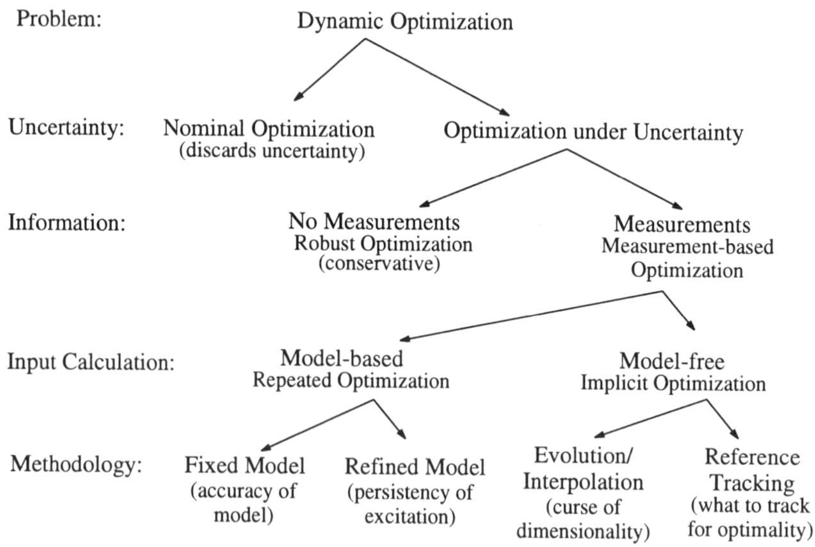

Figure 9.2. Dynamic optimization scenarios with, in parentheses, the corresponding major disadvantage [73].

surements at the end of a batch run are available, they could be used in determining the optimal operation policy of the next batch. Consider the kth run of a batch process where process measurements from the previous $(k-1)$ batches and measurements up to the current time t_l of the kth batch are available. The optimal input policy for the remaining time interval $[t_l, t_f]$ of the kth batch can be determined by solving the optimization problem:

$$\min_{\mathbf{u}^k_{[t_l,t_f]}} \quad J^k = L(\mathbf{x}^k(t_f, \boldsymbol{\theta})) \tag{9.13}$$

$$\begin{aligned} \text{such that} \quad & \dot{\mathbf{x}}^k = \mathbf{F}(\mathbf{x}^k, \boldsymbol{\theta}, \mathbf{u}^k) + \mathbf{d}^k(t), \qquad \mathbf{x}^k(0) = \mathbf{x}^k_0 \\ & \mathbf{y}^k = \mathbf{H}(\mathbf{x}^k, \boldsymbol{\theta}) + \mathbf{v}^k(t) \\ & \mathbf{S}(\mathbf{x}^k, \boldsymbol{\theta}, \mathbf{u}^k) \leq \mathbf{0}\ , \qquad \mathbf{T}(\mathbf{x}^k(t_f, \boldsymbol{\theta})) \leq \mathbf{0} \\ \text{given} \quad & \mathbf{y}^j(i)\ , i = 1, N \text{ for } j = 1, k-1 \text{ and } i = 1, l \text{ for } j = k \end{aligned}$$

where the superscript ^{k} denotes the kth batch run, $\mathbf{x}^k(t), \mathbf{u}^k(t), \mathbf{y}^k(t), \mathbf{d}^k(t)$, and $\mathbf{v}^k(t)$ denote the state, input, output, disturbance, and measurement noise vectors, respectively. $\mathbf{S}()$ is a vector of path constraints, and $\mathbf{T}()$ is a vector of terminal constraints. $\mathbf{y}^j(i)$ denotes the ith measurement vector collected during the jth batch run, and N the total number of measurements during a run. The optimization utilizes information from the previous $k-1$ batch runs and measurements up to time t_l of the current batch to reduce uncertainty in the parameter vector $\boldsymbol{\theta}$ and to determine the optimal input policy for the remainder of the current batch run k.

The optimization approaches that rely on process measurements to update the inputs can be divided into two main groups: model-based techniques and model-free techniques [73]. Model-based techniques use the mathematical model of the batch process to predict the evolution of the run, compute the cost sensitivity with respect to input variations, and update the inputs. Measurement information is used to improve the estimates of the state variables and parameters. The estimation and optimization tasks are repeated over time (as frequently as at each sampling time), yielding significant computational burden. In this *repeated optimization* approach the model can be fixed or refined during the batch run and its optimization. If the model is fixed, a higher level of model accuracy is necessary. If the model parameters are known with accuracy and uncertainty is caused by disturbances, the fixed model can yield satisfactory results. If model refinement such as estimation of model parameters is carried out during the run, the initial model may not need to have high accuracy. The tradeoff is heavier computational burden and addition of persistent excitation to input signals in order to generate data rich in dynamic information for more reliable model identification. Unfortunately, the requirement for sufficient excitation in inputs may conflict with optimal value of inputs.

Model-free optimization relies on measurements from batch runs for updating the inputs to achieve optimality without using a model or an explicit numerical optimization procedure. These *implicit optimization* schemes use either the deviation from a reference trajectory or measurement information to update the inputs. The reference-based techniques update the inputs by using feedback controllers to track the reference trajectories. Reference (optimal) trajectories are usually computed using a nominal model (See Section 7.2). Uncertainty in the model may cause significant deviation of the actual optimal (unknown) trajectories from the nominal ones computed by the model. Data-based techniques compute the inputs directly by using measurement information from past and current batch runs. A reliable historical database is needed to implement this approach.

The type of measurements (off-line taken at the end of the batch run or on-line during the progress of the batch) indicate the type of optimization sought. Off-line end-of-batch measurements lead to *batch-to-batch optimization* where process knowledge obtained in earlier batches enable update of the operating strategy of the current run, approaching an optimal solution as information from additional batch runs are used. The availability of on-line measurements during the run enable the use of an *on-line optimization* approach. On-line measurement-based optimization schemes have many similarities to model-predictive control. The Iterative Learning Control approach (Section 7.6) integrates MPC and batch-to-batch optimization [327, 328, 682]. The integrated methodology is capable of eliminating persisting errors from previous runs and responds to new disturbances that occur in the current run [100, 323, 327]. The differences between measurement-based optimization and MPC are discussed and an extensive list of references for measurement-based optimization studies is given in [73]. Table 9.1 summarizes the classification of measurement-based optimization methods in [73] and provides additional references.

An *invariant-based optimization* approach is proposed in [73] to identify the important characteristics of optimal trajectories of a batch run that are invariant under uncertainty and provide them as reference to feedback controllers. The method consists of three steps: state-dependent parameterization of inputs, selection of signals that are invariant under uncertainty, and tracking the invariant by using process measurements. The state-dependent parameterization is related to the characteristics of the optimal solution: switching times of inputs (related to the concept of process landmarks) and the types of input arcs that occur between switching times. The two types of input arcs are singular arcs where the input lies in the interior of the feasible region and nonsingular arcs where the inputs are determined by a path

Table 9.1. MBO methods specifically designed to compensate uncertainty [73].

Methodology	Batch-to-batch optimization (Off-line measurements)	On-line optimization (On-line measurements)
Model-based Fixed model	[131, 683, 684]	[2, 6, 380]
Model-based Refined model	[152, 157, 178, 317] [365, 369, 497]	[100, 142, 177, 323] [321, 430, 529]
Model-free Evolution Interpolation	[108, 687]	[155, 307, 486] [537, 596, 673]
Model-free Reference tracking	[536, 562]	[158, 186, 312] [532, 559, 585] [602, 612, 623]

constraint. The structure of the optimal solution is determined by the type and sequence of arcs, and switching times. This can be based on experiential knowledge of plant personnel, analytical expressions for optimal inputs, or inspection of the solution from numerical optimization. Uncertainty affects the numerical values of optimal inputs, but the necessary conditions for optimality remain invariant. This fact is exploited to identify the invariants and the measurements to track the invariants by use of feedback. The proposed approach is effective when the optimization potential stems from meeting path and/or terminal constraints of a batch run [73].

9.4 Integrated Supervisory KBS for On-line Process Supervision

The integration of various tasks for fault-tolerant optimal operation of batch processes is closer to realization because of the availability of software environments for efficient real-time operations and powerful computer hardware. The first integration problem (horizontal integration) focuses on the coordination of monitoring, diagnosis, and control tasks, and their supervision for enhanced decision making and intervention. The second integration problem (vertical integration) focuses on the coordination of technical tasks related to process operations (monitoring, diagnosis and control) with higher level management tasks (supply chain management and plantwide optimization).

Horizontal integration can be formulated as a fault-tolerant control problem where appropriate control policies are formulated and implemented in real time in response to equipment failures and disturbances to prevent operational losses or process shutdown. Consider as an example the failure of a sensor that provides critical information to the control system. The monitoring system will detect an abnormality in process operation either because the change in sensor readings will be significantly different from their expected values or the controller acting on erroneous information will cause significant changes in some process variables. This will trigger diagnostic activities either by an automated fault diagnosis system or by plant personnel. Actuator faults or disturbances will also follow this detect-diagnose-decide-intervene sequence. In a fault-tolerant environment, these activities will be carried out automatically under the supervision of a supervisory real-time knowledge-based system.

Basila and co-workers have developed such a supervisory real-time KBS (MOBECS) for retuning or restructuring multivariable feedback controllers for a tubular packed-bed reactor system [43, 274, 275]. MOBECS (Model-Object Based Expert Control System) was developed initially for a single-input single-output control system [43], then extended to multivariable processes to provide fault-tolerant, minimally conservative robust control by using advanced multivariable control techniques. MOBECS is capable of emulating the steps typically carried out in redesigning the multivariable control system and bumplessly implementing the new control law with the entire control system remaining under automatic control. Control system redesign efforts can be initiated by plant personnel or MOBECS can be instructed to assess process performance automatically and take the appropriate controller redesign actions. If controller restructuring is necessary, MOBECS initiates the development of a new process model by using closed-loop data collected at the current operating point. The new model is used in developing a new controller with improved performance.

Powerful real-time KBS development environments such as G2 of Gensym enable the development of more sophisticated supervisory systems for automating and integrating all process supervision and control activities. The building blocks for a real-time supervisory KBS for monitoring and fault diagnosis of multivariable batch fermentation processes are discussed in Section 8.4.2. They illustrate how the KBS coordinates and enhances the interface between detection of abnormality in process operation and fault diagnosis. Additional modules for model development and control system design/modification can be added to implement fault-tolerant control of batch processes.

The real-time supervisory KBS can be vertically interfaced with various software tools that fulfill supply chain management, tracking of process equipment maintenance and repair, and plantwide optimization. Supply chain management interface would transmit consumption levels of raw materials, specific properties (supplier, impurity level, critical characteristics) of feed materials, production schedules and forecasts on product availability. Updating of equipment maintenance and repair records would reduce the surprise and cost of emergency repairs. Automated logging of equipment faults discovered, repairs made, and parts replaced would provide a health record for various equipment, forecast the parts that may be replaced in the near future and restock them to minimize emergency ordering and downtimes caused by waiting for their shipment. Interface with plantwide optimization and planning software will reduce the constraints and time losses in running batch campaigns with many fermentation and storage vessels, increase effective use of every process operation from raw materials to final products, and prevent raw material, resource, and product shortages. This interface will also provide appropriate financial data to management for higher level decision making.

Appendix

	$\bar{X}$ and R Charts				$\bar{X}$ and s Charts			
	Chart for Averages ($\bar{X}$)	Chart for Ranges (R)			Chart for Averages ($\bar{X}$)	Chart for Standard Deviations (s)		
Subgroup Size	Factors for Control Limits	Divisors for Estimate of Standard Deviation	Factors for Control Limits		Factors for Control Limits	Divisors for Estimate of Standard Deviation	Factors for Control Limits	
n	A_2	d_2	D_3	D_4	A_3	c_4	B_3	B_4
2	1.880	1.128	-	3.267	2.659	0.7979	-	3.267
3	1.023	1.693	-	2.574	1.954	0.8862	-	2.568
4	0.729	2.059	-	2.282	1.628	0.9213	-	2.266
5	0.577	2.326	-	2.114	1.427	0.9400	-	2.089
6	0.483	2.534	-	2.004	1.287	0.9515	0.030	1.970
7	0.419	2.704	0.076	1.924	1.182	0.9594	0.118	1.882
8	0.373	2.847	0.136	1.864	1.099	0.9650	0.185	1.815
9	0.337	2.970	0.184	1.816	1.032	0.9693	0.239	1.761
10	0.308	3.078	0.223	1.777	0.975	0.9727	0.284	1.716
11	0.285	3.173	0.256	1.744	0.927	0.9754	0.321	1.679
12	0.266	3.258	0.283	1.717	0.886	0.9776	0.354	1.646
13	0.249	3.336	0.307	1.693	0.850	0.9794	0.382	1.618
14	0.235	3.407	0.328	1.672	0.817	0.9810	0.406	1.594
15	0.223	3.472	0.347	1.653	0.789	0.9823	0.428	1.572
16	0.212	3.532	0.363	1.637	0.763	0.9835	0.448	1.552
17	0.203	3.588	0.378	1.622	0.739	0.9845	0.466	1.534
18	0.194	3.640	0.391	1.608	0.718	0.9854	0.482	1.518
19	0.187	3.689	0.403	1.597	0.698	0.9862	0.497	1.503
20	0.180	3.735	0.415	1.585	0.680	0.9869	0.510	1.490
21	0.173	3.778	0.425	1.575	0.663	0.9876	0.523	1.477
22	0.167	3.819	0.434	1.566	0.647	0.9882	0.534	1.466
23	0.162	3.858	0.443	1.557	0.633	0.9887	0.545	1.455
24	0.157	3.895	0.451	1.548	0.619	0.9892	0.555	1.445
25	0.153	3.931	0.459	1.541	0.606	0.9896	0.565	1.435

$\text{UCL}_{\bar{X}}, \text{LCL}_{\bar{X}} = \bar{\bar{X}} \pm \text{A}_2 \bar{R}$

$\text{UCL}_R = \text{D}_4 \bar{R}$

$\text{LCL}_R = \text{D}_3 \bar{R}$

$\hat{\sigma} = \bar{R}/\text{d}_2$

$\text{D}_3 = 1 - 3\text{d}_3/\text{d}_2$

$\text{UCL}_{\bar{X}}, \text{LCL}_{\bar{X}} = \bar{\bar{X}} \pm \text{A}_3 \bar{s}$

$\text{UCL}_s = \text{B}_4 \bar{s}$

$\text{LCL}_s = \text{B}_3 \bar{s}$

$\hat{\sigma} = \bar{s}/\text{c}_4$

$\text{D}_4 = 1 + 3\text{d}_3/\text{d}_2$

Bibliography

[1] R Aarts, A Suvarinta, P Rauman-Aalto, and P Linko. An expert system in enzyme production control. *Food Biotechnol*, 4:301–315, 1990.

[2] O Abel, A Helbig, W Marquardt, H Zwick, and T Daszkowski. Productivity optimization of an industrial semi-batch polymerization reactor under safety constraints. *J.Pr cess Contr.*, 10(4):351–362, 2000.

[3] B Abraham and A Chuang. Outlier detection and time series modeling. *Technometrics*, 31:241–248, 1989.

[4] E-M Abulesz and G Lyberatos. Periodic optimization of continuous microbial growth processes. *Biotechnol. Bioengng*, 29:1059–1065, 1987.

[5] E-M Abulesz and G Lyberatos. Periodic operation of a continuous culture of baker's yeast. *Biotechnol. Bioengng*, 34:741–749, 1989.

[6] M Agarwal. Feasibility of on-line reoptimization in batch processes. *Chem.Eng.Comm.*, 158:19–29, 1997.

[7] P Agrawal, C Lee, HC Lim, and D Ramkrishna. Theoretical investigations of dynamic behavior of isothermal continuous stirred tank biological reactors. *Chemical Eng. Sci.*, 37:453–462, 1982.

[8] S Aiba and M Shoda. Reassessment of the product inhibition in alcohol fermentation. *J. Ferment. Technol.*, 47:790–794, 1969.

[9] S Aiba, M Shoda, and M Nagatani. Kinetics of product inhibition in alcohol fermentation. *Biotechnol. Bioengng.*, 10:845–864, 1968.

[10] M Aitkin and GT Wilsom. Mixture models, outliers, and the EM algorithm. *Technometrics*, 22:325–331, 1980.

[11] S Albert and RD Kinley. Multivariate statistical monitoring of batch processes: an industrial case study of fermentation supervision. *Trends in Biotechnology*, 19(2):53–62, 2001.

[12] JS Albuquerque and LT Biegler. Data reconciliation and gross error detection for dynamic systems. *AIChE J*, 42:2841, 1996.

[13] J Alford, C Cairney, R Higgs, M Honsowetz, V Huynh, A Jines, D Keates, and C Skelton. Real-awards from artificial intelligence. *InTech*, April:52–55, 1999.

[14] J Alford, C Cairney, R Higgs, M Honsowetz, V Huynh, A Jines, D Keates, and C Skelton. Online expert-system applications use in fermentation plants. *InTech*, July:50–54, 1999.

[15] H Allende and S Heiler. Recursive generalized M estimates for autoregressive moving-average models. *J Time Series*, 13:1–18, 1992.

[16] LC Alwan and HV Roberts. Time-series modeling for statistical process control. *Journal of Business and Economic Statistics*, 6:87–95, 1988.

[17] BDO Anderson and JB Moore. *Optimal Filtering*. Prentice-Hall, Englewood Cliffs, NJ, 1979.

[18] TW Anderson. *Introduction to Multivariate Statistical Analysis*. Wiley, New York, 2nd edition, 1984.

[19] JF Andrews. A mathematical model for the continuous culture of microorganisms utilizing inhibitory substrates. *Biotechnol. Bioengng.*, 10:707–723, 1968.

[20] FJ Anscombe. Rejection of outliers. *Technometrics*, 2:123–147, 1960.

[21] M Aoki. *State Space Modeling of Time Series*. Springer-Verlag, New York, 2nd edition, 1990.

[22] Y Arkun and WH Ray, editors. *Proceedings of Chemical Process Control - CPC IV*. AIChE, New York, 1991.

[23] JA Asenjo. Process kinetics and bioreactor design for the direct bioconversion of cellulose into microbial products. *Biotechnol. Bioengng. Symp.*, 13:449–456, 1983.

[24] JA Asenjo and C Jew. Primary metabolite or microbial protein from cellulose: conditions, kinetics and modeling of the simultaneous saccharification and fermentation to citric acid. *Ann. N.Y. Acad. Sci.*, 413:211–217, 1983.

[25] AJ Assis and RM Filho. Soft sensors development for on-line bioreactor state estimation. *Comp. Chem. Engng.*, 24:1099–1103, 2000.

[26] B Atkinson and F Mavituna. *Biochemical Engineering and Biotechnology Handbook.* Stockton Press, New York, 1991.

[27] M Aynsley, A Hofland, GA Montague, D Peel, and AJ Morris. A real-time knowledge based system for the operation and control of a fermentation plant. In *Proc. American Control Conference*, pages 1992–1997, San Diego, CA, 1990.

[28] M Aynsley, A Hofland, AJ Morris, GA Montague, and C Di Massimo. Artificial intelligence and the supervision of bioprocesses (real-time knowledge-based systems and neural networks). *Adv. in Biochem. Eng. Biotech.*, 48:1–27, 1993.

[29] MJ Bagajewicz. *Process Plant Instrumentation: Design and Upgrade.* C.H.I.P.S. Books, Weimar, Texas, 2001.

[30] JE Bailey. Periodic operation of chemical reactors: A review. *Chem. Engng Commun.*, 1:111–124, 1973.

[31] JE Bailey. Toward a science of metabolic engineering. *Science*, 252:1668–1675, 1991.

[32] JE Bailey. Mathematical modelling and analysis in biochemical engineering: Past accomplishments and future opportunities. *Biotechnology Progress*, 14:8–20, 1998.

[33] JE Bailey, MA Hjortso, SB Lee, and F Srinec. Kinetics of product formation and plasmid segregation in recombinant microbial populations. *Ann. N. Y. Acad. Sci.*, 413:71–87, 1983.

[34] JE Bailey and FJM Horn. Comparison between two sufficient conditions for improvement of an optimal steady-state process by periodic operation. *J. Optim. Theor. Applic.*, 18:378–384, 1971.

[35] JE Bailey and DF Ollis. *Biochemical Engineering Fundamentals.* McGraw Hill, New York, 2nd edition, 1986.

[36] RK Bajpai and M Reuss. A mechanistic model for penicillin production. *J. Chem. Technol. Biotechnol.*, 30:332–344, 1980.

[37] A Bakhtazad, A Palazoglu, and JA Romagnoli. Detection and classification of abnormal process situations using multidimensional wavelet domain hidden markov trees. *Comp Chem Engng*, 24:769–775, 2000.

[38] BR Bakshi. Multiscale PCA with application to multivariate statistical monitoring. *AIChE Journal*, 44(7):1596–1610, 1998.

[39] BR Bakshi, G Locher, G Stephanopoulos, and G Stephanopoulos. Analysis of operating data for evaluation, diagnosis and control of batch operations. *Journal of Process Control*, 4(4):179–194, 1994.

[40] A Banerjee, Y Arkun, B Ogunnaike, and R Pearson. Estimation of non-linear systems using linear multiple models. *AIChE Journal*, 43:1204–1226, 1997.

[41] V Barnett. The study of outliers: Purpose and model. *Appl. Statist.*, 27:242–250, 1978.

[42] V Barnett and T Lewis. *Outliers in Statistical Data.* John Wiley, New York, 1978.

[43] MR Basila Jr., G Stefanek, and A Cinar. A model-object based supervisory expert system for fault tolerant chemical reactor control. *Comp Chem Engng*, 14(4/5):551–560, 1990.

[44] M Basseville. Detecting changes in signals and systems - a survey. *Automatica*, 24:309–326, 1988.

[45] M Basseville and IV Nikiforov. *Detection of abrubt changes: Theory and application.* Prentice Hall, New Jersey, 1993.

[46] D Batty and MS Kamel. Automating knowledge acquisition:A propositional approach to representing expertise alternative to repertory grid technique. *IEEE Trans. Knowledge and Data Engng*, 7(1):53–67, 1995.

[47] CD Bazua and CR Wilke. Ethanol effects on the kinetics of a continuous fermentation with *Saccharomyces cerevisiae*. *Biotechnol. Bioengng Symp.*, 7:105–118, 1977.

[48] S Becker. Unsupervised learning procedures for neural networks. *Int J Neural Syst*, 2:17–33, 1991.

[49] R Bellman and S Dreyfus. *Applied Dynamic Programming.* Princeton Univ. Press, New Jersey, 1962.

[50] A Benveniste, M Basseville, and G Moustakides. The asymptotic local approach to change detection and model validation. *IEEE Trans. Automatic Control*, AC-32:583–592, 1987.

[51] R Berber. Control of batch reactors: A review. *IChemE Proceedings, Part A*, 74:3–20, 1996.

[52] A Berry. Improving production of aromatic compounds in *Escherichia coli* by metabolic engineering. *Trends in Biotechnology*, 14:219–259, 1996.

[53] DP Bertsekas. *Dynamic Programming and Optimal Control.* Athena Scientific, Belmont, MA, 2nd edition, 2000.

[54] LT Biegler and JB Rawlings. Optimization Approaches to Nonlinear Model Predictive Control. In *Proceedings of Chemical Process Control - CPC IV*, pages 543–571, New York, 1991. AIChE.

[55] `http://turnbull.dcs.st-and.ac.uk/history/Mathematicians/Birkhoff.html`. [Accessed 26 November 2002].

[56] GJ Birky and TJ McAvoy. A general framework for creating expert systems for control system design. *Comp Chem Engng*, 14(7):713–728, 1990.

[57] G Birol. *Fermentation Characteristics of Four Genetically Engineered Saccharomyces cerevisiae Strains: Ph.D. Thesis.* Boğaziçi University, Istanbul, 1997.

[58] G Birol, İ Birol, B Kirdar, and Zİ Önsan. Modeling of recombinant yeast cells: Reduction of phase space. In SF Barrett and CHG Wright, editors, *Biomedical Sciences Instrumentation Volume 34. Presented at: Copper Mountain, CO*, Research Triangle Park, 1998. ISA.

[59] G Birol, İ Birol, B Kirdar, and Zİ Önsan. Investigating the fermentation dynamics structure of recombinant *saccharomyces cerevisiae*, ypb-g. *Computers and Chemical Engineering*, To Appear.

[60] G Birol, Zİ Önsan, B Kırdar, and SG Oliver. Ethanol production and fermentation characteristics of recombinant *Saccharomyces cerevisiae* strains grown on starch. *Enzyme and Microbial Technol.*, 22(8):672–677, 1998.

[61] G Birol, C Undey, and A Cinar. A modular simulation package for fed-batch fermentation: Penicillin production. *Comp. Chem. Eng.*, 26(11):1553–1565, 2002.

[62] G Birol, C Undey, SJ Parulekar, and A Cinar. A comparative study on the modeling of penicillin fermentation. In *AIChE Annual Meeting*, Dallas, TX, 1999.

[63] G Birol, C Undey, SJ Parulekar, and A Cinar. A morphologically structured model for penicillin production. *Biotechnol. Bioeng.*, 77(5):538–552, 2002.

[64] İ Birol, SJ Parulekar, and F Teymour. Effect of environment partitioning on the survival and coexistence of autocatalytic replicators. *Phys. Rev. E*, 66(5):051916, 2002.

[65] İ Birol and F Teymour. Statics and dynamics of multiple autocatalytic reactions. *Physica D*, 144:279–297, 2000.

[66] S Bittani, G Fronza, and G Guardabassi. Periodic control: a frequency domain approach. *IEEE Trans. Autom. Control*, AC-18:33–38, 1973.

[67] HW Blanch and DS Clark. *Biochemical Engineering.* Marcel Dekker, New York, 1997.

[68] F Blomer and HO Gunther. LP-based heuristics for scheduling chemical batch processes. *International J. of Production Research*, 38(5):1029–1051, 2000.

[69] G Boente and R Fraiman. Robust nonparametric regression estimation for dependent observations. *Ann. Statist.*, 17:1242–1256, 1989.

[70] D Bonné and SB Jørgensen. Development of learning control for reproducible and high quality operation of batch processes. In *IFAC DYCOPS6*, pages 449–454, Cheju Island, Korea, 2001.

[71] D Bonvin. Optimal operation of batch reactors–a personal view. *Journal of Process Control*, 5–6(8):355–368, 1998.

[72] D Bonvin, P de Valliere, and DWT Rippin. Application of estimation techniques to batch reactors – I. Modeling thermal effects. *Computers Chem. Engng*, 13:1–9, 1989.

[73] D Bonvin, B Srinivasan, and D Ruppen. Dynamic optimization in the batch chemical industry. In *Preprints of Chemical Process Control Conference VI*, pages 283–307, Arizona, January, 2001.

[74] W Borzani, RE Gregori, and MCR Vairo. Response of a continuous anaerobic culture to periodic variation of the feeding mash concentration. *Biotechnol. Bioengng*, 18:623–632, 1976.

[75] JD Boskovic and KS Narendra. Comparison of linear, nonlinear and neural network based adaptive controllers for a class of fed-batch fermentations. *Automatica*, 31(6):817–840, 1995.

[76] GEP Box. Some theorems on quadratic forms applied in the study of analysis of variance problems: Effect of inequality of variancc in onc-way classification. *The Annals of Mathematical Statistics*, 25:290–302, 1954.

[77] GEP Box and NR Draper. *Empirical Model Building and Response Surfaces.* Wiley, New York, 1987.

[78] GEP Box, WG Hunter, and JS Hunter. *Statistics for experimenters.* Wiley, New York, 1978.

[79] GEP Box, GM Jenkins, and GC Reinsel. *Time series analysis - Forecasting and Control.* Prentice-Hall, Inc., Englewood Cliffs, NJ, 3rd edition, 1994.

[80] DR Brillinger. Discussion on linear functional relationships (by p. sprent). *J. R. Statist. Soc. B*, 28:294–294, 1966.

[81] EH Bristol. On a new measure of interactions for multivariable process control. *IEEE Trans. Auto. Control*, AC-11:133, 1966.

[82] R Bro and AK Smilde. Centering and scaling in component analysis. J Chemometrics, Submitted.

[83] RW Brockett. Volterra series and geometric control theory. *Automatica*, 12:167–176, 1976.

[84] AE Bryson and YC Ho. *Applied Optimal Control.* Hemisphere, Washington, D.C., 1975.

[85] IY Caldwell and APJ. Trinci. The growth unit of the mould *Geotrichum candidum. Arch. Microbiol.*, 88:1–10, 1973.

[86] DC Cameron and F Chaplen. Developments in metabolic engineering. *Current Opinions in Biotechnology*, 8:175–180, 1997.

[87] DC Cameron and IT Tong. Cellular and metabolic engineering. *Appl. Biochem. Biotech.*, 38:105–140, 1993.

[88] R Carlson. Preludes to a screening experiment: A tutorial. *Chemometrics and Intelligent Laboratory Systems*, 14:103–114, 1992.

[89] `http://turnbull.dcs.st-and.ac.uk/history/Mathematicians/Cartwright.html`. [Accessed 26 November 2002].

[90] M Casdagli, T Sauer, and JA Yorke. Embedology. *J. Stat. Phys.*, 65:579–616, 1991.

[91] I Chang, CG Tiao, and C Chen. Estimation of time series parameters in the presence of outliers. *Technometrics*, 30:193–204, 1988.

[92] B Chaudhuri and JM Modak. Optimization of fed-batch bioreactor using neural network model. *Bioprocess Engineering*, 19:71–79, 1998.

[93] S Chen and SA Billings. Modeling and analysis of nonlinear time series. *Int. J. Control*, 49:2151–2171, 1989.

[94] S Chen and SA Billings. Orthogonal least squares methods and its application to non-linear system identification. *Int. J. Control*, 50:1873–1896, 1989.

[95] S Chen and SA Billings. Representations of nonlinear systems: The NARMAX model. *Int. J. Control*, 49:1013–1032, 1989.

[96] S Chen and L Lou. Joint estimation of model parameters and outlier effects in time series. *J. Amer. Statist. Assoc.*, 88:284–297, 1993.

[97] Y Cheng, W Karjala, and DM Himmelblau. Resolving problems in closed loop nonlinear process identification using IRN. *Comp. and Chem. Engng.*, 20(10):1159–1176, 1996.

[98] A Cheruy. Software sensors in bioprocess engineering. *Journal of Bioengineering*, 52:193–199, 1997.

[99] LH Chiang and RD Braatz. *Fault Detection and Diagnosis in Industrial Systems.* Springer-Verlag, London, UK, 2001.

[100] IS Chin, KS Lee, and JH Lee. A technique for integrating quality control, profile control and constraint handling for batch processes. *Ind. Eng. Chem. Res*, 39:693–705, 2000.

[101] EY Chow and AS Willsky. Analytical redundancy and the design of robust failure detection systems. *IEEE Trans. Automatic Control*, AC-29(1):603–614, 1984.

[102] A Cinar, J Deng, SM Meerkov, and X Shu. Vibrational control of an exothermic reaction in a CSTR: Theory, simulations, experiments. *AIChE J*, 33:353–365, 1987.

[103] A Cinar, J Deng, SM Meerkov, and X Shu. Vibrational stabilization of a chemical reactor: An experimental study. *IEEE Trans. Autom. Control*, "AC-32"(4):348–352, 1987.

[104] A Cinar, K Rigopoulos, X Shu, and SM Meerkov. Vibrational control of chemical reactors: Stability and conversion improvement in an exothermic CSTR. *Chem. Eng. Comm.*, 59:299–308, 1987.

[105] RN Clark. Detecting instrument malfunction in control systems. *IEEE Trans. Aerospace and Electronic Systems*, AES-11:465–473, 1975.

[106] RN Clark, DC Fosth, and WM Walton. Instrument fault detection. *IEEE Trans. Aerospace and Electronic Systems*, AES-14:456–465, 1978.

[107] DW Clarke and C Mohtadi. Properties of generalized predictive control. *Automatica*, 25:859–875, 1989.

[108] TL Clarke-Pringle and JF MacGregor. Optimization of molecular weight distribution using batch-to-batch adjustments. *Ind. Eng. Chem. Res.*, 37:36603669, 1998.

[109] P Collet and J-P Eckmann. Properties of continuous maps of the interval to itself. In K Osterwalder, editor, *Mathematical Problems in Theoretical Physics*, New York, 1979. Springer-Verlag.

[110] P Collet and J-P Eckmann. *Iterated Maps on the Interval as Dynamical Systems.* Birkhäuser, Boston, 1980.

[111] Western Electric Company. *Statistical Quality Control Handbook.* AT&T Technologies, Indianapolis, 1984.

[112] CL Cooney and F. Acevedo. Theoretical conversion yields for penicillin synthesis. *Biotechnol. Bioeng.*, 19:1449–1462, 1977.

[113] CM Crowe, YA Garcia Campos, and A Hrymak. Reconciliation of process flow rates by matrix projection. I. The linear case. *AIChE J*, 29:818, 1983.

[114] FX Cunningham and E Gantt. Genes and enzymes of carotenoid biosynthesis in plants. *Molecular Biology*, 49:557–583, 1998.

[115] M Cutlip. Concentration forcing of catalytic surface rate processes. *AIChE J*, 25:502–508, 1979.

[116] I Daubechies. *Ten Lectures on Wavelets.* Society for Industrial and Applied Mathematics, Pennsylvania, 1992.

[117] ME Davies. Noise reduction schemes for chaotic time series. *Physica D*, 79:174–192, 1994.

[118] BS Dayal and JF MacGregor. Improved PLS algorithms. *J of Chemometrics*, 11:73–85, 1997.

[119] S de Jong and HAL Kiers. Principal covariates regression Part I. Theory. *Chemometrics Intell. Lab. Syst.*, 14:155–164, 1992.

[120] P de Valliere and D Bonvin. Application of estimation techniques to batch reactors – II. Experimental studies in state and parameter estimation. *Computers Chem. Engng*, 13:11–20, 1989.

[121] M deBoor. *A Practical Guide to Splines.* Springer-Verlag, New York, 1978.

[122] J DeCicco and A Cinar. Empirical modeling of systems with output multiplicities by multivariate additive narx models. *Industrial and Engineering Chemistry Research*, 39(6):1747–1755, 2000.

[123] JR Deller, JG Proakis, and JHL Hansen. *Discrete-Time Processing of Speech Signals.* Wiley-IEEE Press, New York, 1999.

[124] GH Denis and RL Kabel. The effect on conversion of flow rate variations in a heterogeneous catalytic reactor. *AIChE J*, 16:972–978, 1970.

[125] C Di Massimo, GA Montague, MJ Willis, MT Tham, and AJ Morris. Towards improved penicillin fermentation via artificial neural networks. *Computers Chem. Engng.*, 16(4):283–291, 1992.

[126] M Dixon and EC Webb. *Enzymes.* Academic Press, New York, NY, 3rd edition, 1979.

[127] RA Dixon and CJ Arntzen. Transgenic plant technology is entering the era of metabolic engineering. *Trends in Biotechnology*, 11:441–444, 1997.

[128] EJ Doedel. Auto: A program for the automatic bifurcation analysis of autonomous systems. In *Proc. 10th Manitoba Conf. on Num. Math. and Comp., Univ. of Manitoba, Winnipeg, Canada*, pages 265–284, 1981.

[129] D Dong and TJ McAvoy. Multi-stage batch process monitoring. In *Proc. American Control Conference*, pages 1857–1861, Seattle, WA, 1995.

[130] D Dong and TJ McAvoy. Batch tracking via nonlinear principal component analysis. *AIChE Journal*, 42:2199–2208, 1996.

[131] D Dong, TJ McAvoy, and E Zafiriou. Batch-to-batch optimization using neural networks. *Ind. Eng. Chem. Res.*, 35:2269–2276, 1996.

[132] DL Donoho and M Johnstone. Wavelet shrinkage: Asymptotia? *J.R. Stat. Soc. B*, 57:301, 1995.

[133] JS Dordick. `www.rpi.edu/dept/chem-eng/Biotech-Environ/FERMENT/batchb.htm`. [Accessed 26 November 2002].

[134] AW Dorsey and JH Lee. Monitoring of batch processes through state-space models. In *IFAC DYCOPS6*, pages 245–250, Cheju Island, Korea, 2001.

[135] JM Douglas. *Process Dynamics and Control, Vol. II*, page 288. Prentice-Hall, Englewood Cliffs, NJ, 1972.

[136] JM Douglas and TG Dorawala. Complex reactions in oscillating reactors. *AIChE J.*, 17:974–981, 1971.

[137] JJ Downs and EF Vogel. A plant-wide industrial control problem. In *AIChE Annual Meeting*, Chicago, IL, 1990.

[138] F Doymaz, JA Romagnoli, and A Palazoglu. A strategy for detection and isolation of sensor faults and process upsets. *Chemometrics and Intelligent Laboratory Systems*, 55:109–123, 2001.

[139] RO Duda and PE Hart. *Pattern Classification and Scene Analysis.* John Wiley, New York, 1973.

[140] EJ Dudewicz and SN Mishra. *Modern Mathematical Statistics.* John Wiley, New York, 1988.

[141] J Dun and E Gulari. Rate and selectivity modification in Fischer-Tropsch synthesis over charcoal supported Molybdenum by forced concentration cycling. In *AIChE Annual Meeting*, Chicago, IL, 1985.

[142] JW Eaton and JB Rawlings. Feedback control of nonlinear processes using on-line optimization techniques. *Comp. Chem. Eng.*, 14:469–479, 1990.

[143] L Eggeling, H Sahm, and AA de Graaf. Quantifying and directing metabolic flux: application to amino acid overproduction. *Adv. Biochem Eng. Biotechnol.*, 54:1–30, 1996.

[144] AP Engelbrecht. A new pruning heuristic based on variance analysis of sensitivity information. *IEEE Trans. on Neural Networks*, 12(6):1386–1399, 2001.

[145] L Eriksson, E Johansson, N Kettaneh-Wold, and S Wold. *Multi- and Megavariate Data Analysis.* Umetrics Academy, Umeå, Sweden, 2001.

[146] B Ermentrout. *Simulating, Analyzing, and Animating Dynamical Systems: A Guide to XPPAUT for Researchers and Students.* SIAM, Philadelphia, 2002.

[147] B.F̃uchssteiner et al. *MuPAD Tutorial.* Birkhäuser, Basel, 1994.

[148] M Evans, N Hastings, and B Peacock. *Statistical Distributions.* John Wiley, New York, 1993.

[149] WR Farmer and JC Liao. Progress in metabolic engineering. *Current Opinion in Biotechnology*, 7:198–204, 1996.

[150] MJ Feigenbaum. Quantitative universality for a class of nonlinear transformations. *J. Stat. Phys.*, 19:153–180, 1978.

[151] DA Fell. Metabolic control analysis-A survey of its theoretical and experimental development. *Biochem. J.*, 152:313–330, 1992.

[152] C Filippi-Bossy, J Bordet, J Villermaux, S Marchal-Brassely, and C Georgakis. Batch reactor optimization by use of tendency models. *Comp. Chem. Eng.*, 13:35–47, 1989.

[153] RA Fisher. The statistical utilization of multiple measurements. *Annals of Eugenics*, 8:376–386, 1938.

[154] A Fleming. On the antibacterial action of cultures of a penicillium, with special reference to their use in the isolation of *B. influenza. Brit. J. Exp. Path.*, 10:226–236, 1929.

[155] J Flores-Cerrillo and JF MacGregor. Control of particle size distributions in emulsion semibatch polymerization using mid-course correction policies. *Ind. Engng. Chem. Res.*, 41:1805–1814, 2002.

[156] BA Foss, TA Johnson, and AV Sorensen. Nonlinear model predictive control using local models- applied to a batch fermentation process. *Control Eng. Prac.*, 3:389–396, 1995.

[157] J Fotopoulos, C Georgakis, and HG Stenger. Uncertainty issues in the modeling and optimisation of batch reactors with tendency modeling. *Chem. Engng. Sci.*, 49:5533–5548, 1994.

[158] F Fournier, MA Latifi, and G Valentin. Methodolgy of dynamic optimization and optimal control of batch electrochemical reactors. *Chem. Engng. Sci.*, 54:2707–2714, 1999.

[159] AJ Fox. Outliers in time series. *J. R. Statist. Soc. B*, 34:340–363, 1972.

[160] I Frank. A nonlinear PLS model. *Chemometrics and Intelligent Laboratory Systems*, 8:109–119, 1990.

[161] I Frank and L Keller. Sensitivity discriminating observer design for instrument failure detection. *IEEE Trans. Aerospace and Electronic Systems*, AES-16:460–467, 1980.

[162] P Frank. Fault diagnosis in dynamic systems using analytical and knowledge-based redundancy. *Automatica*, 26:459–474, 1990.

[163] PM Frank and J Wünnenberg. Robust fault diagnosis using unknown input observers schemes. In RJ Patton, PM Frank, and R Clark, editors, *Fault Diagnosis in Dynamical Systems*, pages 47–98. Prentice Hall, 1989.

[164] MP Franklin. Selecting defining contrasts and confounded effects in p^{n-m} factorial experiments. *Technometrics*, 27:321–326, 1985.

[165] AM Fraser and HL Swinney. Independent coordinates for strange attractors from mutual information. *Phys. Rev. A*, 33:1134–1140, 1986.

[166] AG Fredrickson and HM Tsuchiya. Chapter 7 - Microbial Kinetics and Dynamics. In L Lapidus and NR Amundson, editors, *Chemical Reactor Theory. A Review.* Prentice-Hall, Inc., New York, NY, 1977.

[167] JE Freund. *Mathematical Statistics.* Prentice Hall, Englewood Cliffs, New Jersey, 1971.

[168] B Friedland. Maximum likelihood estimation of a process with random transitions (failures). *IEEE Trans. Automatic Control*, AC-24:932–937, 1979.

[169] JH Friedman. Multivariate adaptive regression splines. *Ann. Statist.*, 19:1–144, 1991.

[170] PC Fu and JP Barford. A hybrid neural network-first principles approach for modelling of cell metabolism. *Comp. Chem. Engng.*, 20(6-7):951–958, 1996.

[171] K Fukunaga. *Statistical Pattern Recognition.* Academic Press, San Diego, CA, 1990.

[172] RA Gabel and RA Roberts. *Signals and Linear Systems.* John Wiley, New York, 3rd edition, 1980.

[173] JPD Gagnepain and DE Seborg. Analysis of process interactions with applications to multiloop control system design. *Ind. Eng. Chem. Proc. Des. Dev.*, 21:5, 1982.

[174] CE Garcia and M Morari. Internal Model Control - 1. A Unifying Review and Some New Results. *Ind. Eng. Chem. Process Des. Dev.*, 21:308, 1982.

[175] CE Garcia and AM Morshedi. Quadratic Programming Solution of Dynamic Matrix Control (QDMC). *Chem. Eng. Commun.*, 46:73, 1986.

[176] CE Garcia, DM Prett, and M Morari. Model Predictive Control: Theory and Practice - A Survey. *Automatica*, 25:335, 1989.

[177] G Gattu and E Zafiriou. A methodology for on-line setpoint modification for batch reactor control in the presence of modeling error. *Chem. Eng. Journal*, 75(1):21–29, 1999.

[178] M Ge, QG Wang, MS Chin, TH Lee, CC Hang, and KH Teo. An effective technique for batch process optimization with application for batch process optimization with application to crystallization. *Trans. IChemE.*, 78A:99–106, 2000.

[179] P Geladi. Analysis of multi–way (multi–mode) data. *Chemometrics Intelligent Lab. Systems*, 7:11–30, 1989.

[180] P Geladi and BR Kowalski. An example of 2–block predictive partial least–squares regression with simulated data. *Analytica Chimica Acta*, 185:19–32, 1986.

[181] P Geladi and BR Kowalski. Partial least-squares regression: A tutorial. *Analytica Chimica Acta*, 185:1–17, 1986.

[182] A Gelb. *Applied Optimal Estimation.* MIT Press, Cambridge, MA, 1974.

[183] J Gensel. Integrating constraints in an object-based knowledge representation system. In M Meyer, editor, *Constraint Processing: Selected Papers (Lecture Notes in Computer Science, No 923)*, pages 67–77. Springer-Verlag, 1995.

[184] G2 reference manual, 2002. Gensym Corporation, Cambridge, MA.

[185] `http://www.gensym.com/product`. [Accessed 26 November 2002].

[186] C Gentric, F Pla, MA Latifi, and JP Corriou. Optimization and nonlinear control of a batch emulsion polymerization. *Chem. Engng. Journal*, 75(1):31–46, 1999.

[187] J Gertler and D Singer. Augmented models for statistical fault diagnosis in complex dynamical systems. In *Proc. American Control Conference*, pages 317–322, Boston, MA, 1985.

[188] JJ Gertler. Analytical redundancy methods in fault detection and isolation - survey and synthesis. In *Prep. IFAC Safeprocess Conference*, pages 9–22, Baden-Baden, GE, 1991.

[189] JJ Gertler. *Fault detection and diagnosis in engineering systems.* Marcel Dekker, New York, NY, 1998.

[190] TK Ghose and RD Tyagi. Rapid ethanol fermentation of cellulose hydrolysate. II. Product and substrate inhibition and optimization of fermenter design. *Biotechnol. Bioeng.*, 21:1401–1420, 1979.

[191] J Giarratano and G Riley. *Expert Systems: Principles and Programming.* PWS Publishing Co., Boston, MA, 3rd edition, 1998.

[192] JW Gilley and HR Bungay. Frequency response analysis of dilution rate effects on yeast growth. *Biotechnol. Bioeng.*, 10:99–101, 1968.

[193] J Glassey, G Montague, and P Mohan. A case study in industrial bioprocess advisory system development. In *Proc. International Symposium on Advanced Control of Chemical Processes (ADCHEM 2000)*, Pisa, ITALY, 2000.

[194] J Glassey, GA Montague, AC Ward, and BV Kara. Artificial neural network based design procedures for enhancing fermentation development. *Biotechnol. Bioeng.*, 44:397–405, 1994.

[195] J Gleick. *Chaos: Making a New Science.* Penguin, New York, 1988.

[196] A. L. Goldberger and D. R. Rigney. Nonlinear dynamics at the bedside. In L. Glass, P. Hunter, and A. McCulloch, editors, *Theory of Heart: Biomechanics, Biophysics, and Nonlinear Dynamics of Cardiac Function*, New York, 1991. Springer-Verlag.

[197] K Gollmer and C Posten. Supervision of bioprocesses using a dynamic time warping algorithm. *Control Engineering Practice*, 4(9):1287–1295, 1996.

[198] C Goutis. A fast method to compute orthogonal loadings partial least squares. *J of Chemometrics*, 11:33–38, 1997.

[199] L Gregersen and SB Jorgensen. Supervision of fed-batch fermentations. *Chemical Engng J*, 75:69–76, 1999.

[200] L Gregersen and SB Jørgensen. Identification of linear models for batch control and optimization. In *IFAC DYCOPS6*, pages 449–454, Cheju Island, Korea, 2001.

[201] FE Grubbs. Procedures for detecting outlying observations in samples. *Technometrics*, 11:1–21, 1969.

[202] G Guardabassi, A Locatelli, and S Rinaldi. Survey paper: status of periodic optimization of dynamical systems. *J. Optim. Theor. Applic.*, 14:1–20, 1974.

[203] Guay, M. Personal communication, 2000.

[204] J Guckenheimer, A Back, J Guckenheimer, M Myers, Wicklin F, and Worfolk P. dstool: Computer assisted exploration of dynamical systems. *Notices of the American Mathemtical Society*, 39:303–309, 1992.

[205] MA Guerreiro, SR Andrietta, and F Maugeri. Expert system for the design of an industrial fermentation plant for the production of alcohol. *J Chem Tech Biotechnol*, 68:163–170, 1997.

[206] SP Gurden, JA Westerhuis, R Bro, and AK Smilde. A comparison of multiway regression and scaling methods. *Chemometrics Intell. Lab. Syst.*, 59(1-2):121–136, 2001.

[207] F Gustafsson. *Adaptive Filtering and Change Detection*. John Wiley, New York, 2000.

[208] R Guthke, W Schmidt-Heck, and M Pfaff. Knowledge acquisition and knowledge based control in bioprocess engineering. *J Biotech*, 65:37–46, 1998.

[209] H Haario and V-M Taavitsainen. Nonlinear data analysis. II. Examples on new link functions and optimization aspects. *Chemometrics and Intelligent Laboratory Systems*, 23:51–64, 1994.

[210] R Haber and H Unbehauen. Structure identification of nonlinear dynamic systems – A survey on input/output approaches. *Automatica*, 26:651–677, 1990.

[211] V Haggan and T Ozaki. Modeling non-linear random vibrations using an amplitude dependent autoregrcssive time series model. *Biometrika*, 68:186–196, 1981.

[212] AC Hahn. *Forecasting, Structural Time Series Models and the Kalman Filter.* Cambridge University Press, New York, NY, 1989.

[213] GJ Hahn. Some things engineers should know about experimental design. *Journal of Quality Technology*, 9:13–20, 1977.

[214] GJ Hahn and WQ Meeker. *Statistical Intervals. A Guide to Practitioners.* John Wiley, New York, 1991.

[215] H Haken. At least one lyapunov exponent vanishes if the trajectory of an attractor does not contain a fixed point. *Phys. Lett. A*, 94:71–72, 1983.

[216] P Hall and MC Jones. Adaptive M-estimation in nonparametric regression. *Ann. Statist.*, 18:1712–1728, 1990.

[217] FR Hampel, EM Ronchetti, PJ Rousseeuw, and WA Stahel. *Robust Statistics: The Approach Based on Influence Functions.* Wiley, New York, 1986.

[218] J Hansen and MC Kiellandbrandt. Modification of biochemical pathways in industrial yeasts. *Journal of Biotechnology*, 49:1–12, 1996.

[219] W Härdle and T Gasser. Robust non-parametric function fitting. *J. R. Statist. Soc. B*, 46:42–51, 1984.

[220] TJ Harris and WH Ross. Statistical process control procedures for correlated observations. *Canadian J. of Chemical Eng.*, 69:48, 1991.

[221] A Hart. *Knowledge Acquisition for Expert Systems.* Kogan Page, London, United Kingdom, 1989.

[222] AC Harvey and RG Pierse. Estimating missing observations in economic time series. *J. Amer. Statist. Ass.*, 79:125–131, 1984.

[223] V Hatzimankatis and JE Bailey. MCA has more to say. *J. Theor. Biol.*, 182:233–242, 1996.

[224] DM Hawkins. The detection of errors is multivariate data using principal components. *J. Amer. Statist. Ass.*, 69:340–344, 1974.

[225] DM Hawkins. *Identification of Outliers.* Chapman-Hall, London, 1980.

[226] S Haykin. *Neural Networks.* Prentice-Hall, Upper Saddle River, New Jersey, 2nd edition, 1999.

[227] DO Hebb. *The Organization of Behavior: A Neurophysiological Theory.* Wiley, New York, 1949.

[228] R Hegger, H Kantz, and T Schreiber. Practical implementation of nonlinear time series methods: The tisean package. *Chaos*, 9:413–435, 1999.

[229] R Heinrich and TA Rapoport. A linear steady-state treatment of enzymatic chains. *Eur. J. Biochem.*, 42:89–95, 1974.

[230] B Hendricks, RA Korus, and RC Heimsch. Propionic acid production by bacterial fermentation. *Biotechnol. Bioeng. Symp.*, 15:241–245, 1985.

[231] M Hénon. A two dimensional mapping with a strange attractor. *Comm. Math. Phys.*, 50:69–78, 1976.

[232] R Henrion. N-way principal component analysis. Theory, algorithms and applications. *Chemo. Intell. Lab. Sys.*, 25:1–23, 1994.

[233] B. Henry, N. Lovell, and F. Camacho. Nonlinear dynamics time series analysis. In M Akay, editor, *Nonlinear Biomedical Signal Processing*, New York, 2001. IEEE.

[234] MA Henson. Nonlinear model predictive control: Current status and future directions. *Comp. Chem. Engng.*, 23:187–202, 1998.

[235] DM Himmelblau. *Fault Detection and Diagnosis in Chemical and Petrochemical Processes.* Elsevier Science, New York, 1978.

[236] Hitzmann, B and A Lübbert and K Schügerl. An expert system approach for the control of a bioprocess. I:Knowledge representation and processing. *Biotech. Bioengng*, 39:33–43, 1992.

[237] U Hoffmann and H-K Schadlich. The influence of reaction orders and of changes in the total number of moles on the conversion in a periodically operated CSTR. *Chem. Engng Sci.*, 41:2733–2738, 1986.

[238] CP Hollenberg and AW Strasser. Improvement of baker's and brewer's yeast by gene technology. *Food Biotechnol.*, 4:527–534, 1990.

[239] H Honda, T Mano, M Taya, K Shimizu, M Matsubara, and T Kobayashi. A general framework for the assessment of extractive fermentations. *Chem. Engng Sci.*, 42:493–498, 1987.

[240] FJM Horn and RC Lin. Periodic processes: a variational approach. *Ind. Engng Chem. Proc. Des. Dev.*, 6:21–30, 1967.

[241] IM Horowitz. *Synthesis of Feedback Systems.* Academic Press, London, 1963.

[242] JC Hoskins and DM Himmelblau. Artificial neural network models of knowledge representation in chemical engineering. *Computers Chem. Engng.*, 12(9-10):881–890, 1988.

[243] A Hoskuldsson. PLS regression methods. *Journal of Chemometrics*, 2:211, 1988.

[244] H Hotelling. Analysis of a complex of statistical variables into principal components. *J. Educ. Psychol.*, 24:417, 1933.

[245] PJ Huber. *Robust Statistics.* Wiley, New York, 1981.

[246] R Hudlet and R Johnson. Linear discrimination and some further results on best lower dimensional representations. In V Rzyin, editor, *Classification and Clustering*, pages 371–394. Academic Press, Inc., New York, NY, 1977.

[247] CR Hutchinson. Drug synthesis by genetically engineered microorganisms. *Bio/Technology*, 12:375–380, 1994.

[248] M Ignova, J Glassey, AC Ward, and GA Montague. Multivariate statistical methods in bioprocess fault detection and performance testing. *Trans Inst MC*, 19(5):387–404, 1997.

[249] M Ikeda and R Katsumata. Metabolic engineering to produce tyrosine or phenylalanine in a tryptophan-producing *Corynebacterium glutamicum* strain. *Appl. Environ. Microbiol.*, 58:781–785, 1992.

[250] LO Ingram and T Conway. Expression of different levels of ethanologenic enzymes from Zymomonas mobilis in recombinant strains of *Escherichia coli. Appl. Environ. Microbiol.*, 54:397–404, 1988.

[251] R Isermann. Process fault detection based on modelling and estimation methods. *Automatica*, 20:387–404, 1984.

[252] F Itakura. Minimum prediction residual principle applied to speech recognition. *IEEE Trans.*, ASSP-23(1):67–72, 1975.

[253] JE Jackson. Principal components and factor analysis : Part I - Principal Components. *Journal of Quality Technology*, 12(4):201–213, 1980.

[254] JE Jackson. *A Users Guide to Principal Components.* Wiley, New York, 1991.

[255] JE Jackson and GS Mudholkar. Control procedures for residuals associated with principal components analysis. *Technometrics*, 21:341–349, 1979.

[256] AK Jain, RR Hudgins, and PL Silveston. Influence of forced feed composition cycling on the rate of ammonia synthesis over an industrial iron catalyst. *Can. J. Chem. Engng*, 61:824–832, 1983.

[257] AH Jazwinski. *Stochastic Processes and Filtering Theory.* Academic Press, New York, 1970.

[258] Q Jiang and M Bagajewicz. On a strategy of serial identification with collective compensation for multiple gross error estimation in linear steady state reconciliation. *Ind & Eng Chem Research*, 38:2119–2128, 1999.

[259] Q Jiang, M Sanchez, and M Bagajewicz. On the performance of principal components analysis in multiple gross error identification. *Ind & Eng Chem Research*, 38:2005–2012, 1999.

[260] TA Johansen and R Murray-Smith. *Multiple model approaches to modelling and control.* Taylor & Francis Inc., Bristol, PA, 1997.

[261] A Johnson. The control of fed-batch fermentation processes - A Survey. *Automatica*, 23(6):691–705, 1987.

[262] RA Johnson and DW Wichern. *Applied Multivariate Statistical Analysis.* Prentice-Hall, Englewood Cliffs, NJ, 4th edition, 1998.

[263] IT Jolliffe. *Principal Component Analysis.* Springer Verlag, New York, 1986.

[264] H Jorgensen, J Nielsen, J Villadsen, and H Mollgaard. Metabolic flux distributions in *Penicillium chrysogenum* during fed-batch cultivations. *Biotechnol. Bioeng.*, 46:117–131, 1995.

[265] MR Juba and JW Hamer. Progress and challenges in batch process control. In M Morari and TJ McAvoy, editors, *Chemical Process Control (CPC III)*, pages 139–183. CACHE, Austin, TX and Elsevier, Amsterdam, 1986.

[266] Eaton JW. Gnu octave. `http://www.octave.org/`. [Accessed 26 November 2002].

[267] H Kacser and JA Burns. The control of flux. *Symp. Soc. Exp. Biol.*, 27:65–104, 1973.

[268] MN Karim and SL Rivera. Artificial neural networks in bioprocess state estimation. *Adv. in Biochem. Eng. Biotech.*, 46:1–33, 1992.

[269] TW Karjala and DM Himmelblau. Dynamic data rectification by recurrent neural networks and the extended kalman filter. *AIChE J*, 42:2225, 1996.

[270] A Kassidas, JF MacGregor, and PA Taylor. Synchronization of batch trajectories using dynamic time warping. *AIChE Journal*, 44(4):864–875, 1998.

[271] SM Kay. *Fundamentals of Statistical Signal Processing: Detection Theory*. Prentice Hall, New Jersey, 1998.

[272] DB Kell and HV Westerhoff. Metabolic control theory: Its role in microbiology and biotechnology. *FEMS Microbiol. Rew*, 39:305–320, 1986.

[273] ST Kellogg, DK Chatterjee, and AM Chakrabarty. Plasmid assisted molecular breeding; new technique for enhanced biodegradation of persistent toxic chemicals. *Science*, 214:1133–1135, 1981.

[274] SJ Kendra, MR Basila, and A Cinar. Intelligent process control with supervisory knowledge-based systems. *IEEE Control Systems*, 14:37–47, 1994.

[275] SJ Kendra, MR Basila, and A Cinar. Intelligent process control with supervisory knowledge-based systems. In SG Tzafstas, editor, *Methods and Applications of Intelligent Control*, pages 139–171. Kluwer Academic Publishers, 1997.

[276] MB Kennel, R Brown, and HDI Abarbanel. Determining embedding dimension for phase-space reconstruction using a geometrical construction. *Phys. Rev. A*, 45:3403–3411, 1992.

[277] N Kettaneh-Wold, JF MacGregor, B Dayal, and S Wold. Multivariate design of process experiments. *Chemometrics and Intelligent Laboratory Systems*, 23:39–50, 1994.

[278] C Khosla and JE Bailey. Heterologous expression of a bacterial hemoglobin improves the growth properties of recombinant *Escherichia coli*. *Nature*, 331:633–635, 1988.

[279] DH Kil and FB Shin. *Pattern Recognition and Prediction with Applications to Signal Characterization.* AIP Press, Woodbury, NY, 1996.

[280] RE King. *Computational Intelligence in Control Engineering.* Marcel Dekker, Inc, New York, NY, 1999.

[281] G Kitagawa. On the use of aic for the detection of outliers. *Technometrics*, 21:193–199, 1979.

[282] S Klamt. Fluxanalyzer: A graphical interface for metabolic flux analysis (stoichiometric analysis and determination of flux distributions in metabolic networks). `www.mpi-magdeburg.mpg.de/research/project_a/pro_a5a/mfaeng/intro.html`, 2000. [Accessed 26 November 2002].

[283] R Kohn and CF Ansley. Estimation, prediction, and interpolation for arima models with missing data. *J. Amer. Statist. Ass.*, 81:751–761, 1986.

[284] `http://turnbull.dcs.st-and.ac.uk/history/Mathematicians/Kolmogorov.html`. [Accessed 26 November 2002].

[285] KB Konstantinov and T Yoshida. Physiological state control of fermentation processes. *Biotech. Bioengng.*, 33:1145–1156, 1989.

[286] KB Konstantinov and T Yoshida. An expert approach for control of fermentation processes as variable structure plants. *J Ferment Bioengng*, 70(1):48–57, 1990.

[287] KB Konstantinov and T Yoshida. A knowledge-based pattern recognition approach for real-time diagnosis and control of fermentation processes as variable structure plants. *IEEE Trans. Systems, Man, Cyber.*, 21(4):908–914, 1991.

[288] KB Konstantinov and T Yoshida. Knowledge-based control of fermentation processes. *Biotech. Bioengng.*, 39(5):479–486, 1992.

[289] KB Konstantinov and T Yoshida. Real-time qualitative analysis of the temporal shapes of (bio)process variables. *AIChE Journal*, 38(11):1703–1715, 1992.

[290] MJ Korenberg. Orthogonal identification of nonlinear difference equation models. In *Midwest Symp. on Circuits and Systems*, Louisville, KY, 1985.

[291] KA Kosanovich, KS Dahl, and MJ Piovoso. Improved process understanding using multiway principal component analysis. *Ind. Eng. Chem. Res.*, 35:138–146, 1996.

[292] F Kosebalaban and A Cinar. Integration of multivariate spm and fdd by parity space technique for a food pasteurization process. *Comp Chem Engng*, 25:473–391, 2001.

[293] B Kosko. *Neural Networks and Fuzzy Systems.* Prentice-Hall, Englewood Cliffs, NJ, 1992.

[294] B Kosko. *Fuzzy Engineering.* Prentice-Hall, Englewood Cliffs, NJ, 1996.

[295] EJ Kostelich and T Schreiber. Noise reduction in chaotic time series data: A survey of common methods. *Phys. Rev. E*, 48:1752–1763, 1993.

[296] T Kourti, J Lee, and JF MacGregor. Experiences with industrial applications of projection methods for multivariate statistical process control. *Comp. and Chem. Engng.*, 20(Suppl. A):745, 1996.

[297] T Kourti and JF MacGregor. Process analysis, monitoring and diagnosis using multivariate projection methods. *Chemometrics and Intelligent Laboratory Systems*, 28:3–21, 1995.

[298] T Kourti, P Nomikos, and JF MacGregor. Analysis, monitoring and fault diagnosis of batch processes using multiblock and multiway PLS. *Journal of Process Control*, 5(4):277–284, 1995.

[299] K Kovarova-Kovar, S Gehlen, A Kunze, T Keller, R von Daniken, M Kolb, and APGM van Loon. Application of model-predictive control based on artificial neural networks to optimize the fed-batch process for riboflavin production. , *J. Biotechnol*, 79:39–52, 2000.

[300] DJ Kozub and JF MacGregor. State estimation for semi-batch polymerization reactors. *Chem. Eng. Science*, 47:1047–1062, 1992.

[301] MA Kramer. Autoassociative neural networks. *Comp. Chem. Engng.*, 16(4):313–328, 1992.

[302] MA Kramer and JA Leonard. Diagnosis using backpropagation neural networks-Analysis and criticism. *Comp. Chem. Engng.*, 14:1323, 1990.

[303] MA Kramer and BL Palowitch. A rule-based approach to fault diagnosis using sign directed graph. *AIChE J*, 33:130, 1987.

[304] JV Kresta, JF MacGregor, and TE Marlin. Multivariate statistical monitoring of process operating performance. *Canadian Journal of Chemical Engineering*, 69:35–47, 1991.

[305] PM Kroonberg. *Three Mode Principal Component Analysis: Theory and Applications.* DWSO Press, Leiden, 1983.

[306] B Krose and P van der Smagt. `http://www.robotic.dlr.de/Smagt/books`. [Accessed 26 November 2002].

[307] M Krothapally, B Bennett, W Finney, and S Palanki. Experimental implementation of an on-line optimization scheme to batch PMMA synthesis. *ISA Trans.*, 38:185–198, 1999.

[308] WJ Krzanowski. Between-groups comparison of principal components. *J. of Amer. Stat. Assn.*, 74:703–707, 1979.

[309] WJ Krzanowski. Cross-validation choice in principal component analysis. *Biometrics*, 43:575–584, 1987.

[310] V Kumar, S Ramakrishnan, TT Teeri, JKC Knowles, and BS Hartey. *Saccharomyces cerevisiae* cells secreting an *Aspergillus niger* beta-galactosidase grown on whey permeate. *Bio/Technology*, 71:766–770, 1992.

[311] YuA Kuznetsov and VV Levitin. *CONTENT: A Multiplatform Environment for Analyzing Dynamical Systems.* Dynamical Systems Lab., Centrum voor Wiskunde en Informatica, Amsterdam, 1996.

[312] N Lakshmanan and Y Arkun. Estimation and control of batch processes using multiple models. *Int J of Control*, 72(7/8):659–675, 1999.

[313] OE Lanford. A computer-assisted proof of the feigenbaum conjectures. *Bull. Amer. Math. Soc*, 6:427–434, 1982.

[314] J Lapointe, B Marcos, M Veillette, and G Laflamme. BIOEXPERT-An expert system for waste water treatment process diagnosis. *Computers Chem. Engng.*, 13:619, 1989.

[315] WE Larimore. System identification, reduced-order filtering and modeling via canonical variate analysis. In *Proc. of Automatic Control Conf.*, page 445, 1983.

[316] WE Larimore. Canonical variate analysis in identification, filtering, and adaptive control. In *Proc. of IEEE Conf. on Decision and Control*, page 596, 1990.

[317] MV Le Lann, M Cabassud, and G Casamatta. Modeling, optimization, and control of batch chemical reactors in fine chemical production. In *IFAC DYCOPS5*, pages 751–760, Corfu, Greece, 1998.

[318] CK Lee and JE Bailey. Modification of consecutive-competitive reaction selectivity by periodic operation. *Ind. Engng Chem. Proc. Des. Dev.*, 19:160–166, 1980.

[319] CK Lee, SYS Yeung, and JE Bailey. Experimental studies of a consecutive-competitive reaction in steady state and forced periodic CSTRs. *Can. J. Chem. Engng*, 58:212–218, 1980.

[320] J Lee, K Jung, SH Choi, and H Kim. Combination of the tod and the tol pathways in redesigning a metabolic route of *Pseudomonas putida* for the mineralization of a benzene, toluene and p-xylene mixture. *Appl. Environ. Microbiol.*, 61:2211–2217, 1995.

[321] J Lee, KS Lee, JH Lee, and S Park. An on-line batch span minimization and quality control strategy for batch and semi-batch processes. In *IFAC ADCHEM'00*, page 705712, Pisa, Italy, 2000.

[322] J Lee and SJ Parulekar. Periodic operation of continuous recombinant cultures improves antibiotic selection. *Chem. Engng Sci.*, 51:217–231, 1996.

[323] JH Lee, KS Lee, and WC Kim. Model-based iterative learning control with quadratic criterion for time-varying linear systems. *Automatica*, 36:641–657, 2000.

[324] JH Lee, M Morari, and CE Garcia. State-space interpretation of model predictive control. *Automatica*, 30(4):707–717, 1994.

[325] JM Lee, JF Pollard, and GA Coulman. Ethanol fermentation with cell recycling: Computer simulation. *Biotechnol. Bioeng.*, 25:497–511, 1983.

[326] KJ Lee, DE Tribe, and PL Rogers. Ethanol production by *Zymomonas mobilis* in continuous culture at high glucose concentrations. *Biotechnol. Lett.*, 1:421–426, 1979.

[327] KS Lee, IS Chin, HJ Lee, and JH Lee. Model predictive control technique combined with iterative learning for batch processes. *AIChE J.*, 45:2175–2187, 1999.

[328] KS Lee and JH Lee. Model-based refinement of input trajectories for batch and other transient processes. In *AIChE Annual Meeting*, Chicago, IL, 1996.

[329] SB Lee and JE Bailey. A mathematical model for lambda dv plasmid replication: Analysis of wild-type plasmid. *Plasmid*, 11(2):151–165, 1984a.

[330] SB Lee and JE Bailey. Analysis of growth rate effects on productivity of recombinant *Escherichia coli* populations using molecular mechanism models. *Biotechnol. Bioeng.*, 26:6673, 1984b.

[331] SY Lee and ET Papoutsakis. *Metabolic Engineering.* Marcel Dekker, New York, 1999.

[332] B Lefrancois. Detecting over-influential observations in time series. *Computers Chem. Engng.*, 78:91–99, 1991.

[333] B Lennox, GA Montague, HG Hiden, G Kornfeld, and PR Goulding. Process monitoring of an industrial fed-batch fermentation. *Bioeng Biotechnol*, 74(2):125–135, 2001.

[334] J Leonard and MA Kramer. Improvement of the backpropagation algorithm for training neural networks. *Computers Chem. Engng.*, 14(3):337–341, 1990.

[335] J Leonard, MA Kramer, and LH Ungar. A neural network architecture that computes its own reliability. *Computers Chem. Engng.*, 16(9):819–835, 1992.

[336] IJ Leontaritis and SA Billings. Input-output parametric models for nonlinear systems. *Int. J. Ctrl.*, 41:303–344, 1985.

[337] D Leung and J Romagnoli. An integration mechanism for multivariate knowledge-based fault diagnosis. *J Process Control*, 12:15–26, 2002.

[338] O Levenspiel. The Monod equation: A revisit and a generalization to product inhibition situations. *Biotechnol. Bioeng.*, 22:1671–1687, 1980.

[339] S Li, KY Lim, and DG Fisher. A state-space formulation of model predictive control. *AIChE J.*, 35:241–249, 1989.

[340] W. Liebert and H. G. Schuster. Proper choice of the time delay for the analysis of chaotic time series. *Phys. Lett. A*, 142:107–111, 1989.

[341] MJ Liebman, TF Edgar, and LS Lasdon. Efficient data reconciliation and estimation for dynamic processes using nonlinear programming techniques. *Chem. Eng. Sci*, 16:963, 1992.

[342] C Lin and CS George Lee. *Neural Fuzzy Systems*. Prentice-Hall PTR, Upper Saddle River, NJ, 1996.

[343] F Lindgren, P Geladi, S Rännar, and S Wold. Interactive variable selection (IVS) for PLS. Part I.Theory and algorithms. *J of Chemometrics*, 8:349–363, 1994.

[344] `http://turnbull.dcs.st-and.ac.uk/history/Mathematicians/Littlewood.html`. [Accessed 26 November 2002].

[345] W Liu. An extended Kalman filter and neural network cascade fault diagnosis strategy for glutamic acid fermentation process. *Artificial Intelligence in Engineering*, 13:131–140, 1999.

[346] L Ljung. *System Identification: Theory for the user*. Prentice-Hall, Englewood Cliffs, NJ, 2nd edition, 1999.

[347] L Ljung and T Glad. *Modeling of Dynamic Systems*. Prentice-Hall, Englewood Cliffs, New Jersey, 1994.

[348] C Loeblein, JD Perkins, B Srinivasan, and D Bonvin. Economic performance analysis in the design of on-line batch optimization systems. *J Proc Cont*, 9:61–78, 1999.

[349] A Lorber, L Wangen, and B Kowalski. A theoretical foundation for the PLS algorithm. *J of Chemometrics*, 1:19–31, 1987.

[350] EN Lorenz. Deterministic nonperiodic flow. *J. Atmospheric Science*, 20:130–141, 1963.

[351] CA Lowry and DC Montgomery. A review of multivariate control charts. *IIE Transactions*, 27:800–810, 1995.

[352] CA Lowry, WH Woodall, CW Champ, and SE Rigdon. A multivariate exponentially weighted moving average control chart. *Technometrics*, 34(1):46–53, 1992.

[353] R Luo, M Misra, SJ Qin, R Barton, and DM Himmelblau. Sensor fault detection via multiscale analysis and parametric statistical inference. *Ind. Eng. Chem. Res.*, 37:1024–1032, 1998.

[354] H Lütkepohl. *Introduction to Multiple Time Series Analysis*. Springer, Berlin, Germany, 1991.

[355] JF MacGregor, C Jaeckle, C Kiparissides, and M Koutoudi. Process monitoring and diagnosis by multiblock PLS methods. *AIChE Journal*, 40(5):826–838, 1994.

[356] JF MacGregor, DJ Kozub, A Penlidis, and AE Hamielec. State estimation of polymerization reactors. In *Preprints IFAC Symp. on Dynamics and Control of Chemical Reactors and Distillation Columns*, pages 147–152, Bournemouth, U.K., 1986.

[357] JJ MacQuitty. Impact of biotechnology on the chemical industry. *ACS Sympos. Ser.*, 362:11–29, 1988.

[358] C Maffezoni. Hamilton-Jacobi theory for periodic control problems. *J. Optim. Theor. Applic.*, 14:21–29, 1974.

[359] RS Mah, GM Stanley, and DM Downing. Reconciliation and rectification of process flow and inventory data. *Ind. Eng. Chemistry, Process Design*, 15:175–183, 1976.

[360] Y Maki and KA Loparo. A neural-network approach to fault detection and diagnosis in industrial processes. *IEEE Trans. Cont. Syst. Techn.*, 5(6):529–541, 1997.

[361] ER Malinowski. Statistical F-tests for abstract factor analysis and target testing. *J. of Chemometrics*, 3:49–60, 1988.

[362] SG Mallat. A theory for multiresolution signal decomposition: The wavelet representation. *IEEE Trans.Pattern Anal. Mach. Intell.*, 11:674–693, 1989.

[363] EC Malthouse, AC Tamhane, and RSH Mah. Nonlinear partial least squares. *Comp. Chem. Engng.*, 21(8):875–890, 1997.

[364] Maple v, release 7, 2001.

[365] S Marchal-Brassely, J Villermaux, JL Houzelot, and JL Barnay. Optimal operation of a semibatch reactor by self-adaptive models for temperature and feedrate profiles. *Chem. Engng. Sci.*, 47:2445–2450, 1992.

[366] TE Marlin. *Process Control.* McGraw Hill, New York, 1995.

[367] PZ Marmarelis and VZ Marmarelis. *Analysis of Physiological Systems.* Plenum Press, New York, 1978.

[368] H Martens and T Næs. *Multivariate Calibration.* Wiley, New York, 1989.

[369] EC Martinez. Batch process modeling for optimization and reinforcement learning. *Comp. Chem. Engng.*, 24:1187–1193, 2000.

[370] RL Mason, RF Gunst, and JL Hess. *Statistical Design and Analysis of Experiments.* Wiley, New York, 1989.

[371] M Masuda, S Takamatu, N Nishimura, S Komatsubara, and T Tosa. Improvement of culture conditions for l-proline production by a recombinant strain of *Serratia marcescens. Appl. Biochem. Biotechnol.*, 43:189–197, 1993.

[372] MathWorks. *The Matlab, Version 6.1.* The MathWorks, Inc., 2001.

[373] Matlab Signal Processing Toolbox (Version 5). User's Guide, The Mathworks, Inc. Natick, MA, 2000.

[374] M Matsumura, T Imanaka, T Yoshida, and H Tagushi. Modelling of *Cephalosporin C* production and application to fed-batch cultivation. *J. Ferment. Technol.*, 59:115–123, 1981.

[375] RM May. Simple mathematical models with very complicated dynamics. *Nature*, 26:459–467, 1976.

[376] TJ McAvoy, Y Arkun, and E Zafiriou. *Model-based Process Control: Proceeding of the IFAC Workshop.* Pergamon Press, Oxford, 1989.

[377] WS McCulloch and W Pitts. A logical calculus of the ideas immanent in nervous activity. *Bulletin of Mathematical Biophysics*, 5:115–133, 1943.

[378] GJ McLachlan. *Discriminant Analysis and Statistical Pattern Recognition.* John Wiley & Sons, New York, 1992.

[379] RG McMillan. Tests for one or two outliers in normal samples with unknown variance. *Technometrics*, 13:87–100, 1971.

[380] ES Meadows and JB Rawlings. Model identification and control of a semibatch chemical reactor. In *Proc. American Control Conference*, pages 249–255, Boston, MA, 1991.

[381] ES Meadows and JB Rawlings. Model predictive control. In MA Henson and DE Seborg, editors, *Nonlinear Process Control*, pages 233–310. Prentice Hall, Upper Saddle River, NJ, 1997.

[382] RD Megee, S Kinoshita, AG Fredrikson, and HM Tsuchiya. Differentiation and product formation in molds. *Biotechnol. Bioeng.*, 12:771–801, 1970.

[383] RK Mehra and S Mahmoud. Model Algorithmic Control. In PB Deshpande, editor, *Distillation Dynamics and Control*, chapter 15. ISA, Research Triangle Park, NC, 1985.

[384] RK Mehra et. al. Model Algorithmic Control Using IDCOM for the F100 Jet Engine Multivariable Control Design Problem. In RJ Patton, PM Frank, and R Clark, editors, *Alternatives for Linear Multivariable Control*, pages 47–98. NEC, Chicago, IL, 1978.

[385] GR Meira and AF Johnson. Molecular weight distribution control in continuous "living" polymerizations through periodic operation of the monomer feed. *Polym. Engng Sci*, 21:415–423, 1981.

[386] RM Merem. Cellular engineering. *Ann. Biomed. Eng.*, 19:529–545, 1991.

[387] LD Mermelstein, ET Papoutsakis, DJ Petersen, and GN Bennett. Metabolic engineering of *Clostridium acetobutylicum ATCC 824* for increased solvent production by enhancement of acetone formation enzyme activities using a synthetic acetone operon. *Biotechnol. Bioeng.*, 42(9):1053–1060, 1993.

[388] P Miller and RE Swanson. Contribution plots: The missing link in multivariate quality control. In *37th Annual Fall Technical Conf., ASQC*, Rochester, NY, 1993.

[389] P Miller, RE Swanson, and CF Heckler. Contribution plots: The missing link in multivariate quality control. *Int J of Applied Mathematics and Computer Science*, 8(4):775–792, 1998.

[390] ML Minsky. *Computation: Finite and Infinite Machines.* Prentice-Hall, Englewood Cliffs, NJ, 1967.

[391] T Moberg, JS Ramberg, and RH Randles. An adaptive multiple regression procedure based on M-estimators. *Technometrics*, 22:213–224, 1980.

[392] JM Modak and HC Lim. Feedback optimization of fed-batch fermentation. *Biotechnol. Bioengng.*, 30:528–540, 1987.

[393] A Modi and R Musier. Systematic batch process development and analysis via an advanced modeling and simulation tool. *Aspen Tech. Inc.*, 1998.

[394] RR Mohler. *Bilinear Control Processes.* Academic Press, New York, 1973.

[395] HG Monbouquette, GD Sayles, and DF Ollis. Immobilized cell biocatalyst activation and pseudo-steady-state behavior: model and experiment. *Biotechnol. Bioeng.*, 35:609–629, 1990.

[396] G Montague. *Monitoring and Control of Fermenters.* IChemE, Galliard Ltd. United Kingdom, 1997.

[397] G Montague and J Morris. Neural-network contributions in biotechnology. *TIBTECH*, 12:312–324, 1994.

[398] GA Montague, AJ Morris, AR Wright, M Aynsley, and A. Ward. Growth monitoring and control through computer-aided on-line mass balancing in fed-batch penicillin fermentation. *Can. J. Chem. Eng.*, 64:567–580, 1986.

[399] JL Montesinos, C Campmajo, J Iza, F Valero, J Lafuente, and C Sola. Use of intelligent system to monitor and control fermentation processes. Application to lipase production by *Candida rugosa*. *Process Control and Quality*, 5:237–244, 1994.

[400] DC Montgomery. *Introduction to Statistical Quality Control.* Wiley, New York, 1991.

[401] DC Montgomery. *Design and Analysis of Experiments.* Wiley, New York, 5th edition, 2000.

[402] DC Montgomery and CM Mastrangelo. Some statistical process control methods for autocorrelated data. *Journal of Quality Technology*, 23:179, 1991.

[403] M Morari and E Zafiriou. *Robust Process Control.* Prentice-Hall, Englewood Cliffs, NJ, 1989.

[404] N Murata, S Yoshizawa, and S Amari. Network information criterion-Determining the number of hidden units for an artificial neural network model. *IEEE Trans. on Neural Networks*, 5(3):865–872, 1994.

[405] KR Muske and JB Rawlings. Model predictive control with linear unstable processes. *J. Proc. Cont.*, 3:85–96, 1993.

[406] C Myers, LR Rabiner, and AE Rosenberg. Performance tradeoffs in dynamic time warping algorithms for isolated word recognitions. *IEEE Trans. on Acoustics, Speech and Signal Process.*, 6(28):623–635, 1980.

[407] RH Myers, AI Khuri, and WH Carter. Response surface methodology. *Technometrics*, 31:137–157, 1989.

[408] T Naes and T Isaksson. Splitting of calibration data by cluster analysis. *J Chemometrics*, 5:49–65, 1991.

[409] S Narasimhan and RSH Mah. Generalized likelihood ratio method for gross error identification. *AIChE Journal*, 33:1514–1521, 1987.

[410] S Narasimhan and RSH Mah. Generalized likelihood ratios for gross error identification in dynamic processes. *AIChE Journal*, 34:1321, 1988.

[411] A Negiz. *Statistical dynamic modeling and monitoring methods for multivariable continuous processes.* PhD thesis, Illinois Inst. of Tech., Dept. of Chemical and Env. Eng., Chicago-IL, USA, 1995.

[412] A Negiz and A Cinar. On the detection of multiple sensor abnormalities in multivariable processes. In *Proc. American Control Conference*, pages 2364–2369, 1992.

[413] A Negiz and A Cinar. A parametric approach to statistical monitoring of processes with autocorrelated observations. In *AIChE Annual Meeting*, Miami, FL, 1995.

[414] A Negiz and A Cinar. Statistical monitoring of strongly autocorrelated processes. In *Proc. of Joint Statistical Meeting*, Chicago, IL, 1996.

[415] A Negiz and A Cinar. Statistical monitoring of multivariable dynamic processes with state–space models. *AIChE Journal*, 43(8):2002–2020, 1997.

[416] A Negiz and A Cinar. PLS, balanced and canonical variate realization techniques for identifying varma models in state space. *Chemometrics Intell. Lab. Syst.*, 38:209–221, 1997.

[417] A Negiz and A Cinar. Monitoring of multivariable dynamic processes and sensor auditing. *Journal of Process Control*, 8(5-6):375–380, 1998.

[418] D Neogi and C Schlags. Multivariate statistical analysis of an emulsion batch process. *Ind. Eng. Chem. Res.*, 37(10):3971–3979, 1998.

[419] S Newhouse, D Ruelle, and F Takens. Occurrence of strange axiom-a attractors near quasiperiodic flow on t^m, $m \leq 3$. *Commun. Math. Phys.*, 64:35–40, 1978.

[420] GC Newton, LA Gould, and JF Kaiser. *Analytical Design of Feedback Controls.* Wiley, New York, 1957.

[421] HT Nguyen and M Sugeno. *Fuzzy Systems: Modeling and Control.* Kluwer Academic Publishers, Boston, MA, 1998.

[422] HT Nguyen and EA Walker. *A First Course in Fuzzy Logic.* CRC Press, Boca Raton, FL, 1999.

[423] J Nielsen. A simple morphologically structured model describing the growth of filamentous microorganisms. *Biotechnol. Bioeng.*, 41:715–727, 1993.

[424] J Nielsen. *Physiological Engineering Aspects of Penicillium chrysogenum.* World Scientific, Singapore, 1997.

[425] J Nielsen and P Krabben. Hyphal growth and fragmentation of *Penicillium chrysogenum* in submerged cultures. *Biotechnol. Bioeng.*, 46:588–598, 1995.

[426] J Nielsen and J Villadsen. *Bioreaction Engineering Principles.* Plenum Press, New York, 1994.

[427] R Nikoukhah. Innovations generation in the presence of unknown inputs: Application to robust failure detection. *Automatica*, 30:1851–1867, 1994.

[428] NJ Nilsson. *Artificial Intelligence: A New Synthesis.* Morgan Kaufmann Publishers, San Francisco, CA, 1998.

[429] E Nisenfeld. *Batch Control.* ISA PGS Series, 1996.

[430] M Noda, T Chida, S Hasebe, and I Hashimoto. On-line optimization system of pilot scale multi-effect batch distillation system. *Comp. Chem. Engng.*, 24:1577–1583, 2000.

[431] EJ Noldus. Periodic optimization of a chemical reactor system using perturbation methods. *J. Engng Math.*, 11:49–66, 1977.

[432] P Nomikos. Detection and diagnosis of abnormal batch operations based on multiway principal components analysis. *ISA Trans.*, 35:259–266, 1996.

[433] P Nomikos and JF MacGregor. Monitoring batch processes using multiway principal component analysis. *AIChE Journal*, 40(8):1361–1375, 1994.

[434] P Nomikos and JF MacGregor. Multi-way partial least squares in monitoring batch processes. *Chemometrics and Intell. Lab. Syst.*, 30:97–108, 1995.

[435] P Nomikos and JF MacGregor. Multivariate SPC charts for monitoring batch processes. *Technometrics*, 37:41–59, 1995.

[436] A Norvilas, A Negiz, J DeCicco, and A Cinar. Intelligent process monitoring by interfacing knowledge–based systems and multivariate statistical monitoring. *Journal of Process Control*, 10(4):341–350, 2000.

[437] NM Nounou and B Bakshi. On-line multiscale filtering of random and gross errors without process models. *AIChE Journal*, 45(5):1041–1058, 1999.

[438] BA Ogunnaike and WH Ray. *Process Dynamics, Modeling and Control*. Oxford University Press, New York, NY, 1994.

[439] Committee on New Sensor Technologies. *Expanding the Vision of Sensor Materials*. The National Academy of Sciences, `http://www.nap.edu/openbook/0309051754/html/1.html`, 1995. [Accessed 26 November 2002].

[440] DG O'Neill and G Lyberatos. Feedback identification of continuous microbial growth systems. *Biotechnol. Bioengng*, 28:1323–1333, 1986.

[441] D O'Shaughnessy. Speaker recognition. *IEEE ASSP Magazine*, 4(3):4–17, 1986.

[442] F Ozgulsen, R Adomaitis, and A Cinar. A numerical method for determining optimal parameter values in forced periodic operation. *Chem. Eng. Sci.*, 47:605–613, 1992.

[443] F Ozgulsen and A Cinar. Forced periodic operation of tubular reactors. *Chem. Eng. Sci.*, 49(20):3409–3419, 1994.

[444] F Ozgulsen, S Kendra, and A Cinar. Nonlinear predictive control of periodically forced chemical reactors. *AIChE J*, 39:589–598, 1993.

[445] F Ozgulsen, K Rigopoulos, and A Cinar. A comparative study of tools for assessing the effects of forced periodic operation of catalytic reactors. *Chem. Engng Commun.*, 112:85–104, 1992.

[446] TW Parsons. *Voice and Speech Processing*. McGraw Hill, New York, 1987.

[447] SJ Parulekar. Analytical optimization of some single-cycle and repeated fed-batch fermentations. *Chem. Eng. Sci.*, 47:4077–4097, 1992.

[448] SJ Parulekar. Analysis of forced periodic operations of continuous bioprocesses - Single input variations. *Chem. Eng. Sci.*, 53:2481–2502, 1998.

[449] SJ Parulekar. Analysis of forced periodic operations of continuous bioprocesses - Multiple input variations. *Chem. Eng. Sci.*, 55:513–533, 2000.

[450] SJ Parulekar. Forced periodic operations of continuous recombinant cell cultures subject to antibiotic selection pressure. *Chem. Eng. Sci.*, 56:6463–6484, 2001.

[451] SJ Parulekar and J Lee. Structure analysis of continuous cultures subject to periodic medium tuning. *Chem. Eng. Sci.*, 48:3007–3035, 1993.

[452] SJ Parulekar and HC Lim. Modeling, optimization and control of semi-batch bioreactors. *Advances in Biochemical Engineering/Biotechnology*, 32:207–258, 1985.

[453] SJ Parulekar and HC Lim. Dynamics of continuous commensalistic cultures - I. Multiplicity and local stability characteristics and bifurcation analysis. *Chem. Eng. Sci.*, 41:2605–2616, 1986.

[454] SJ Parulekar, RS Waghmare, and HC Lim. Yield optimization for multiple reactions. *Chem. Eng. Sci.*, 43:3077–3091, 1988.

[455] SJ Parulekar and D Wei. Periodic operation of continuous fermentations subject to periodic variation of dilution rate. In *Proc. American Control Conference*, volume 2, pages 1118–1123, Atlanta, GA, 1988.

[456] PM Frank Patton, RJ and R Clark, editors. *Fault Diagnosis in Dynamical Systems*. Prentice Hall, Englewood Cliffs, NJ, 1989.

[457] RJ Patton. Robustness in model-based fault diagnosis - the 1995 situation. In *Prep IFAC Workshop on On-Line Fault Detection in Chemical Process Industries*, pages 55–77, 1995.

[458] RJ Patton and SM Kangethe. Robust fault diagnosis using eigenstructure assignment. In RJ Patton, PM Frank, and R Clark, editors, *Fault Diagnosis in Dynamical Systems*, pages 99–154. Prentice Hall, Englewood Cliffs, NJ, 1989.

[459] GC Paul and CR Thomas. A structured model for hyphal differentiation and penicillin production using *Penicillium chrysogenum*. *Biotechnol. Bioeng.*, 51:558–572, 1996.

[460] S Pavlou, IG Kevrekidis, and G Lyberatos. On the coexistence of competing microbial species in a chemostat under cycling. *Biotechnol. Bioeng.*, 35:224–232, 1990.

[461] S Pazoutova, J Votruba, and Z Rehacek. A mathematical model for growth and alkaloid production in the submerged culture of *Calviceps Purpurea*. *Biotechnol. Bioeng.*, 23:2837, 1981.

[462] K Pearson. On lines and planes of closest fit to systems of points in space. *Philos. Mag.*, 2:559, 1901.

[463] W Pedrycz and F Gomide, editors. *An Introduction to Fuzzy Sets*. MIT Press, Cambridge, MA, 1998.

[464] Y Peng, A Youssouf, P Arte, and M Kinneart. A complete procedure for residual generation and evaluation with application to a heat exchanger. *IEEE Trans. Control Systems Technology*, 5(6):542–554, 1997.

[465] WA Perkins and A Austin. Adding temporal reasoning to expert-system-building environments. *IEEE Expert*, February:23–30, 1990.

[466] TF Petti and PS Dhurjati. A coupled knowledge based system using fuzzy optimization for advisory control. *AIChE J*, 38(9):1369–1378, 1992.

[467] TF Petti, J Klein, and PS Dhurjati. Diagnostic model processor:Using deep knowledge for process fault diagnosis. *AIChE J*, 36(4):565–575, 1990.

[468] AM Picket, MJ Bazin, and HH Topiwala. Growth and composition of *Escherichia coli* subjected to square-wave perturbations in nutrient supply: effect of varying frequencies. *Biotechnol. Bioengng.*, 21:1043–1055, 1979.

[469] AM Picket, MJ Bazin, and HH Topiwala. Growth and composition of *Escherichia coli* subjected to square-wave perturbations in nutrient supply: effect of varying amplitudes. *Biotechnol. Bioengng.*, 22:1213–1224, 1980.

[470] A Pinches and LJ Pallent. Rate and yield relationships in the production of Xanthan gum by batch fermentations using complex and chemically defined growth media. *Biotechnol. Bioengng.*, 28:1484–1496, 1986.

[471] SJ Pirt and RC Righoletto. Effect of growth rate on the synthesis of penicillin by penicillium chrysogenum in batch and chemostat cultures. *Appl. Microbiol.*, 15:1284–1290, 1967.

[472] PLS Toolbox for Matlab (Version 1.2). User's Guide, 2001. Eigenvector Research (`http://www.eigenvector.com`), Seattle, WA.

[473] `http://turnbull.dcs.st-and.ac.uk/history/Mathematicians/Poincare.html`. [Accessed 26 November 2002].

[474] Y Poirier, DE Dennis, K Klomparens, and C Somerville. Polyhydroxybutyrate, a biodegradable thermoplastic, produced in transgenic plants. *Science*, 96:73–80, 1992.

[475] MN Pons. *Bioprocess Monitoring and Control.* John Wiley and Sons, New York, 1992.

[476] P Prescott. A review of some robust data analyses and multiple outlier detection procedures. *Bulletin in Appl. Statist.*, pages 141–158, 1980.

[477] DM Prett and CE Garcia. *Fundamental Process Control.* Butterworths, Stoneham, 1988.

[478] DM Prett, CE Garcia, and BL Ramaker. *The Second Shell Process Control Workshop.* Butterworths, Boston, 1990.

[479] MB Priestley. *Nonlinear and Nonstationary Time Series Analysis.* Academic Press, London, 1988.

[480] D Psichogios, R de Veaux, and L Ungar. Nonparametric system identification: A comparison of MARS and neural networks. In *Proc. American Control Conference*, pages 1436–1440, 1992.

[481] DC Psichogios and LH Ungar. A hybrid neural network-first principles approach to process modeling. *AIChE J*, 38(10):1499–1511, 1992.

[482] DC Psichogios and LH Ungar. SVD-NET:An algorithm that automatically selects network structure. *IEEE Trans. on Neural Networks*, 5(3):513–515, 1994.

[483] TE Quantrille and YA Liu. *Artificial Intelligence in Chemical Engineering.* Academic Press, San Diego, CA, 1991.

[484] L Rabiner and B Juang. *Fundamentals of Speech Recognition.* Prentice-Hall, Englewood Cliffs, New Jersey, 1993.

[485] LR Rabiner, AE Rosenberg, and SE Levinson. Consideration in dynamic time warping algorithms for discrete word recognition. *IEEE Trans. on Acoustics, Speech and Signal Process.*, 6(26):575, 1978.

[486] S Rahman and S Palanki. State feedback synthesis for on-line optimization in the presence of measurable disturbances. *AIChE Journal*, 42:2869–2882, 1996.

[487] A Raich and A Cinar. Multivariate statistical methods for monitoring continuous processes: Assessment of discrimination power of disturbance models and diagnosis of multiple disturbances. *Chemometrics and Intelligent Laboratory Systems*, 30:37–48, 1995.

[488] A Raich and A Cinar. Statistical process monitoring and disturbance diagnosis in multivariable continuous processes. *AIChE Journal*, 42(4):995–1009, 1996.

[489] A Raich and A Cinar. Diagnosis of process disturbances by statistical distance and angle measures. *Comp and Chem Engng*, 21(6):661–673, 1997.

[490] GK Raju and CL Cooney. Active learning from process data. *AIChE J*, 44(10):2199–2211, 1998.

[491] C Ramamurthi, C Situ, and W Bequette. Control relevant dynamic data reconciliation and parameter estimation. *Comp and Chem Engng*, 17:41–59, 1991.

[492] TS Ramesh, SK Shum, and JF Davis. A structured framework for efficient problem solving in diagnostic expert systems. *Comp and Chem Engng*, 12(9/10):891–902, 1988.

[493] JO Ramsay. Principal differential analysis: Data reduction by differential operators. *Journal of the Royal Statistical Society*, 58(Series B):495–508, 1996.

[494] JO Ramsay. Estimating smooth monotone functions. *Journal of the Royal Statistical Society - Series* B, 60:365–375, 1998.

[495] JO Ramsay and BW Silverman. *Functional Data Analysis*. Springer-Verlag, New York, NY, 1997.

[496] S Rannar, JF MacGregor, and S Wold. Adaptive batch monitoring using hierarchical PCA. *Chemometrics Intell. Lab. Syst.*, 41:73–81, 1998.

[497] A Rastogi, J Fotopoulos, C Georgakis, and HG Stenger. The identification of kinetic expression and the evolutionary optimization of specialty chemical batch reactors using tendency models. *AIChE J*, 47(9-11):2487–2492, 1992.

[498] WH Ray. *Advanced Process Control.* McGraw-Hill, New York, NY, 1981.

[499] R Reed. Pruning algorithms-a survey. *IEEE Trans Neural Networks*, 4:740–747, 1993.

[500] GC Reinsel. *Elements of Multivariate Time Series Analysis.* Springer, New York, NY, 2nd edition, 1997.

[501] G Reklaitis, A Sunol, D Rippin, and O Hortacsu. Batch Processing Systems Engineering: Current Status and Future Directions. In GV Reklaitis, A Sunol, D Rippin, and O Hortacsu, editors, *Batch Processing Systems Engineering: Fundamentals and Applications for Chemical Engineering*, pages 309–330. Springer-Verlag, Berlin, Germany, 1996.

[502] E Rich and K Knight. *Artificial Intelligence.* McGraw Hill, New York, NY, 1992.

[503] JA Richalet and B Bimonet. Identification des Systems Discrets Linearines Monovariables Par Minimisation d'Une Distance de Structure. *Elec. Lett*, 4:24, 1968.

[504] JA Richalet, F Lecamus, and P Hummel. New trends in identification, minimization of a structural distance, weak topology. In *Second IFAC Symposium on Identification*, Prague, 1970.

[505] JA Richalet, A Rault, and R Pouliquen. *Identification des Processus par la Methode du Modele.* Gordon and Breach, 1970.

[506] JA Richalet, A Rault, JD Testud, and J Papon. Model Predictive Heuristic Control: Applications to Industrial Processes. *Automatica*, 14:413, 1978.

[507] NL Ricker. The use of quadratic programming for constrained internal model control. *Ind. Eng. Chem. Process Des. Dev.*, 24:925, 1985.

[508] A Rigopoulos, Y Arkun, and F Kayihan. Full CD profile control of sheet forming processes using adaptive PCA and reduced order MPC design. In *Proc. of ADCHEM'97*, page 396, 1997.

[509] K Rigopoulos, X Shu, and A Cinar. Vibrational control of an exothermic CSTR: Stabilization by multiple input oscillations. *AIChE J.*, 34:2041–2051, 1988.

[510] DWT Rippin. Control of batch processes. In *Proc. IFAC Symp. Dynamics and Control of Reactors and Distillation Columns (DYCORD+'89)*, pages 131–141, Maastricht, The Netherlands, 1989.

[511] J Rissanen. *Stochastic Complexity in Statistical Inquiry.* World Scientific, Singapore, 1989.

[512] Beckman RJ and RD Cook. Outlier .... s. *Technometrics*, 25:119–149, 1983.

[513] IR Roda, J Comas, M Poch, M Sanchez-Marre, and U Cortes. Automatic knoowledge acquisition from complex processes for the development of knowledge-based systems. *Ind. Eng. Chem. Res.*, 40:3353–3360, 2001.

[514] JA Roels. *Energetics and Kinetics in Biotechnology.* Elsevier, New York, 1983.

[515] F Rojo, DH Pieper, KH Engesser, HM Knackmuss, and KN Timmis. Assemblage of orthocleavage route for simultaneous degradation of chloro- and methylaromatics. *Science*, 235:1395–1398, 1987.

[516] DK Rollins and JF Davis. Unbiased estimation of gross errors in process measurements. *AIChE J*, 38:563–572, 1992.

[517] DK Rollins and S Devanathan. Unbiased estimation in dynamic data reconciliation. *AIChE J*, 39:1330, 1993.

[518] JA Romagnoli and MC Sanchez. *Data Processing and Reconciliation for Chemical Process Operations.* Academic Press, 2000.

[519] JA Romagnoli and G Stephanopoulos. Rectification of process measurement data in the presence of gross errors. *Chem Eng Sci*, 36:1849, 1981.

[520] F Rosenblatt. The perceptron: A probabilistic model for information storage and organization in the brain. *Psychological Review*, 65:386–408, 1958.

[521] F Rosenblatt. *Principles of Neurodynamics.* Spartan Books, Washington D.C., 1962.

[522] SG Rothwell, EB Martin, and AJ Morris. Comparison of methods for handling unequal length batches. In *IFAC DYCOPS5*, pages 66–71, Corfu, Greece, 1998.

[523] PJ Rousseeuw and Leroy C. *Robust Regression and Outlier Detection.* Wiley, New York, 1987.

[524] PJ Rousseeuw and BCV Zomeren. Unmasking multivariate outliers and leverage points. *J. Amer. Statist. Ass.*, 85:633–639, 1990.

[525] L Ruan and XD Chen. Comparison of several periodic operations of a continuous fermentation process. *Biotechnol. Prog.*, 12:286–288, 1996.

[526] D Ruelle. Deterministic chaos: The science and the fiction. *Proc. R. Soc. London A*, 427:241–248, 1990.

[527] DE Rumelhart and JL McClelland, eds. *Parallel Distributed Processing: Explorations in the Microstructure of Cognition*, volume 1. MIT Press, Cambridge, MA, 1986.

[528] GC Runger and FB Alt. Choosing principal components for multivariate statistical process control. *Commun. Statis.-Theory Meth.*, 25(5):909–922, 1996.

[529] D Ruppen, D Bonvin, and DWT Rippin. Implementation of adaptive optimal operation for a semi-batch reaction system. *Comp. Chem Engng.*, 22:185–189, 1998.

[530] EL Russell, LH Chiang, and RD Braatz. *Data-driven Techniques for Fault Detection and Diagnosis in Chemical Processes.* Springer-Verlag, London, UK, 2000.

[531] SA Russell, P Kesavan, JH Lee, and BA Ogunnaike. Recursive data-based prediction and control of product quality for batch and semi-batch processes applied to a nylon 6, 6 autoclave. *AIChE Journal*, 44:2442–2464, 1998.

[532] I Saenz de Buruaga, A Echevarria, PD Armitage, JC de la Cal, JR Leiza, and JM Asua. On-line control of a semi-batch emulsion polymerization reactor based on calorimetry. *AIChE Journal*, 43(4):1069–1081, 1997.

[533] H Sakoe and S Chiba. Dynamic programming algorithm optimization for spoken word recognition. *IEEE Trans. on Acoustics, Speech and Signal Process.*, 2(26):43–49, 1978.

[534] T Sastri. A recursive estimation algorithm for adaptive estimation and parameter change detection of time series models. *J. Op. Res. Soc.*, 37:987–999, 1986.

[535] T Sauer, J Yorke, and M Casdagli. Embedology. *J. Stat. Phys.*, 65:579–616, 1991.

[536] GW Scheid, SJ Qin, and TJ Riley. Run-to-run optimization, monitoring, and control on a rapid thermal processor. In *AIChE Annual Meeting*, Dallas, TX, 1999.

[537] B Schenker and M Agarwal. On-line optimized feed switching in semi-batch reactors using semi-empirical dynamic models. *Control Eng. Practice*, 8(12):1393–1403, 2000.

[538] W Schmid. Outliers in a multivariate autoregressive moving-average process. *Stochastic Process and Their Appl.*, 36:117–133, 1990.

[539] W Schmid. MARS: A tutorial. *Journal of Chemometrics*, 6:199–216, 1992.

[540] HG Schuster. *Deterministic Chaos.* Physik-Verlag, Weinheim, 1984.

[541] DE Seborg, TF Edgar, and DA Mellichamp. *Process Dynamics and Control.* Wiley, New York, 1989.

[542] WA Shewhart. *Economic control of quality of manufactured product.* Van Nostrand, NJ, 1931.

[543] K Shimizu. A tutorial review on bioprocess systems engineering. *Computers Chem. Engng.*, 20(6-7):915–941, 1996.

[544] K Shimizu and M Matsubara. A solvent screening criterion for multicomponent extractive fermentation. *Chem. Engng Sci.*, 42:499–504, 1987.

[545] ML Shuler and F Kargi. *Bioprocess Engineering Basic Concepts.* Prentice Hall, New Jersey, 1992.

[546] ML Shuler and F Kargi. *Bioprocess Engineering Basic Concepts.* Prentice Hall, New Jersey, 2nd edition, 2002.

[547] HF Silverman and DP Morgan. The application of dynamic programming to connected speech recognition. *IEEE ASSP Magazine*, 7(7):6–25, 1990.

[548] SIMCA (Version 6.0). User's Guide, 2001. UMETRICS AB (`http://www.umetrics.com`), Umeå, Sweden.

[549] D Sincic and JE Bailey. Optimal periodic control of activated sludge processes - I. Results for the base case with Monod/decay Kinetics. *Wat. Res.*, 12:47–53, 1978.

[550] D Sincic and JE Bailey. Analytical optimization and sensitivity analysis of forced periodic chemical processes. *Chem. Engng Sci.*, 35:1153–1161, 1980.

[551] `http://turnbull.dcs.st-and.ac.uk/history/Mathematicians/Smale.html`. [Accessed 26 November 2002].

[552] AK Smilde. Three–way analysis. Problems and prospects. *Chemometrics Intelligent Lab. Systems*, 15:143–157, 1992.

[553] AK Smilde and DA Doornbos. Three–way methods for the calibration of chromatographic systems :Comparing PARAFAC and three–way PLS. *Journal of Chemometrics*, 5:345, 1991.

[554] AK Smilde and AL Kiers. Multiway covariates regression models. *Journal of Chemometrics*, 13:31–48, 1999.

[555] AK Smilde, Z Wang, and B Kowalski. Theory of medium–rank second-order calibration with restricted-Tucker models. *Journal of Chemometrics*, 8:21–36, 1994.

[556] AK Smilde, JA Westerhuis, and R Boque. Multiway multiblock component and covariates regression models. *Journal of Chemometrics*, 14:301–331, 2000.

[557] P Smyth. Hidden markov models for fault detection in dynamic systems. *Pattern Recognition*, 27:149–164, 1994.

[558] T Södersrtöm and P Stoica. *System Identification.* Prentice-Hall, Englewood Cliffs, New Jersey, 1989.

[559] M Soroush and C Kravaris. Nonlinear control of a batch polymerization reactor: An experimental study. *AIChE J*, 38(9):1429–1448, 1992.

[560] Special Issue of *IEEE Tran. Automatic Control*, December 1971.

[561] JJ Spitz, DC Chappelear, and RL Laurence. Periodic operation of a bulk styrene polymerization. *Chem. Engng Prog. Symp. Ser.*, 72:74, 1977.

[562] B Srinivasan, E Visser, and D Bonvin. Optimization-based control with imposed feedback structures. *Control Eng. Practice*, in press, 2001.

[563] GM Stanley and RSH Mah. Estimation of flows and temperatures in process networks. *AIChE J*, 23:642, 1977.

[564] G Stephanopoulos. Metabolic fluxes and metabolic engineering. *Metabolic Engineering*, 1:1–11, 1999.

[565] G Stephanopoulos, AA Aristidou, and J Nielsen. *Metabolic Engineering Principles and Methodologies.* Academic Press, USA, 1998.

[566] G Stephanopoulos and AJ Sinskey. Metabolic engineering methodologies and future prospects. *Trends in Biotechnology*, 11:392–396, 1993.

[567] G Stephanopoulos and JJ Vallino. Network rigidity and metabolic engineering in metabolite overproduction. *Science*, 252:1675–1681, 1991.

[568] ML Stephens, C Christensen, and G Lyberatos. Plasmid stabilization of an *Escherichia coli* culture through cycling. *Biotechnol. Prog.*, 8:1–4, 1992.

[569] ML Stephens and G Lyberatos. Effect of cycling on final mixed culture fate. *Biotechnol. Bioengng*, 29:672–678, 1987.

[570] ML Stephens and G Lyberatos. Effect of cycling on the stability of plasmid-bearing microorganisms in continuous culture. *Biotechnol. Bioengng*, 31:464–469, 1988.

[571] LE Sterman and BE Ydstie. The steady-state process with periodic perturbations. *Chem. Engng Sci.*, 45:721–736, 1990.

[572] LE Sterman and BE Ydstie. Unsteady-state multivariable analysis of periodically perturbed systems. *Chem. Engng Sci.*, 45:737–749, 1990.

[573] LE Sterman and BE Ydstie. Periodic forcing of the CSTR: an application of the generalized π-criterion. *AIChE J*, 37:986–996, 1991.

[574] JP Steyer, I Queinnec, and D Simoes. Biotech: A real-time application of artificial intelligence for fermentation processes. *Control Eng. Practice*, 1(2):315–321, 1993.

[575] RW Stieber and P Gerhardt. Dialysis continuous process for ammonium lactate fermentation: Simulated and experimental dialysate-feed, immobilized-cell systems. *Biotechnol. Bioengng.*, 23:535–549, 1981.

[576] L Stols and MI Donnelly. Production of succinic acid through overexpression of NAD+-dependent malic enzyme in an *Escherichia coli* mutant. *Appl. Environ. Microbiol.*, 63:2695–2701, 1997.

[577] CL Stork and BR Kowalski. Distinguishing between process upsets and sensor malfunctions using sensor redundancy. *Chemometrics and Intelligent Lab. Sys.*, 46:117–131, 1999.

[578] CL Stork, DJ Veltcamp, and BR Kowalski. Identification of multiple sensor disturbances during process monitoring. *Analytical Chemistry*, 69:5031–5036, 1997.

[579] W Sun, A Palazoglu, A Bakhtazad, and JA Romagnoli. Process trend analysis using wavelet domain hidden markov models. Submitted to AIChE J, 2002.

[580] T Suzuki, Y Sakino, M Nakajima, H Asama, T Fujii, K Sato, H Kaetsu, and I Endo. A novel man-machine interface for a bioprocess expert system constructed for cooparative decision making and operation. *J Biotechnol*, 52:277–282, 1997.

[581] V-M Taavitsainen and P Korhonen. Nonlinear data analysis with latent variables. *Chemometrics and Intelligent Laboratory Systems*, 14:185–194, 1992.

[582] F Takens. Detecting strange attractors in turbulence. *Lecture Notes in Math.*, 898, 1981.

[583] M Tantirungkij, T Seki, and T Yoshida. Genetic improvement of *Saccharomyces cerevisiae* for ethanol production from xylose. *Ann. NY Acad. Sci.*, 721:138–147, 1994.

[584] E Tatara and A Cinar. An intelligent system for multivariate statistical process monitoring and diagnosis,. *ISA Trans.*, 41:255–270, 2002.

[585] P Terwiesch and M Agarwal. On-line correction of pre-determined input profiles for batch reactors. *Comp. Chem. Engng.*, 18:S433–S437, 1994.

[586] MT Tham and A Parr. Succeed at on-line validation and reconstruction of data. *Chem. Eng. Prog.*, 46:90, 1994.

[587] J Thibault, V vanBreusegem, and A Cheruy. On-line prediction of fermentation variables using neural networks. *Biotechnol. Bioeng.*, 36:1041–1048, 1990.

[588] ML Thompson and MA Kramer. Modeling chemical processes using prior knowledge and neural networks. *AIChE J*, 40(8):1328–1338, 1994.

[589] AN Tikhonov, AB Vasileva, and AG Sveshikov. *Differential Equations.* Springer-Verlag, New York, 1984.

[590] KN Timmis, F Rojo, and JL Ramos. Prospects for laboratory engineering of bacteria to degrade pollutants. *Basic Life Science*, 45:61–79, 1988.

[591] H Tong and CM Crowe. Detection of gross errors in data by principal components analysis. *AIChE J*, 41:1712, 1995.

[592] H Tong and CM Crowe. Detecting persistent gross errors by sequential analysis of principal components. *AIChE J*, 43:1242–1249, 1997.

[593] IT Tong, HH Liao, and DC Cameron. 1,3-propanediol production by *Escherichia coli* expression genes from Klebsiella pneumonia dha regulon. *Appl. Environ. Microbiol.*, 57:3541–3546, 1991.

[594] ND Tracy, JC Young, and RL Mason. Multivariate control charts for individual observations. *Journal of Quality Control*, 24(2):88–95, 1992.

[595] RS Tsay. Time series model specification in the presence of outliers. *J. Amer. Statist. Ass.*, 81:132–141, 1986.

[596] AY Tsen, SS Yang, DSH Wong, and B Joseph. Predictive control of quality in a batch polymerization using a hybrid artificial neural network model. *AIChE J*, 42:435–455, 1996.

[597] LR Tucker. Some mathematical notes on three–mode factor analysis. *Psychometrika*, 31:279–311, 1966.

[598] WT Tucker, FW Faltin, and SA Vander Wiel. Algorithmic statistical process control : An elaboration. *Technometrics*, 35(4):363–375, 1993.

[599] JW Tukey. *Exploratory Data Analysis.* Addison-Wesley, Reading, Mass., 1977.

[600] RD Tyagi and TK Ghose. Batch and multistage continuous ethanol fermentation of cellulose hydrolysate and optimum design of fermenter by graphical analysis. *Biotechnol. Bioengng.*, 22:1907–1928, 1980.

[601] VK Tzouanas, C Georgakis, WL Luyben, and LH Ungar. Expert multivariable control:Part I-Structure and design methodology. *Ind. Eng. Chem. Res.*, 29:382–389, 1990.

[602] O Ubrich, B Srinivasan, F Stossel, and D Bonvin. Optimization of a semi-batch reaction system under safety constraints. In *Proc. European Control Conf.*, pages F306.1–6, Karlsruhe, Germany, 1999.

[603] C Undey, G Birol, I Birol, and A Cinar. An educational simulation package for penicillin fermentation. In *AIChE Annual Meeting*, Los Angeles, CA, 2000.

[604] C Undey, I Boz, E Oztemel, and A Cinar. Statistical monitoring of multistage batch processes. In *AIChE Annual Meeting*, Dallas, TX, 1999.

[605] C Undey and A Cinar. Statistical monitoring of multiphase, multistage batch processes. *IEEE Control Systems Mag.*, 22(5):40–52, 2002.

[606] C Undey, S Ertunc, and A Cinar. An integrated framework for online monitoring and product quality prediction in real-time for batch fermentations. In *AIChE Annual Meeting*, Indianapolis, IN, 2002.

[607] C Undey, E Tatara, BA Williams, G Birol, and A Cinar. A hybrid supervisory knowledge-based system for monitoring penicillin fermentation. In *Proc. American Control Conf.*, volume 6, pages 3944–3948, Chicago, IL, 2000.

[608] C Undey, E Tatara, BA Williams, G Birol, and A Cinar. On-line real-time monitoring of penicillin fermentation. In *Proc. International Symposium on Advanced Control of Chemical Processes (ADCHEM 2000)*, pages 243–248, Pisa, ITALY, 2000.

[609] C Undey, BA Williams, and A Cinar. Monitoring of batch pharmaceutical fermentations: Data synchronization, landmark alignment, and real-time monitoring. In *Proc. 15th IFAC World Congress on Automatic Control*, Barcelona, Spain, 2002.

[610] MLR Vairo, W Borzani, and P Magalhaes. Response of a continuous anaerobic culture to variations in the feeding rate. *Biotechnol. Prog.*, 19:595–598, 1977.

[611] JJ Vallino and G Stephanopoulos. Carbon flux distributions at the pyruvate branch point in *Corynebacterium glcutamicum* during lysine overproduction. *Biotechnol. Prog.*, 10:320–326, 1994.

[612] JF van Impe and G Bastin. Optimal adaptive control of fed-batch fermentation processes. *Control Engng. Practice*, 3(7):939–954, 1995.

[613] P van Overschee and B de Moor. N4SID : Subspace algorithms for the identification of combined deterministic-stochastic systems. *Automatica*, 30:75, 1994.

[614] SA Vander Wiel, WT Tucker, FW Faltin, and N Doganaksoy. Algorithmic statistical process control : Concepts and application. *Technometrics*, 34(3):286–297, 1992.

[615] J Varner and D Ramkrishna. The nonlinear analysis of cybernetic models. Guidelines for model formulation. *J. of Biotechnol.*, 71:67–104, 1999.

[616] A Vasilache, B Dahhou, G Roux, and G Goma. Classification of fermentation process models using recurrent neural networks. *Int J of Systems Sci*, 32(9):1139–1154, 2001.

[617] V Venkatasubramanian. CATDEX: An expert system for diagnosing a fluidized catalytic cracking unit. In *Knowledge-based systems in Chemical Engineering*, chapter 2. CACHE Corp., Austin, TX, 1988.

[618] V Venkatasubramanian and K Chan. A neural network methodology for process fault diagnosis. *AIChE J*, 35(12):1993–2002, 1989.

[619] V Venkatasubramanian and SH Rich. An object-oriented two-tier architechture for integrating compiled and deep-level knowledge for process diagnosis. *Comp. Chem. Engng*, 12(9/10):903–921, 1988.

[620] V Venkatasubramanian, R Vaidyanathan, and Y Yamamoto. Process fault detection and diagnosis using neural networks-I.Steady-state processes. *Comp. Chem. Engng*, 14(7):699–712, 1990.

[621] M Verhaegen and P Dewilde. Subspace model identification. part i: The output error state space model identification class of algorithms. *Int. J. Contr.*, 56:1187–1210, 1992.

[622] Z Verwater-Lukszo. A practice approach to recipe improvement and optimization in the batch process industry. *Comp. in Industry*, 36:279–300, 1998.

[623] E Visser. *A Feedback-based Implementation Scheme for Batch Process Optimization.* PhD thesis, Swiss Federal Institute of Technology, Lausanne, Switzerland, 1999.

[624] V Volterra. *Theory of Functionals and Integro-Differential Equations.* Dover, New York, 1959.

[625] A Wald. *Sequential Analysis.* Wiley, New York, 1947.

[626] RE Walpole, RH Myers, and SL Myers. *Probability and Statistics for Engineers and Scientists.* Prentice-Hall, Upper Saddle River, NJ, 6th edition, 1998.

[627] Z Wang, C Di Massimo, MT Tham, and AJ Morris. Procedure for determining the topology of multilayer feedforward neural networks. *Neural Networks*, 7(2):291–300, 1994.

[628] Z Wang, TM Tham, and AJ Morris. Multilayer feedforward neural networks: a canonical form approximation of nonlinearity. *Int J of Control*, 56(3):655, 1992.

[629] LE Wangen and BR Kowalski. A multiblock partial least squares algorithm for investigating complex chemical systems. *Journal of Chemometrics*, 3:3–20, 1988.

[630] K Watanabe, S Hirota, L Hou, and DM Himmelblau. Diagnosis of multiple simultaneous fault via hierarchical artificial neural networks. *AIChE Journal*, 40(5):839–848, 1994.

[631] K Watanabe, I Matsuura, M Abe, M Kubota, and DM Himmelblau. Incipient fault diagnosis of chemical processes via artificial neural networks. *AIChE Journal*, 35(11):1803–1812, 1989.

[632] N Watanabe, H Kurimoto, M Matsubara, and K Onogi. Periodic control of continuous stirred tank reactors - II. Cases of a nonisothermal single reactor. *Chem. Engng Sci*, 37:745–752, 1982.

[633] N Watanabe, S Ohbayashi, and H Kurimoto. Application of the infinite-frequency Pi criterion to a periodically operated isothermal CSTR. *Chem. Engng Sci*, 45:2984–2986, 1990.

[634] N Watanabe, K Onogi, and M Matsubara. Periodic control of continuous stirred tank reactors - I. The Pi criterion and its applications to isothermal cases. *Chem. Engng Sci*, 36:809–818, 1981.

[635] M Weighell, EB Martin, M Bachmann, AJ Morris, and J Friend. Multivariate statistical process control applied to an industrial production facility. In *Proc. of ADCHEM'97*, pages 359–364, 1997.

[636] JB Welles and HW Blanch. The effect of discontinuous feeding on ethanol production by *Saccharomyces cerevisiae*. *Biotechnol. Bioengng*, 18:129–132, 1976.

[637] PJ Werbos. *Beyond regression: new tools for prediction and analysis in the behavioral sciences.* PhD thesis, Harvard University, in Applied Mathematics, 1984.

[638] JA Westerhuis and PMJ Coenegracht. Multivariate modelling of the pharmaceutical two-step process of wet granulation and tableting with multiblock partial least squares. *Journal of Chemometrics*, 11:379–392, 1997.

[639] JA Westerhuis, SP Gurden, and AK Smilde. Generalized contribution plots in multivariate statistical process monitoring. *Chemometrics Intell. Lab. Syst.*, 51:95–114, 2000.

[640] JA Westerhuis, T Kourti, and JF MacGregor. Analysis of multiblock and hierarchical PCA and PLS models. *Journal of Chemometrics*, 12:301–321, 1998.

[641] JA Westerhuis, T Kourti, and JF MacGregor. Comparing alternative approaches for multivariate statistical analysis of batch process data. *Journal of Chemometrics*, 13:397–413, 1999.

[642] A Whitaker. . *Process Biochem.*, 15:10, 1980.

[643] B Widrow and ME Hoff. Adaptive switching circuits. *IRE WESCON Convention Record*, pages 96–104, 1960.

[644] N Wiener. *Non-linear Problems in Random Theory.* MIT Press, Cambridge, MA, 1958.

[645] WE Wiesel. Continuous time algorithm for lyapunov exponents. I. *Phys. Rev. E*, 47:3686–3691, 1993.

[646] B Williams and A Cinar. Semi batch process monitoring by using three–way chemometric methods. In *AIChE Annual Meeting*, Miami Beach, FL, 1998.

[647] BA Williams, C Undey, and A Cinar. Physiological phase detection of batch bioprocesses with unequal run lengths. In *8th International Conference on Computer Applications in Biotechnology (CAB 8)*, June 24-27, Quebec City, Quebec, Canada, 2001.

[648] T Williams, C Kelley, J Campbell, D Kotz, and R Lang. *GNUPLOT— An Interactive Plotting Program*, 31 August 1990.

[649] MJ Willis, C Di Massimo, GA Montague, MT Tham, and AJ Morris. Artificial neural networks in process engineering. *IEE Proceedings-D*, 138(3):256–266, 1992.

[650] AS Willsky. A survey of design methods for failure detection in dynamic systems. *Automatica*, 12:601–611, 1976.

[651] AS Willsky and HL Jones. A generalized likelihood ratio approach to state estimation in linear systems subject to abrupt changes. In *Proc. IEEE Conf. Decision and Control*, page 846, 1974.

[652] AS Willsky and HL Jones. A generalized likelihood ratio approach to detection and estimation of jumps in linear systems. *IEEE Trans. Automatic Control*, AC-21:108–112, 1992.

[653] HD Wilson and RG Rinker. Concentration forcing in ammonia synthesis. *Chem. Engng Sci*, 37:343–355, 1982.

[654] JD Windass, MJ Worsey, EM Pioli, D Pioli, PT Barth, KT Atherton, and EC Dart. Improved conversion of methanol to single-cell protein by *Methyllophilus methylotrophus. Nature*, 287:396–401, 1980.

[655] BM Wise and NB Gallagher. The process chemometrics approach to process monitoring and fault detection. *Journal of Process Control*, 6(6):329–348, 1996.

[656] BM Wise, NL Ricker, and DJ Veltkamp. Upset and sensor fault detection in multivariable processes. In *AIChE Annual Meeting, Paper 164b*, San Francisco, CA, 1989.

[657] MF Witcher. PhD thesis, University of Massachusetts, Amherst, MA, 1977.

[658] S Wold. Spline functions in data analysis. *Technometrics*, 16:1–11, 1974.

[659] S Wold. Cross-validatory estimation of the number of components in factor and principal components analysis. *Technometrics*, 20(4):397–405, 1978.

[660] S Wold. Nonlinear partial least squares modelling: II. Spline inner relation. *Chemometrics and Intelligent Laboratory Systems*, 14:71–84, 1992.

[661] S Wold, P Geladi, K Esbensen, and J Ohman. Multi-way principal component and PLS analysis. *Journal of Chemometrics*, 1:41–56, 1987.

[662] S Wold, S Hellberg, T Lundstedt, M Sjostrom, and H Wold. PLS modeling with latent variables in two or more dimensions. In *Proc. Symp. on PLS Model Building: Theory and Application*, Frankfurt, Germany, Sept. 1987.

[663] S Wold, N Kettaneh, H Friden, and A Holmberg. Modelling and diagnostics of batch processes and analogous kinetic experiments. *Chemometrics Intelligent Lab. Systems*, 44:331–340, 1998.

[664] S Wold, N Kettaneh-Wold, and B Skagerberg. Nonlinear PLS modeling. *Journal of Chemometrics*, 7:53–65, 1989.

[665] S Wold, N Kettaneh-Wold, and K Tjessem. Hierarchical multiblock PLS and PC models, for easier model interpretation, and as an alternative to variable selection. *Journal of Chemometrics*, 10:463–482, 1996.

[666] S Wold, H Martens, and H Wold. MULDAST Proc. ed. by S. Wold. In *Technical Report*, Research Group of Chemometrics, Umea University, Umea, Sweden, 1984.

[667] S Wold, A Ruhe, H Wold, and WJ Dunn. The collinearity problem in linear regression. Partial least squares PLS approach to generalized inverses. *SIAM J. Sci. Stat. Comput.*, 3(5):735, 1984.

[668] S Wold, M Sjostrom, and L Eriksson. PLS-regression: a basic tool of chemometrics. *Chemometrics Intelligent Lab. Systems*, 58:109–130, 2001.

[669] A Wolf, JB Swift, HL Swinney, and JA Vastano. Determining lyapunov exponents from a time series. *Physica D*, 16:285–317, 1985.

[670] S Wolfram. *The Mathematica Book.* Wolfram Research, Inc., Champaign, 2002.

[671] JC Wong, KA McDonald, and A Palazoglu. Classification of process trends based on fuzzified symbolic representation and hidden markov models. *Journal of Process Control*, 8(5–6):395–408, 1998.

[672] WH Woodall and MM Ncube. Multivariate CUSUM quality control procedures. *Technometrics*, 27:285–292, 1985.

[673] Y Yabuki and JF MacGregor. Product quality control in semi-batch reactors using mid-course correction policies. In *IFAC ADCHEM'97*, pages 189–194, Banf, Canada, 1997.

[674] H Yang, U Reichl, R King, and ED Gilles. Mathematical model for apical growth, septation and branching of mycelial microorganisms. *Biotechnol. Bioeng.*, 39(1):49–58, 1992.

[675] H Yang, U Reichl, R King, and ED Gilles. Measurement and simulation of the morphological development of filamentous microorganisms. *Biotechnol. Bioeng.*, 39(1):44–48, 1992.

[676] YT Yang, GN Bennett, and KY San. Genetic and metabolic engineering. *EJB: Electronic Journal of Biotechnology [online]*, 3(1):Available at: `www.ejb.org/content/vol1/issue3/full/3/bip/` ISSN: 0717–3458, issue of Dec 1998.

[677] E Yashchin. Performance of Cusum control schemes for serially correlated observations. *Technometrics*, 35:37–52, 1993.

[678] SYS Yeung, D Sincic, and JE Bailey. Optimal periodic control of activated sludge processes: II. Comparison with conventional control for structured sludge kinetics. *Wat. Res.*, 14:77–83, 1980.

[679] Y You and M Nikolaou. Dynamic process modeling with recurrent neural networks. *AIChE Journal*, 39(10):1654–1667, 1993.

[680] L Zadeh. Fuzzy sets. *Inf Cont*, 8:338–353, 1965.

[681] E Zafiriou. On the robustness of model predictive controllers. In *Proceedings of Chemical Process Control - CPC IV*, pages 359–364, New York, 1991. AIChE.

[682] E Zafiriou, RA Adomaitis, and G Gattu. Approach to run-to-run control for rapid thermal processing. In *Proceedings of the American Control Conference*, page 1286, Seattle, 1995.

[683] E Zafiriou, HW Chion, and RA Adomaitis. Nonlinear model-based run-to-run control for rapid thermal processing with unmeasured variable estimation. *Electr chem.Soc.Proc.*, 95(4):18–31, 1995.

[684] E Zafiriou and JM Zhu. Optimal control of semibatch processes in the presence of modeling error. In *Proceedings of the American Control Conference*, pages 1644–1649, San Diego, CA, 1990.

[685] TC Zangirolami, CL Johensen, J Nielsen, and SB Jorgensen. Simulation of penicillin production in fed-batch cultivations using a morphologically structured model. *Biotechnol. Bioeng.*, 56(6):593–604, 1997.

[686] Q Zhang, M Basseville, and A Benveniste. Early warning of slight changes in systems. *Automatica*, 30:95–113, 1994.

[687] LL Zheng, TJ McAvoy, Y Huang, and G Chen. Application of multivariate statistical analysis in batch processes. *Ind. Eng. Chem. Res.*, 40:1641–1649, 2001.

[688] DO Zines and PL Rogers. A chemostat study of ethanol inhibition. *Biotechnol. Bioengng*, 13:293–308, 1971.

Index